VOLUME SIX HUNDRED AND FORTY EIGHT

METHODS IN ENZYMOLOGY

Enzymatic Plastic Degradation

METHODS IN ENZYMOLOGY

VOLUME SIX HUNDRED AND FORTY EIGHT

METHODS IN ENZYMOLOGY

Enzymatic Plastic Degradation

Edited by

GERT WEBER

Macromolecular Crystallography, Helmholtz-Zentrum Berlin für Materialien und Energie, Berlin, Germany

UWE T. BORNSCHEUER

Department of Biotechnology & Enzyme Catalysis, Institute of Biochemistry, University of Greifswald, Greifswald, Germany

REN WEI

Department of Biotechnology & Enzyme Catalysis; Junior Research Group Plastic Biodegradation, Institute of Biochemistry, University of Greifswald, Greifswald, Germany

ELSEVIER

ACADEMIC PRESS
An imprint of Elsevier

Academic Press is an imprint of Elsevier
50 Hampshire Street, 5th Floor, Cambridge, MA 02139, United States
525 B Street, Suite 1650, San Diego, CA 92101, United States
The Boulevard, Langford Lane, Kidlington, Oxford OX5 1GB, United Kingdom
125 London Wall, London, EC2Y 5AS, United Kingdom

First edition 2021

Notices
Knowledge and best practice in this field are constantly changing. As new research and experience broaden our understanding, changes in research methods, professional practices, or medical treatment may become necessary.

Practitioners and researchers must always rely on their own experience and knowledge in evaluating and using any information, methods, compounds, or experiments described herein. In using such information or methods they should be mindful of their own safety and the safety of others, including parties for whom they have a professional responsibility.

To the fullest extent of the law, neither the Publisher nor the authors, contributors, or editors, assume any liability for any injury and/or damage to persons or property as a matter of products liability, negligence or otherwise, or from any use or operation of any methods, products, instructions, or ideas contained in the material herein.

ISBN: 978-0-12-822012-2
ISSN: 0076-6879

For information on all Academic Press publications
visit our website at https://www.elsevier.com/books-and-journals

Publisher: Zoe Kruze
Editorial Project Manager:
Federico Paulo S. Mendoza
Production Project Manager: James Selvam
Cover Designer: Mark Rogers

Typeset by SPi Global, India

Contents

Contributors

Luc Avérous
BioTeam/ICPEES-ECPM, UMR CNRS 7515, Université de Strasbourg, Strasbourg Cedex 2, France

Christoffel P.S. Badenhorst
Department of Biotechnology & Enzyme Catalysis, Institute of Biochemistry, University of Greifswald, Greifswald, Germany

Gert-Jan Bekker
Institute for Protein Research, Osaka University, Suita, Osaka, Japan

Lars M. Blank
iAMB—Institute of Applied Microbiology, ABBt—Aachen Biology and Biotechnology, RWTH Aachen University, Aachen, Germany

Uwe T. Bornscheuer
Department of Biotechnology & Enzyme Catalysis, Institute of Biochemistry, University of Greifswald, Greifswald, Germany

Daniel Breite
Leibniz Institute of Surface Engineering (IOM), Leipzig, Germany

María José Cárdenas Espinosa
Helmholtz Centre for Environmental Research—UFZ, Department of Environmental Biotechnology, Leipzig, Germany

Michael Carus
nova-Institut für politische und ökologische Innovation GmbH, Hürth, Germany

Zhuozhi Chen
School of Life Sciences, Tianjin University, Tianjin, China

Jennifer Chow
Department of Microbiology and Biotechnology, University of Hamburg, Hamburg, Germany

Andrea Colina Blanco
Helmholtz Centre for Environmental Research—UFZ, Department of Environmental Biotechnology, Leipzig, Germany

Craig S. Criddle
Department of Civil & Environmental Engineering, Stanford University, Stanford, CA, United States

Yinglu Cui
CAS Key Laboratory of Microbial Physiological and Metabolic Engineering, State Key Laboratory of Microbial Resources, Institute of Microbiology, Chinese Academy of Sciences, Beijing, China

Lara Dammer
nova-Institut für politische und ökologische Innovation GmbH, Hürth, Germany

Dominik Danso
Department of Microbiology and Biotechnology, University of Hamburg, Hamburg, Germany

Jinfeng Ding
College of Environmental Science and Engineering, Ocean University of China, Qingdao, China

Christian Eberlein
Helmholtz Centre for Environmental Research—UFZ, Department of Environmental Biotechnology, Leipzig, Germany

Tobias J. Erb
SYNMIKRO Research Center; Max Planck Institute for Terrestrial Microbiology, Marburg, Germany

Patricia Falkenstein
Institut für Analytische Chemie, Universität Leipzig, Leipzig, Germany

Fenglei Gao
Key Laboratory of Marine Chemistry Theory and Technology (Ocean University of China), Ministry of Education, Qingdao, China

Leonie G. Graf
Macromolecular Crystallography, Institute of Biochemistry, University of Greifswald, Greifswald, Germany

Georg M. Guebitz
Institute for Environmental Biotechnology, IFA Tulln, University of Natural Resources and Life Sciences, Vienna, Austria

Hermann J. Heipieper
Helmholtz Centre for Environmental Research—UFZ, Department of Environmental Biotechnology, Leipzig, Germany

Henric M.T. Hintzen
iAMB—Institute of Applied Microbiology, ABBt—Aachen Biology and Biotechnology, RWTH Aachen University, Aachen, Germany

Kazumi Hiraga
Kyoto Institute of Technology, Kyoto, Japan

Rong Ji
State Key Laboratory of Pollution Control and Resource Reuse, School of the Environment, Nanjing University, Nanjing, China

Yu Ji
Institute of Biotechnology, RWTH Aachen University, Aachen, Germany

Katsumasa Kamiya
Education Development Center, Kanagawa Institute of Technology, Atsugi, Japan

Narutoshi Kamiya
Graduate School of Simulation Studies, University of Hyogo, Kobe, Japan

Dai-ichiro Kato
Department of Science, Graduate School of Science and Engineering, Kagoshima University, Kagoshima, Japan

Fusako Kawai
Graduate School of Life and Environmental Sciences, Kyoto Prefectural University, Kyoto, Japan

Yasuyuki Kawashima
Department of Applied Chemistry, Graduate School of Engineering, University of Hyogo, Himeji, Japan

Lars Krause
nova-Institut für politische und ökologische Innovation GmbH, Hürth, Germany

Weidong Liu
Department of Biotechnology & Enzyme Catalysis, Institute of Biochemistry, University of Greifswald, Greifswald, Germany; Industrial Enzymes National Engineering Laboratory, Tianjin, Institute of Industrial Biotechnology, Chinese Academy of Sciences, Tianjin, China

Yi Lu
Institute of Biotechnology, RWTH Aachen University, Aachen, Germany

Yini Ma
State Key Laboratory of Pollution Control and Resource Reuse, School of the Environment, Nanjing University, Nanjing, China

Audrey Magnin
BioTeam/ICPEES-ECPM, UMR CNRS 7515, Université de Strasbourg, Strasbourg Cedex 2, France

Uwe G. Maier
Laboratory for Cell Biology, University of Marburg; SYNMIKRO Research Center, Marburg, Germany

Jörg Matysik
Institut für Analytische Chemie, Universität Leipzig, Leipzig, Germany

Ingrid Eileen Meyer Cifuentes
Junior Research Group Microbial Biotechnology, Leibniz Institute DSMZ German Collection of Microorganisms and Cell Cultures, Braunschweig, Germany

Emil A.P. Michels
Macromolecular Crystallography, Institute of Biochemistry, University of Greifswald, Greifswald, Germany

Daniel Moog
Laboratory for Cell Biology, University of Marburg; SYNMIKRO Research Center, Marburg, Germany

Keisuke Nagai
Department of Applied Chemistry, Graduate School of Engineering, University of Hyogo, Himeji, Japan

Seiji Negoro
Department of Applied Chemistry, Graduate School of Engineering, University of Hyogo, Himeji, Japan

Nobutaka Numoto
Medical Research Institute, Tokyo Medical and Dental University (TMDU), Bunkyo-ku, Tokyo, Japan

Kohei Oda
Kyoto Institute of Technology, Kyoto, Japan

Masayuki Oda
Graduate School of Life and Environmental Sciences, Kyoto Prefectural University, Kyoto, Japan

Taku Ohki
Department of Applied Chemistry, Graduate School of Engineering, University of Hyogo, Himeji, Japan

Basak Öztürk
Junior Research Group Microbial Biotechnology, Leibniz Institute DSMZ German Collection of Microorganisms and Cell Cultures, Braunschweig, Germany

Gottfried J. Palm
Macromolecular Crystallography, Institute of Biochemistry, University of Greifswald, Greifswald, Germany

Pablo Pérez-García
Department of Microbiology and Biotechnology, University of Hamburg, Hamburg, Germany

Lara Pfaff
Department of Biotechnology & Enzyme Catalysis, Institute of Biochemistry, University of Greifswald, Greifswald, Germany

Eric Pollet
BioTeam/ICPEES-ECPM, UMR CNRS 7515, Université de Strasbourg, Strasbourg Cedex 2, France

Hendrik Puetz
Institute of Biotechnology, RWTH Aachen University, Aachen, Germany

Achim Raschka
nova-Institut für politische und ökologische Innovation GmbH, Hürth, Germany

Karl-Heinz Rexer
Department for Evolutionary Ecology of Plants, University of Marburg, Marburg, Germany

Doris Ribitsch
Institute for Environmental Biotechnology, IFA Tulln, University of Natural Resources and Life Sciences, Vienna; ACIB—Austrian Centre of Industrial Biotechnology, Graz, Austria

Dietmar Schlosser
Department of Environmental Microbiology, Helmholtz Centre for Environmental Research—UFZ, Leipzig, Germany

Ulrich Schwaneberg
Institute of Biotechnology, RWTH Aachen University; DWI-Leibniz Institute for Interactive Materials, Aachen, Germany

Naoki Shibata
Department of Picobiology, Graduate School of Life Science, University of Hyogo, Ako-gun, Hyogo, Japan

Yasuteru Shigeta
Center for Computational Sciences, University of Tsukuba, Tsukuba, Japan

Pia Skoczinski
nova-Institut für politische und ökologische Innovation GmbH, Hürth, Germany

Chen Song
Institut für Analytische Chemie, Universität Leipzig, Leipzig, Germany

Wolfgang R. Streit
Department of Microbiology and Biotechnology, University of Hamburg, Hamburg, Germany

Chengjun Sun
Key Lab of Marine Eco-Environmental Science and Technology, Marine Bioresource and Environment Research Center, First Institute of Oceanography, Ministry of Natural Resources; Laboratory of Marine Drugs and Bioproducts, Pilot National Laboratory for Marine Science and Technology, Qingdao, China

Jinyuan Sun
CAS Key Laboratory of Microbial Physiological and Metabolic Engineering, State Key Laboratory of Microbial Resources, Institute of Microbiology, Chinese Academy of Sciences; University of Chinese Academy of Sciences, Beijing, China

Masahiro Takeo
Department of Applied Chemistry, Graduate School of Engineering, University of Hyogo, Himeji, Japan

Ikuo Taniguchi
International Institute for Carbon-Neutral Energy Research, Kyushu University, Fukuoka, Japan

Lili Tian
State Key Laboratory of Pollution Control and Resource Reuse, School of the Environment, Nanjing University; Institute of Agricultural Resources and Environment, Jiangsu Academy of Agricultural Sciences, Nanjing, China

Till Tiso
iAMB—Institute of Applied Microbiology, ABBt—Aachen Biology and Biotechnology, RWTH Aachen University, Aachen, Germany

Zefang Wang
School of Life Sciences, Tianjin University, Tianjin, China

Gert Weber
Macromolecular Crystallography, Institute of Biochemistry, University of Greifswald, Greifswald; Macromolecular Crystallography, Helmholtz-Zentrum Berlin, Berlin, Germany

Ren Wei
Department of Biotechnology & Enzyme Catalysis; Junior Research Group Plastic Biodegradation, Institute of Biochemistry, University of Greifswald, Greifswald, Germany

Gina Welsing
iAMB—Institute of Applied Microbiology, ABBt—Aachen Biology and Biotechnology, RWTH Aachen University, Aachen, Germany

Birger Wolter
iAMB—Institute of Applied Microbiology, ABBt—Aachen Biology and Biotechnology, RWTH Aachen University, Aachen, Germany

Bian Wu
CAS Key Laboratory of Microbial Physiological and Metabolic Engineering, State Key Laboratory of Microbial Resources, Institute of Microbiology, Chinese Academy of Sciences, Beijing, China

Wei-Min Wu
Department of Civil & Environmental Engineering, Stanford University, Stanford, CA, United States

Yunjie Xiao
School of Life Sciences, Tianjin University, Tianjin, China

Kengo Yasuhira
Department of Applied Chemistry, Graduate School of Engineering, University of Hyogo, Himeji, Japan

Yelwin Yew
Biotechnology, University of Applied Sciences Bremerhaven, Bremerhaven, Germany

Shosuke Yoshida
Institute for Research Initiatives, Nara Institute of Science and Technology & Division of Biological Science, Nara Institute of Science and Technology, Ikoma, Nara, Japan

Jan Zarzycki
Max Planck Institute for Terrestrial Microbiology, Marburg, Germany

Hongli Zhang
Department of Microbiology and Biotechnology, University of Hamburg, Hamburg, Germany

Preface

With the onset of large-scale synthetic polymer (plastic) production around 80 years ago, humanity has experienced a technological quantum leap, fueled by these durable, multifaceted, and versatile materials. A plethora of different polymer types have successfully conquered nearly every niche in our economy. These polymers are supplemented with a wide range of additives, for example, to extend durability, to establish superior material properties or add colors. This has led to the omnipresence of synthetic polymers in our daily lives. The ever-increasing production scales have amounted to nearly 10 billion metric tons of plastic synthesized up to now. Irrespective of the life cycle duration, the majority of dumped plastic are not covered by sustainable waste management strategies. For example, synthetic polymers in landfills may take several hundred years to decompose. Lastly, mismanaged plastic waste has led to a worldwide pollution of terrestrial and marine habitats where plastic debris of virtually any size is continuously accumulating. Micro- and nanoplastics are distributed in the air and can thus reach even remote places on this planet, for example glaciers, or arctic regions. Aiming at a sustainable plastic disposal, recycling procedures were established around 30 years ago. To date, mechanical or chemical recycling strategies of waste plastics require still technology upgradation in order to balance adequate product quality, commercial viability, and sustainability. Further difficulties arise from polymer mixtures that require elaborate sorting, and also primarily from additives which cannot be easily extracted in most mechanical recycling processes thereby causing product downgrade. Alternatively, chemical recycling methods address the recovery of low-molecular building blocks after depolymerizing the waste plastics. However, as long as virgin plastics are still being produced using inexpensive crude oil, the chemical recycling strategies will have a limited economic viability in most cases.

In the past 15 years, enzymes have been considered as a promising alternative to the chemical methods for plastic recycling. Enzymatic plastic degradation may enable a full decomposition of selected polymer types at low energy costs. Unspecific microbial reactions toward synthetic polymers have been reported since about 40 years. More recently, biotechnological methods allowed identification and engineering of plastic degrading enzymes. In most of these cases, the enzyme activities rely on the similarity of the chemical nature of natural and synthetic substrates, for example, ester bonds in plant cutin and synthetic polyesters. Through a specific enzymatic depolymerization, a full recovery of the building blocks from certain polymer types such as polyesters or polyamides is feasible. In contrast, vinyl polymers like polystyrene (PS) or polypropylene cannot be broken down to specific small molecules due to their highly recalcitrant carbon–carbon backbones. In 2016, the discovery of a naturally occurring enzyme system which is specific for polyethylene terephthalate (PET) degradation in the bacterium *Ideonella sakaiensis* triggered extensive research activities on plastic degrading enzymes, some of which have recently been identified in microbial metagenomes from various environments. Over the last years, an arsenal of valuable methods to identify, characterize, and improve plastic degrading enzymes were developed. With this compilation of articles from leading researchers in this field, it is our goal to make a large set of very useful and broadly applicable methods available at a high level of detail to foster future research initiatives on enzymatic plastic degradation and recycling.

Our *Methods in Enzymology* volume on *Enzymatic Plastic Degradation* is organized as follows. Chapter 1 introduces the current status and pictures future developments of industrially synthesized polymers, also in the light of biodegradability. In Chapter 2, methods for sampling and analysis of microplastics in sea or fresh water are detailed. Nature is a valuable resource for enzymes that potentially degrade synthetic polymers. In Chapter 3, cultivation methods for aerobic and anaerobic microbial consortia are described while Chapter 4 focuses on the cultivation of filamentous fungi that employ Fenton chemistry for the biodegradation of sulfonated PS. However, plastic degradation activity in nature is not limited to free-living fungi and bacteria. In Chapter 5, methods to investigate the decomposition of highly recalcitrant vinyl polymers by insect larvae are presented. As described in Chapter 6, the ^{14}C isotope may be successfully employed as a tracer to assess the deterioration of PS. The methodology to identify potential enzyme candidates from large metagenomic datasets is presented in Chapter 7. In Chapter 8, the authors have compiled a comprehensive method collection on the identification and characterization of actinomycete cutinases. In Chapter 9, the approach for the isolation and characterization of MHETase and PETase from *Ideonella sakaiensis* is presented in detail. In the light of maturing bioinformatic approaches for enzyme enhancement, Chapter 10 describes a strategy for a significant improvement of PETase through a combination of bioinformatics and activity assessment. Chapter 11 contains a comprehensive description of solution and solid nuclear-magnetic-resonance procedures for characterizing PET polymers that have undergone enzymatic degradation. A fluorimetric high-throughput screening method for polyester hydrolase activity using PET nanoparticles is covered by Chapter 12. Enzymes active on natural polymers often contain substrate-binding modules fostering elevated activity. In Chapter 13, the authors present the corresponding method to identify and engineer polymer-binding anchor peptides for an enhanced enzymatic plastic degradation exemplified for polyester hydrolases. Likewise, in Chapter 14, methods to introduce naturally occurring hydrophobic polymer-binding modules to enhance the adsorption of polyester hydrolases for a consequently improved depolymerization activity are presented. Apart from polyesters, polyurethane is another prominent polymer type with hydrolysable backbones which can be specifically degraded by enzymes. In Chapter 15, the authors describe pioneering methods to characterize the enzymatic degradation of polyurethanes. Structural information of enzymes, ideally bound to the substrate, is a prerequisite for understanding their mechanisms and improving their catalytic performance. In Chapter 16, the detailed methodology underlying the first substrate-bound structures of PETase and MHETase is presented at a high level of detail. Due to the chemical similarity to polypeptides, nylon is the next polymer group which can potentially be depolymerized by enzymatic hydrolysis. Chapter 17 comprehensively details a large set of methods dedicated to the enzymatic degradation of nylon oligomers. Genetic engineering methods of pseudomonads for the biosynthesis of a fatty acid derivative by metabolizing PET monomers are presented in Chapter 18. The authors of Chapter 19 detail the methodology for the cultivation of novel microbial strains able to grow on the building blocks of polyurethane. The efficiency of plastic degrading enzymes can be further enhanced when employed as whole cell catalysts. In Chapter 20, protocols for constructing and culturing microalgae capable of secreting recombinant PETase are presented and finally, in Chapter 21, methods to engineer yeast as a cellular chassis for surface-displayed PETase are detailed.

We were very pleased when encouraged to compile this volume for the famous *Methods in Enzymology* series, pioneering a collection of protocols on enzymatic plastic degradation. We thank all authors of this volume for their highly valuable contributions which were provided in a timely manner despite the still ongoing COVID-19 pandemic. This volume greatly benefited from the contributions of the external reviewers Imke Lang, Christian Feiler, Christoffel P.S. Badenhorst, Elvis Genbo Xu, and

Weiliang Dong and we are very grateful for their assistance during the review process. We also thank the Elsevier staff, in particular Zoe Kruze, Shellie Bryant, and James Selvam for their excellent support. With this 648th volume of *Methods in Enzymology*, it is our goal to provide a valuable methodological resource which will support current and seed novel research activities in this field and lastly help to establish the role of enzymes in a sustainable circular plastics economy. We hope that you, the readers, will enjoy this collection of articles covering the expanding field of enzymatic plastic degradation.

Gert Weber
Uwe T. Bornscheuer
Ren Wei

CHAPTER ONE

Current status and future development of plastics: Solutions for a circular economy and limitations of environmental degradation

Pia Skoczinski*, Lars Krause, Achim Raschka, Lara Dammer, and Michael Carus
nova-Institut für politische und ökologische Innovation GmbH, Hürth, Germany
*Corresponding author: e-mail address: pia.skoczinski@nova-institut.de

Contents

Abstract

Plastics are the most advanced materials, regarding their application and property range, with an annual worldwide demand of over 400 million tonnes, but have been facing strong public opposition lately.

Methods in Enzymology, Volume 648
ISSN 0076-6879
https://doi.org/10.1016/bs.mie.2020.11.001

In the debate on sustainable plastics, plastic waste and plastic littering, bio-based and especially biodegradable plastics have gained attention as potential solution to these problems. This has been accompanied by misunderstanding, false and inadequate labeling of plastics as well as insufficient communication with the public and authorities. The plastic waste and littering problems are addressed by already implemented and planned bans on single-use plastics, improved regulations for labeling and communication of biodegradable plastics, and enhancement of plastic recycling. The future will show how successful these strategies are. If plastics are produced with renewable carbon from biomass, direct CO_2 utilization and recycling, the production can be completely independent from fossil resources in a long-term future perspective which supports the circular renewable carbon economy.

1. Introduction

The circular economy as a regenerative, economic system aims to minimize emissions, use of resources, energy, as well as the production of waste by closing the loops of energy and material use (Geissdoerfer, Savaget, Bocken, & Hultink, 2017).

Plastics as a part of this economy have been facing strong public opposition lately, although they are the most advanced materials concerning application range, processing properties, and efficiency with an annual worldwide demand of over 400 million tonnes (Carus, Dammer, Raschka, Skoczinski, & vom Berg, 2020). If plastics are produced with renewable carbon from biomass, direct CO_2 utilization and recycling (Carus & Raschka, 2018), the plastic production can be completely independent from fossil resources in a long-term future perspective which supports a circular renewable carbon economy. In the debate on sustainable plastics, plastic waste and plastic littering, bio-based and especially biodegradable plastics have gained attention as potential solution to these problems. This has been accompanied by misunderstanding, false and inadequate labeling of plastics as well as insufficient communication with the public in general and with ministries of environment, non-governmental organization and waste management companies specifically, which have not been strategically involved.

The following sections will focus on terminologies around plastics, limitations in biodegradation, the current plastic market, disposal and recycling as well as political regulations for plastics to mitigate waste and littering. Thus, providing a comprehensive view on a current reassessment and evolving plastic market, strengthening a circular economy.

2. Definitions

In course of implementing a circular economy and a sustainable production of all kind of products, several terms have been provided especially concerning origin and properties of plastics, e.g. bio-based plastics, biodegradable plastics, compostable plastics, and recyclable plastics. Although, some of these descriptions have been clearly defined by the International Organization for Standardization (ISO) or the European Committee for Standardization (CEN), their use combined with a lack of explanation and understanding have led to misinterpretation, confusion and misunderstanding of products linked to these terms. Therefore, clear definitions and their right use and explanation are indispensable.

2.1 Polymer or plastic—What is the difference?

The terms polymer and plastic have been defined in ISO 472:2013 to avoid improper use particularly when the detailed definition plays a role in any kind of certification e.g. biodegradability (ISO, 2013). Within this definition the ISO differentiates between different kind of polymers, based on their chemical composition and the final plastic product:

A regular *polymer* is defined as a molecule that can be described by only one species of constitutional unit in a single sequential arrangement. A constitutional unit means the repeating unit within the polymer, this can be composed of one single monomer (homopolymer) or two or more monomers (copolymer).

A *homopolymer* is derived from a single species of monomer, e.g. polyethylene (PE).

A *copolymer* on the other hand is derived from more than one species of a monomer, e.g. polyethylene terephthalate (PET), with the two different monomers ethylene glycol and terephthalic acid.

A *plastic* is a material that contains a high polymer as an essential ingredient and that can be shaped by flow at some stage in its processing into finished products.

A high polymer is a large molecule with a molecular weight greater than 10,000 g/mol (units; Da), that is usually but not always composed of repeating units of low-molecular weight, e.g. ethylene or propylene. As additives as such are not included in ISO 472:2013, the following explanation focuses on plasticizers and fillers as additives, that are explicitly defined within ISO 472:2013: During the plastic processing it is sometimes necessary *to plasticize* the polymer by chemical modification or the addition of plasticizers to make the polymer softer, more flexible and/or workable. On the other hand, the plastic can be modified regarding its strength, permanence, working properties or other qualities by adding a *filler*, a relatively inert solid material.

2.2 What is bio-based?

The CEN has exactly defined the term bio-based for all kinds of products named as such. Nevertheless, the term bio-based and its use sometimes leads to confusion and discussions. The Technical Committee 411 is specifically dedicated to "Bio-based products" within the CEN and established the EN 16575:2014 with the following definitions (CEN, 2014):

Bio-based is defined as derived from biomass and the term *bio-based content* is defined as the fraction of a product that is derived from biomass. Based on this a *bio-based product* can be wholly or partly derived from biomass. Additionally, also *bio-based carbon* as carbon derived from biomass and *bio-based carbon content* as the fraction of carbon derived from biomass in a product have been defined.

The measurement of the bio-based content and the bio-based carbon content are also determined by specific standards. The fundamental difference in the measurement is that, bio-based carbon content specifically and only measures the carbon content in the bio-based product while the bio-based content also includes other elements such as oxygen, hydrogen or nitrogen.

The *bio-based carbon content* is determined by using the radiocarbon analysis (CSN EN 16640) (CSN, 2017) and is declared as "the percentage of the *total carbon* content of the product." The *bio-based content* is also determined by the radiocarbon analysis (CSN EN 16640) (CSN, 2017), but also includes an elemental analysis to also measure the amount of bio-based derived other elements, such as oxygen or hydrogen (CSN EN 16785-1) (CSN, 2015), and therefore refers not only to the bio-based carbon content but to the *overall* bio-based content of the product.

3. Biodegradation and biodegradability—Explanations and limits

First of all, it has to be strictly pointed out that the term biodegradability is not linked to a product being bio-based. The term bio-based refers to the feedstock biomass from which the product originates, while biodegradability is an end-of-life option. A bio-based product is not necessarily biodegradable, also fossil-based products can be biodegradable. While biodegradation describes a complex biogeochemical mechanism, the term biodegradability describes the potential of a product to undergo the process of biodegradation. Biodegradation, specifically concerning plastics has been defined in ISO 472:2013 as degradation caused by biological activity, especially by enzymatic action, leading to a significant change in the chemical structure of a material (ISO, 2013). Based on the fact that the term "significant change in the chemical structure of a material" is not defined in more detail caused misinterpretation and false use of the description biodegradation and the linked statement for a plastic of being biodegradable (Narancic et al., 2018).

From a scientific point of view *biodegradation* can be defined as an aerobic or anaerobic biochemical process mediated by microorganisms that break down materials to CO_2, methane, water, and biomass via enzymatic hydrolysis and metabolization. This biotic degradation and assimilation process by microorganisms can be enhanced and facilitated by prior abiotic degradation, including disintegration and fragmentation of the material through environmental impacts as UV radiation, wind and temperature (Fig. 1).

Therefore, a plastic is only fully biodegraded if it is completely converted into CO_2, water, and biomass. All prior steps, abiotic degradation and incomplete biotic degradation, just lead to a disintegrated, fragmented plastic that then remains as macro-, micro-, or nanoplastic or an even smaller portion, the plastic monomer, in the environment. Successful biodegradation depends on the microbial activity and related specific environmental aspects as the temperature, humidity or available nutrients. So not every plastic will biodegrade in different environments in a similar way since the biodegradability of a certain plastic is always linked to specific biogeochemical condition. Therefore, the certificate for biodegradability is always

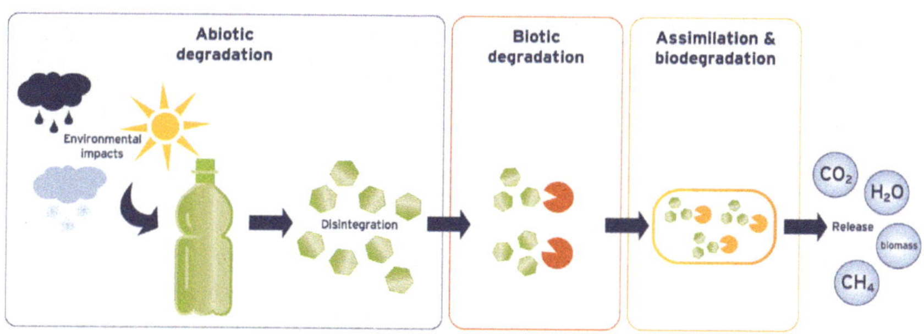

Fig. 1 Overview of different steps for biodegradation.

marked with the addition of the specific environment, e.g. OK biodegradable MARINE (Table 1).

To test biodegradation of plastics and certify them as biodegradable within a particular environment, different testing conditions for a variety of environments have been established (Table 1).

Standardized testing conditions and certificates exists for biodegradation in marine, fresh water, soil, home composting, and industrial composting environments. These standardized testing conditions might not exactly fit all real environmental conditions, but are necessary to perform the biodegradation test in the first place and yield certifications:

(1) Biodegradation testing of plastics is often performed after milling, showing the inherent nature of the material to biodegrade.

(2) The temperature for these tests is chosen for shortening of the testing time and a linear progression of biodegradation is assumed.

Table 1 Overview of different environments for biodegradation, testing conditions and certifications.

	Testing conditions			
Environment	**Temperature (°C)**	**Time for 90% biodegradation**	**Certification**	**Certifier**
Marine	30	Max. 6 months	OK biodegradable MARINE	TÜV AUSTRIA
Fresh water	21	Max. 56 days	OK biodegradable WATER	TÜV AUSTRIA
Soil	25	Max. 2 years	OK biodegradable SOIL	TÜV AUSTRIA
			DIN-Geprüft biodegradable in soil	DIN Certco
Home composting	28	Max. 12 months	OK biodegradable compost HOME	TÜV AUSTRIA
			DIN-Geprüft Home Compostable	DIN Certco
Landfill	No standard specifications or certification scheme available, since this is not a preferred end-of-life option			
Anaerobic digestion	52/37	Standard specification not yet available, but 90% generally considered as completely biodegradable		
Industrial composting	58	Max. 6 months	Standard: EN 13432	

(3) The time of the test until about 90% of the maximum level of biodegradation has been reached is the biodegradation phase.
(4) The maximum level of measurable biodegradation is 90%. By biodegrading plastics, microorganisms always metabolize them, increasing their biomass and thereby releasing CO_2. This CO_2 is the measurable parameter for biodegradation and reflects the 90%, while the non-measurable remaining 10% are related to the accumulated biomass during the process.

In actual real environments the same level of biodegradation will be achieved, although possibly within a different timeframe (Narancic et al., 2018; nova-Institut, IKT Stuttgart, OWS, & HYDRA Marine Sciences, 2020).

Since plastics ending up in landfill is not a preferred end-of-life option, no standard specification or certification scheme is available here. Dependent on the microbial presence anaerobic digestion can take place at 52 °C for thermophilic microorganisms or at 37 °C for mesophilic microorganisms. A standard specification for anaerobic digestion is not yet available, but 90% are generally considered as completely biodegradable.

3.1 Certified biodegradable polymers and derived plastics

The certification of biodegradable polymers and plastics is usually performed with the final sold end product (prepared and applied for the abovementioned testing conditions). In most cases, this is not the synthesized pure polymer, but the plastic that also might include plasticizers, additives or fillers. Therefore, biodegradability is only guaranteed if all additives and organic fillers are biodegradable, too. Certain processing steps to the final product, such as dying might prevent their biodegradation in specific environments.

Due to this fact, no general statement and certification for a plastic derived from a specific biodegradable polymer can be made, because the specific product is certified and not the polymer in general. The certification is always dependent on the specific grade of the polymer and the entire composition of the plastic. Nevertheless, it was possible to establish generalized claims for the most commonly used biodegradable polymers (Table 2) (nova-Institut et al., 2020).

It has to be clearly emphasized here that biodegradability depends on the complex biogeochemical conditions at each testing site and that these generalized statements can only serve as approximations which need to be confirmed for each individual specific product based on these biodegradable polymers by standardized testing under lab conditions. The behavior of

Table 2 Biodegradable polymers in various environments.

Polymer	Proven biodegradability in the following environments according to standards/ certification systems	Proven biodegradability in the following environments under certain conditions, longer time frames or only for certain polymer grades
Cellulose (lignin <5%)	Marine	
	Fresh water	
	Soil	
	Home composting	
	Landfill	
	Anaerobic digestion	
	Industrial composting	
Cellulose acetate (CA) and other derivatives		Marine
		Fresh water
		Soil
		Home composting
		Landfill
		Anaerobic digestion
		Industrial composting
Lignin, wood		Marine
		Fresh water
		Soil
		Home composting
		Landfill
		Anaerobic digestion
		Industrial composting
Poly(butylene-*co*-adipate terephthalate) (PBAT)	Industrial composting	Soil
		Home composting
Polybutylene succinate (PBS)	Industrial composting	

Table 2 Biodegradable polymers in various environments.—cont'd

Polymer	Proven biodegradability in the following environments according to standards/ certification systems	Proven biodegradability in the following environments under certain conditions, longer time frames or only for certain polymer grades
Polybutylene succinate-*co*-adipate (PBSA)	Soil	
	Home composting	
	Industrial composting	
Polyhydroxybutyrate (PHB) and copolymers	Marine	
	Fresh water	
	Soil	
	Home composting	
	Landfill	
	Anaerobic digestion	
	Industrial composting	
Polylactic acid (PLA)	Industrial composting	Anaerobic digestion
Starch and other natural polymers	Marine	
	Fresh water	
	Soil	
	Home composting	
	Landfill	
	Anaerobic digestion	
	Industrial composting	

Adapted from nova-Institut, IKT Stuttgart, OWS, & HYDRA Marine Sciences. (2020). Biodegradable polymers in various environments. Retrieved from http://bio-based.eu/graphics/#biodegPoster: nova-Institut (Ed.), Hürth, Germany.

the tested plastic can vary, depending on the temperature, the available nutrients and oxygen, as well as the microbial activity at each testing site. Furthermore, other factors as the size of the plastic, the grade, the specific chemical constitution and intrinsic factors can impact the biodegradability. The influence of the latter two can be explained using polylactic acid (PLA) as an example. The rate of biodegradation is dependent on the crystallinity of the polymer. PLA with a higher content of its D-lactic acid monomer in PLA shows higher biodegradability as PLA with only L-lactic acid as a monomer. Because D-lactic acid reduces the crystallinity by inducing twists in the otherwise very regular L-PLA chain. This decrease in crystallinity can increase its biodegradability (Kale et al., 2007).

4. Current plastic market and production

Worldwide plastic production in 2018/2019 was around 400 million tonnes, with fossil-based plastics being the major part of 90% (Carus et al., 2020). Recycled plastics are already making up 9% of the worldwide production, while bio-based and CO_2-based plastic are having a share of 1 and 0.1% respectively (Fig. 2).

The average compound annual growth rate (CAGR) for the whole market is expected to be at 4% until 2025. Fossil-based plastics are estimated to grow with a CAGR of 3–4%, the growth for bio-based plastics is estimated to be in the same range as fossil-based ones with 3%. The largest increase is expected for recycled plastics with a CAGR of over 6%. Although the use of CO_2 for polycarbonate production is established and several other technologies for plastic production from CO_2 have been demonstrated, a valid CAGR for this group of plastics cannot be estimated yet, as the market is still young and under current dynamic development.

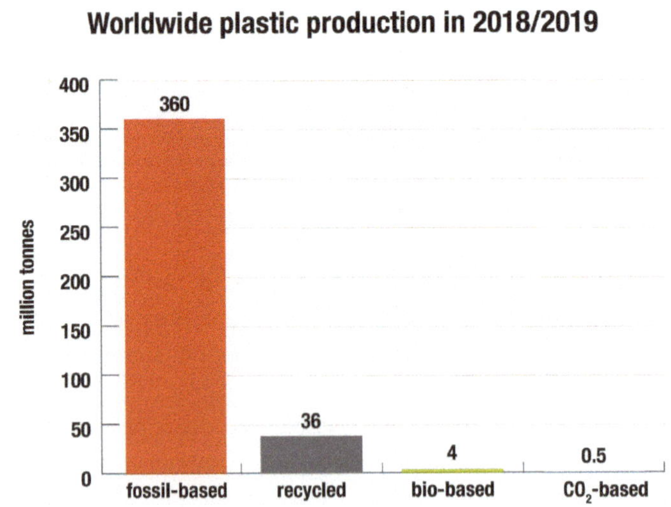

Fig. 2 Worldwide plastic production in 2018/2019 (Carus et al., 2020).

All plastics produced today find their use in a wide range of applications as for agriculture, in automotive industry, building and construction, consumer goods, electrics and electronics, packaging, and textiles.

4.1 Fossil-based plastics

Fossil-based plastics with the largest use in 2018/2019 were polyethylene (PE), polypropylene (PP) and polyvinyl chloride (PVC) (Fig. 3). Although this distribution is specifically for the European demand of fossil-based plastics, it is assumed to be similar in a worldwide perspective (Geyer, Jambeck, & Law, 2017). With 30% PE makes up the majority of the European demand and constitutes on the one hand low-density (PE-LD) and linear low-density PE (PE-LLD) with 18% and on the other hand high-density (PE-HD) and medium-density (PE-MD) with 12%. While PE-LD and PE-LLD are mainly used for bags, trays and containers, as well as agricultural films and food packaging, PE-HD and PE-MD are used for toys and milk bottles, pipes.

The share of PP in the European fossil-based plastic demand is 19%, it is mainly used for several kinds of food packaging. With 10%, PVC is mainly used in the construction sector in window frames or floor and wall covering. Polyurethanes (PUR) and polyethylene terephthalate (PET) share 8%, respectively. PURs are commonly used in a foam formulation for building insulations, in pillows or mattresses and PET for the production of all kind of bottles for water, soft drinks, juices, and cleaners. Polystyrene (PS) together

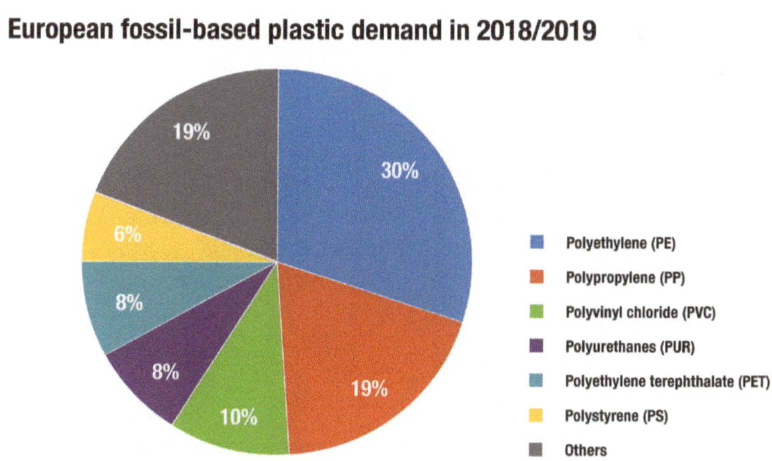

Fig. 3 European fossil-based plastic demand in 2018/2019. *Adapted from Plastics Europe. (2019). Plastics—The facts 2019. Retrieved from https://www.plasticseurope.org/de/resources/publications/1804-plastics-facts-2019.*

with expanded polystyrene (EPS) is making up 6% and are used for food packaging, building insulation or in electrical and electronic equipment. Acrylonitrile butadiene styrene (ABS), polybutylene terephthalate (PBT), polycarbonate (PC) and polymethyl methacrylate (PMMA) add up as others to 19% of the European demand (Plastics Europe, 2019).

The major application for fossil-based plastics in Europe is in all kind of packaging, with 40% (Fig. 4). Here PE, PP, and PET have the largest share. Packaging is followed by building and construction with 20%, mainly made up by the use of PVC. The automotive industry shares 10%, electrical and electronic equipment 6%, household, leisure and sports 4%, agriculture 3% and others as hub caps, optical fibers and eyeglasses lenses etc. add up to 17% (Plastics Europe, 2019).

4.2 Bio-based plastics

The major share within the worldwide bio-based plastic demand in 2018/2019 was coming from bio-based epoxy resins and cellulose acetate (CA), with 27% and 20%, respectively (Fig. 5). Bio-based epoxy resins belong to the class of smart drop-ins and are made from glycerol as a by-product in biodiesel production (Carus, Dammer, Puente, Raschka, & Arendt, 2017). Epoxy resins in general are mainly used in the automotive industry and building and construction. Cellulose acetate (CA) finds its main applications in the textile segment, mainly fibers (cigarette filters) and in the consumer goods sector.

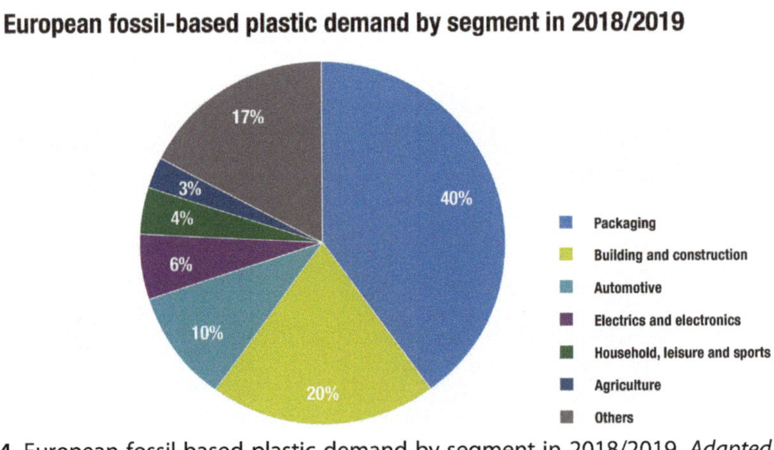

Fig. 4 European fossil-based plastic demand by segment in 2018/2019. *Adapted from Plastics Europe. (2019). Plastics—The facts 2019. Retrieved from https://www.plasticseurope.org/de/resources/publications/1804-plastics-facts-2019.*

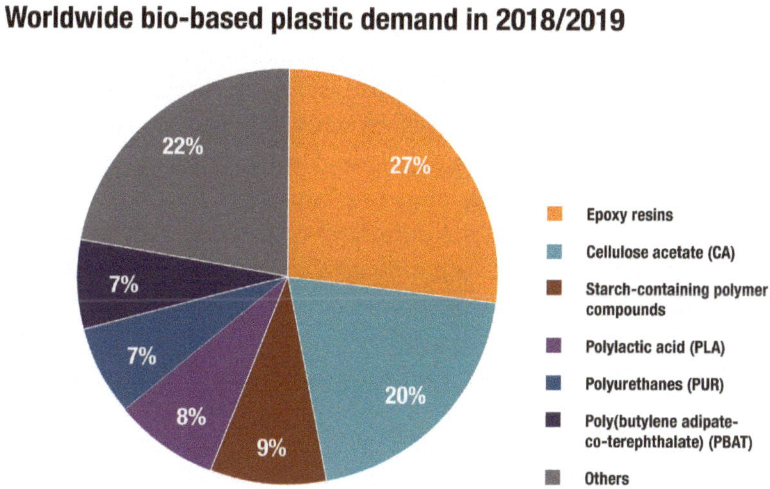

Fig. 5 Worldwide bio-based plastic demand in 2018/2019. *Adapted from Skoczinski, P., Chinthapalli, R., Carus, M., Baltus, W., de Guzman, D., Käb, H., et al. (2020). Bio-based building blocks and polymers—Global capacities, production and trends 2019–2024. Retrieved from http://bio-based.eu/markets/#TRPolymerdata2020: nova-Institut (Ed.), Hürth, Germany.*

These two major bio-based plastics are followed by starch-containing polymer compounds with 9%, mainly used in flexible packaging, and bio-based PUR with 8% and PLA with 7%. While the bio-based PUR, based on natural oil polyols (NOPs) and bio-based polyester and polyether polyols, is mainly used in the automotive industry and building and construction similarly as its fossil-based counterpart, PLA shows a wide variety in applications and is used in all segments except the automotive industry. Here, PLA is only used in small portions, mainly as PLA fibers (included in the textiles segment) for under the hood applications. Poly(butylene adipate-co-terephthalate) (PBAT) that is currently equally produced using fossil- and bio-based feedstocks makes up 7% and is used for flexible and rigid packaging as well as for agricultural applications as mulching films. Others include 10 bio-based plastics that add up to 22% in the worldwide demand. This include the bio-based drop-in plastics PE, PP and PET, the smart drop-in polyamides (PA), polybutylene succinate (PBS) and polytrimethylene terephthalate (PTT) and the dedicated plastics, that do not have commercially available fossil-based counterparts, including aliphatic polycarbonates (APC), casein polymers, polyethylene furanoate (PEF) and polyhydroxyalkanoates (PHAs).

Different from the European fossil-based plastic demand per segment where packaging shows by far the highest share, the worldwide bio-based plastic demand by segment shows that packaging with 24% and textiles including woven, non-woven and fibers with 22% have a quite similar share in application for bio-based plastics (Fig. 6). Packaging includes 13% for flexible and 11% for rigid packaging as food service ware. These two major applications are followed by the automotive industry with 14%, building and construction and consumer goods with 13%, respectively, and electrics and electronics with 5%, as well as agricultural applications with 3%. The remaining 6% of the bio-based plastic demand is used by other applications like coatings, adhesives or medical devices (Skoczinski et al., 2020).

4.3 Biodegradable plastics

A few of the abovementioned bio-based plastics, as well as fossil-based PBAT and PBS, are also biodegradable and are explicitly shown here and sum up to almost 2 million tonnes in 2018/2019. It has to be clearly pointed out here that the following numbers are derived from the basic bio-based polymer that is used to produce the respective plastic. As mentioned above, the successful biodegradation is dependent on the specific composition of the final plastic and the environment for biodegradation (Table 2).

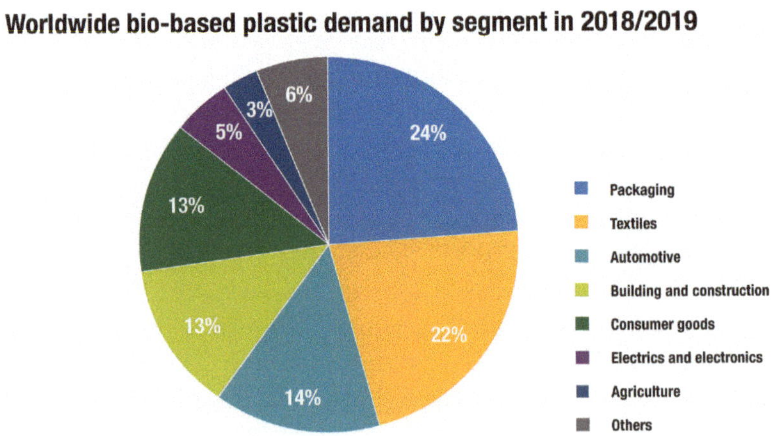

Fig. 6 Worldwide bio-based plastic demand by segment in 2018/2019. *Adapted from Skoczinski, P., Chinthapalli, R., Carus, M., Baltus, W., de Guzman, D., Käb, H., et al. (2020). Bio-based building blocks and polymers—Global capacities, production and trends 2019–2024. Retrieved from http://bio-based.eu/markets/#TRPolymerdata2020: nova-Institut (Ed.), Hürth, Germany.*

The major worldwide share for biodegradable plastics is made up by cellulose acetate (CA) with 46%, followed by starch-containing polymer compounds with 22%, PLA with 16% and PBAT with 14%. Casein polymers, PBS and PHA add up to the 2% of other biodegradable plastics in their worldwide demand (Fig. 7).

The major applications for biodegradable plastics are textiles with 40% and flexible and rigid packaging with 33% (Fig. 8). While textiles are mainly covered by cellulose acetate (cigarette filters and wipes), PLA is the major biodegradable plastic used in packaging. A share of 8% is demanded by consumer goods and 7% by agricultural applications, here also PLA finds a major application share.

Others include automotive, coating, adhesives, building and construction and electrics and electronics and make up 12% of the biodegradable plastics demand. For application in the automotive industry PBS is mainly used, as well as PBAT for coatings and adhesives. The major application share in building and construction is again coming from PLA (Skoczinski et al., 2020).

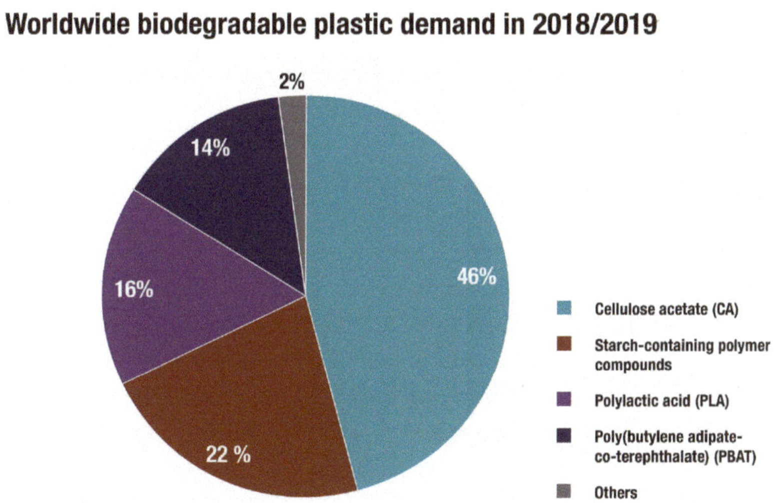

Fig. 7 Worldwide biodegradable plastic demand in 2018/2019. *Adapted from Skoczinski, P., Chinthapalli, R., Carus, M., Baltus, W., de Guzman, D., Käb, H., et al. (2020). Bio-based building blocks and polymers—Global capacities, production and trends 2019–2024. Retrieved from http:/bio-based.eu/markets/#TRPolymerdata2020: nova-Institut (Ed.), Hürth, Germany.*

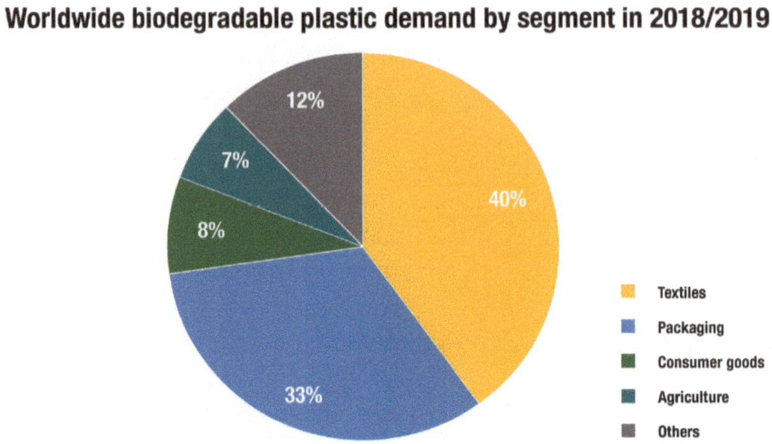

Fig. 8 Worldwide biodegradable plastic demand by segment in 2018/2019. *Adapted from Skoczinski, P., Chinthapalli, R., Carus, M., Baltus, W., de Guzman, D., Käb, H., et al. (2020). Bio-based building blocks and polymers—Global capacities, production and trends 2019–2024. Retrieved from http://bio-based.eu/markets/#TRPolymerdata2020: nova-Institut (Ed.), Hürth, Germany.*

4.4 CO_2-based plastics

CO_2 as a renewable carbon feedstock is currently emerging. More and more technologies have been established to produce a wide range of chemical products, as well as fuels from CO_2. The utilization of CO_2 within polymers is by far the longest established and the most production-volume-related advanced technology. CO_2 can be directly used for the production of different polycarbonates as polyethylene carbonate (PEC) and polypropylene carbonate (PPC) as well for the synthesis of polyol precursors that are then further used to synthesize CO_2-based polyurethanes (Hilken, 2019). Overall, the amount of CO_2 incorporated is currently between 3 (PC) over 20% (PUR) to nearly 50% (PEC and PPC). Other technologies that indirectly use CO_2 via intermediates or fermentative processing steps for polymer synthesis are developing and currently under research (Raschka, Skoczinski, Ravenstijn, & Carus, 2019). For instance, the PHAs and PLAs could be possibly produced from CO_2 via fermentative microbial conversion (Ravenstijn, 2017). Also, electrochemical conversions of CO_2 into polymer building blocks such as ethylene are under development (Tufa et al., 2020). Additionally, CO_2 is currently investigated to be used for the production of textile fibers (Manvi, 2020).

5. Plastic recycling

Plastic recycling can be divided into four types which are the primary, secondary, tertiary, and quaternary recycling (ASTM D7209) (ASTM, 2006).

Both, *primary and secondary recycling* are mechanical recycling methods (e.g. melting and extrusion) in which plastic wastes are processed into a recyclate without significantly changing the molecular structure of the polymer. For extrusion, the material needs to be melted and therefore the mechanical recycling is limited to thermoplastics. The primary recycling is mainly applied on production- or pre-consumer waste with a known homogeneous composition (Achilias et al., 2012). In contrast to primary recycling, the secondary recycling is applied on post-consumer waste which is often contaminated. Therefore the feedstock needs to be pre-processed via collection, sorting, cleaning, drying, and chipping before it can be pelletized or extruded (Sethi, 2016). In comparison to virgin material, the recyclate is often characterized by a lower quality, impurities, and contamination with additives and hazardous substances which cannot be removed via mechanical recycling methods. Other limitations are the presence of water which could induce unwanted chain-scission reactions or trace acidic impurities which may alter the molecular structure (Achilias et al., 2012). Critical factors for secondary recycling are therefore heterogeneity, cleanliness, contamination, as well as the polymer degradation with each recycling loop (Sethi, 2016). Nevertheless, the method is characterized by low costs delivering reliable results depending on the feedstock and pre-processing.

Tertiary recycling is often capable to process a broader range of different plastics and is not only limited to thermoplastics. It may include different mechanisms which are falling under the terms of advanced or chemical recycling but this may vary since it strongly depends on their respective definitions. Depending on the definition three basic mechanisms can be covered which are the dissolution, depolymerization, and conversion. Via dissolution, the contaminations can be removed from plastics without changing the molecular structure of the polymer. The method is based on selectively dissolving and thus separating a targeted polymer from plastic waste in a specified solvent, followed by the removal of non-dissolved components (e.g. additives, pigments, other non-target polymers) and the precipitation of the target polymer from the solvent. In contrast to that, the depolymerization breaks the molecular structure of a polymer into its building blocks such as monomers, dimers, or oligomers. After breakdown, the obtained building blocks often need to be purified from the residual plastic components before new polymers can be synthesized via polymerization. In general, the depolymerization can be initiated through heat (thermal depolymerization), as well as heat combined with chemicals (solvolysis) or enzymes (enzymolysis). Although the enzymolysis is still at an early developmental stage (Wei et al., 2020) several advantages to solvolysis and thermal depolymerization can be expected such as relative mild reaction conditions (e.g. atmospheric pressure and room temperature) and the potential of fine-tuning via enzyme and metabolic engineering. The conversion of plastic waste is carried out via thermochemical methods such as pyrolysis or gasification. These methods are breaking polymers into simpler molecules which are chemically different to the polymers building units. The obtained products are often liquids or gases which can be used as feedstock for the production of plastics, chemicals, and fuels (Sethi, 2016) whereby further upstream processing steps such as refining, reforming, and purification might be necessary.

Quaternary recycling represents the energy recovery from waste via incineration. Plastic wastes are characterized by a high caloric value which can be converted into heat energy during incineration. The produced CO_2 can be captured and recycled as secondary valuable material together with the remaining components which are not burned such as metals and minerals. In comparison to the other methods, the quaternary recycling accepts a large range of feedstocks including municipal solid waste without the need for excessive pre-processing.

6. Plastic waste and disposal

In 2018, the amounts of worldwide post-consumer plastic waste have reached 250 million tonnes, from which 69% were collected. 29% of this collected waste ended up in managed landfill, 20% were sorted for recycling and 20% were used for energy recovery. The remaining 31%, that counts for 77 million tonnes of waste were disposed inappropriately or dumped irresponsibly (Fig. 9) (Conversio Market & Strategy, 2019).

From a European perspective, the production of post-consumer plastic waste was 30 million tonnes in 2018. 97% were collected, used for energy recovery (42%), recycled (32%) and ended up in managed landfill (24%). The amount of waste that was disposed inappropriately, dumped irresponsibly or leaked from the waste stream was 1 million tonnes, which corresponds to over 2% (Fig. 10) (Conversio Market & Strategy, 2019; Plastics Europe, 2019).

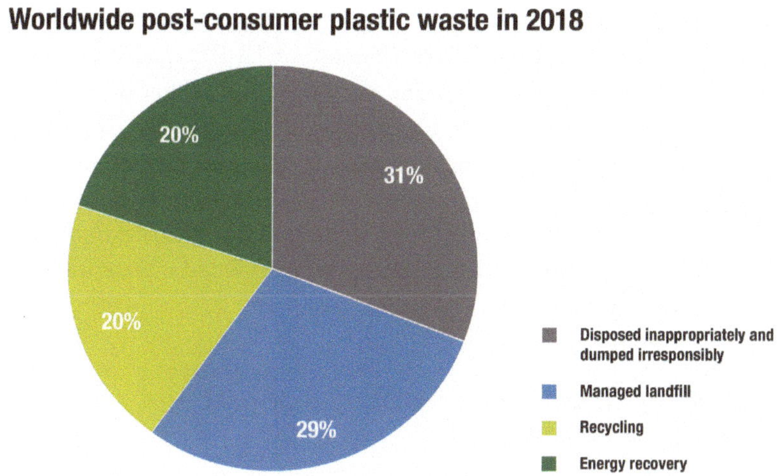

Fig. 9 Worldwide post-consumer plastic waste in 2018. *Data from Conversio Market & Strategy. (2019). Global plastic flow 2018. Retrieved from https:/www.k-online.de/cgi-bin/md_k/lib/pub/object/downloadfile.cgi/Global_Plastics_Flow_Summary_Oct14_2019.pdf?oid=133807&lang=1&ticket=g_u_e_s_t.*

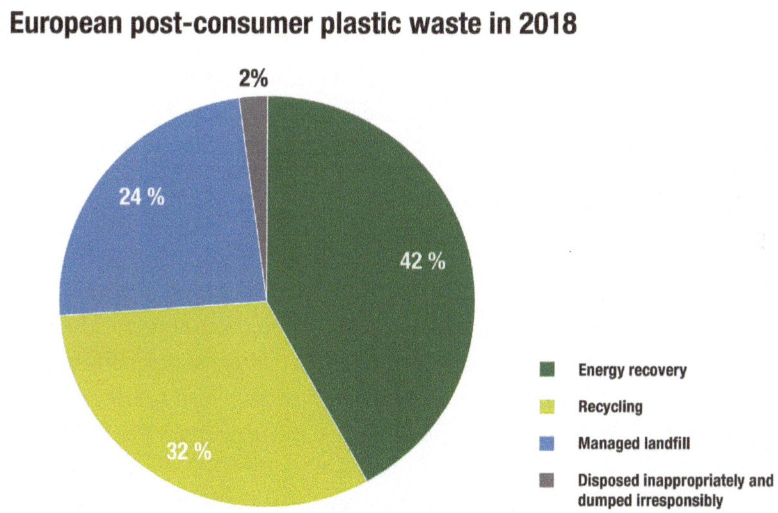

Fig. 10 European post-consumer plastic waste in 2018. *Data (rounded) from Plastics Europe. (2019). Plastics—The facts 2019. Retrieved from https://www.plasticseurope.org/de/resources/publications/1804-plastics-facts-2019 and Conversio Market & Strategy. (2019). Global plastic flow 2018. Retrieved from https:/www.k-online.de/cgi-bin/md_k/lib/pub/object/downloadfile.cgi/Global_Plastics_Flow_Summary_Oct14_2019.pdf?oid=133807&lang=1&ticket=g_u_e_s_t.*

Post-consumer plastic waste collection within the European Union is mainly performed by door-to-door systems. In 18 EU-countries, plastic is collected by this system, four of these countries collect plastic as a separate fraction while in the other countries the plastic is co-mingled together with one (e.g. metal) or more other fractions (European Commission – DG ENV & Copenhagen Resource Institute, 2015). Depending on the member state, these systems vary from one bin to up to six bins or sacks. Besides door-to-door systems additional collection systems are offered such as bring points or civic amenity sites.

7. Mismanagement of plastic disposal and recycling—Plastic littering

Over 30% of the worldwide produced plastic waste is disposed inappropriately and dumped irresponsibly. This accounts for nearly 80 million tonnes of plastic waste ending up in the environment worldwide (Fig. 9) (Conversio Market & Strategy, 2019). Around 25 million tonnes of this waste (10% of world-wide post-consumer plastic waste (Borrelle et al., 2020)) are ending up in the marine environment, the rest remains on land in soils, lakes and rivers. These plastics continue as macro- or microplastics in the environment and can cause severe impacts on ecological systems.

These facts are not only valid for fossil-based plastics, but also for bio-based and biodegradable plastics (Narancic & O'Connor, 2019). Even if a plastic is made from biomass, this does not necessarily mean that nature can handle its remains. As previously explained, being bio-based is not necessarily linked to biodegradability. And even though a plastic is biodegradable, it does not have this ability in general in every environment. Biodegradability depends on complex biogeochemical conditions. The behavior of the plastic can vary, depending on the temperature, the available nutrients and oxygen, as well as the microbial activity at each testing site. Furthermore, other factors as the size of the plastic, the grade, the specific chemical constitution and intrinsic factors can impact the biodegradability (Wei et al., 2020).

Some plastics are labeled biodegradable without being tested and certified as such. Other plastics are certified as being biodegradable in a particular environment, but the lack of explanation and understanding at consumer side leads to inappropriate disposal. An example for this is the certification of being "compostable." This certified biodegradability is linked to being compostable in an industrial composting facility, the conditions here are quite different from those in home compost. But this difference is not clear to the majority of consumers that finally are responsible for the proper disposal. Additionally, so called oxo-degradable plastics have been launched to the market. These plastics include additives, which lead to the fragmentation of the plastic material into micro-fragments through oxidation, or to chemical decomposition. In principle, through a prior fragmentation by oxidation subsequent biodegradation can possibly be enhanced but from experience this just leads to microplastics that pollute the environment.

To properly manage conscious use and disposal of plastic, the correct utilization, communication and understanding of their definitions, certifications and labels have to be ensured and improved. Alongside with this improved recycling technologies, standardized technologies for microplastic leakage as well as an increase in waste management infrastructure especially in developing countries is essential (for detailed scenarios see (Lau et al., 2020; Lebreton & Andrady, 2019)).

8. Political actions against plastic littering

In 2018, the European Commission launched the European Bioeconomy Strategy (European Commission, 2018). This strategy clearly pointed out the significance of bio-based materials for a sustainable and future-oriented European economy, but lacked dedicated political regulations and support mechanisms or incentives for bio-based plastics. This is similar for the Circular Economy Action Plan. This plan was first published in 2015, has been implemented since then and was updated in 2020 (European Commission, 2020). It includes the European Plastics Strategy with different actions for expansion and separation of plastic waste collection, standardization, plastic bag legislation, increase of using recycled plastics etc. as well as some tentative plans on bio-based and biodegradable plastics and a legislative initiative on reducing single-use items.

The establishment of a clear regulatory framework for plastics with biodegradable properties is of great importance to overcome the abovementioned problems of false labeling and misunderstanding. These problems are, among others, specifically addressed in the Plastics Strategy.

The legislative initiative on reducing single-use items is also known as the Single-Use Plastics Directive (SUPD) (European Parliament and Council, 2019) and currently under heavy discussion. This Directive targets all kinds of plastics and does not make any exceptions for bio-based and/or biodegradable plastics, only for "natural polymers." Several measures are included as improved labeling and consumer communication, step-wise reduction in consumption or a full ban of certain single-use products. The background for this legislation was the majority of marine beach litter items being single-use plastic ones (Addamo, Laroche, & Hanke, 2017). The aim of the Directive is the substitution of such single-use items with multi-use ones or a complete change in consumption behavior. Therefore, the following products are banned: Cotton bud sticks, cutlery (forks, knives, spoons, chopsticks), plates, straws, beverage stirrers, sticks to be attached to and to support balloons, food containers made of expanded polystyrene and all products made from oxo-degradable plastics.

At the point of writing this chapter, the last details of implementing the SUPD are still being defined. A final version was first expected in September 2020, but was postponed several times, now to the end of 2020, due to intensive and extensive discussions on the details. One major controversial point are the exemptions from the ban, most prominently with what constitutes a natural polymer exactly (nova-Institut, 2019).

On a worldwide level, many other countries and regions are legislating to ban the manufacture, import free distribution and/or use of single-use plastics (UNEP, 2018). Since the legislative landscape is changing rapidly, even such recent sources as the UNEP (2018) report are not fully up-to-date anymore and it is not possible to present a full overview within the scope of this chapter. It should be noted, though, that regional legislation in this case is often more important than national ones.

9. Conclusion and future development

Plastics are by far the most advanced materials when it comes to the spectrum of properties and applications. In addition, they can be brought into any conceivable form with the greatest efficiency. Nevertheless, plastics have recently gained a more harmful than indispensable image based on plastic waste and littering. These significant problems can and must be solved urgently, as it is already on the political agenda, specifically in the European Union but also from a worldwide point of view. Already implemented and planned bans on single-use plastics as well as improved regulations for labeling and communication of biodegradable plastics and enhancement of plastic recycling target these problems and the future will show how reasonable and successful these strategies are. Biodegradable plastics, for example, should only be used when the collection of other plastics is not possible and an uncontrolled release into the environment is inevitable e.g. agricultural foils or clips for vines in wine growing.

Although the public might assume that it would be manageable to just switch to other materials than plastics, it is not. Because suitable metals and also minerals as sand for cement production are already becoming scarce worldwide. Resources for plastic production on the other hand are infinite when they come from renewable carbon. Renewable carbon entails all carbon sources that avoid or substitute the use of any additional fossil carbon from the geosphere. This way renewable carbon can come from the biosphere, namely biomass, from the atmosphere via direct CO_2 utilization or from the technosphere via recycling.

Renewable carbon circulates between those atmospheres and thereby creating a carbon circular economy (Carus & Raschka, 2018). The transition to renewable carbon in the plastics industry is the equivalent to decarbonization in the energy sector and the most promising long-term vision for mitigating climate change, because it completely avoids the further use of fossil resources.

If the future plastic industry is built on renewable carbon, the worldwide plastic production in 2050 might be independent of further extracted fossil carbon and entirely based on biomass, CO_2, and recycling (Fig. 11).

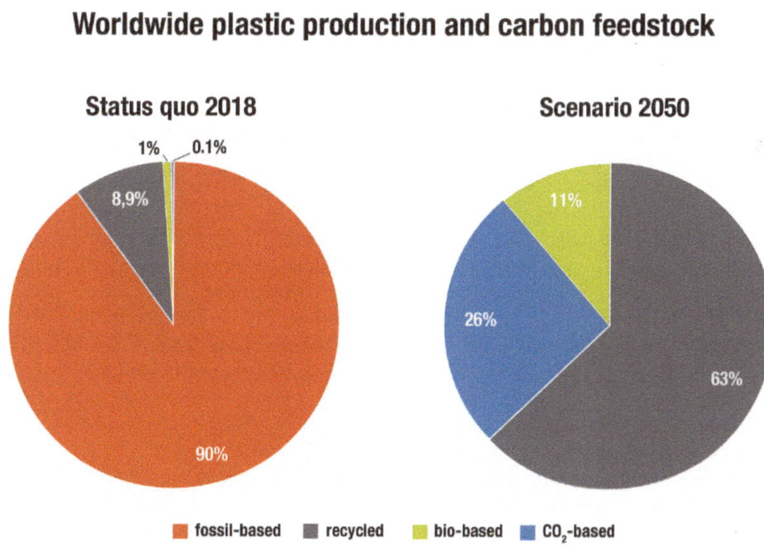

Fig. 11 Worldwide plastic production and carbon feedstock—Status quo in 2018 and future scenario for 2050.

10. Methods

The content within this chapter is based on desk research that includes market analysis, results of market expert interviews, as well as literature research and analysis of recent political drafts, papers and directives.

References

Achilias, D. S., Andriotis, L., Koutsidis, I. A., Louka, D. A., Nianias, N. P., Siafaka, P., et al. (2012). Recent advances in the chemical recycling of polymers (PP, PS, LDPE, HDPE, PVC, PC, nylon, PMMA). In *Material recycling—Trends and perspectives*. IntechOpen. https://doi.org/10.5772/33457.

Addamo, A., Laroche, P., & Hanke, G. (2017). *Top marine beach litter items in Europe. A review and synthesis based on beach litter data*. Retrieved from Luxembourg: Publications Office of the European Union. https://publications.jrc.ec.europa.eu/repository/bitstream/JRC108181/technical_report_top_marine_litter_items_eur_29249_en_pdf.pdf.

ASTM. (2006). Standard guide for waste reduction, resource recovery, and use of recycled polymeric materials and products (ASTM D7209) (withdrawn 2015). In *American society for testing and materials*. USA: ASTM International.

Borrelle, S. B., Ringma, J., Law, K. L., Monnahan, C. C., Lebreton, L., McGivern, A., et al. (2020). Predicted growth in plastic waste exceeds efforts to mitigate plastic pollution. *Science*, *369*(6510), 1515–1518. https://doi.org/10.1126/science.aba3656.

Carus, M., Dammer, L., Puente, Á., Raschka, A., & Arendt, O. (2017). *Bio-based drop-in, smart drop-in and dedicated chemicals*. Retrieved from Hürth, Germany: nova-Institut. http://bio-based.eu/technology/#RtBDropin.

Carus, M., Dammer, L., Raschka, A., Skoczinski, P., & vom Berg, C. (2020). *nova-Paper#12: Renewable carbon—Key to a sustainable and future-oriented chemical and plastic industry*. Retrieved from Hürth, Germany: nova-Institut. http://bio-based.eu/nova-papers/#novapaper12.

Carus, M., & Raschka, A. (2018). *nova-Paper#10: Renewable carbon is key to a sustainable and future-oriented chemical industry*. Retrieved from Hürth, Germany: nova-Institut. http://bio-based.eu/nova-papers/#novapaper10.

CEN. (2014). Bio-based products (EN 16575:2014). In *European committee for standardization—Technical committe 411 (CEN/TC 411)*. Belgium: CEN.

Conversio Market & Strategy. (2019). *Global plastic flow 2018*. Retrieved from https://www.k-online.de/cgi-bin/md_k/lib/pub/object/downloadfile.cgi/Global_Plastics_Flow_Summary_Oct14_2019.pdf?oid=133807&lang=1&ticket=g_u_e_s_t.

CSN. (2015). Bio-based products—Bio-based content—Part 1: Determination of the bio-based content using the radiocarbon analysis and elemental analysis (CSN EN 16785-1). In *Commonwealth standards network—European standards*. European Standards.

CSN. (2017). Bio-based products—Bio-based carbon content—Determination of the bio-based carbon content using the radiocarbon method (CSN EN 16640). In *Commonwealth standards network—European standards*. European Standards.

European Commission. (2018). *Communication from the commission to the European parliament, the council, the European economic and social committee and the committee of the regions. A sustainable bioeconomy for Europe: Strengthening the connection between economy, society and the environment. COM(2018) 673 final. Brussels, Belgium*. Retrieved from https://eur-lex.europa.eu/legal-content/EN/TXT/?uri=CELEX%3A52018DC0673.

European Commission. (2020). *Communication from the commission to the European parliament, the council, the European economic and social committee and the committee of the regions. A new circular economy action plan for a cleaner and more competitive Europe. Brussels, Belgium*. Retrieved from https://eur-lex.europa.eu/legal-content/EN/TXT/?qid=1583933814386&uri=COM:2020:98:FIN.

European Commission—DG ENV, & Copenhagen Resource Institute. (2015). *Assessment of separate collection schemes in the 28 capitals of the EU*. European Commission. Retrieved from https://ec.europa.eu/environment/waste/studies/pdf/Separate%20collection_Final%20Report.pdf.

European Parliament and Council. (2019). *Directive (EU) 2019/904 of the European parliament and of the council of 5 June 2019 on the reduction of the impact of certain plastic products on the environment*. Brussels, Belgium: Official Journal of the European Union. Retrieved from https://eur-lex.europa.eu/legal-content/EN/TXT/?uri=uriserv:OJ.L_.2019.155.01.0001.01.ENG&toc=OJ:L:2019:155:TOC.

Geissdoerfer, M., Savaget, P., Bocken, N. M. P., & Hultink, E. J. (2017). The circular economy—A new sustainability paradigm? *Journal of Cleaner Production*, *143*, 757–768. https://doi.org/10.1016/j.jclepro.2016.12.048.

Geyer, R., Jambeck, J. R., & Law, K. L. (2017). Production, use, and fate of all plastics ever made. *Science Advances*, *3*(7), e1700782. https://doi.org/10.1126/sciadv.1700782.

Hilken, P. (2019). Brighter use of CO_2 for broad range of applications. In *7th conference on carbon dioxide as feedstock for fuels, chemistry and polymers*. Germany: Cologne.

ISO. (2013). Plastics—Vocabulary (ISO 472:2013). In *International Organization for Standardization*. Switzerland: ISO Copyright Office.

Kale, G., Kijchavengkul, T., Auras, R., Rubino, M., Selke, S. E., & Singh, S. P. (2007). Compostability of bioplastic packaging materials: An overview. *Macromolecular Bioscience*, 7(3), 255–277. https://doi.org/10.1002/mabi.200600168.

Lau, W. W. Y., Shiran, Y., Bailey, R. M., Cook, E., Stuchtey, M. R., Koskella, J., et al. (2020). Evaluating scenarios toward zero plastic pollution. *Science*, *369*(6510), 1455–1461. https://doi.org/10.1126/science.aba9475.

Lebreton, L., & Andrady, A. (2019). Future scenarios of global plastic waste generation and disposal. *Palgrave Communications*, *5*(1). https://doi.org/10.1057/s41599-018-0212-7.

Manvi, K. P. (2020). Carbon dioxide: A raw material for textile industry. In *8th conference on carbon dioxide as feedstock for fuels, chemistry and polymers*. Germany: Cologne.

Narancic, T., & O'Connor, K. E. (2019). Plastic waste as a global challenge: Are biodegradable plastics the answer to the plastic waste problem? *Microbiology (Reading)*, *165*(2), 129–137. https://doi.org/10.1099/mic.0.000749.

Narancic, T., Verstichel, S., Reddy Chaganti, S., Morales-Gamez, L., Kenny, S. T., De Wilde, B., et al. (2018). Biodegradable plastic blends create new possibilities for end-of-life management of plastics but they are not a panacea for plastic pollution. *Environmental Science & Technology*, *52*(18), 10441–10452. https://doi.org/10.1021/acs.est.8b02963.

nova-Institut. (2019). *Which polymers are "natural polymers" in the sense of the single-use plastic ban?*. Retrieved from Hürth, Germany: nova-Institut. http://bio-based.eu/policy/#LetterDGEnvironment.

nova-Institut, IKT Stuttgart, OWS, & HYDRA Marine Sciences. (2020). *Biodegradable polymers in various environments*. Retrieved from Hürth, Germany: nova-Institut. http://bio-based.eu/graphics/#biodegPoster.

Plastics Europe. (2019). *Plastics—The facts* (p. 2019). Retrieved from https://www.plasticseurope.org/de/resources/publications/1804-plastics-facts-2019.

Raschka, A., Skoczinski, P., Ravenstijn, J., & Carus, M. (2019). *Carbon dioxide (CO2) as chemical feedstock for polymers—Technologies, polymers, developers and producers*. Retrieved from Hürth, Germany: nova-Institut. http://bio-based.eu/markets/#TRCCU2019.

Ravenstijn, J. (2017). *Bio-based polymers, a revolutionary change*. Retrieved from Hürth, Germany: nova-Institut. http://bio-based.eu/markets/#TRPolymerRevolution.

Sethi, B. (2016). Methods of recycling. In R. Francis (Ed.), *Recycling of polymers: Methods, characterization and applications* (pp. 55–93). Weinheim, Germany: Wiley-VCH.

Skoczinski, P., Chinthapalli, R., Carus, M., Baltus, W., de Guzman, D., Käb, H., et al. (2020). *Bio-based building blocks and polymers—Global capacities, production and trends 2019–2024*. Retrieved from Hürth, Germany: nova-Institut. http://bio-based.eu/markets/#TRPolymerdata2020.

Tufa, R. A., Chanda, D., Ma, M., Aili, D., Demissie, T. B., Vaes, J., et al. (2020). Towards highly efficient electrochemical CO2 reduction: Cell designs, membranes and electrocatalysts. *Applied Energy*, *277*, 1–40. https://doi.org/10.1016/j.apenergy.2020.115557.

UNEP. (2018). *Legal limits on single-use plastic and microplastics: A global review of national laws and regulations*. Retrieved from https://wedocs.unep.org/bitstream/handle/20.500.11822/27113/plastics_limits.pdf?sequence=1&isAllowed=y. https://wedocs.unep.org/bitstream/handle/20.500.11822/27113/plastics_limits.pdf?sequence=1&isAllowed=y.

Wei, R., Tiso, T., Bertling, J., O'Connor, K., Blank, L. M., & Bornscheuer, U. T. (2020). Possibilities and limitations of biotechnological plastic degradation and recycling. *Nature Catalysis*, *3*(11), 867–871. https://doi.org/10.1038/s41929-020-00521-w.

CHAPTER TWO

Methods for microplastic sampling and analysis in the seawater and fresh water environment

Chengjun Sun[a,b,*], Jinfeng Ding[c], and Fenglei Gao[d]

[a]Key Lab of Marine Eco-Environmental Science and Technology, Marine Bioresource and Environment Research Center, First Institute of Oceanography, Ministry of Natural Resources, Qingdao, China
[b]Laboratory of Marine Drugs and Bioproducts, Pilot National Laboratory for Marine Science and Technology, Qingdao, China
[c]College of Environmental Science and Engineering, Ocean University of China, Qingdao, China
[d]Key Laboratory of Marine Chemistry Theory and Technology (Ocean University of China), Ministry of Education, Qingdao, China
*Corresponding author: e-mail address: csun@fio.org.cn

Contents

Abstract

Microplastic (plastic smaller than 5 mm in size) is ubiquitous around the world both in the ocean and the freshwater system. Due to their potential serious negative impact on marine organisms and human beings, marine microplastics have attracted worldwide attention in the past decade. Information and knowledge of the spatial and temporal distribution of marine microplastics are crucial for accurately assessing our current and

Methods in Enzymology, Volume 648
ISSN 0076-6879
https://doi.org/10.1016/bs.mie.2020.12.009

future environmental health conditions. This is also important for developing mitigation plans and measures to protect our environment. Since the measured microplastic pollution level is closely related to the sampling methods and identification techniques, it is important to employ standardized sampling and analysis operation procedures for cross-comparison. In this chapter, we present the basic sampling, sample pretreatment and microplastic identification techniques involved in microplastic pollution assessment and discuss the adaptability of different sampling and pretreatment methods. The pros and cons of different techniques are also discussed.

1. Introduction

Microplastics (defined as particles <5 mm in diameter) (Thompson et al., 2004) have gained global attention due to their ubiquitous distribution and potential risks to bio-organisms and human beings. They come from two sources (Auta, Emenike, & Fauziah, 2017). Primary microplastics are directly manufactured. Secondary microplastics are defined as plastic particles or debris originated from the degradation or fragmentation of large plastics through various physical, chemical, or biological processes. Microplastics have been frequently found in various media including rivers, lakes, seawater, sediments, biota, sea salt, waste water, soil, and air (de Souza Machado et al., 2019; Ding et al., 2019; Eo et al., 2018; Kataoka, Nihei, Kudou, & Hinata, 2018; Kim, Lee, Kim, & Kim, 2018; Liu et al., 2019; Park et al., 2020; Wang et al., 2019; Xiong et al., 2018; Zheng et al., 2019). Geographically they have been found in coastal waters, open oceans, and the polar regions (Desforges, Galbraith, Dangerfield, & Ross, 2014; Eriksen et al., 2014, 2013; Obbard, 2018; Song et al., 2018). Once microplastics are released into or formed in the environment, they are very difficult to be removed. A lot of studies have shown that microplastics are bioavailable to animals and plants (de Souza Machado et al., 2019; Desforges, Galbraith, & Ross, 2015; Ding et al., 2020; Wright, Thompson, & Galloway, 2013). Even though the toxicity of microplastic under environment-related abundance is still under investigation, their potential effects on organisms are not negligible.

The monitoring of microplastics in various environments can provide basic data for determining and assessing their pollution status. There are multiple guides on sampling and analysis (Galgani et al., 2013; GESAMP, 2015; Masura, Baker, Foster, & Arthur, 2015). The reported mean abundance of microplastics in seawater worldwide ranged from 4.8×10^{-6} items/m^3 in the eastern equatorial Pacific (Spear, Ainley, & Ribic, 1995) to 10^{2-3} items/L in the northeast Pacific (Brandon, Freibott, & Sala, 2020), showing a maximum difference of 12 orders of magnitude. Worth noticing is that microplastic abundance tends to increase significantly when going down to smaller sizes (Cózar et al., 2014; Isobe, Uchida, Tokai, & Iwasaki, 2015; Song et al., 2014). Summary of published data shows that mean microplastic abundances in water samples are 0.00028, 0.22, 0.96, 93, and 2444 items/m^3 with mesh sizes of 900–1000, 450–505, 280–350, 120–250, and 10–80 μm, respectively. The eight orders of magnitude difference observed here is largely due to different sampling methods and different water volumes sampled. To obtain reliable data, Liu et al. (2019) recommended using at least 8 m^3 of water samples. Unfortunately, this volume is not possible under a lot of conditions and the time involved can be tremendous. Thus, when comparing abundance, it is important to have comparable methods and consider the lower end of microplastic size during sampling and detection. Most water monitoring studies use Neuston or Manta trawl with mesh sizes of 20–1000 μm to collect microplastics in the surface waters. Other bulk water collection methods with different pore sized membranes have also been used. Subsurface waters are collected by pumps during ship cruising (Desforges et al., 2014) or by submersible pumps place at different depths (Zobkov, Esiukova, Zyubin, & Samusev, 2018). There are also methods for sediment and air (Enyoh, Verla, Verla, Ibe, & Amaobi, 2019; Prata, Costa, Duarte, & Rocha-Santos, 2019). In order to acquire comparable data, monitoring studies require similar or comparable sampling methods. A lack of standard and comparable sampling and analysis protocols has hindered large scale microplastic distribution depicting (Law, 2017; Maximenko et al., 2019; Zhao, Zhu, Gao, & Li, 2018).

In addition, appropriate qualitative and quantitative analyses are also important for reliable microplastic data acquisition. Conventionally, visual sorting with naked eyes or a stereomicroscope has been used to identify large microplastics in the 1–5 mm size range. However, this method cannot be applied to small micro- or nano-sized plastics. Limited by instrumentation, the complexity of microplastics in size, shape, and color, and analytical methods, the most acceptable and practical method so far is micro-FTIR. For the review on microplastic identification method, please refer to Shim, Hong, and Eo (2017).

We focus the current article on water microplastic sampling and sample treatment and discuss the pros and cons of different methods, hoping to provide guidance to researchers sampling under different water conditions in different geographical regions. We also hope this will help to improve the standardization of existing methods and protocols.

2. Water sample collection methods

Different water collection methods can be selected depending on the location and characteristics of the water body, instrumentation and vessel availability, weather condition, and time available. Commonly used methods include towing (both low and high speed, 2–8 knots), bulk sampling at different water depths by submersible pumps or Rosetta sampler, manual sampling with bucket and hand net, and etc. Towing is the most frequently used method for large-scale surface water sampling in both marine and freshwater environments (Dris et al., 2018, Liu, Wu, et al. (2019); Shim, Hong, & Eo, 2018). The difference in towing lies mostly in the size of net mesh and net opening. Medium to small scale samplings by different tools using different pre-concentrating methods or direct filtration have been used for both surface water and water at different depths. Method to choose varies by lab and person but a golden rule is the larger the sampled volume, the better the result can represent the real microplastic abundance.

2.1 Towing

Towing using a net with a mesh size of about 330 μm is conventionally applied for plankton sampling. This method is useful for large-scale surface water microplastic sampling and has helped to provide historical microplastic data. The disadvantage is that it can miss the microplastics of smaller size, especially those smaller than 100 μm, due to the relatively large mesh size.

2.1.1 Materials

- Vessel equipped at the bow side with a boom which can lift at least 3.5 m
- Manta net with collection tube attached: opening: 90–100 cm × 40–50 cm; mesh size: about 330 μm for both monitoring and research purpose, mesh size can go down to 100–150 μm for research purpose; net length: 3 m
- A flow meter that can be attached to the net
- Seawater rinsing hose, prefer using filtered seawater
- 2 L beaker
- Stainless steel tweezer
- 500 mL wide-mouth glass bottles
- Cotton lab coat
- Procedural blank membranes of choice (glass fiber membrane, 0.45 μm nitrocellulose membrane, 2–10 μm metal membrane)

Stainless steel sieves of 5, 1 and 300 μm mesh sizes
- Drying oven (60 °C)
- Analytical balance (0.1 mg)

2.1.2 Sample collection procedures

2.1.2.1 Cruise at a peed between 1 and 3 knots under a sea condition lower than Beaufort level 3. Try to avoid rough water conditions.

2.1.2.2 Attach the Manta net to the boom, close the valve at the end of the collection tube and use the boom on board to first lift and move the Manta net away from the ship and then lower the Manta net to the surface of the water. Make sure the net opening is just above the water surface (Fig. 1). The boom is usually installed toward the bow of the vessel to minimize the wicking effect. Record starting position, wind speed and direction, wave height, temperature, cruise direction, and any other information available.

2.1.2.3 Start towing against the wave for 15–30 min. Constantly watch the net condition. The net should stay just above the water surface all the time. Sampling should be stopped if the net is clogged or torn.

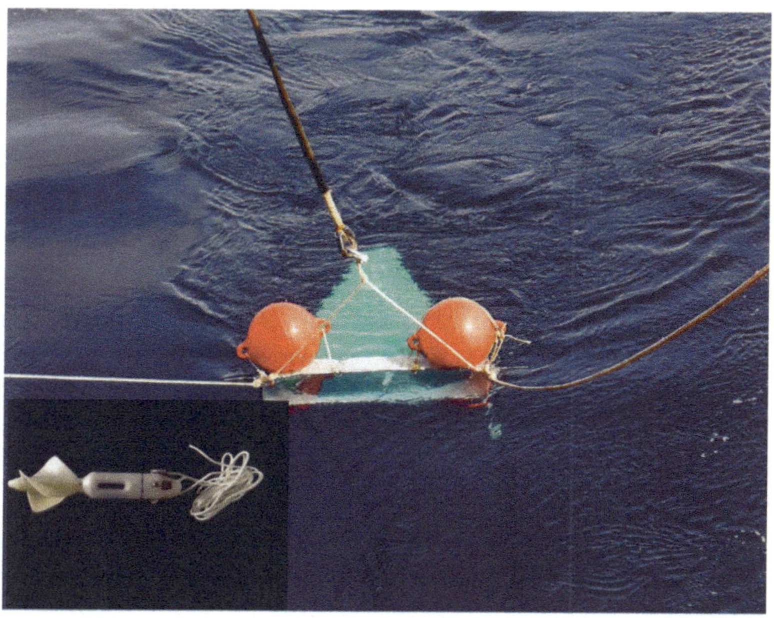

Fig. 1 A modified Manta trawl in-work with a flowmeter attached.

Restart sampling after the net is changed. Make sure to record the new position and time.

2.1.2.4 When towing is done, lift the boom and retrieve the net onboard. Make sure the collection tube is at least 30 cm up in the air. Remove and record the value of the flow meter. Operator handling the net should stay at a downwind position.

2.1.2.5 Use the rinsing hose to flush the outer surface of the net from top-down with seawater (do not let the rinsing seawater enter the net). This step will ensure the sample attached to the inner wall of the net is flushed to the bottom collection tube.

2.1.2.6 Prepare the 2 L beaker and place it close to the end of the collection tube. Carefully and slowly open the valve at the end of the collection tube. Collect all the content in the collection tube. If there are large plastic pieces, use a tweezer to retrieve them and place them together with the rest of the sample. If there are large non-plastic materials in the collection tube that do not fit in the 500 mL glass bottle, use the tweezer to retrieve those pieces and wrap them in aluminum foil before taking them to the lab for further handling. Or carefully and thoroughly rinse the surface of these pieces, pool the rinse off liquid with the collected sample, and discard these large non-plastic pieces.

2.1.2.7 Close the valve at the end of the collection tube, rinse the collection tube with pure distilled water in a rinse bottle. Collect and pool the liquid with the content above. Repeat this step at least two times to ensure all the content in the collection tube is rinsed off and collected.

2.1.2.8 If the on-board filtration system is not available, transfer all the content from the beaker to the wide-mouth glass bottle. Seal the bottle and store it at 4 °C till further processing in the lab. When on-board filtration is available, There is no need to transfer the content from the beaker to the wide-mouth glass bottle. Cover the beaker with aluminum foil and continue with the following further separation steps.

2.1.2.9 Depending on the total amount of particles in the above sample, the collected sample can either sequentially go through 5, 1, and 0.3 mm (Fig. 2) sieve or be directly filtrated (Fig. 3). If there are a lot of visible large pieces, sequential sieving is often used. The

Fig. 2 Sieves sequentially stacked together. Mesh sizes (from top down) are 5, 1, and 0.3 mm, respectively.

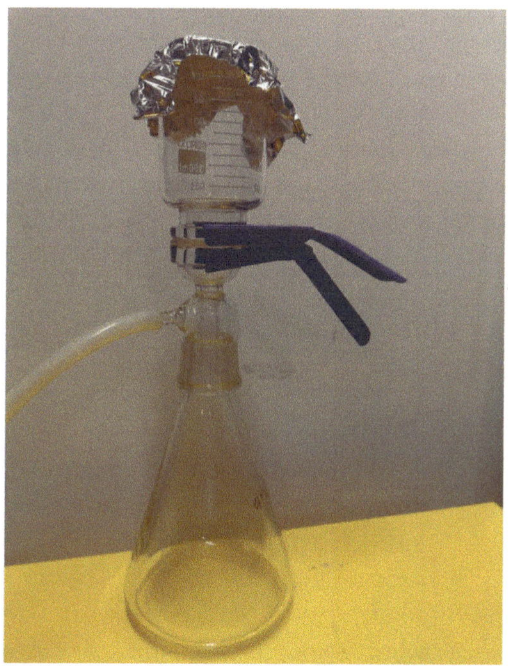

Fig. 3 Glass filter set used for filtration.

5 mm sieve is usually first used to separate any pieces larger than 5 mm from those smaller than 5 mm. Plastic and suspected plastic pieces larger than 5 mm can be manually picked from this fraction, placed in an aluminum-foil-covered glass beaker or petri dish, and put in the drying oven. The total number of microplastics should be recorded. Fractions collected on the 1 mm sieve and 0.3 mm sieves can go through the same hand-picking and drying procedure. For research purpose, we recommend substituting the 0.3 mm sieve with direct filtration using a 0.45 μm membrane. For open ocean water and other clean water samples without much visible microplastic pieces, the content in the beaker can be filtrate directly onto a 0.45 μm membrane for further pre-treatment and identification. In some cases, one might encounter zooplankton in the sample. Try to separate these zooplankton under a dissecting microscope if possible.

2.1.2.10 Dry the microplastic in the oven at 60 °C until constant weight and weigh the microplastics. Record the total weight of each microplastic fraction. For the samples with very few visible microplastic pieces, weight recording might not be possible. Use the total number of identified microplastic to report the data. All suspected microplastic pieces should go through the following sample pre-treatment and identification steps.

Note: When high speed towing is possible, one can employ the AVANI (All-purpose Velocity Accelarated Net Instrument) trawl (Eriksen et al., 2018) and tow at a speed up to 8 knots. Towing time will depend on the actual trawl parameters and total area intended to cover.

2.2 Direct pumping

Direct pumping can be used both on moving and stationary vessels wherever there is a pump on-board. A submersible pump can be used for collecting water samples at different depths. Direct pumping can be used to collect a large volume of water sample in tens or hundreds of liters with more freedom in sample volume and mesh size choices. A principal rule is to collect as much water as possible while using a mesh size as small as practical. This method is particularly useful for collecting surface water samples along the path of a cruising vessel and a large water volume with a submersible pump.

2.2.1 Materials

- A submersible pump installed on-board or be used independently
- A 20 μm mesh sized concentrating net with a 100 mL concentrating tube and valve at the end with a supporting frame (Fig. 4). Mesh sizes can vary from 10 to 330 μm
- A stainless steel frame to support the concentrating net
- A clamp to secure the incoming seawater hose to the concentration net
- A flow meter
- A glass filtration set
- 0.45 μm filter membrane
- A 2 L glass beaker
- 500 mL wide-mouth glass bottles
- cotton lab coat
- Procedural blank membranes and filter membrane of choice (glass fiber membrane, 0.45 μm nitrocellulose membrane, 2–10 μm metal membrane)
- Stainless steel sieves of 5, 1 and 300 μm mesh sizes
- Drying oven (60 °C)
- Watch glasses

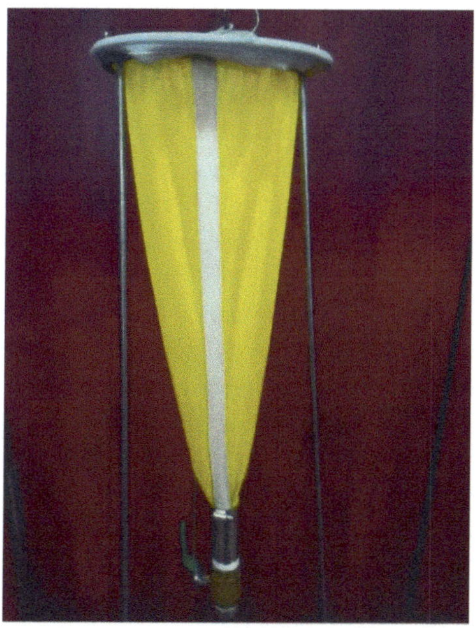

Fig. 4 A 20 μm mesh sized concentrating net standing on a metal frame.

2.2.2 Sample collection procedure

2.2.2.1 Place the submersible pump at the right water depth. Turn on the pump and let it run for at least 10 min to rinse the pipeline. Keep the pump running at a flow rate no higher than 10 L/min or measure the flow with a flow meter attached to the outlet of the pump. This flow rate is adjustable depending on the speed the pre-concentrating net can handle. For the pump installed on-board, simply turn on the pump and let the water run for at least 10 min before starting water collection.

2.2.2.2 Close the collection tube valve at the end of the concentrating tube. Connect the pump outlet to the concentrating net or simply secure the pump outlet close to the mouth of the concentrating net and start sampling. Record the sampling starting time and position.

2.2.2.3 Let the water flow until at least 500 L sample is processed. A larger volume is preferred. Constantly watch the net to avoid overflow. Record sampling end time, position, and total volume processed.

2.2.2.4 Rinse the inner surface of the net with Milli-Q H_2O from top-down. Then cover the concentrating net mouth with aluminum foil. Rinse the outer surface of the net from top-down with filtered seawater or Milli-Q water (do not let the rinse water enter the net from the mouth). Make sure the sample attached to the inner wall of the net is flushed to the collection tube.

2.2.2.5 Place the glass beaker on the floor or bench close to the end of the collection tube. Carefully and slowly open the valve at the end of the collection tube. Collect all the content in the collection tube. Use a tweezer to remove large pieces if needed.

2.2.2.6 Close the collection tube valve, rinse the collection tube from outside with 50 mL pure distilled water. Collect and pool the liquid with the content into the beaker. Repeat the rinsing step twice.

2.2.2.7 Either transfer all the content to the wide-mouth glass bottle for storage or cover the beaker with aluminum foil and bring the beaker back to the lab for filtration. Use a 5 mm sieve to separate or tweezer to pick out larger particles. Filtrate the rest of the content onto a 0.45 μm nitrocellulose membrane or 2–10 μm metal membrane with a glass filter unit. Microplastic pieces that can be hand-picked can be separately wrapped and stored in aluminum foil.

2.2.2.8 Store the membrane in between two pieces of watch glasses, seal the glasses, and prepare for further treatment and identification.

Note: A blank procedural membrane should be prepared during each sampling. Put a wet blank membrane on one piece of watch glass and place it right beside the concentrating net for the duration of sampling. Seal the blank membrane with another piece of watch glass at the end of sampling to serve as a procedural blank.

2.3 Bulk sampling

Bulk sampling is used for collecting water samples with a finite volume at specific sites or depths. It can be done with a Rosette sampler, large volume water sampler, manual surface water sampling with a stainless steel bucket, hand net, or other devices. Bulk sampling usually collects a smaller volume than direct pumping. The advantage of this method is that it can sample at places where other methods are not feasible, for example, a Rosette sampler or large volume water sampler for deep water, and manual surface water sampling for where no other water sampling devices are available.

2.3.1 Materials

- Rosetta sampler, large volume water sampler or a stainless steel bucket
- Optional: a 20 μm mesh sized concentrating net with a 100 mL concentrating tube and valve at the end. Mesh sizes can vary from 10 μm to 330 μm
- Optional: a frame to support the concentrating net
- A glass filtration set
- 0.45 μm nitrocellulose filter membrane
- Cotton lab coat
- Stainless steel tweezer
- Watch glasses
- Optional: one 2 L glass beaker
- Optional: 500 mL wide-mouth glass bottles
- cotton lab coat
- Procedural blank membranes and filter membrane of choice (glass fiber membrane, 0.45 μm pore size nitrocellulose membrane, 2–10 μm pore size metal membrane)

2.3.2 Sample collection procedure

2.3.2.1 Follow standard Rosetta sampler or large volume sample collection protocol, collect seawater samples at different depths. Retrieve the sampler on board (Fig. 5).

Fig. 5 A Rosetta sampler retrieved on board.

2.3.2.2 Option #1: close the collection tube valve at the end of the concentrating net. Place the concentrating net right below the Rosetta sampler bottle. Slowly open the sampler bottle and let all the collected water pass through the concentrating net to concentrate. Each sampler bottle usually holds 5–20 L of water. Follow steps 2.2.2.4–2.2.2.8 to finish sample collection and pre-concentrating.

2.3.2.3 Option #2: For volume-limited samples, use clean glass bottles to collect a certain volume of water from the sampler. Cap the bottle and bring it back to the lab. Directly filtrate all the water collected using a glass filter unit with the membranes of pore size of choice.

2.3.2.4 Store the filter membrane in between two pieces of watch glasses and get ready for further processing.

Note: For deep water sampling or under special occasions, if the total volume of water sampled is less than 5 L, we recommend direct filtration onto metal membrane or 0.45 μm nitrocellulose filter membrane to collect all possible microplastic pieces.

3. Sample pretreatment before identification

Whether the sample needs pre-treatment before identification or not depends on how much organic material there is in the sample. Some samples can be directly loaded onto the FT-IR or μ-FT-IR or Raman for identification because there is little or no interfering organics. This is the case for some of the non-weathered microplastic pieces larger than 1 mm and membranes with little residue. In most cases, the samples need further clean up to minimize organic material interference before identification either because there is a lot of plankton or the microplastic surface is covered with organic materials. It is well known that the surfaces of microplastics are covered with microbes in the ocean (Oberbeckmann & Labrenz, 2020). Occasionally, when the whole sample is filtrated onto a small pore-sized membrane, there is too much organic material that further separation steps are needed after organic material removal. Since actual samples vary greatly in the amount, size, and composition of insoluble material, the protocol to choose will depend on the nature of the samples. If there is a large amount of insoluble material, please refer to the NOAA guide (Masura et al., 2015). Otherwise, we recommend the following protocol.

3.1 Materials

- Sonicator
- 30% H_2O_2
- Optional: 0.05 M $FeSO_4$, 10% KOH, both filtered through 0. 45 μm membrane
- 250 mL and 500 mL beaker
- Milli-Q water or distilled pure water
- Stainless steel tweezer
- Glass separation funnel
- Glass filtration system
- 0.45 μm membrane
- Solid NaCl
- 1.6 g/mL $ZnCl_2$
- Optional: sodium iodide (1.4 g/mL), lithium metatungstate (1.62 g/mL)

3.2 Sample pre-treatment procedure

Sample pre-treatment serves the purpose of removing the interfering organics either on the surface of the microplastics or mixed with the microplastics. Both 30% H_2O_2 and 10% KOH are commonly used to either oxidize or degrade the organics. It needs to be noted that some colors in the microplastics might fade or disappear when using 30% H_2O_2.

3.2.1 Transfer all the content on the 1 mm, 0.3 mm sieve and membrane to clean 250 mL beakers. Special care is needed when transferring the material on the membrane to the beaker. Make sure everything on the membrane is rinsed off with Milli-Q H_2O. Check the rinsed membrane under a stereomicroscope to ensure complete transfer.

3.2.2 Cover the beaker loosely with aluminum foil. Place the beaker in an oven set at 60 °C overnight to dry.

3.2.3 Slowly add 20–50 mL of preheated (60 °C) 30% H_2O_2 into the beaker. Cover the beaker with aluminum foil. Wait until no violent bubbling. Sonicate for 2–5 min. Adding 0.05 M $FeSO_4$ to a final concentration of 0.025 M is optional. In our experience, we found adding Fe (II) can sometimes make the reaction too violent which increases the risk of losing samples. In cases with the zooplankton-containing sample, 10% KOH works better than H_2O_2.

3.2.4 Place the beaker in the incubate at 60 °C overnight.

3.2.5 Check the content in the beaker and make sure there is no visible organic material. If there is still organic material, add more 30% H_2O_2 depending on how much organic material is left and continue oxidation at 60 °C until no visible organic material.

3.2.6 Separate the microplastic after oxidation

3.2.6.1 Option #1: For density separation with NaCl, add 3.6 g solid NaCl to each 10 mL solution in the beaker or add until NaCl can no longer dissolve in a clean hood. Then transfer all the solution to a separation funnel. Rinse the beaker with saturated NaCl solution and pool the wash solution to the separation funnel. Settle overnight and then separate the solid in the top layer into a new beaker. Dry and store the top layer solid in a foil-covered beaker or Petri dish.

3.2.6.2 Option #2: Place the beaker into another clean larger glass container or beaker in a clean hood. Add 3.6 g solid NaCl to each 10 mL solution in the beaker or add until NaCl can no longer dissolve. Keep gently adding filtered saturated NaCl solution along the side of the beaker until all insoluble material is overflown. Collect all the overflow and rinse the outside of the beaker with Milli-Q water to ensure all insoluble material is collected. Dry the sample in the larger beaker or filtrate it onto a 0.45 μm membrane.

3.2.6.3 Option #3: For the sample with no visible microplastics or very few pieces of microplastics, take the beaker out from the oven and filtrate the solution through a 0.45 μm membrane when the solution is still warm. Filtrate without cooling enables easier pass-through of the dissolved organics and leave less organic material on the membrane.

Note: Other solutions can be used for density separation, such as zinc chloride (1.6 g/mL) solution, sodium iodide (1.4 g/mL), lithium metatungstate (1.62 g/mL).

4. Sample identification

4.1 Visual inspection with a stereomicroscope

Visual sorting is often used before the spectroscopic method to save time. Large microplastic pieces (>1 mm) can often be identified by the naked eye even though spectroscopic identification is still needed to confirm the polymer type because visual identification is very subjective. The size of the microplastic is usually measured under a stereomicroscope. For smaller sized microplastics or suspected microplastics, we recommend keeping them on the membrane and mark their positions under the stereomicroscope before loading the membrane onto a spectrometer for polymer identification (Zheng et al., 2019). To date, spectroscopic identification of polymer type is a must to ensure data quality.

Epifluorescence and staining techniques have been used to identify microplastics in a sample (Brandon et al., 2020; Shim et al., 2017). There are also other labeling techniques. The application of these methods is limited by the complexity of natural samples.

4.2 Microplastic polymer composition identification

There are different methods for polymer identification. Non-destructive methods include FT-IR or μ-FT-IR and Raman spectrometry. Destructive method includes Pyrolysis-GC/MS. There are also other complementary methods including SEM and AFM. While FT-IR seems to be the most widely accepted method, most of these methods are complementary to each other. Please refer to reviews by Dris et al. (2018), Nguyen et al. (2019), and Shim et al. (2017), for a detailed description of each method. We recommend using μ-FT-IR if it is possible.

5. Data reporting

Calculate the total volume of sampled water based on the formula the flow meter gives. Usually, the formula is as following:

$V = R \times S$, in which V is the total volume of water sampled, R is the reading on the flowmeter, S is the opening area of the net which is by multiplying the length and width of the opening. For some programmable flow meter, reading can be set to directly reflect total volume.

Divide the number of total microplastic pieces or the total weight of microplastics by the total volume of sampled water to calculate the microplastic abundance. Report microplastic abundance data in both weight/m^3 and number/m^3. For the samples with non-weighable microplastic, use only number/m^3 to report the data. Some towing reports have used weight/km^2 and number/ km^2. This can be done by dividing the number of total microplastic pieces or the total weight of microplastic by the total water area sampled. The total area is calculated by multiplying the net width and the distance of sampling. It is always favorable to report as many units as possible to enable data comparison.

6. Blank and contamination control

With the increasing amount of microplastic found in various media, extreme precautions need to be taken in the whole process of sampling, sample treatment, and sample analysis to avoid contamination. General precautions like wearing cotton lab-coats, using non-plastic containers or tools, using filtered solutions, minimize airborne contaminants, and etc. should be taken at all possible times. Control experiments and procedural blanks should be conducted whenever possible.

7. Summary

Accurate quantification of microplastics is crucial for their risk assessment. There are multiple steps involved in the quantification process in place of various environments and media involved. With a quickly evolving field and a vast difference in geographical position, water characteristics, weather conditions, time availability, instrumentation and vessel accessibility, manpower, and other factors, the lack of standardization in each step will contribute greatly to largely varied data. While protocol standardization is a priority toward comparable data, we need to recognize and acknowledge the complexity of the nature of microplastic. While it might be hard to find a protocol that fits all conditions, we should always observe the best practice to minimize variation and inaccuracy, maximize consistency, and improve comparability.

Acknowledgments

This work was supported by the Basic Scientific Fund for National Public Research Institutes of China (No. 2020Q10), Investigation and Evaluation of Microplastics in Seawater (No. ZY0720007), China Ocean Mineral Resources R & D Association (No. DY135-E2-1-07, DY135-E2-4-05) , Basic Scientific Fund for International Marine Debris & Microplastic Cooperation (No. QY0920009).

References

Auta, H. S., Emenike, C. U., & Fauziah, S. H. (2017). Distribution and importance of microplastics in the marine environment: A review of the sources, fate, effects, and potential solutions. *Environment International, 102*, 165–176.

Brandon, J. A., Freibott, A., & Sala, L. M. (2020). Patterns of suspended and salp-ingested microplastic debris in the North Pacific investigated with epifluorescence microscopy. *Limnology and Oceanography Letters, 5*, 46–53.

Cózar, A., Echevarria, F., Gonzalez-Gordillo, J. I., Irigoien, X., Ubeda, B., Hernandez-Leon, S., et al. (2014). Plastic debris in the open ocean. *Proceedings of the National Academy of Sciences of the United States of America, 111*, 10239–10244.

Desforges, J. W., Galbraith, M., Dangerfield, N., & Ross, P. S. (2014). Widespread distribution of microplastics in subsurface seawater in the NE Pacific Ocean. *Marine Pollution Bulletin, 79*, 94–99.

Desforges, J. P., Galbraith, M., & Ross, P. S. (2015). Ingestion of microplastics by zooplankton in the Northeast Pacific Ocean. *Archives of Environmental Contamination and Toxicology, 69*, 320–330.

de Souza Machado, A. A., Lau, C. W., Kloas, W., Bergmann, J., Bachelier, J. B., Faltin, E., et al. (2019). Microplastics can change soil properties and affect plant performance. *Environmental Science & Technology, 53*, 6044–6052.

Ding, J., Jiang, F., Li, J., Wang, Z., Sun, C., Wang, Z., et al. (2019). Microplastics in the coral reef systems from Xisha Islands of South China Sea. *Environmental Science & Technology, 53*, 8036.

Ding, J., Li, J., Sun, C., Jiang, F., He, C., Zhang, M., et al. (2020). An examination of the occurrence and potential risks of microplastics across various shellfish. *Science of the Total Environment, 739*, 139887.

Dris, R., Imhof, H. K., Löder, M. G. J., Gasperi, J., Laforsch, C., & Tassin, B. (2018). Microplastic contamination in freshwater systems: Methodological challenges, occurrence and sources. In E. Zeng (Ed.), *Microplastic Contamination in Aquatic Environments* (pp. 51–93). Elsevier Inc.

Enyoh, C. E., Verla, A. W., Verla, E. N., Ibe, F. C., & Amaobi, C. E. (2019). Airborne microplastics: A review study on method for analysis, occurrence, movement and risks. *Environmental Monitoring and Assessment, 191*, 668.

Eo, S., Hong, S. H., Song, Y. K., Lee, J., Lee, J., & Shim, W. J. (2018). Abundance, composition, and distribution of microplastics larger than 20 μm in sand beaches of South Korea. *Environmental Pollution, 238*, 894–902.

Eriksen, M., Lebreton, L. C. M., Carson, H. S., Thiel, M., Moore, C. J., Borerro, J. C., et al. (2014). Plastic pollution in the world's oceans: More than 5 trillion plastic pieces weighing over 250,000 tons afloat at sea. *PLoS One, 9*, 0111913.

Eriksen, M., Liboiron, M., Kiessling, T., Charron, L., Alling, A., Lebreton, L., et al. (2018). Microplastic sampling with the AVANI trawl compared to two neuston trawls in the bay of Bengal and South Pacific. *Environmental Pollution, 232*, 430–439.

Eriksen, M., Maximenko, N., Thiel, M., Cummins, A., Lattin, G., Wilson, S., et al. (2013). Plastic pollution in the South Pacific subtropical gyre. *Marine Pollution Bulletin, 68*, 71–76.

Galgani, F., Hanke, G., Werner, S., Oosterbaan, L., Nilsson, P., Fleet, D., et al. (2013). *Guidance on monitoring of marine litter in European seas*. European Commission http://hdl.handle.net/10508/1649.

GESAMP. (2015). Sources, fate and effects of microplastics in the marine environment. In *Reports and Studies. IMO/FAO/UNESCO-IOC/UNIDO/WMO/IA EA/UN/UNEP/UNDP Joint Group of Experts on the scientific aspects of marine environmental protection, GESAMP*.

Isobe, A., Uchida, K., Tokai, T., & Iwasaki, S. (2015). East Asian seas: A hot spot of pelagic microplastics. *Marine Pollution Bulletin, 101*, 618–623.

Kataoka, T., Nihei, Y., Kudou, K., & Hinata, H. (2018). Assessment of the sources and inflow processes of microplastics in the river environments of Japan. *Environmental Pollution, 244*.

Kim, J. S., Lee, H. J., Kim, S. K., & Kim, H. J. (2018). Global pattern of microplastics (MPs) in commercial food-grade salts: Sea salt as an indicator of seawater MP pollution. *Environmental Science and Technology, 52*, 12819–12828.

Law, K. L. (2017). Plastics in the marine environment. *Annual Review of Marine Science, 9*, 205–229.

Liu, K., Wu, T., Wang, X., Song, Z., Zong, C., Wei, N., et al. (2019). Consistent transport of terrestrial microplastics to the ocean through atmosphere. *Environmental Science and Technology, 53*.

Liu, K., Zhang, F., Song, Z., Zong, C., Wei, N., & Li, D. (2019). A novel method enabling the accurate quantification of microplastics in the water column of deep ocean. *Marine Pollution Bulletin, 146*, 462–465.

Masura, J., Baker, J., Foster, G., & Arthur, C. (2015). Laboratory methods for the analysis of microplastics in the marine environment: Recommendations for quantifying synthetic particles in waters and sediments. In *NOAA technical memorandum NOS-OR&R-48*.

Maximenko, N., Corradi, P., Law, K. L., Van Sebille, E., Garaba, S. P., Lampitt, R. S., et al. (2019). Toward the integrated marine debris observing system. *Frontiers in Marine Science, 6*, 447.

Nguyen, B., Claveau-Mallet, D., Hernandez, L. M., Xu, E. G., Farner, J. M., & Tufenkji, N. (2019). Separation and analysis of microplastics and nanoplastics in complex environmental samples. *Accounts of Chemical Research, 52*, 858–866.

Obbard, R. W. (2018). Microplastics in polar regions: The role of long range transport. *Current Opinion in Environmental Science & Health, 1*, 24–29.

Oberbeckmann, S., & Labrenz, M. (2020). Marine microbial assemblages on microplastics: Diversity, adaptation, and role in degradation. *Annual Review of Marine Science, 12*, 209–232.

Park, H. J., Oh, M. J., Kim, P. G., Kim, G., Jeong, D. H., Ju, B. K., et al. (2020). National reconnaissance survey of microplastics in municipal wastewater treatment plants in Korea. *Environmental Science & Technology, 54*, 1503–1512.

Prata, J. C., Costa, J. P. D., Duarte, A. C., & Rocha-Santos, T. (2019). Methods for sampling and detection of microplastics in water and sediment: A critical review. *TrAC Trends in Analytical Chemistry, 110*, 150–159.

Shim, W. J., Hong, S. H., & Eo, S. (2017). Identification methods in microplastic analysis: A review. *Analytical Methods, 9*, 1384–1391.

Shim, W. J., Hong, S. H., & Eo, S. (2018). Marine microplastics: Abundance, distribution, and composition. In E. Zeng (Ed.), *Microplastic Contamination in Aquatic Environments* (pp. 1–26). Elsevier Inc.

Song, Y. K., Hong, S. H., Eo, S., Jang, M., Han, G. M., Isobe, A., et al. (2018). Horizontal and vertical distribution of microplastics in Korean coastal waters. *Environmental Science and Technology*, *52*, 12188–12197.
Song, Y. K., Hong, S. H., Jang, M., Kang, J. H., Kwon, O. Y., Han, et al. (2014). Large accumulation of micro-sized synthetic polymer particles in the sea surface microlayer. *Environmental Science & Technology*, *48*, 9014–9021.
Spear, L. B., Ainley, D. G., & Ribic, C. A. (1995). Incidence of plastic in seabirds from the tropical pacific, 1984–1991: Relation with distribution of species, sex, age, season, year and body weight. *Marine Environmental Research*, *40*, 123–146.
Thompson, C. R., Olsen, Y., Mitchell, R. P., Davis, A., Rowland, S. J., John, A. W., et al. (2004). Lost at sea: Where is all the plastic? *Science*, *304*, 838.
Wang, J., Liu, X., Li, Y., Powell, T., Wang, X., Wang, G., et al. (2019). Microplastics as contaminants in the soil environment: A mini-review. *Science of the Total Environment*, *691*, 848–857.
Wright, S. L., Thompson, R. C., & Galloway, T. S. (2013). The physical impacts of microplastics on marine organisms: A review. *Environmental Pollution*, *178*, 483–492.
Xiong, X., Zhang, K., Chen, X., Shi, H., Luo, Z., & Wu, C. (2018). Sources and distribution of microplastics in China's largest inland Lake - Qinghai lake. *Environmental Pollution*, *235*, 899–906.
Zhao, S., Zhu, L., Gao, L., & Li, D. (2018). Limitations for microplastic quantification in the ocean and recommendations for improvement and standardization. In E. Zeng (Ed.), *Microplastic Contamination in Aquatic Environments* (pp. 27–49). Elsevier Inc.
Zheng, Y., Li, J., Cao, W., Liu, X., Jiang, F., Ding, J., et al. (2019). Distribution characteristics of microplastics in the seawater and sediment: A case study in Jiaozhou BayChina. *Science of the Total Environment*, *674*, 27–35.
Zobkov, M. B., Esiukova, E. E., Zyubin, A. Y., & Samusev, I. G. (2018). Microplastic content variation in water column: The observations employing a novel sampling tool in stratified Baltic Sea. *Marine Pollution Bulletin*, *138*, 193–205.

CHAPTER THREE

Exploring microbial consortia from various environments for plastic degradation

Ingrid Eileen Meyer Cifuentes and Basak Öztürk*
Junior Research Group Microbial Biotechnology, Leibniz Institute DSMZ German Collection of Microorganisms and Cell Cultures, Braunschweig, Germany
*Corresponding author: e-mail address: basak.oeztuerk@dsmz.de

Contents

Abstract

Many complex natural and synthetic compounds are degraded by microbial assemblages rather than single strains, due to usually limited metabolic capacities of single organisms. It can therefore be assumed that plastics can be more efficiently degraded by microbial consortia, although this field has not been as widely explored as plastic degradation by individual strains. In this chapter, we present some of the current studies on this topic and methods to enrich and cultivate plastic-degrading microbial consortia from aquatic and terrestrial ecosystems, including substrate preparation and biodegradation assessment. We focus on both conventional and biodegradable plastics as potential growth substrates. Cultivation methods for both aerobic and anaerobic microorganisms are presented.

Methods in Enzymology, Volume 648
ISSN 0076-6879
https://doi.org/10.1016/bs.mie.2020.12.005

1. Introduction

Biodegradation of hydrocarbons and xenobiotics by microbial consortia is a widely studied phenomenon. Microorganisms acting in a consortium can have higher biodegradation efficiencies than individual strains, either by having a direct role in biodegradation or removal of potentially toxic degradation intermediates (Dejonghe et al., 2003; Pelz et al., 1999; Yu et al., 2019). Consortium members can also improve biodegradation indirectly, for example by secreting metabolites that induce co-metabolic degradation (Chen & Aitken, 1999) or metabolic cross-feeding (Hu et al., 2020; Sorensen, Ronen, & Aamand, 2002). In terms of plastic degradation, most studies have so far focused on the isolation of single strains with degradative capability. Examples from the degradation of other pollutants, however, show that consortia have a great potential in plastic waste biodegradation and bioremediation.

Several studies have been published on the biodegradation of plastics by bacterial and fungal consortia. Vargas-Suárez et al. reported the biodegradation of polyester polyurethane (PUR) and polyacrylic polymers by enrichment cultures, where the degradation was more efficient in the consortium than by individual community members (Vargas-Suárez, Fernández-Cruz, & Loza-Tavera, 2019). Similarly, a polyurethane-acrylate copolymer with additives was degraded by a bacterial enrichment culture, which hosted a broad range of genes and pathways related to xenobiotic degradation (Gaytán et al., 2020). In both cases, polymers were supplied as suspensions, and cultures were started from deteriorated PUR foam pieces from a landfill site. Zafar et al. reported the deterioration of PUR coupons by fungal compost communities (Zafar et al., 2014), but it is unclear whether the fungal strains perform the degradation individually, or as a consortium. Among other polymers for which degradation by consortia have been proposed are polyethylene and polypropylene (Skariyachan et al., 2018), weathered polyethylene (W-PE) (Syranidou et al., 2017) and polystyrene (PS) (Syranidou et al., 2017). It is worth noting, however, that in these studies the degradation was mainly determined by measuring weight loss and by monitoring changes in Fourier transform infrared spectroscopy (FTIR) FTIR spectra which can also result from the breakdown of chemical additives (Danso, Chow, & Streit, 2019). Synergistic microbial degradation was also shown with the biodegradable plastic polycaprolactone (PCL). This polymer was completely degraded by a two-member thermophilic consortium with one facilitating the other by removing degradation intermediates and thus maintaining an optimum pH (Nakasaki, Matsuura, Tanaka, & Sakai, 2006). For other common polymers such as polyamides or poly(ethylene terephthalate) (PET), degradation by consortia has so far not been demonstrated.

In this chapter we provide an overview of current studies that involve plastic degradation by microbial consortia under both aerobic and anaerobic conditions. In addition, we describe methods to prepare plastic substrates for enrichment cultures, to quantify biodegradation and cultivate enrichment cultures from different environmental samples.

2. Enrichment and cultivation of aerobic plastic-degrading consortia

Enrichment of plastic-degrading consortia from environmental samples follows the same general principles that are used to enrich and isolate other xenobiotic-degrading microorganisms from the environment. Samples are generally acquired from plastic-contaminated environments such as landfills and recycling facilities (Espinosa et al., 2020; Gaytán et al., 2020; Yoshida et al., 2016). However, various other environmental samples have been used for enrichment studies as summarized below.

2.1 Soil and compost

Soil and compost are the best-studied environments for plastic biodegradation. Enrichment cultures can be started using soil/compost as inoculum (Teeraphatpornchai et al., 2003; Yoshida et al., 2016), or plastic pieces which have been pre-incubated or partially degraded in these environments (Hu, Thumarat, Zhang, Tang, & Kawai, 2010; Vargas-Suárez et al., 2019). Yoshida et al., added small quantities of yeast extract to the defined medium in the initial stages of the enrichment of PET-degrading consortia (Yoshida et al., 2016), which can accelerate degradation by increasing biomass and supplying nutrients. If the elimination of non-degraders is desired, the culture can be transferred into a medium with the plastic as the only carbon source.

Another approach to generate plastic-degrading consortia from soil is to isolate individual strains with catabolic potential and combine them in a synthetic consortium. Esmaeili et al. isolated one fungal and one bacterial candidate from landfill samples pre-incubated with low density polyethylene (LDPE), based on their capacity to degrade ethylene oligomers. When inoculated in soil microcosms, these strains improved LDPE degradation compared to non-inoculated samples (Esmaeili, Pourbabaee, Alikhani, Shabani, & Esmaeili, 2013).

2.2 Freshwater and seawater

Microbial consortia associated with aquatic plastic debris have been extensively studied (Kesy, Oberbeckmann, Kreikemeyer, & Labrenz, 2019; Zettler, Mincer, & Amaral-Zettler, 2013). The enrichment of plastic degrading bacteria from marine samples would follow the same principles as for soil and landfills, with the growth media being adjusted to the respective inoculum source. The following protocol describes a procedure for the enrichment of consortia from marine eulittoral samples.

2.2.1 Protocol for the enrichment of marine aerobic consortia

The following protocol illustrates the steps to enrich aerobic plastic-degrading bacteria from marine eulittoral samples. PCL pellets (Sigma Aldrich ref. no 440744 (average $M_n \sim 80{,}000$), 704,105 (average $M_n \sim 45{,}000$)) are used as an example substrate, but can be replaced by other target polymers. If thin films are used instead of pellets, the substrate preparation steps can be omitted.

Plastic substrate sterilization is usually accomplished by submerging the plastic in 70% ethanol and drying completely under sterile conditions. UV sterilization should be used with caution: while some polymers such as PE may be rendered more amenable to biodegradation with a UV-pretreatment (Orr, Hadar, & Sivan, 2004), the pretreatment can reduce the biodegradability of others such as poly(butylene adipate terephthalate) (De Hoe et al., 2019; Kijchavengkul, Auras, Rubino, Ngouajio, & Fernandez, 2008) and polyethylene terephthalate (PET) (Falkenstein et al., 2020).

2.2.1.1 Materials, equipment and reagents

2.2.1.1.1 Media The following tables list the components of a marine minimal medium (Gade, Stührmann, Reinhardt, & Rabus, 2005a) with added N and vitamins, which can be used for plastic enrichment cultures (Tables 1–5).

2.2.1.1.2 Equipment

1. Analytical balance
2. 100 mL beaker
3. 150 mm petri dishes
4. Laminar flow hood
5. 10 mm mesh sieve
6. 250 mL Erlenmeyer flask
7. Forceps

Table 1 Artificial seawater (Kester, Duedall, Connors, & Pytkowicz, 1967).

Substance	Weight (g/L)
NaCl	23.9
$MgCl_2 \times 6H_2O$	5.5
$Na_2SO_4 \times 7H_2O$	4
KCl	0.67
KBr	0.1
$NaHCO_3$	0.2
H_3BO_3	0.03
$CaCl_2 \times 2H_2O$	1.5
$SrCl_2 \times 6H_2O$	0.015
NaF	0.005

Table 2 Hutner's Basal Salts (Cohen-Bazire, Sistrom, & Stanier, 1957).

Substance	Amount /L
Nitrilotriacetate (NTA)	10 g
$MgSO_4 \times 7H_2O$	29.7 g
$CaCl_2 \times 2H_2O$	3.34 g
$NaMoO_4 \times 2H_2O$	12.67 mg
$FeSO_4 \times 7H_2O$	99 mg
metal salt solution '44'	50 mL

Dissolve NTA by adding an equimolar amount of NaOH, before adding the rest of the ingredients (Inglett, Reddy, & Corstanje, 2005).

8. Vortex
9. Orbital shaker
10. 0.22 μM syringe filter

2.2.1.1.3 Chemicals

1. Plastic as pellets or films
2. Dichloromethane (≥99.5%, J.T. Baker-Fischer Scientific)
3. Ethanol diluted in Milli-Q water to 70% (Absolute, Sigma-Aldrich)
4. Tryptone pancreatic digest of casein (Merck, ref. no T9410)
5. Glycerol (Sigma-Aldrich, ref. no G9012)

Table 3 The metal salts solution '44' (Schlesner, 1994).

Substance	Weight (mg/L)
EDTA	250
$ZnSO_4 \times 7H_2O$	1095
$FeSO_4 \times 7H_2O$	500
$MnSO_4 \times H_2O$	154
$CuSO_4 \times 5H_2O$	39.2
$CoCl_2 \times 6H_2O$	20.3
$Na_2B_4O_7 \times 10H_2O$	7.7

Dissolve EDTA by adding an equimolar amount of NaOH, before adding the rest of the ingredients.

Table 4 Vitamin solution no. 6 (Staley, 1968).

Substance	Weight (g/L)
Biotin	2
Pyridoxin hydrochloride	10
Thiamin hydrochloride	5
Ca-pantothenate	5
para-aminobenzoic	5
Folic acid	2
Riboflavin	5
Nicotine amide	5
Vitamin B_{12}	0.1

Store at 4 °C in the dark.

6. D(+)-Trehalose dihydrate min. 98% (Carl Roth GmbH, ref. no 5151.1)
7. Tryptic soy broth (Sigma-Aldrich, ref. no T8907)
8. Sterile DMSO (Sigma-Aldrich, ref. no D2650)

All other media components should be at the highest purity grade possible.

2.2.1.2 Protocol

1. To prepare PCL films: weigh 300 mg of PCL pellets in a beaker, add 30 mL of dichloromethane under a fume hood. Stir to dissolve completely

Table 5 Marine minimal medium (Gade, Stührmann, Reinhardt, & Rabus, 2005b).

Solution	**mL/L**
Artificial seawater	500
Hutner's basal salts	20
Vitamin solution no. 6	10
NH_4Cl (1 M)	1
K_2HPO_4 (1 M)	0.1
dd H_2O	470

Sterilize by autoclaving at 121 °C for 20 min. Underlined solutions are filter-sterilized and added aseptically to the medium after the other components have been autoclaved and cooled.

2. Pour in a 150 mm glass Petri dish and cover with a lid, leaving a very small opening to let the solvent evaporate (the slower it evaporates, the more even the film will be in thickness).
3. Weight the desired plastic film to have a final concentration of 100 mg carbon/L. The carbon content of plastics without additives can be calculated from the molecular formula. For plastics with additives, elemental analysis should be used
4. Sterilize the films by submerging them in 70% ethanol. Let them stand for 20 min. After sterilization, dry completely under a laminar flow hood
5. To prepare the marine inoculum: mix 1 g of aerobic sediment with 10 mL of seawater sample. Vortex at maximum speed for 30 s to dislocate sediment-attached bacteria. Let settle for 2 min. If the sediment contains large particles, it can be sieved through a 10 mm-mesh sieve
6. In a 250 mL Erlenmeyer flask, combine 49 mL or minimal medium and 1 mL of marine inoculum. Add the sterile PCL film with flamed forceps. To speed up the enrichment process, tryptone can be added to a final concentration of 0.01%. For polymers that are more difficult to degrade, this amount can be increased to 0.1%.
7. Shake at 100 rpm at 20 °C until the film starts to disintegrate
8. Transfer 1 mL of the culture supernatant into 49 mL of minimal medium amended with a fresh plastic film piece, without any extra C source. Incubate as in the previous step
9. Keep reinoculating into fresh medium with plastic, until the time it takes from inoculation to plastic disintegration is stable among at least three transfers

10. Store cultures in 20% glycerol or DMSO-TT (Table 6). Harvest the biomass from the desired culture volume by centrifugation, and resuspend in the desired storage solution, using 1 mL storage solution per 10 mL of initial culture volume. Flash-freeze 1 mL aliquots in liquid N_2 and store at −80 °C

Note

[1] Dichloromethane is an organochloride that has serious health effects if inhaled. Therefore, it should be handled with nitrile gloves at a fume hood.

The salinity of the minimal medium, as well as the incubation temperature needs to be optimized according to the physical parameters of the inoculum source. After a stable culture is obtained, the mineralization of the film can be measured as stated in the next section.

2.3 Methods to quantify plastic aerobic mineralization

Biodegradation of plastic materials into CO_2 and biomass can be monitored by measuring oxygen demand as per ISO 18830:2016 (ISO, 2016), or CO_2 release as per ISO 19679:2020 and OECD 301 (ISO, 2020; OECD, 1992). The latter method is more accurate, and can be performed in most laboratories. The cultures are grown in a closed vessel with an alkaline trap (commonly NaOH, KOH or $Ba(OH)_2$) to convert CO_2 into bicarbonate salt (Fig. 1), which can then be measured by titration (Briassoulis, Pikasi, Papardaki, & Mistriotis, 2020) or using a total organic carbon measuring device.

Instead of using a liquid trap, Yoshida et al., used Ascarite® CO_2 absorbent to trap the released CO_2 upon PET degradation. The amount of CO_2 was then determined by periodically weighing the absorbent (Yoshida et al., 2016).

Table 6 DMSO-TT storage solution (Kerckhof et al., 2014).

Component	Amount
DMSO	5 mL
Trehalose	1 g
Tryptone soy broth	0.3 g
Double-distilled (dd) H_2O	95 mL

Dissolve the trehalose and the tryptone soy broth in 50 mL of dd H_2O. Fill up to 95 mL with ddH_2O and sterilize through a 0.22 μM syringe filter. Add 5 mL of sterile DMSO aseptically.

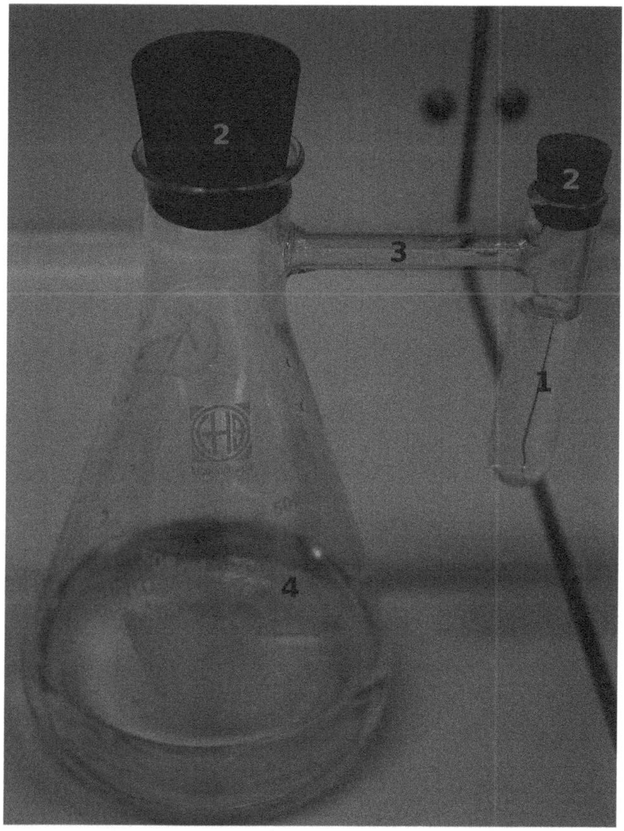

Fig. 1 Model mineralization flask. The side arm that contains the alkaline solution (1) and the flask are sealed tight with rubber stoppers (2). It is important that the connecting tube (3) is long enough so that the alkaline solution doesn't spill into the culture medium (4) during shaking.

The percentage of biodegradation is calculated as:

$$\%\text{degradation} = \frac{mg\ CO_2\ produced}{ThCO_2 \times mg\ test\ substance\ added} \times 100\ (\text{OECD}, 1992).$$

where $ThCO_2$ is the theoretical CO_2 in mg that would evolve if the entire C content of the test substance is converted to CO_2. The total C content of the plastic material is required for this calculation. As plastic materials may contain unknown organic additives, the C content may not always match the content calculated from the molecular formula, and needs to be determined empirically for accurate degradation curves. The use of carbon-isotope labeled plastics is the most specific method to trace CO_2 resulting from plastic mineralization, as has been recently demonstrated (Yang et al., 2015; Zumstein et al., 2018). This method is mainly limited by the availability and cost of labeled polymers.

Although CO_2 evolution is the most direct method to measure plastic mineralization, conventional plastics may require a very long time to be fully oxidized to CO_2 (Chamas et al., 2020). Other methods such as Fourier-transform infrared spectroscopy and nuclear magnetic resonance and thermal analysis to assess molecular-level changes in the plastic structure can be used to complement standard CO_2 evolution tests (Brandon et al., 2018; Chamas et al., 2020; Kim et al., 2020), although these methods alone don't provide adequate proof that the plastic is being converted to biomass and CO_2.

3. Enrichment and cultivation of anaerobic plastic-degrading consortia

Similar to oxic environments, plastic accumulation has been also found in natural and artificial anoxic settings. In the oceans, the presence of plastics in the anoxic oceanic zones depends on the plastic density. Some plastics float and accumulate at the surface and others sink to the sea floor (Bergmann & Klages, 2012). In terrestrial environments, microplastics have been found in anoxic soils such as wetlands and floodplains (Inglett et al., 2005; Scheurer & Bigalke, 2018; Townsend, Lu, Sharley, & Pettigrove, 2019).

The main difference between aerobic and anaerobic plastic degraders is that: (1) Anaerobic growth requires the presence of electrons acceptors (e.g., nitrate, iron, sulfate) other than oxygen (Moodie & Ingledew, 1990) and (2) within a consortium, the process of anaerobic growth involves not only the formation of CO_2 but also a mixture of CH_4 and CO_2 known as biogas (Fig. 2), besides other gases and water (Bátori, Åkesson, Zamani, Taherzadeh, & Sárvári Horváth, 2018; Quecholac-Piña, Hernández-Berriel, Mañón-Salas, Espinosa-Valdemar, & Vázquez-Morillas, 2020). In this regard anaerobic degradation is advantageous: Not only the monomers of the biodegraded polymers can be recovered but also the biogas produced can be collected and used to create energy (Capodaglio, Callegari, & Lopez, 2016).

In the following sections we give an insight to different approaches used to enrich soil and sludge anaerobic plastic degrading consortia. After these sections, a detailed methodology is described for the preparation of anaerobic marine media and the enrichment of anaerobic marine bacteria in the presence of a polybutylene adipate terephthalate (PBAT) film.

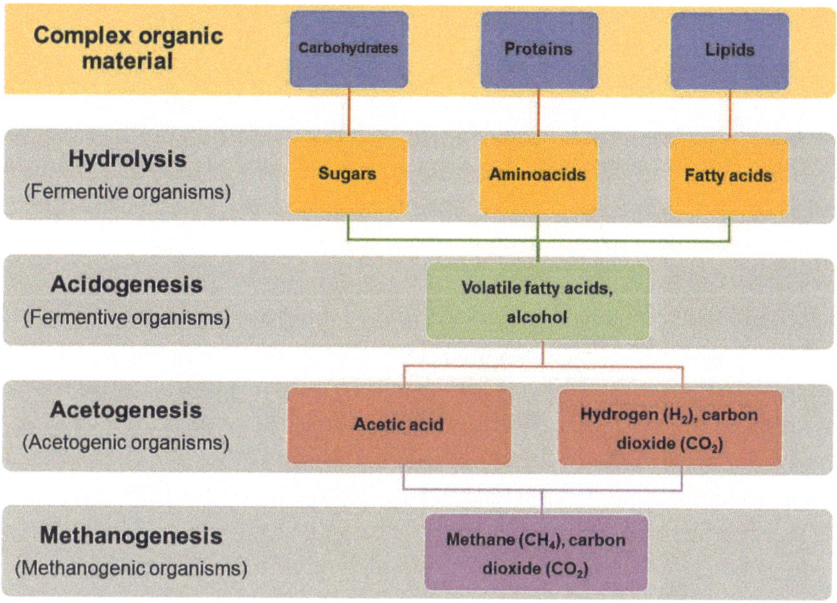

Fig. 2 A simplified diagram of the anaerobic digestion process. The five main steps of the process are framed in gray boxes: Hydrolysis, acidogenesis, acetogenesis and methanogensis and the main products are shown for each step.

3.1 Soil and landfills

To date, a limited number of studies describe soil anaerobic bacteria that can degrade plastic. Similar to studies concerning aerobic microbial community, anaerobic enrichment cultures can be started using soil/compost mixture (e.g., garden soil mixed with kitchen and yard wastes). These types of inocula can be then transferred to artificial systems that mimic landfill conditions (Björn, Hörsing, Karlsson, Mersiowsky, & Ejlertsson, 2007; Ishigaki et al., 2004) and are incubated under optimized parameters of temperature, oxygen, pH, aeration, etc. Typically, incubation temperatures are adjusted within a range from mesophilic to thermophilic, obtaining commonly higher degradation rates at thermophilic conditions (Björn et al., 2007).

In these systems, a few polymers have been tested so far, such as non-biodegradable plastics (PVC) and commercialized bioplastics. Within this group, *co*-polyesters of polyhydroxybuyrate and valerate (PHBV), PCL, blends of starch and polyvinyl alcohol and cellulose acetate have been tested (Björn et al., 2007; Ishigaki et al., 2004).

3.2 Anaerobic digesters and fermentation tanks

The metabolic capacities of plastic-degrading anaerobic microbial communities from anaerobic digesters and fermentation tasks are more widely studied than those from other environments. This is likely because these systems provide better conditions to promote plastic-degrading microbial communities (e.g., optimized temperatures, moisture, and supply of organic materials) and in more controlled environments.

There are several studies on plastic degradation using sludge samples recovered from anaerobic digesters of wastewater treatment plants and fermentation tanks. These samples are further incubated anaerobically in anoxic serum bottles (Miller & Wolin, 1974) or in controlled bioreactors (Mauerhofer et al., 2019). Within these systems, the communities are incubated from mesophilic to thermophilic (commonly 37 to 55 °C) conditions. In controlled bioreactors, degradation normally takes place faster than under natural conditions.

In these systems the degradation of diverse plastics has been tested, including poly(3-hydroxybutyrate-*co*-3-hydroxyhexanoate) with high and low molecular weight (Wang et al., 2018), polyhydroxybutyrate (PHB), PHBV and PCL (Abou-Zeid, Müller, & Deckwer, 2001; Yagi, Ninomiya, Funabashi, & Kunioka, 2014), PLA and poly-l-lactides (Itävaara, Karjomaa, & Selin, 2002; Yagi, Ninomiya, Funabashi, & Kunioka, 2009), PBS (Dvorackova, Svoboda, Kostka, & Pekarova, 2015; Yagi et al., 2014) and poly(ethylene terephthalate-*co*-lactate) copolyesters (Hermanová, Šmejkalová, Merna, & Zarevúcka, 2015) among many others. The degradation of these plastics was observed and linked to the presence of anaerobic communities.

3.3 Marine ecosystem

In contrast to the numerous studies of anaerobic communities in landfills and digesters, studies on marine anaerobic bacteria are rather scarce. In a nutshell, the approach to grow marine bacteria involves filling up sterile containers with sediment samples collected from the marine environment. The enrichment cultures are then set up by transferring the samples into anoxic artificial sea water previously sparged with a mixture of CO_2 and N_2 (20:80 vol/vol) (Köpke, Wilms, Engelen, Cypionka, & Sass, 2005). This medium, contrary to oxic media, contains a reducing agent (e.g., Na_2S) (Plugge, 2005) that keeps the media anoxic/reduced and sodium bicarbonate (Delgado, Parameswaran, Fajardo-Williams, Halden, & Krajmalnik-Brown, 2012) as a pH buffer.

In addition, large volumes of marine media can be prepared by using Widdel flasks (Brandis-Heep, 2020; Laso-Pérez, Krukenberg, Musat, & Wegener, 2018; F Widdel, 1980).

3.3.1 Protocol for the enrichment of marine anaerobic consortia

The following protocol describes in detail the steps needed to enrich anaerobic marine plastic degrading consortia from marine sediments.

3.3.1.1 Materials, equipment and reagents

3.3.1.1.1 Media The following tables list the components of an anaerobic marine medium that can be used to enrich general anaerobic marine communities.

First dissolve $FeCl_2$ in HCl, then dilute in water, add and dissolve the other salts. Finally make up to 1 L and autoclaved separately from the other solutions.

3.3.1.1.2 Equipment

1. 1 L Erlenmeyer flask
2. 2 L Widdel flask (ILMABOR-TGI)
3. Stirring plate
4. 5 mL sterile serological pipette
5. pH Meter
6. 125 mL Wheaton serum bottles (Sigma-Aldrich)
7. Butyl rubber septum stoppers (Ochs laborbedarf)
8. Aluminum caps (Ochs laborbedarf)
9. 0.45 × 25 mm disposable-sterile needle (B. Braun Sterican)
10. 0.2 μm sterile filter (Minisart, Sartorius Stedim Biotech)
11. Spatula
12. 6 m-long aluminum tubes
13. Vinyl anaerobic chamber (Coy lab products)
14. Analytical scale
15. 50 mL Falcon tubes

3.3.1.2 Protocol

3.3.1.2.1 Media Preparation

1. Prepare the anaerobic marine basal media (Table 7) in an Erlenmeyer flask and transfer it into a 2 L Widdel flask (Fig. 3) containing a magnetic stirring rod

Table 7 Artificial sea water (Widdel, 1980).

Substance	**Amount/2 L**
NaCl	40.0 g
$MgCl_2 \times H_2O$	6.0 g
$CaCl_2 \times 2H_2O$	0.3 g
KCl	1 g
NH_4Cl	0.5 g
KH_2PO_4	0.4 g
dd H_2O	1930 mL

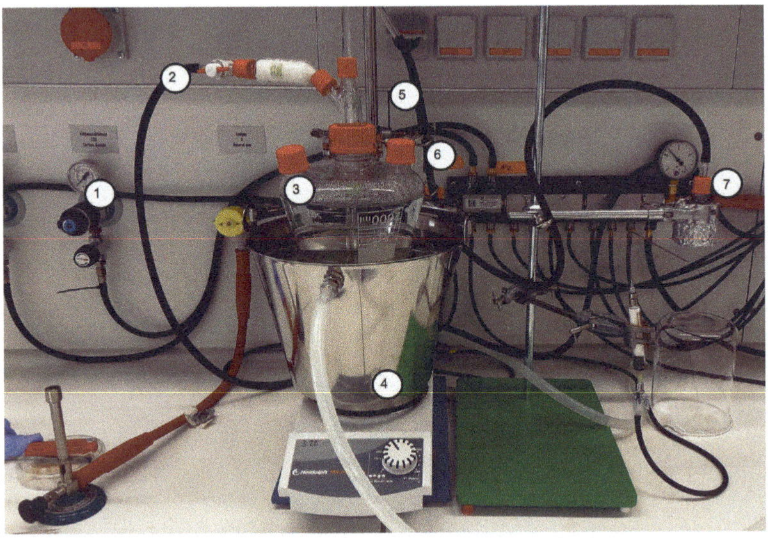

Fig. 3 Widdel-flask set up for preparation of anaerobic media. A Duran flask containing the basal salts of the anoxic media is autoclaved and sparged with a mixture of CO_2 and N_2. During sparging the temperature is controlled in a water bath. At lower temperatures, additional solutions (e.g., vitamins) can be added through the side lid. The media can be then released through an outflow tube and collected into serum bottles. Components: 1. Gas pressure regulator; 2. Gas inflow tube; 3. Side lid; 4. Water bath; 5. Media outflow tube; 6. Hoffman clamps; 7. Bottle-filling system.

2. Autoclave for 30 min, 120 °C. Immediately after autoclaving, place the Widdel flask on a stirring plate inside an empty water bath bucket (Fig. 3, component 4).
3. Gas the medium for 1 h with a mixture of CO_2 (20%) and N_2 (80%) through the gas inflow tube (Fig. 3, component 1). To allow gas exchange, keep one of the side lids loose (Fig. 3, component 3).

4. After 1 h, fill the water bath to cool down the media to room temperature while stirring. Let it cool for 1 h. During this time, keep the flow of CO_2 (20%) and N_2 (80%) running
5. When the medium has reached room temperature, inject the solutions (Table 8–13) according to Table 14 from one of the Widdel's side lids. Stir for 20 min at 250 rpm
6. With a sterile pipette, take 5 mL of medium and measure the pH. The pH can be adjusted, if needed, by adding 1 M HCl or 0.5 M Na_2CO_3
7. Dispense the medium (50 mL) in previously autoclaved 125 mL serum bottles through the bottle filling system of the Widdel flask (Fig. 3, component 7). Dispense the medium under a CO_2 (20%) and N_2 (80%) atmosphere
8. After disposal, gas the headspace for 10 s and seal the bottles with butyl rubber septum stoppers and aluminum caps

Table 8 Trace element solution SL-10 (Widdel, Kohring, & Mayer, 1983).

Substance	Amount/L
HCl (25%; 7.7 M)	10.0 mL
$FeCl_2 \times H_2O$	1.5 g
$ZnCl_2$	70.0 mg
$MnCl_2 \times 4H_2O$	100.0 mg
H_3BO_3	6.0 mg
$CoCl_2 \times 6H_2O$	190.0 mg
$CuCl_2 \times 2H_2O$	2.0 mg
$NiCl_2 \times 6H_2O$	24.0 mg
$Na_2MoO_4 \times 2H_2O$	36.0 mg
dd H_2O	990.0 mL

Table 9 Tungstate/Selenite solution (F Widdel, 1980).

Substance	Amount/0.1 L
NaOH	0.5 g
$Na_2SeO_3 \times 5H_2O$	3.0 mg
$Na_2WO_4 \times 2H_2O$	4.0 mg
Distilled water	100.0 mL

Autoclave separately from the other solutions.

Table 10 Vitamin solution (Friedrich Widdel et al., 1983).

Substance	Amount/0.2 L
para-Aminobenzoic acid	10.0 mg
D(+)-Biotin	2.0 mg
Nicotinic acid	20.0 mg
Calcium pantothenate	5.0 mg
Pyridoxamine dihydrochloride	50.0 mg
Thiamine dihydrochloride	10.0 mg
Cyanocobalamin	10 mg
Distilled water	200.0 mL

Filter sterilized the vitamins and store at 4 °C.

Table 11 Sodium bicarbonate.

Substance	Amount/0.06 L
$NaHCO_3$	5.04 g
Distilled water	60.0 mL

$NaHCO_3$ is dissolved in 60 mL distilled water that has been previously CO_2 saturated. Autoclave separately from the other solutions.

Table 12 Sodium sulfide solution.

Substance	Amount/0.025 L
$Na_2S \times 9H_2O$	3.0 g
Distilled and boiled water	25.0 mL

Headspace is gassed with N_2 and the solution is autoclaved separately from the other solutions.

Table 13 Na-resazurin solution.

Substance	Amount/0.1 L
Na-resazurin	40 mg
Distilled water	100.0 mL

Table 14 Solutions are added in the following order:

Supplement	Addition volume (mL/2L medium)
Sodium bicarbonate	60
Trace element solution SL-10	2
Tungstate/Selenite solution	2
Vitamin solution	1
Sodium sulfide solution	4

9. Gas pressure can be released from the cultures by using a 0.45 × 25 mm disposable-sterile needle inserted in the stopper while gassing 20% CO_2 and 80% N_2 through a second disposable-sterile needle attached to a sterile 0.2 μm filter

3.3.1.2.2 Sample Collection and Preparation

1. Samples collection: Anaerobic samples can be collected, e.g., from marine tidal flats. At low tide, aluminum tubes are inserted in the sediment and anaerobic samples can be scratched out from the undisturbed sediment fraction with a sterile spatula (Köpke et al., 2005). The sediments can be then stored in sterile anoxic vessels
2. Sample preparation:
   a. Place the collected samples in a vinyl anaerobic chamber equipped with all necessary accessories (e.g., 3 door airlocks, catalyst boxes, atmosphere filtration system and an O_2/H_2 control monitor).
   b. Weigh 1 g of sediment and transfer it to an anoxic 50 mL falcon tube. Prepare several samples to have enough replicates
   c. Outside the chamber and under a sterile environment, transfer the sediments with a small and previously flamed spatula into 125 mL serum bottles. The serum bottles contain 50 mL of anoxic marine medium. Inoculation should be carried out under a CO_2 (20%) and N_2 (80%) atmosphere
   d. After inoculation, add the sterile plastic film (Protocol 2.2.1.2, steps 1–4). Gas the headspace for 10 s and seal the bottles with butyl rubber septum stoppers and aluminum caps
   e. Gas pressure should be released from the cultures by using a 0.45 × 25 mm disposable-sterile needle inserted in the stopper while gassing (20% CO_2 and 80% N_2) through a second disposable-sterile needle attached to a sterile 0.2 μm filter
   f. Incubate cultures at 20 °C, shaking at 70 rpm

As for aerobic marine enrichments, the salinity and temperature can be optimized according to the inoculum source. Plastic disintegration can be tracked visually, and/or assessed as described below.

3.4 Methods to quantify plastic anaerobic mineralization

Several methods similar to those used for aerobic bacteria are also commonly used in anaerobic environments to assess polymer biodegradation. The most typical and common methods to assess plastic degradation by anaerobic bacteria are weight loss tests, tensile strength changes, FTIR analysis, among others. However, a method unique to anaerobic systems is the detection of cumulative biogas formation (CH_4, CO_2 and H_2). These detection methods rely on the growth of anaerobic communities, composed of bacteria and methanogenic archaea, to produce CO_2 and CH_4. Currently, the most prevailing methods used for detecting biogases are gas chromatography systems coupled to flame ionizing detectors, thermal conductivity detectors, etc., and volumetric by liquid displacement methods (Guwy, 2004). In both cases, the implementation of closed systems is required. GC detection requires the collection of biogas samples from the headspaces of cultures/reactors, while volumetric methods require the use of displacement measurement devices (Fig. 4A). Moreover, the addition of NaOH or KOH solution traps allows the capture of CO_2 and consequently the separation from CH_4 (Fig. 4B). Gases, other than CO_2 and CH_4, (e.g., H_2S) can be also separately measured (Mohee, Unmar, Mudhoo, & Khadoo, 2008; Yagi et al., 2009).

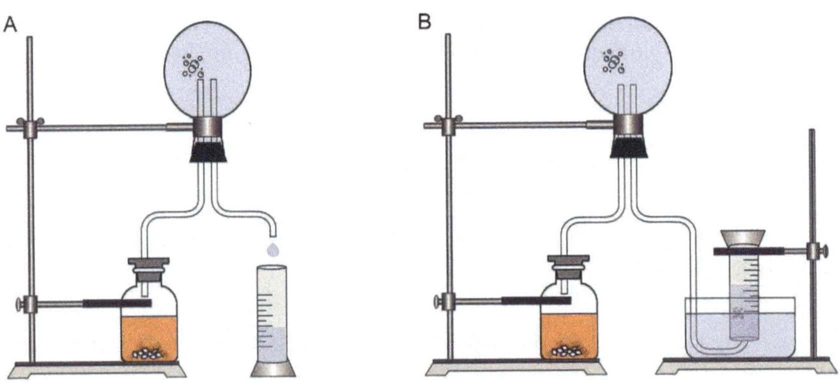

Fig. 4 Schematic diagram of the liquid displacement method. Two alternatives for liquid displacement method are shown. (A) The simplified version of this method uses a gas trap containing, e.g., NaOH and a graduated cylinder to measure the displaced liquid. (B) In a more elaborated system, the CO_2 can be trapped in, e.g., KOH or NaOH solution while remaining gases (e.g., CH_4) are volumetrically measured with a graduated cylinder.

4. Future perspectives

Plastic biodegradation by microbial consortia in aerobic and anaerobic systems is a currently understudied field with great biotechnological potential. Natural and synthetic consortia can be used to enhance biodegradation of compostable plastics in industrial composters, or degrade conventional plastics into monomers for recycling. To better understand how microorganisms work synergistically to degrade a complex plastic substrate, the current degradation studies should be enhanced by meta(omics) analyses and enzyme characterization.

References

Abou-Zeid, D.-M., Müller, R.-J., & Deckwer, W.-D. (2001). Degradation of natural and synthetic polyesters under anaerobic conditions. *Journal of Biotechnology*, *86*(2), 113–126.

Bátori, V., Åkesson, D., Zamani, A., Taherzadeh, M. J., & Sárvári Horváth, I. (2018). Anaerobic degradation of bioplastics: A review. *Waste Management*, *80*, 406–413.

Bergmann, M., & Klages, M. (2012). Increase of litter at the Arctic deep-sea observatory HAUSGARTEN. *Marine Pollution Bulletin*, *64*(12), 2734–2741.

Björn, A., Hörsing, M., Karlsson, A., Mersiowsky, I., & Ejlertsson, J. (2007). Impacts of temperature on the leaching of organotin compounds from poly(vinyl chloride) plastics—A study conducted under simulated landfill conditions. *Journal of Vinyl and Additive Technology*, *13*(4), 176–188.

Brandis-Heep, A. (2020). In A. Brandis-Heep, E. Kothe, T. Zimmermann, & A. Störiko (Eds.), *Isolierung und Kultivierung von Bakterien BT - Methoden der Mikrobiologie: Ein Praxishandbuch* (pp. 1–62). Berlin, Heidelberg: Springer Berlin Heidelberg.

Brandon, A. M., Gao, S.-H., Tian, R., Ning, D., Yang, S.-S., Zhou, J., et al. (2018). Biodegradation of polyethylene and plastic mixtures in mealworms (larvae of Tenebrio molitor) and effects on the gut microbiome. *Environmental Science and Technology*, *52*(11), 6526–6533.

Briassoulis, D., Pikasi, A., Papardaki, N. G., & Mistriotis, A. (2020). Aerobic biodegradation of bio-based plastics in the seawater/sediment interface (sublittoral) marine environment of the coastal zone – Test method under controlled laboratory conditions. *Science of the Total Environment*, *722*, 137748.

Capodaglio, A. G., Callegari, A., & Lopez, M. V. (2016). European framework for the diffusion of biogas uses: Emerging technologies, acceptance, incentive strategies, and institutional-regulatory support. *Sustainability (Switzerland)*, *8*(4), 298.

Chamas, A., Moon, H., Zheng, J., Qiu, Y., Tabassum, T., Jang, J. H., et al. (2020). Degradation rates of plastics in the environment. *ACS Sustainable Chemistry & Engineering*, *8*(9), 3494–3511.

Chen, S.-H., & Aitken, M. D. (1999). Salicylate stimulates the degradation of high-molecular weight polycyclic aromatic hydrocarbons by Pseudomonas saccharophila P15. *Environmental Science & Technology*, *33*(3), 435–439.

Cohen-Bazire, G., Sistrom, W. R., & Stanier, R. Y. (1957). Kinetic studies of pigment synthesis by non-sulfur purple bacteria. *Journal of Cellular and Comparative Physiology*, *49*(1), 25–68.

Danso, D., Chow, J., & Streit, W. R. (2019). Plastics: Environmental and biotechnological perspectives on microbial degradation. *Applied and Environmental Microbiology, 85*(19), e01095–19. American Society for Microbiology.

De Hoe, G. X., Zumstein, M. T., Getzinger, G. J., Rüegsegger, I., Kohler, H.-P. E., Maurer-Jones, M. A., et al. (2019). Photochemical transformation of poly(butylene adipate-co-terephthalate) and its effects on enzymatic hydrolyzability. *Environmental Science & Technology, 53*(5), 2472–2481.

Dejonghe, W., Berteloot, E., Goris, J., Boon, N., Crul, K., Maertens, S., et al. (2003). Synergistic degradation of linuron by a bacterial consortium and isolation of a single linuron-degrading *Variovorax* strain. *Applied and Environmental Microbiology, 69*(3), 1532–1541.

Delgado, A. G., Parameswaran, P., Fajardo-Williams, D., Halden, R. U., & Krajmalnik-Brown, R. (2012). Role of bicarbonate as a pH buffer and electron sink in microbial dechlorination of chloroethenes. *Microbial Cell Factories, 11*(1), 128.

Dvorackova, M., Svoboda, P., Kostka, L., & Pekarova, S. (2015). Influence of biodegradation in thermophilic anaerobic aqueous conditions on crystallization of poly(butylene succinate). *Polymer Testing, 47*, 59–70.

Esmaeili, A., Pourbabaee, A. A., Alikhani, H. A., Shabani, F., & Esmaeili, E. (2013). Biodegradation of low-density polyethylene (LDPE) by mixed culture of Lysinibacillus xylanilyticus and Aspergillus Niger in soil. *PLoS One, 8*(9), e71720.

Espinosa, M. J. C., Blanco, A. C., Schmidgall, T., Atanasoff-Kardjalieff, A. K., Kappelmeyer, U., Tischler, D., et al. (2020). Toward biorecycling: Isolation of a soil bacterium that grows on a polyurethane oligomer and monomer. *Frontiers in Microbiology, 11*, 404. Frontiers.

Falkenstein, P., Gräsing, D., Bielytskyi, P., Zimmermann, W., Matysik, J., Wei, R., et al. (2020). UV pretreatment impairs the enzymatic degradation of polyethylene terephthalate. *Frontiers in Microbiology, 11*, 689.

Gade, D., Stührmann, T., Reinhardt, R., & Rabus, R. (2005a). Growth phase dependent regulation of proteincomposition in Rhodopirellula baltica. *Environmental Microbiology*, 7(8), 1074–1084.

Gade, D., Stührmann, T., Reinhardt, R., & Rabus, R. (2005b). Growth phase dependent regulation of protein composition in Rhodopirellula baltica. *Environmental Microbiology*, 7(8), 1074–1084.

Gaytán, I., Sánchez-Reyes, A., Burelo, M., Vargas-Suárez, M., Liachko, I., Press, M., et al. (2020). Degradation of recalcitrant polyurethane and xenobiotic additives by a selected landfill microbial community and its biodegradative potential revealed by proximity ligation-based metagenomic analysis. *Frontiers in Microbiology, 10*, 2986.

Guwy, A. J. (2004). Equipment used for testing anaerobic biodegradability and activity. *Reviews in Environmental Science & Bio/Technology, 3*(2), 131–139.

Hermanová, S., Šmejkalová, P., Merna, J., & Zarevúcka, M. (2015). Biodegradation of waste PET based copolyesters in thermophilic anaerobic sludge. *Polymer Degradation and Stability, 111*, 176–184.

Hu, X., Thumarat, U., Zhang, X., Tang, M., & Kawai, F. (2010). Diversity of polyester-degrading bacteria in compost and molecular analysis of a thermoactive esterase from *Thermobifida alba* AHK119. *Applied Microbiology and Biotechnology, 87*(2), 771–779.

Hu, B., Wang, M., Geng, S., Wen, L., Wu, M., Nie, Y., et al. (2020). Metabolic exchange with non-alkane-consuming *Pseudomonas stutzeri* SLG510A3-8 improves n-alkane biodegradation by the alkane degrader *Dietzia sp.* strain DQ12-45-1b. (A. J. M. Stams, Ed.). *Applied and Environmental Microbiology, 86*(8). e02931–19.

Inglett, P. W., Reddy, K. R., & Corstanje, R. (2005). *Anaerobic soils.* Elsevier.

Ishigaki, T., Sugano, W., Nakanishi, A., Tateda, M., Ike, M., & Fujita, M. (2004). The degradability of biodegradable plastics in aerobic and anaerobic waste landfill model reactors. *Chemosphere, 54*(3), 225–233.

ISO. (2016). *ISO 18830:2016 Plastics—Determination of aerobic biodegradation of non-floating plastic materials in a seawater/sandy sediment interface—Method by measuring the oxygen demand in closed respirometer*. ISO.

ISO. (2020). *ISO 19679:2020 Plastics—Determination of aerobic biodegradation of non-floating plastic materials in a seawater/sediment interface—Method by analysis of evolved carbon dioxide*. ISO.

Itävaara, M., Karjomaa, S., & Selin, J.-F. (2002). Biodegradation of polylactide in aerobic and anaerobic thermophilic conditions. *Chemosphere*, *46*(6), 879–885.

Kerckhof, F.-M., Courtens, E. N. P., Geirnaert, A., Hoefman, S., Ho, A., Vilchez-Vargas, R., et al. (2014). Optimized cryopreservation of mixed microbial communities for conserved functionality and diversity. *PLoS One*, *9*(6), e99517.

Kester, D. R., Duedall, I. W., Connors, D. N., & Pytkowicz, R. M. (1967). Preparation of artificial seawater. *Limnology and Oceanography*, *12*(1), 176–179.

Kesy, K., Oberbeckmann, S., Kreikemeyer, B., & Labrenz, M. (2019). Spatial environmental heterogeneity determines young biofilm assemblages on microplastics in Baltic Sea mesocosms. *Frontiers in Microbiology*, *10*, 1665.

Kijchavengkul, T., Auras, R., Rubino, M., Ngouajio, M., & Fernandez, R. (2008). Assessment of aliphatic–aromatic copolyester biodegradable mulch films. Part II: Laboratory simulated conditions. *Chemosphere*, *71*, 1607–1616.

Kim, H. R., Lee, H. M., Yu, H. C., Jeon, E., Lee, S., Li, J., et al. (2020). Biodegradation of polystyrene by Pseudomonas sp. isolated from the gut of superworms (larvae of Zophobas atratus). *Environmental Science & Technology*, *54*(11), 6987–6996.

Köpke, B., Wilms, R., Engelen, B., Cypionka, H., & Sass, H. (2005). Microbial diversity in coastal subsurface sediments: A cultivation approach using various electron acceptors and substrate gradients. *Applied and Environmental Microbiology*, *71*(12). 7819 LP–7830.

Laso-Pérez, R., Krukenberg, V., Musat, F., & Wegener, G. (2018). Establishing anaerobic hydrocarbon-degrading enrichment cultures of microorganisms under strictly anoxic conditions. *Nature Protocols*, *13*(6), 1310–1330.

Mauerhofer, L.-M., Pappenreiter, P., Paulik, C., Seifert, A. H., Bernacchi, S., & Rittmann, S. K.-M. R. (2019). Methods for quantification of growth and productivity in anaerobic microbiology and biotechnology. *Folia Microbiologica*, *64*(3), 321–360.

Miller, T. L., & Wolin, M. J. (1974). A serum bottle modification of the Hungate technique for cultivating obligate anaerobes. *Applied Microbiology*, *27*(5). 985 LP–987.

Mohee, R., Unmar, G. D., Mudhoo, A., & Khadoo, P. (2008). Biodegradability of biodegradable/degradable plastic materials under aerobic and anaerobic conditions. *Waste Management*, *28*(9), 1624–1629.

Moodie, A. D., & Ingledew, W. J. (1990). In A. H. Rose, & D. W. Tempest (Eds.), *Vol. 31. Microbial anaerobic respiration* (pp. 225–269).

Nakasaki, K., Matsuura, H., Tanaka, H., & Sakai, T. (2006). Synergy of two thermophiles enables decomposition of poly-ε-caprolactone under composting conditions. *FEMS Microbiology Ecology*, *58*(3), 373–383.

OECD. (1992). *Test no. 301: Ready biodegradability. OECD guidelines for the testing of chemicals, Section 3*. OECD.

Orr, I., Hadar, Y., & Sivan, A. (2004). Colonization, biofilm formation and biodegradation of polyethylene by a strain of *Rhodococcus ruber*. *Applied Microbiology and Biotechnology*, *65*(1), 97–104.

Pelz, O., Tesar, M., Wittich, R.-M., Moore, E. R. B., Timmis, K. N., & Abraham, W.-R. (1999). Towards elucidation of microbial community metabolic pathways: Unravelling the network of carbon sharing in a pollutant-degrading bacterial consortium by immunocapture and isotopic ratio mass spectrometry. *Environmental Microbiology*, *1*(2), 167–174.

Plugge, C. M. B. (2005). Anoxic media design, preparation, and considerations. *Environmental Microbiology*, *397*, 3–16.

Quecholac-Piña, X., Hernández-Berriel, M.d. C., Mañón-Salas, M.d. C., Espinosa-Valdemar, R. M., & Vázquez-Morillas, A. (2020). Degradation of plastics under anaerobic conditions: A short review. *Polymers, 12*(1), 109.

Scheurer, M., & Bigalke, M. (2018). Microplastics in Swiss floodplain soils. *Environmental Science & Technology, 52*(6), 3591–3598. United States.

Schlesner, H. (1994). The development of media suitable for the microorganisms morphologically resembling Planctomyces spp, Pirellula spp, and other Planctomycetales from various aquatic habitats using dilute media. *Systematic and Applied Microbiology, 17*(1), 135–145.

Skariyachan, S., Patil, A. A., Shankar, A., Manjunath, M., Bachappanavar, N., & Kiran, S. (2018). Enhanced polymer degradation of polyethylene and polypropylene by novel thermophilic consortia of *Brevibacillus* sps. and *Aneurinibacillus* sp. screened from waste management landfills and sewage treatment plants. *Polymer Degradation and Stability, 149*, 52–68.

Sorensen, S. R., Ronen, Z., & Aamand, J. (2002). Growth in coculture stimulates metabolism of the phenylurea herbicide isoproturon by Sphingomonas sp. strain SRS2. *Applied and Environmental Microbiology, 68*(7), 3478–3485.

Staley, J. T. (1968). *Prosthecomicrobium* and *Ancalomicrobium*: New prosthecate freshwater bacteria. *Journal of Bacteriology, 95*(5), 1921–1942.

Syranidou, E., Karkanorachaki, K., Amorotti, F., Franchini, M., Repouskou, E., Kaliva, M., et al. (2017). Biodegradation of weathered polystyrene films in seawater microcosms. *Scientific Reports*, 7(1), 17991.

Syranidou, E., Karkanorachaki, K., Amorotti, F., Repouskou, E., Kroll, K., Kolvenbach, B., et al. (2017). Development of tailored indigenous marine consortia for the degradation of naturally weathered polyethylene films. *PLoS One, 12*(8), e0183984.

Teeraphatpornchai, T., Nakajima-Kambe, T., Shigeno-Akutsu, Y., Nakayama, M., Nomura, N., Nakahara, T., et al. (2003). Isolation and characterization of a bacterium that degrades various polyester-based biodegradable plastics. *Biotechnology Letters, 25*(1), 23–28.

Townsend, K. R., Lu, H.-C., Sharley, D. J., & Pettigrove, V. (2019). Associations between microplastic pollution and land use in urban wetland sediments. *Environmental Science and Pollution Research, 26*(22), 22551–22561.

Vargas-Suárez, M., Fernández-Cruz, V., & Loza-Tavera, H. (2019). Biodegradation of polyacrylic and polyester polyurethane coatings by enriched microbial communities. *Applied Microbiology and Biotechnology, 103*(7), 3225–3236.

Wang, S., Lydon, K. A., White, E. M., Grubbs, J. B., 3rd, Lipp, E. K., Locklin, J., et al. (2018). Biodegradation of poly(3-hydroxybutyrate- co-3-hydroxyhexanoate) plastic under anaerobic sludge and aerobic seawater conditions: Gas evolution and microbial diversity. *Environmental Science & Technology, 52*(10), 5700–5709.

Widdel, F. (1980). *Anaerober Abbau von Fettsäuren und Benzoesäure durch neu isolierte Arten Sulfat-reduzierender Bakterien.* Georg-August-Universität zu Göttingen.

Widdel, F., Kohring, G.-W., & Mayer, F. (1983). Studies on dissimilatory sulfate-reducing bacteria that decompose fatty acids. *Archives of Microbiology, 134*(4), 286–294.

Yagi, H., Ninomiya, F., Funabashi, M., & Kunioka, M. (2009). Anaerobic biodegradation tests of poly(lactic acid) under mesophilic and thermophilic conditions using a new evaluation system for methane fermentation in anaerobic sludge. *International Journal of Molecular Sciences, 10*(9), 3824–3835.

Yagi, H., Ninomiya, F., Funabashi, M., & Kunioka, M. (2014). Mesophilic anaerobic biodegradation test and analysis of eubacteria and archaea involved in anaerobic biodegradation of four specified biodegradable polyesters. *Polymer Degradation and Stability, 110*, 278–283.

Yang, Y., Yang, J., Wu, W.-M., Zhao, J., Song, Y., Gao, L., et al. (2015). Biodegradation and mineralization of polystyrene by plastic-eating mealworms: Part 1. Chemical and physical characterization and isotopic tests. *Environmental Science & Technology*, *49*(20), 12080–12086. American Chemical Society.

Yoshida, S., Hiraga, K., Takanaha, T., Taniguchi, I., Yamaji, H., Maeda, Y., et al. (2016). A bacterium that degrades and assimilates poly(ethyleneterephthalate). *Science*, *351*(6278), 1196–1199.

Yu, K., Yi, S., Li, B., Guo, F., Peng, X., Wang, Z., et al. (2019). An integrated meta-omics approach reveals substrates involved in synergistic interactions in a bisphenol A (BPA)-degrading microbial community. *Microbiome*, 7(1), 1–13.

Zafar, U., Nzeram, P., Langarica-Fuentes, A., Houlden, A., Heyworth, A., Saiani, A., et al. (2014). Biodegradation of polyester polyurethane during commercial composting and analysis of associated fungal communities. *Bioresource Technology*, *158*, 374–377.

Zettler, E. R., Mincer, T. J., & Amaral-Zettler, L. A. (2013). Life in the "plastisphere": Microbial communities on plastic marine debris. *Environmental Science and Technology*, *47*(13), 7137–7146.

Zumstein, M. T., Schintlmeister, A., Nelson, T. F., Baumgartner, R., Woebken, D., Wagner, M., et al. (2018). Biodegradation of synthetic polymers in soils: Tracking carbon into CO_2 and microbial biomass. *Science Advances*, *4*(7), eaas9024.

CHAPTER FOUR

Cultivation of filamentous fungi for attack on synthetic polymers via biological Fenton chemistry

Dietmar Schlosser*

Department of Environmental Microbiology, Helmholtz Centre for Environmental Research—UFZ, Leipzig, Germany

*Corresponding author: e-mail address: dietmar.schlosser@ufz.de

Contents

Methods in Enzymology, Volume 648
ISSN 0076-6879
https://doi.org/10.1016/bs.mie.2020.12.006

Abstract

Environmental pollution with synthetic polymers (commonly named plastics) nowadays poses serious threats to the environment and human health. Unfortunately, most conventional plastics are highly recalcitrant even under conditions known to be favorable for microbial degradation. Expanding the knowledge regarding opportunities and limitations of the microbial degradability of plastics would largely contribute to the development of adequate decontamination and management strategies for plastic pollution. This chapter provides cultivation approaches to be applied for the characterization of eco-physiologically diverse asco- and basidiomycete fungi with respect to their ability to attack solid and water-soluble synthetic polymers with the help of quinone redox cycling-based Fenton-type reactions, which result in the production of highly reactive hydroxyl radicals. These reactive oxygen species are the strongest oxidants known from biological systems. However, their potential employment by fungi dwelling in diverse habitats as a biodegradation tool to attack synthetic polymers is still insufficiently explored.

1. Introduction

The common term "plastics" refers to a group of solid synthetic polymers, which are poorly bioavailable and particularly resistant to microbial attack (Krueger, Harms, & Schlosser, 2015). In 2015, conventional plastic types such as polyethylene (PE), polypropylene (PP), polyester, polyamide (PA) and acrylic fibers (as a sum; PP&A), polyvinyl chloride (PVC), polyethylene terephthalate (PET), polyurethanes (PUR) and polystyrene (PS; related structural information is given in Fig. 1) accounted for about 32%, 19%, 16%, 10%, 9%, 7% and 6%, respectively, of a global production volume (including additives) of 407 million metric tons MMT (Geyer, Jambeck, & Law, 2017). By contrast, the global bioplastics production is still quite small compared to the global production numbers of conventional plastics. Bioplastics accounted for only about 1% of the total plastic materials produced in 2017, with shares of approximately 47% and 43% for biodegradable and bio-based plastics, respectively (Andreessen & Steinbuchel, 2019; EUBP, 2017).

1.1 Types and biodegradability of plastics

Solid synthetic polymers have a very low bioavailability and are therefore generally poor microbial growth substrates, despite frequently possessing well biodegradable structural elements such as amide and ester bonds, a high energy content, and favorable electron donor quality (Krueger, Harms, et al., 2015). Co-metabolism, which is the prevailing way of altering organic

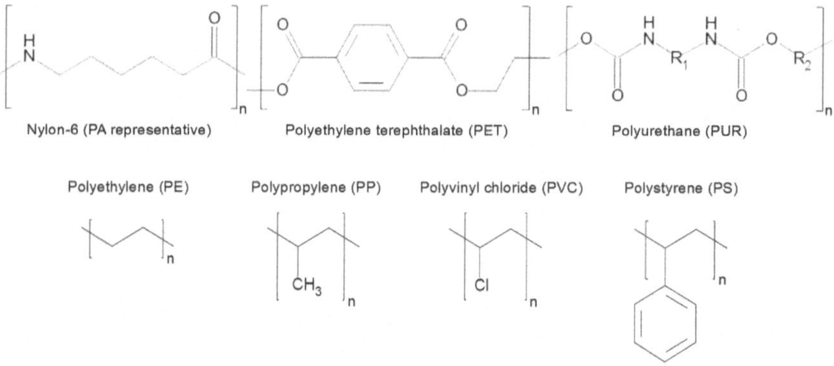

Fig. 1 Chemical structures of important conventional synthetic polymers (plastics). The structures in the upper part of the figure represent compounds with hydrolyzable bonds and those in the lower part plastics with non-hydrolysable bonds in the polymer backbone. Nylon-6 is shown as a PA representative. R_1 stands for the di-isocyanate part in PUR and R_2 for the polyol moiety in PUR (Krueger, Harms, & Schlosser, 2015). *Reprinted by permission from Springer Nature: Springer. Krueger, M.C., Harms, H., & Schlosser, D. (2015) Prospects for microbiological solutions to environmental pollution with plastics.* Applied Microbiology Biotechnology 99 *(21):8857–8874. © Springer-Verlag Berlin Heidelberg 2015.*

chemicals of xenobiotic origin in fungi and is also found among bacteria, may provide less compound-specific alternatives to productive biodegradation of synthetic polymers. However, the presently available information still seems too limited to determine whether co-metabolic or productive degradation of plastics is more important (Krueger, Harms, et al., 2015). The size of synthetic polymers may also represent an obstacle to biodegradation since at least plastic materials above the size of nanoplastics (da Costa, Santos, Duarte, & Rocha-Santos, 2016) would require an extracellular initiation of their breakdown into compounds small enough for cellular uptake and further intracellular catabolism (Harms, Wick, & Schlosser, 2017; Krueger, Harms, et al., 2015).

Plastics can be divided into hydrolysable and non-hydrolysable polymers, based on their respective chemical structures and biochemical mechanisms potentially being able to cleave the polymer backbone (Figs. 1 and 2) (Krueger, Harms, et al., 2015). The biochemical decomposition of naturally occurring polymers such as, e.g., cellulose, chitin, and proteins is usually initiated by various hydrolytic enzymes, which cleave the bonds between the polymer subunits. Similarly, synthetic polymers with hydrolysable structures in their backbones might become depolymerized through initial enzymatic hydrolysis. Enzymes such as esterases, cutinases, and other hydrolases

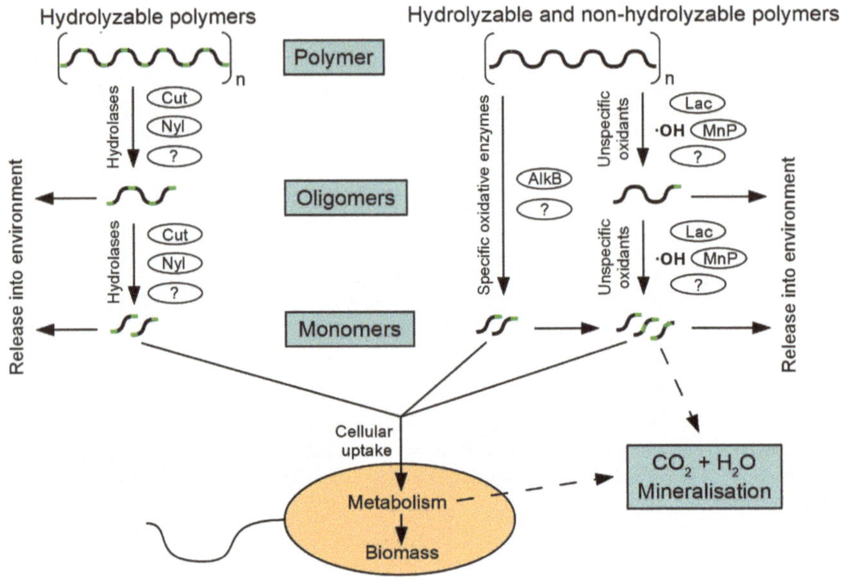

Fig. 2 Proposed pathways of microbial degradation of plastics. Polymer decomposition may be initiated by either hydrolysis (hydrolyzable polymers only) or oxidation (both hydrolyzable and non-hydrolyzable plastics). No experimental evidence exists for potential reductive dehalogenation of halogenated polymers such as PVC as yet, and therefore this type of reaction is not shown. Green lines indicate functional groups (including hydrolyzable bonds linking subunits of hydrolyzable plastics). The question marks denote so-far unknown enzymes. Hydroxyl radicals, which may be produced in Fenton-type reactions, are highlighted in bold. Cut, cutinase; Nyl, nylon hydrolase; AlkB, alkane hydroxylase; Lac, laccase; MnP, manganese peroxidase (Krueger, Harms, & Schlosser, 2015). *Adapted by permission from Springer Nature: Springer, Krueger, M.C., Harms, H., & Schlosser, D. (2015) Prospects for microbiological solutions to environmental pollution with plastics.* Applied Microbiology Biotechnology 99 *(21):8857–8874. © Springer-Verlag Berlin Heidelberg 2015.*

were reported to be involved in these processes (Fig. 2) (Krueger, Harms, et al., 2015). Biodegradable (i.e., compostable) bioplastics all represent hydrolysable plastics. For instance, polyesters like polylactic acid (PLA), polyhydroxyalkanoate (PHA), polybutylene succinate (PBS), and polybutylene adipate terephthalate (PBAT) belong to these types of plastics (EUBP, 2017, 2018). Polyesters such as PET and polymers of the PUR and PA type (Fig. 1) are also potentially hydrolysable (Fig. 2), even though being more recalcitrant and unfortunately also more widespread (Krueger, Harms, et al., 2015).

Non-hydrolysable plastics are lacking of reactive groups in their polymer backbones, which are exclusively built on C—C bonds (Fig. 1). Therefore, redox reactions (including radical-based ones) remain as the only possibility for biochemical polymer decomposition into smaller breakdown products, which then may undergo further microbial degradation (Fig. 2) (Krueger, Harms, et al., 2015). Highly recalcitrant plastics such as PE, PP, PS, and PVC (Fig. 1) share the characteristic of being inert to hydrolysis with lignin (Krueger, Harms, et al., 2015). The latter compound is a major constituent of plant lignocellulose and the second-most abundant natural polymer, which can efficiently be decomposed with the help of extracellular lignin-modifying enzymes and reactive oxygen species produced by particular higher fungi. Nevertheless, non-hydrolysable plastics are obviously even more difficult to degrade than lignin. Bond dissociation energies (BDEs) described for the chemical bonds forming the backbone of lignin range from 160 to 300 kJ/mol for C—O ether bonds and from 240 to 425 kJ/mol (up to exceptional 500 kJ/mol) for C—C bonds. However, plastics like PE, PP, PS, and PVC exhibit BDEs in the range of 330–370 kJ/mol for the C—C bonds of their backbones (350–470 kJ/mol for C—H bonds); with higher BDEs indicating stronger bonds being comparatively less prone to cleavage (Krueger, Harms, et al., 2015). Moreover, particular structural elements of lignin such as partly oxidized moieties make it far more hydrophilic than plastics, and lower the redox potential thus facilitating a successful oxidative attack (Krueger, Harms, et al., 2015).

1.2 Extracellular Fenton-type reactions in filamentous fungi

Redox reactions potentially enabling to attack plastics might be catalyzed by lignin-modifying enzymes (in particular, the multi-copper oxidase laccase and lignin-modifying peroxidases such as lignin peroxidase, manganese peroxidase, and versatile peroxidase) especially known from lignin-degrading basidiomycetes (Krueger, Harms, et al., 2015). These fungi are frequently also referred to as white-rot basidiomycetes (or white-rot fungi) because of their ability to completely degrade lignin to carbon dioxide and water; hereby leaving certain polysaccharide constituents of lignocellulose as a white-grayish residue. Indeed, lignin-modifying enzymes have been implicated in the (albeit mostly limited) oxidation of both hydrolysable and non-hydrolysable plastics to some extent (Deguchi, Kitaoka, Kakezawa, & Nishida, 1998; Fujisawa, Hirai, & Nishida, 2001; Krueger, Harms, et al., 2015; Krueger, Hofmann, Moeder, & Schlosser, 2015).

Alternatively, initial oxidation of polymers may be caused by biochemically produced hydroxyl radicals (Fig. 2). These radicals are the strongest oxidants known from biological systems, possess a redox potential of about +2.7 V, and therefore efficiently oxidize many organic compounds

(Klaning, Sehested, & Holcman, 1985; Krueger, Hofmann, et al., 2015). Hydroxyl radicals produced by certain fungi have been demonstrated to initiate backbone scission in water-soluble polymers such as the recalcitrant ether polyethylene glycol (Kerem, Jensen, & Hammel, 1999) and the polystyrene analog polystyrene sulfonate (PSS; Fig. 3) (Krueger, Hofmann, et al., 2015), and to cause superficial oxidation of solid polystyrene (PS; Fig. 3) (Krueger et al., 2017). Hydroxyl radicals are well known to hydroxylate non-phenolic aromatic structures of xenobiotics (Kramer, Kreisel, Fahr, Kassbohrer, & Schlosser, 2004; Krueger et al., 2017; Schlosser, Fahr, Karl, & Wetzstein, 2000). Beyond reactions potentially initiating backbone scission in polymers solely due to hydroxyl radical attack, the possible hydroxylation of non-phenolic aromatic polymer structures (e.g., aromatic rings in PS and PSS; Fig. 3) by hydroxyl radicals could make such structures accessible to those fungal lignin-modifying enzymes preferably acting on phenolic aromatics (Harms, Schlosser, & Wick, 2011; Krueger et al., 2017; Krueger, Harms, et al., 2015; Krueger, Hofmann, et al., 2015).

Biological hydroxyl radical production has received particular attention in basidiomycetes causing the so-called brown-rot of wood. This type of wood decay involves the utilization of wood polysaccharides for growth and the accompanying partial modification of lignin to a brown residue (Martinez et al., 2009; Yelle, Ralph, Lu, & Hammel, 2008), thereby giving rise to the general designation "brown-rot fungi" (or brown-rot basidiomycetes) for the causative organisms. In brown-rot fungi, wood lignocellulose decomposition can be achieved through the redox cycling of fungal quinones (Kerem et al., 1999; Krueger, Bergmann, & Schlosser, 2016; Krueger, Hofmann, et al., 2015; Suzuki, Hunt, Houtman, Dalebroux, & Hammel, 2006). This process involves the reduction of Fe^{3+} to Fe^{2+} by

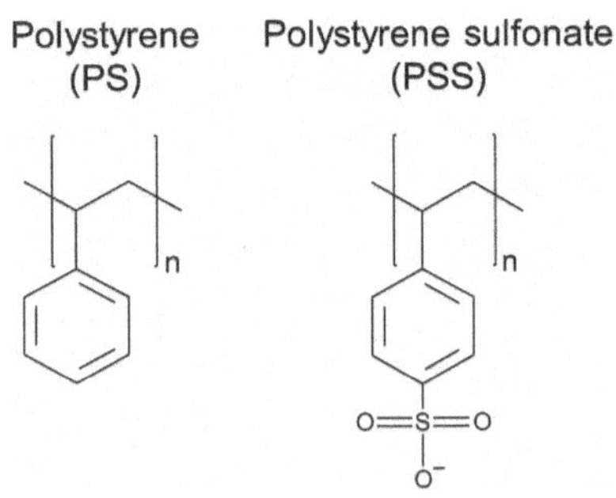

Fig. 3 Chemical structures of the solid PS and the water-soluble structural analog PSS.

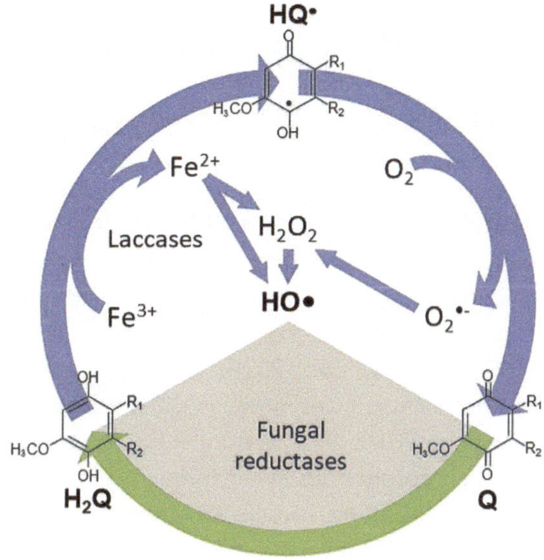

Fig. 4 Fungal quinone redox cycling. Fungal or plant (lignin)-derived diphenols such as various *o*-methoxy-substituted hydroquinones (H_2Q; $R_1 = OCH_3$ and $R_2 = H$, or $R_1 = H$ and $R_2 = OCH_3$) are oxidized by laccases and/or Fe^{3+} to semiquinone radicals (HQ•). These are then further oxidized to quinones (Q) and in turn reduce air oxygen to superoxide anion radicals ($O_2{\bullet}^-$, which give rise to H_2O_2 production via dismutation and/or oxidation of Fe^{2+} to Fe^{3+}. H_2O_2 reacts with Fe^{2+} in a Fenton-type reaction to yield highly reactive hydroxyl radicals (HO•). Quinones produced from semiquinone oxidation are reduced back to diphenols by cell-bound fungal reductases. The figure is composed based on information from the following references: An, Wei and Xiao (2015), Aranda et al. (2010), Gómez-Toribio, Garcia-Martin, Martinez, Martinez and Guillen (2009), Gómez-Toribio, García-Martín, Martínez, Martínez and Guillén (2009), Guillen, Gomez-Toribio, Martinez and Martinez (2000), Guillén, Martínez, Muñoz and Martínez (1997), Kramer et al. (2004), Krueger et al. (2016), Krueger, Hofmann, et al. (2015), Marco-Urrea et al. (2010), Sutzl et al. (2018), Vasiliadou et al. (2019), Wei et al. (2010).

secreted fungal diphenols (e.g., hydroquinones), which are in turn oxidized to semiquinone radicals (Fig. 4). These subsequently reduce molecular oxygen to yield superoxide anion radicals, and hereby are further oxidized to quinones. Superoxide anion radicals can produce H_2O_2 via dismutation and/or oxidation of Fe^{2+} to Fe^{3+}. H_2O_2 reacts with Fe^{2+} in a Fenton-type reaction to yield highly reactive hydroxyl radicals, which decompose especially the crystalline parts of cellulose as a prerequisite for fungal utilization of these only poorly bioavailable cellulose fractions, and concomitantly also modify lignin to some extent. Quinones formed from semiquinone oxidation are reduced back to diphenols by cell-bound fungal reductases to complete the redox cycle (Fig. 4) (Arantes, Jellison, & Goodell, 2012; Jensen Jr., Houtman, Ryan, & Hammel, 2001; Krueger et al., 2017; Krueger, Hofmann, et al., 2015; Suzuki et al., 2006; Yelle et al., 2008).

White-rot basidiomycetes and filamentous fungi (among them environmentally ubiquitous molds, as well as peatland and aquatic isolates) were also reported to drive Fenton chemistry by quinone redox cycling, and to employ this biodegradation tool for the effective depolymerization of the soluble synthetic polymer PSS (Gómez-Toribio, García-Martín, et al., 2009; Guillén et al., 1997; Krueger et al., 2016). Fungi not representing brown-rot basidiomycetes do not seem to synthesize the quinones needed for redox cycling (Krueger et al., 2016; Paszczynski, Crawford, Funk, & Goodell, 1999). Therefore, such compounds were frequently exogenously added to fungal cultures in corresponding studies, in order to reveal redox cycling effects (Aranda et al., 2010; Gómez-Toribio, García-Martín, et al., 2009; Krueger et al., 2016; Marco-Urrea et al., 2010). Lignin-derived diphenols/hydroquinones, e.g., compounds related to the *p*-hydroxyphenyl, guaiacyl, and syringyl units of lignin, might contribute to fungal quinone redox cycling under natural conditions (Gómez-Toribio, García-Martín, et al., 2009; Sutzl et al., 2018).

Extracellular oxidoreductases such as laccases are widespread in the fungal kingdom (Giardina et al., 2010; Rodgers et al., 2010; Theerachat, Guieysse, Morel, Remaud-Simeon, & Chulalaksananukul, 2019), and may potentially be involved in fungal quinone redox cycling in laccase-producing members of several basidio- and ascomycete groups (Fig. 4) (Krueger et al., 2016). Laccases were reported for a few brown-rot basidiomycetes, where they could oxidize diphenols to semiquinones if abiotic diphenol oxidation would be hampered by high concentrations of oxalate; a strong Fe^{3+} chelator produced and secreted by many brown rot basidiomycetes along with diphenols (An et al., 2015; Bissaro, Varnai, Rohr, & Eijsink, 2018; Wei et al., 2009, 2010). Laccases are particularly prominent in white-rot fungi, and were implicated in quinone redox cycling also in this fungal group (Aranda et al., 2010; Gómez-Toribio, Garcia-Martin, et al., 2009; Gómez-Toribio, García-Martín, et al., 2009; Guillén et al., 1997; Guillen, Munoz, Gomez-Toribio, Martinez, & Jesus Martinez, 2000; Vasiliadou et al., 2019). Furthermore, also lignin-modifying peroxidases of white-rot fungi may contribute to quinone redox cycling in combination with laccases or as sole oxidants, since diphenols can easily be oxidized by all of these enzymes (Gómez-Toribio, García-Martín, et al., 2009; Marco-Urrea et al., 2010).

2. Rationale for the cultivation of fungi for Fenton chemistry-dependent attack on synthetic polymers

Brown- and white-rot basidiomycetes, and also eco-physiologically diverse filamentous ascomycetes dwelling in various habitats have previously been demonstrated to be capable of quinone redox cycling under certain conditions resembling those in natural environments (Gómez-Toribio, Garcia-Martin, et al., 2009; Gómez-Toribio, García-Martín, et al., 2009; Krueger et al., 2016; Sutzl et al., 2018). Quinone redox cycling can drive fungal Fenton chemistry and lead to the formation of highly reactive hydroxyl radicals as biodegradation tools, which might attack synthetic polymers and thereby initiating their decomposition (Krueger, Harms, et al., 2015). Superficial oxidation of particularly recalcitrant solid synthetic polymers such as PS by fungal Fenton chemistry-derived hydroxyl radicals was previously demonstrated for the brown-rot fungus *G. trabeum*, albeit signals indicating PS oxidation were rather weak (Krueger et al., 2017). Nevertheless, this type of biochemical attack may be more efficient for other types of solid synthetic polymers, and perhaps not limited to the group of brown-rot fungi. The water-soluble synthetic polymer PSS, a compound with widespread medical and several technical applications, turned out to be vulnerable to quinone-driven fungal Fenton chemistry before (Krueger et al., 2016; Krueger, Hofmann, et al., 2015). Due to the presence of sulfonic acid groups, PSS is water-soluble in contrast to many other polymers and therefore allows circumvent bioavailability limitations caused by solid, insoluble polymers (Krueger, Hofmann, et al., 2015). Despite not directly being comparable with solid polymers with respect to susceptibility to hydroxyl radical attack and further degradation also due to mechanistically important differences in the related chemical structures (Krueger et al., 2017), PSS may be employed as a comparatively easy-to-analyze and time-saving proxy to indicate quinone redox cycling-dependent Fenton-based biodegradation capabilities of fungi (Krueger et al., 2016). In the following, cultivation-based approaches for the characterization of eco-physiologically diverse asco- and basidiomycetes with respect to their ability to attack solid and water-soluble synthetic polymers with the help of quinone redox cycling-based Fenton-type reactions are provided.

3. Materials, equipment and reagents

3.1 Materials

1. Culture flasks (glass; total volume 250 mL)
2. Strains of filamentous fungi intended to be investigated, which may, e.g., be derived from public culture collections of microorganisms or isolated from habitats of interest. For instance, the brown-rot basidiomycetes *Gloeophyllum trabeum* DSM 1398 and *G. striatum* DSM 9592, the white-rot fungus *Trametes versicolor* DSM 11269, and the ascomycetous mold *Penicillium chrysogenum* DSM 848 (all available from the German Collection of Microorganisms and Cell Cultures DSMZ, Braunschweig, Germany) were previously shown to depolymerize synthetic polymers via redox cycling-driven Fenton chemistry (Fig. 5) (Krueger et al., 2016, 2017; Krueger, Hofmann, et al., 2015).

3.2 Equipment

1. Autoclavable sieve (diameter about 6–8 cm)
2. Amicon® stirred cell (200–400 mL) equipped with a 10 kDa nominal MW cutoff polyethersulfone ultrafiltration membrane (Merck Millipore)
3. Bath sonicator
4. Biological safety cabinet (class II or higher)

3.3 Chemicals

1. 2,6-Dimethoxy benzoquinone (2,6-DMBQ; 97%; Sigma-Aldrich)
2. 4-Aminobenzoic acid (Sigma-Aldrich or other suppliers)
3. (−)-Riboflavin (Sigma-Aldrich or other suppliers)
4. Agar (powder, for microbiology; Sigma-Aldrich or other suppliers)
5. $CaCl_2$ (Sigma-Aldrich or other suppliers)
6. $CoSO_4$ (Sigma-Aldrich or other suppliers)
7. $CuSO_4$ (Sigma-Aldrich or other suppliers)
8. Cyanocobalamin (Sigma-Aldrich or other suppliers)
9. DL-α-lipoic acid (Sigma-Aldrich or other suppliers)
10. D-(+)-biotin (Sigma-Aldrich or other suppliers)
11. D-(+)-pantothenic acid calcium salt (Sigma-Aldrich or other suppliers)
12. $FeSO_4$ (Sigma-Aldrich or other suppliers)
13. Folic acid (Sigma-Aldrich or other suppliers)
14. H_3BO_3 (Sigma-Aldrich or other suppliers)

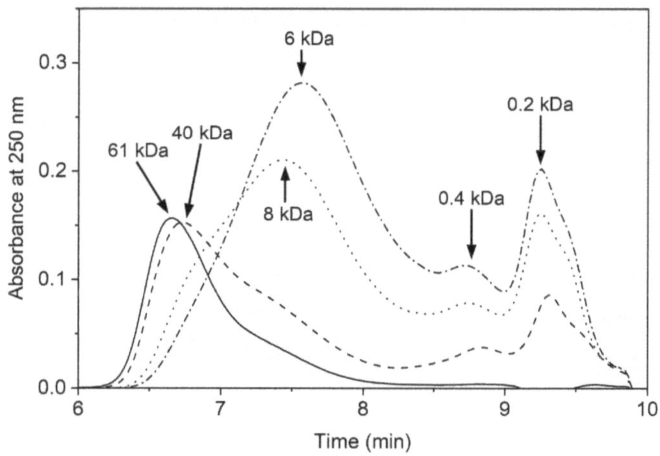

Fig. 5 Typical size exclusion chromatography (SEC) elution profiles of samples derived from the supernatant of a single representative culture of the brown-rot basidiomycete *Gloeophyllum trabeum* in a medium supporting extracellular hydroxyl radical production (herein referred to as Wetzstein medium; please also refer to step 1.2, Section 4.1 below), which had been supplemented with ultrafiltered polystyrene sulfonate (PSS) at the start of cultivation. Samples were taken immediately after the addition of PSS (—), and from culture day 7 (— — —), 14 (······), and 20 (— · — ·), respectively. PSS depolymerization with increasing time is obvious from a continuous shift of the peak corresponding to the major PSS fraction to longer retention times indicating lower molecular masses, and the concomitant appearance of two new peaks with intensities increasing over time in the range of mono- and dimeric compounds (i.e., approximately 0.2 and 0.4 kDa). A direct comparison and quantification is however hampered by a considerably higher absorbance of certain fractions of the partially degraded PSS at around 250 nm, as compared to the undegraded polymer (Krueger, Hofmann, et al., 2015). *Reprinted from: Krueger, M.C., Hofmann, U., Moeder, M., & Schlosser, D. (2015) Potential of wood-rotting fungi to attack polystyrene sulfonate and its depolymerisation by* Gloeophyllum trabeum *via hydroquinone-driven Fenton chemistry. PLoS One 10(7): e0131773. doi:https://doi.org/10.1371/journal.pone.0131773. Copyright: © 2015 Krueger et al. Open access article distributed under the terms of the Creative Commons Attribution License (http://creativecommons.org/licenses/by/4.0/).*

15. $MgSO_4$ (Sigma-Aldrich or other suppliers)
16. $MnSO_4$ (Sigma-Aldrich or other suppliers)
17. Nicotinamide (Sigma-Aldrich or other suppliers)
18. $(NH_4)_6Mo_7O_{24}$ (Sigma-Aldrich or other suppliers)
19. Oxoid CM 57 malt extract broth (Oxoid, Vienna, Austria)
20. Polymer films as desired, e.g., polystyrene calibration films for IR, Ph. Eur. certified (Sigma-Aldrich)

21. Poly(sodium 4-styrene sulfonate) (PSS; MW ~70,000 Da; Sigma-Aldrich)
22. Pyridoxine x HCl (Sigma-Aldrich or other suppliers)
23. Sodium azide (Sigma-Aldrich or other suppliers)
24. Thiamine x HCl (Sigma-Aldrich or other suppliers)
25. $ZnSO_4$ (Sigma-Aldrich or other suppliers)

4. Protocols

4.1 Maintenance of fungal cultures

1.1. Maintain all fungal strains on 1% (m/v) malt extract agar plates (1.5% agar; pH 5.7) at 28 °C.
1.2. Maintain brown-rot basidiomycetes at 28 °C on additional agar plates (1.5% agar) containing a defined mineral medium devoid of a carbon, nitrogen, and phosphate source previously described, which is composed as follows: 0.8 mM $MgSO_4$, 0.2 mM $CaCl_2$, 12 mM H_3BO_3, 0.4 mM $CoSO_4$, 0.2 mM $CuSO_4$, 0.04 mM $(NH_4)_6Mo_7O_{24}$, 2 mM $MnSO_4$, and 0.4 mM $ZnSO_4$ (pH 6), supplemented after autoclaving with 20 μM $FeSO_4$ (freshly prepared) and 1 mL (per liter of medium) of a vitamin stock solution containing (per liter) 20 mg of 4-aminobenzoic acid, 6 mg of D-(+)-biotin, 50 mg of DL-α-lipoic acid, 50 mg of nicotinamide, 10 mg of D-(+)-pantothenic acid calcium salt, 100 mg of pyridoxine x HCl, 50 mg of (−)-riboflavin, 20 mg of folic acid, 10 mg of thiamine × HCl, and 50 mg of cyanocobalamin (Wetzstein, Schmeer, & Karl, 1997; referred to as Wetzstein medium in Krueger, Hofmann, et al., 2015, and below). Add $FeSO_4$ and vitamins as sterile filtered solutions in a biological safety cabinet to the still liquid agar medium, after it had been autoclaved. For agar plate preparation, include 10 g/L milled wheat straw (about 6 mm particle size; autoclave twice separately before mixing with the autoclaved and still liquid agar medium).
1.3. Store well-grown agar plates at 4 °C in the dark.

4.2 Preparation of water-soluble PSS for depolymerization experiments

2.1. Eliminate low molecular mass fractions contained in commercial PSS by ultrafiltration, using an Amicon® stirred cell equipped with a 10 kDa nominal MW cutoff ultrafiltration membrane

4.3 Preparation of solid polymer films for degradation experiments

3.1. Cut solid PS films or films of other plastic types (about 80–100 μm thickness) into square pieces of about 5 cm^2 size.
3.2. Degrease and sterilize polymer films by gentle shaking in 70% (v/v) ethanol for 5–10 min.
3.3. Remove films from the 70% ethanol solution and leave them to air dry under sterile conditions in a biological safety cabinet.
3.4. Store polymer films under sterile conditions until use.

4.4 Depolymerization of water-soluble PSS using brown-rot basidiomycetes

4.1. Culture brown-rot fungi on 1% malt extract agar plates at 28 °C for 1–3 weeks, depending on their growth speed. Adapt the cultivation temperature to meet the requirements of your fungi if needed.
4.2. Use three agar plugs of 1 cm diameter derived from fungal colonies on agar to inoculate 250-mL flasks containing 30 mL Oxoid CM 57 malt extract broth (concentration 5 g/L).
4.3. Incubate the thus obtained fungal precultures without agitation at 28 °C in the dark for 7 days. Adapt the cultivation temperature to meet the requirements of your fungi if needed.
4.4. For degradation experiments, filter liquid precultures over a sterile sieve under sterile conditions (biological safety cabinet).
4.5. Wash the retained mycelium twice with 20 mL liquid Wetzstein degradation medium (supplemented with freshly prepared $FeSO_4$ and vitamins but without agar and wheat straw; see Section 4.1, point 1.2 above).
4.6. Transfer the washed mycelia to new 250-mL flasks containing 25 mL liquid Wetzstein medium (containing $FeSO_4$ and vitamins), which previously had been supplemented with 2 g/L sterile filtered PSS in a biological safety cabinet. Establish at least triplicate cultures for each fungus.
4.7. In order to study effects of $FeSO_4$ (reactant in Fenton-type reactions) on fungal PSS depolymerization, follow step 4.6 above, however, use liquid Wetzstein medium devoid of $FeSO_4$ instead. Establish at least triplicate cultures for each fungus.
4.8. In order to study effects of 2,6-DMBQ (subject to fungal redox cycling driving Fenton chemistry) on fungal PSS depolymerization,

follow step 4.6 (presence of $FeSO_4$) or 4.7 (absence of $FeSO_4$) above and additionally include 250 μM sterile filtered 2,6-DMBQ in the liquid Wetzstein medium. Establish at least triplicate cultures for each fungus.

4.9. In order to distinguish between PSS and its breakdown products and compounds potentially produced by fungi independent from PSS breakdown during analysis (see Section 6.1 below), follow steps 4.6–4.8 above, however, without including PSS in the liquid Wetzstein medium. Establish at least triplicate cultures for each fungus.

4.10. Prepare biologically inactive control cultures for comparison by additionally including 1 g/L sodium azide in liquid Wetzstein media used in steps 4.6–4.9 above. Establish at least triplicate control cultures for each fungus and cultivation variant.

4.11. Incubate fungal cultures for the desired period (up to 3 weeks is recommended) at 28 °C without shaking in the dark. Adapt the cultivation temperature to meet the requirements of your fungi if needed.

4.12. Take initial samples (1 mL; always cell-free culture supernatants from triplicate cultures) before adding fungal mycelia to degradation mixtures, and also immediately following the addition of mycelia to media and thorough mixing. Take further samples at desired time intervals throughout the cultivation period. If necessary, fungal biomass should be removed from samples of culture supernatants by centrifugation (16,000 × g for 15 min at 4 °C). Samples should be stored at −20 °C until analysis.

4.5 Depolymerization of water-soluble PSS using white-rot basidiomycetes and filamentous ascomycetes

5.1. Culture fungal strains on 1% malt extract agar plates at 28 °C for 1–3 weeks, depending on their growth speed. Adapt the cultivation temperature to meet the requirements of your fungi if needed.

5.2. Use three agar plugs of 1 cm diameter derived from fungal colonies on agar to inoculate 250-mL flasks containing 25 mL Oxoid CM 57 malt extract broth (concentration 5 g/L), which had previously been supplemented with 2 g/L sterile filtered PSS and 20 μM sterile filtered $FeSO_4$ (freshly prepared) in a biological safety cabinet. Establish at least triplicate cultures for each fungus.

5.3. In order to study effects of $FeSO_4$ (reactant on Fenton-type reactions) on fungal PSS depolymerization, follow step 5.2 above, however, without including $FeSO_4$ in the liquid Oxoid CM 57 malt extract medium. Establish at least triplicate cultures for each fungus.

5.4. In order to study effects of 2,6-DMBQ (subject to fungal redox cycling driving Fenton chemistry) on fungal PSS depolymerization, follow step 5.2 (presence of $FeSO_4$) or 5.3 (absence of $FeSO_4$) above and additionally include 250 μM sterile filtered 2,6-DMBQ in the liquid Oxoid CM 57 malt extract medium. Establish at least triplicate cultures for each fungus.

5.5. In order to distinguish between PSS and its breakdown products and compounds potentially produced by fungi independent from PSS breakdown during analysis (see Section 6.1 below), follow steps 5.2–5.4 above, however, without including PSS in the liquid Oxoid CM 57 malt extract medium. Establish at least triplicate cultures for each fungus.

5.6. Prepare biologically inactive control cultures for comparison by additionally including 1 g/L sodium azide in the liquid Oxoid CM 57 malt extract media used in steps 5.2–5.5 above. Establish at least triplicate control cultures for each fungus and cultivation variant.

5.7. Follow the steps 4.11 and 4.12 above.

4.6 Investigating biochemical attack on solid polymer films by brown-rot basidiomycetes

6.1. Follow steps 4.1–4.5 above.

6.2. Transfer the washed mycelia to new 250-mL flasks containing 25 mL liquid Wetzstein medium (containing $FeSO_4$ and vitamins), which previously had been supplemented with one sterile PS (or other solid polymer) film (about 5 cm^2 size; 80–100 μm thickness) per flask in a biological safety cabinet. Establish at least triplicate cultures for each fungus or more, depending on the time points considered for sampling (triplicate cultures would have to be harvested per time point of sampling, to be able to analyze three polymer films per sampling point).

6.3. In order to study effects of $FeSO_4$ (reactant on Fenton-type reactions) on fungal attack on polymer films, follow step 6.2 above, however, use liquid Wetzstein medium devoid of $FeSO_4$ instead. Establish at least triplicate cultures for each fungus or more, depending on the time points considered for sampling as explained before.

6.4. In order to study effects of 2,6-DMBQ (subject to fungal redox cycling driving Fenton chemistry) on fungal attack on polymer films, follow step 6.2 (presence of $FeSO_4$) or 6.3 (absence of $FeSO_4$) above and additionally include 250 μM sterile filtered 2,6-DMBQ in the liquid Wetzstein medium. Establish at least triplicate cultures for each fungus or more, depending on the time points considered for sampling as explained before.

6.5. Prepare biologically inactive control cultures for comparison by additionally including 1 g/L sodium azide in liquid Wetzstein media used in steps 6.2–6.4 above. Establish at least triplicate cultures for each fungus or more, depending on the time points considered for sampling as explained before.

6.6. Incubate fungal cultures for the desired period (4–5 weeks is recommended) at 28 °C without shaking in the dark. Adapt the cultivation temperature to meet the requirements of your fungi if needed.

6.7. Harvest three fungal cultures and take initial samples (one polymer film per flask) immediately following the addition of mycelia to media and thorough mixing. Harvest further cultures (always three cultures per sampling point and cultivation variant) and take further samples at desired time intervals throughout the cultivation period. Untreated polymer films should also be analyzed for comparison. Sampled polymer films must be subject to a cleaning and drying procedure described in Section 6.2 below, and then should be stored at a dry, clean and dark place until analysis (also see Sections 6.2 and 6.3 below).

4.7 Investigating biochemical attack on solid polymer films by white-rot basidiomycetes and filamentous ascomycetes

7.1. Culture fungal strains on 1% malt extract agar plates at 28 °C for 1–3 weeks, depending on their growth speed. Adapt the cultivation temperature to meet the requirements of your fungi if needed.

7.2. Use three agar plugs of 1 cm diameter derived from fungal colonies on agar to inoculate 250-mL flasks containing 25 mL Oxoid CM 57 malt extract broth (concentration 5 g/L), which had previously been supplemented with one sterile PS (or other solid polymer) film (about 5 cm^2 size; 80–100 μm thickness) per flask and 20 μM sterile filtered $FeSO_4$ (freshly prepared) in a biological safety cabinet. Establish at least triplicate cultures for each fungus or more, depending on the time points considered for sampling (triplicate cultures would have to be harvested per time point of sampling, to be able to analyze three polymer films per sampling point).

7.3. In order to study effects of $FeSO_4$ (reactant on Fenton-type reactions) on fungal attack on polymer films, follow step 7.2 above, however, use liquid Oxoid CM 57 malt extract medium devoid of $FeSO_4$ instead.

Establish at least triplicate cultures for each fungus or more, depending on the time points considered for sampling as explained before.

7.4. In order to study effects of 2,6-DMBQ (subject to fungal redox cycling driving Fenton chemistry) on fungal attack on polymer films, follow step 7.2 (presence of $FeSO_4$) or 7.3 (absence of $FeSO_4$) above and additionally include 250 μM sterile filtered 2,6-DMBQ in the liquid Oxoid CM 57 malt extract medium. Establish at least triplicate cultures for each fungus or more, depending on the time points considered for sampling as explained before.

7.5. Prepare biologically inactive control cultures for comparison by additionally including 1 g/L sodium azide in liquid Oxoid CM 57 malt extract media used in steps 7.2–7.4 above. Establish at least triplicate cultures for each fungus or more, depending on the time points considered for sampling as explained before.

7.6. Follow steps 6.6 and 6.7 above.

5. Safety considerations

General laboratory safety procedures should be applied. Personal protective materials such as lab coat, gloves and safety glasses should be used during the experiments. Please note that sodium azide is a severe poison and may be fatal in contact with skin or if swallowed (GHS hazard statements: H300, H310, H373, H410). Therefore, appropriate safety measures have to be applied (further information can, e.g., be retrieved from the related safety material data sheet accessible via https://www.sigmaaldrich.com/safety-center.html, followed by a search for "sodium azide"). Biosaftey level 1 applies to laboratory work with the fungi exemplified in Section 3.1. However, readers interested in employing particular fungi should always gather up-to-date information regarding the phylogenetic and taxonomic position of the fungus intended to be used and the applicable legislation concerning its risk group and biosafety level assignment, and follow the related safety measures to be applied.

6. Applicable analytical methods to detect fungal effects on polymers

6.1 Size exclusion chromatography (SEC) for the analysis of water-soluble polymers

SEC may be employed for the analysis of water-soluble polymers such as PSS, and used to detect polymer decomposition resulting from biochemical attack (Fig. 5) (Krueger, Hofmann, et al., 2015). Polymer separation can be achieved using a HEMA MCX 1000 Å column (PSS Polymer Standards Service GmbH, Mainz, Germany) operated at 40 °C, which may be connected to any suitable liquid chromatography system. A running buffer composed of 5 g/l $NaNO_3$, 2 g/l K_2HPO_4 and 20% acetonitrile in distilled water (pH adjusted to 10) and a flow rate of 1 mL/min were previously successfully applied for the separation of PSS (Krueger, Hofmann, et al., 2015). Cell- and particle-free samples from fungal culture supernatants can directly be injected without prior treatment, using an injection volume of 100 μL. With a photodiode array detector (PDA), chromatograms may be recorded over a certain wavelength range (e.g., from 220 to 600 nm, with a monitoring wavelength of 250 nm). Molecular mass calibration can be based on retention times of polystyrene sulfonate sodium salt molecular mass standards (PSS Polymer Standards Service GmbH, Mainz, Germany), e.g., covering a range of 0.208 to 79.9 kDa (Krueger, Hofmann, et al., 2015). Number-average molecular masses (M_n) and weight-average molecular masses (M_w) may, e.g., be calculated from data of chromatograms recorded at 250 nm, using slices corresponding to defined elution time intervals for single fractions (e.g., time intervals of 1.6 s were used for single fractions by Krueger, Hofmann, et al. (2015). In order to avoid potential interferences with unrelated low molecular mass compounds (e.g., fungal metabolites or impurities in the original PSS), only the polymeric fractions of PSS should be considered (Krueger, Hofmann, et al., 2015).

The molar amounts of chain scissions from fungal PSS depolymerization may be calculated as illustrated in the following example, which was taken from Krueger et al. (2017): Given that the untreated PSS (used as a parent compound) has a M_n value of 16 kDa in dissolved state, 2 g/L of dissolved PSS corresponds to a number-average molar concentration of c_n = 2 g/L divided by 16,000 g/mol = 125 μmol/L. Hence, a 50% decrease in M_n (statistically equivalent to one chain scission per PSS molecule) in a reaction volume of 25 mL corresponds to 1 [scission] * 125 μmol/L * 0.025 L = 3.125 μmol of chain scissions (Krueger et al., 2017).

6.2 Water contact angle analysis of solid polymer surfaces

The hydrophilicity of solid polymer film surfaces, which is expected to increase upon biochemical surface oxidation, can be assessed with the help of water contact angle analysis (Fig. 6) (Krueger et al., 2017). Prior to analysis, polymer films removed from fungal cultures are cleaned as follows (Krueger et al., 2017): Films are stored in distilled water at 4 °C overnight

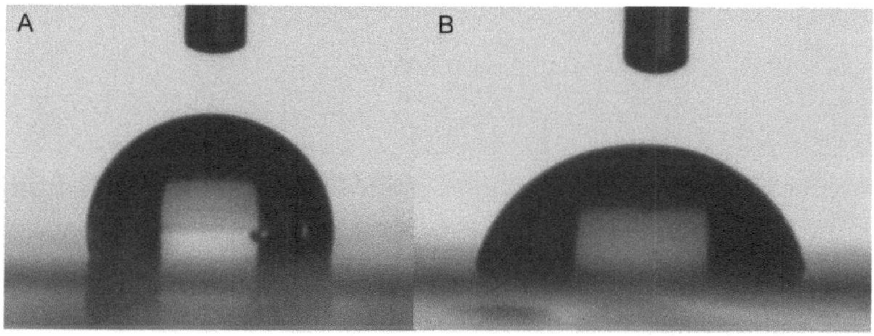

Fig. 6 Representative water contact angle measurement of PS films. (A) Abiotic control. (B) After treatment with the brown-rot fungus *G. trabeum*. The fungal treatment resulted in a significant decrease of the water contact angle as compared to corresponding values observed with abiotic controls ($85.2 \pm 4°$ vs. $101.5 \pm 1.8°$; means ± standard deviations from triplicate cultures; $P < 0.001$ with Student's *t*-test; Krueger et al., 2017), indicating a more hydrophilic surface of PS films due to superficial PS oxidation by fungal hydroxyl radicals (for further details please refer to Krueger et al., 2017). *Figure reprinted from Krueger, M.C., Seiwert, B., Prager, A., Zhang, S., Abel, B., Harms, H., et al. (2017). Degradation of polystyrene and selected analogues by biological Fenton chemistry approaches: opportunities and limitations,* Chemosphere, 173, *520–528. © 2017 Elsevier Ltd., with permission from Elsevier.*

and then subjected to three consecutive incubations of 2h in 2% SDS solution at 50°C, with the last 15min under additional ultrasonication. Thereafter, films are first rinsed with distilled water and then with 70% ethanol for 5min, followed by further 15min incubation in 70% ethanol under shaking at 120, and a final washing step with distilled water. Finally, the films are dried at room temperature for 48h (Krueger et al., 2017).

Water contact angle analysis can be performed, e.g., using a Krüss DSA 100 (Krüss GmbH, Hamburg, Germany) and the Drop Shape Analysis 4 software (Krüss) (Krueger et al., 2017). Water drops may, e.g., be generated at a flow rate of 40mL/min and 3mL volume. Contact angles can, e.g., be measured 0.4s after droplet deposition (10th frame; recording rate 25 frames per second). Several (at least three) positions of each film should be analyzed in order to obtain representative data (Krueger et al., 2017).

6.3 X-ray photoelectron spectroscopy (XPS) analyses of solid polymer surfaces

XPS can be considered to be highly sensitive toward the elemental composition of solid surfaces to a depth of about 10nm, and used to detect related changes potentially caused by fungal biochemical attack on plastics (Fig. 7) (Krueger et al., 2017). Polymer films removed from degradation experiments should rigorously be cleaned prior to analysis, e.g., using the cleaning

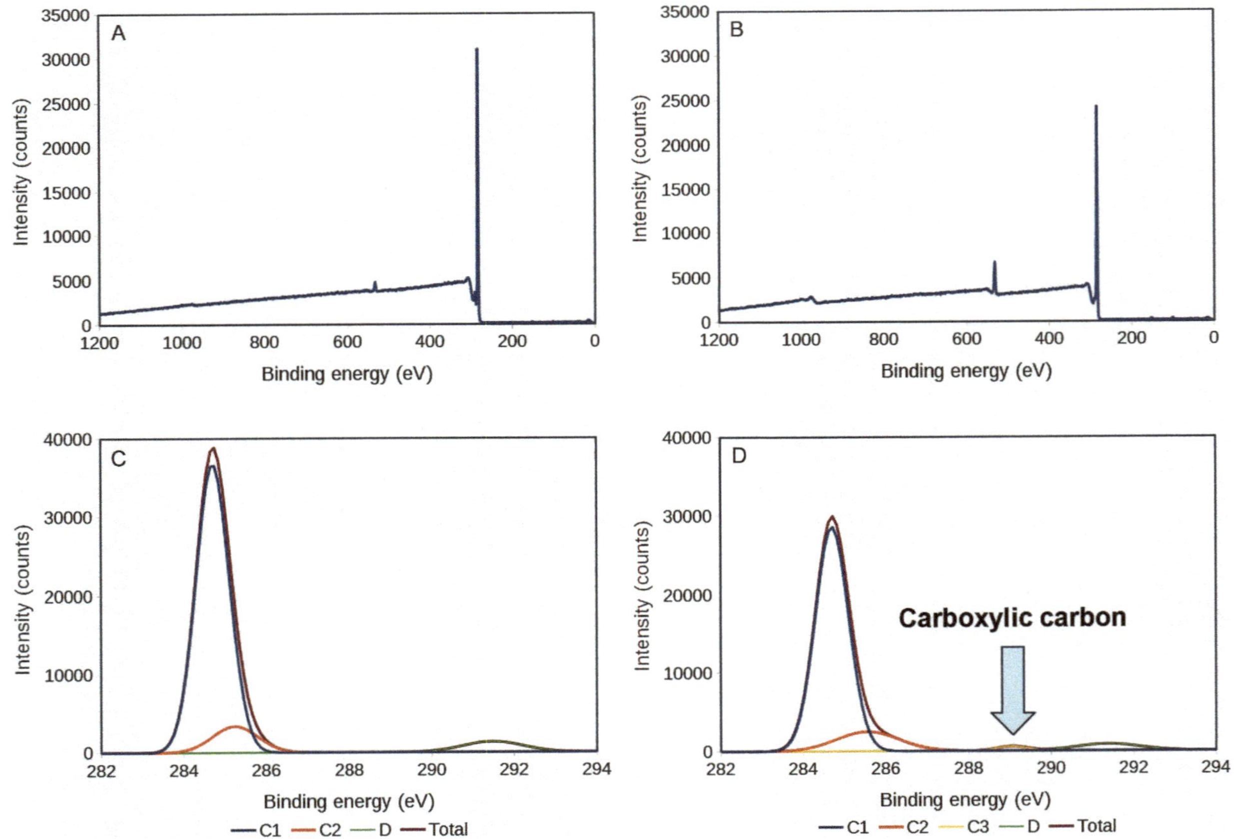

Fig. 7 See figure legend on opposite page.

procedure described in Section 6.2 (Water contact angle analysis) above. XPS analyses can, e.g., be carried out using a Kratos Axis Ultra (Kratos Analytical, Ltd., Manchester, United Kingdom) with a monochromated Al excitation source at 150 W (15 kV, 10 mA), as previously described (Krueger et al., 2017). Surface spectra may, e.g., be collected at 160 eV pass energy with 1.0 eV nominal resolution. High-resolution-scans can be measured by 40 eV pass energy with 0.1 eV nominal resolution. Binding energies could be corrected for static charging of samples by reference to the C1s peak set at a binding energy of 284.7 eV (Schulze, Stoelzer, Striegler, Starke, & Prager, 2015). Each surface should be analyzed at several (at least three) different positions to ensure an adequate statistical coverage of the obtained data. Rejecting measurements that detect elements other than C, O or tiny amounts of Si (e.g., those detecting N) is recommended in order to prevent interfering signals potentially arising from contamination with biomass components or rust precipitates (Krueger et al., 2017).

7. Summary

A more profound knowledge with respect to opportunities and limitations of the microbial degradability of synthetic polymers (frequently referred to as plastics) is highly desirable as it would largely support the development of adequate decontamination and management strategies for environmental pollution with plastics. Such strategies are urgently needed due to growing threats to the environment and human health caused by plastic pollution.

Fig. 7 XPS survey and representative C1s spectra of PS films. (A) Survey spectrum of untreated PS used as a control. (B) Survey spectrum of PS treated with *G. trabeum*. (C) C1s spectrum of (A). (D) C1s spectrum of (B). C1s spectra: Total = total measurement, C1 = aromatic carbon, C2 = aliphatic carbon; C3 = carboxylic carbon (also indicated by arrow in D); D = static discharge. XPS is highly sensitive toward the elemental composition of solid surfaces to a depth of about 10 nm. Significantly elevated oxygen levels of PS film samples derived from treatment with *G. trabeum* (8.1 ± 3.06% for active fungal samples vs. 4.73 ± 2.25% for controls; means ± standard deviations from triplicate cultures; $P = 0.018$ with Student's t-test; Krueger et al., 2017) and the appearance of carboxylic C after fungal treatment (D) advocate for limited superficial PS oxidation by fungal hydroxyl radicals, in line with water contact angle measurement results as exemplified in Fig. 6 (for further details please refer to Krueger et al., 2017). *Figure reprinted from Krueger, M.C., Seiwert, B., Prager, A., Zhang, S., Abel, B., Harms, H., et al. (2017). Degradation of polystyrene and selected analogues by biological Fenton chemistry approaches: opportunities and limitations,* Chemosphere, 173, *520–528. © 2017 Elsevier Ltd., with permission from Elsevier.*

Hydroxyl radicals are the strongest oxidants known from biological systems. In fungi, these reactive oxygen species are produced via extracellular Fenton-type reactions driven by quinone redox cycling. The employment of fungal Fenton-type reactions to attack synthetic polymers has previously been demonstrated to a limited extent. However, the full potential of such reactions for synthetic polymer breakdown and the range of fungi being capable of catalyzing related reactions still remains to be explored. The present chapter provides cultivation-dependent approaches applicable for the characterization of ecophysiologically diverse filamentous fungi with respect to their ability to attack solid and water-soluble synthetic polymers with the help of quinone redox cycling-based Fenton-type reactions, which result in the production of highly reactive hydroxyl radicals potentially serving as biodegradation tools for synthetic polymer breakdown. Additional information concerning examples for applicable analytical methods to detect fungal effects on polymers—by no means exhaustive—is also provided.

Acknowledgments

I want to thank the following people (listed in alphabetical order) for their participation and contributions: Bernd Abel, Michael Bergmann, Hauke Harms, Martina Kolbe, Martin C. Krüger, Kristin Lindstaedt, Nadia Lotti, Katrin Lübke, Andrea Prager, Jana Reichenbach, Madlen Schubert, Bettina Seiwert, Lukas Y. Wick, and Shangwei Zhang. This work was funded by the European Commission under the 7th Framework Programme as a part of the project BIOCLEAN (grant agreement 312100). It was further supported by the Helmholtz Association of German Research Centres and contributes to the Chemicals in the Environment (CITE) Research Programme conducted at the Helmholtz Centre for Environmental Research—UFZ.

References

An, H., Wei, D., & Xiao, T. (2015). Transcriptional profiles of laccase genes in the brown rot fungus *Postia placenta* MAD-R-698. *Journal of Microbiology*, *53*(9), 606–615.

Andreessen, C., & Steinbuchel, A. (2019). Recent developments in non-biodegradable biopolymers: Precursors, production processes, and future perspectives. *Applied Microbiology and Biotechnology*, *103*(1), 143–157.

Aranda, E., Marco-Urrea, E., Caminal, G., Arias, M. E., Garcia-Romera, I., & Guillen, F. (2010). Advanced oxidation of benzene, toluene, ethylbenzene and xylene isomers (BTEX) by *Trametes versicolor*. *Journal of Hazardous Materials*, *181*(1–3), 181–186.

Arantes, V., Jellison, J., & Goodell, B. (2012). Peculiarities of brown-rot fungi and biochemical Fenton reaction with regard to their potential as a model for bioprocessing biomass. *Applied Microbiology and Biotechnology*, *94*(2), 323–338.

Bissaro, B., Varnai, A., Rohr, A. K., & Eijsink, V. G. H. (2018). Oxidoreductases and reactive oxygen species in conversion of lignocellulosic biomass. *Microbiology and Molecular Biology Reviews*, *82*(4), e00029-00018.

da Costa, J. P., Santos, P. S. M., Duarte, A. C., & Rocha-Santos, T. (2016). (Nano)plastics in the environment–sources, fates and effects. *Science of the Total Environment*, *566-567*, 15–26.

Deguchi, T., Kitaoka, Y., Kakezawa, M., & Nishida, T. (1998). Purification and characterization of a nylon-degrading enzyme. *Applied and Environmental Microbiology*, *64*(4), 1366–1371.

EUBP. (2017). *Bioplastics–facts and figures*. European Bioplastics. https://docs.european-bioplastics.org/publications/EUBP_Facts_and_figures.pdf.

EUBP. (2018). *What are bioplastics? European Bioplastics fact sheet*. https://docs.european-bioplastics.org/publications/fs/EuBP_FS_What_are_bioplastics.pdf.

Fujisawa, M., Hirai, H., & Nishida, T. (2001). Degradation of polyethylene and Nylon-66 by the laccase-mediator system. *Journal of Polymers and the Environment*, *9*(3), 103–108.

Geyer, R., Jambeck, J. R., & Law, K. L. (2017). Production, use, and fate of all plastics ever made. *Science Advances*, *3*(7), e1700782.

Giardina, P., Faraco, V., Pezzella, C., Piscitelli, A., Vanhulle, S., & Sannia, G. (2010). Laccases: A never-ending story. *Cellular and Molecular Life Sciences*, *67*(3), 369–385.

Gómez-Toribio, V., Garcia-Martin, A. B., Martinez, M. J., Martinez, A. T., & Guillen, F. (2009). Enhancing the production of hydroxyl radicals by *Pleurotus eryngii* via quinone redox cycling for pollutant removal. *Applied and Environmental Microbiology*, *75*(12), 3954–3962.

Gómez-Toribio, V., García-Martín, A. B., Martínez, M. J., Martínez, A. T., & Guillén, F. (2009). Induction of extracellular hydroxyl radical production by white-rot fungi through quinone redox cycling. *Applied and Environmental Microbiology*, *75*(12), 3944–3953.

Guillen, F., Gomez-Toribio, V., Martinez, M. J., & Martinez, A. T. (2000). Production of hydroxyl radical by the synergistic action of fungal laccase and aryl alcohol oxidase. *Archives of Biochemistry and Biophysics*, *383*(1), 142–147.

Guillén, F., Martı´nez, M. A. J., Muñoz, C., & Martı´nez, A. T. (1997). Quinone redox cycling in the ligninolytic fungus *Pleurotus eryngii* leading to extracellular production of superoxide anion radical. *Archives of Biochemistry and Biophysics*, *339*(1), 190–199.

Guillen, F., Munoz, C., Gomez-Toribio, V., Martinez, A. T., & Jesus Martinez, M. (2000). Oxygen activation during oxidation of methoxyhydroquinones by laccase from *Pleurotus eryngii*. *Applied and Environmental Microbiology*, *66*(1), 170–175.

Harms, H., Schlosser, D., & Wick, L. Y. (2011). Untapped potential: Exploiting fungi in bioremediation of hazardous chemicals. *Nature Reviews Microbiology*, *9*(3), 177–192.

Harms, H., Wick, L. Y., & Schlosser, D. (2017). The fungal community in organically polluted systems. In J. Dighton, & J. F. White (Eds.), *The fungal community: Its organization and role in the ecosystem* (4th ed., pp. 459–469). Boca Raton/FL: CRC Press.

Jensen, K. A., Jr., Houtman, C. J., Ryan, Z. C., & Hammel, K. E. (2001). Pathways for extracellular Fenton chemistry in the brown rot basidiomycete *Gloeophyllum trabeum*. *Applied and Environmental Microbiology*, *67*(6), 2705–2711.

Kerem, Z., Jensen, K. A., & Hammel, K. E. (1999). Biodegradative mechanism of the brown rot basidiomycete *Gloeophyllum trabeum*: Evidence for an extracellular hydroquinone-driven Fenton reaction. *FEBS Letters*, *446*(1), 49–54.

Klaning, U. K., Sehested, K., & Holcman, J. (1985). Standard Gibbs energy of formation of the hydroxyl radical in aqueous solution. Rate constants for the reaction $ClO_2^- + O_3 \leftrightarrows O_3^- + ClO_2$. *The Journal of Physical Chemistry*, *89*(5), 760–763.

Kramer, C., Kreisel, G., Fahr, K., Kassbohrer, J., & Schlosser, D. (2004). Degradation of 2-fluorophenol by the brown-rot fungus *Gloeophyllum striatum*: Evidence for the involvement of extracellular Fenton chemistry. *Applied Microbiology and Biotechnology*, *64*(3), 387–395.

Krueger, M. C., Bergmann, M., & Schlosser, D. (2016). Widespread ability of fungi to drive quinone redox cycling for biodegradation. *FEMS Microbiology Letters*, *363*(11), fnw105.

Krueger, M. C., Harms, H., & Schlosser, D. (2015). Prospects for microbiological solutions to environmental pollution with plastics. *Applied Microbiology and Biotechnology*, *99*(21), 8857–8874.

Krueger, M. C., Hofmann, U., Moeder, M., & Schlosser, D. (2015). Potential of wood-rotting fungi to attack polystyrene sulfonate and its depolymerisation by *Gloeophyllum trabeum* via hydroquinone-driven Fenton chemistry. *PLoS One*, *10*(7), e0131773.

Krueger, M. C., Seiwert, B., Prager, A., Zhang, S., Abel, B., Harms, H., et al. (2017). Degradation of polystyrene and selected analogues by biological Fenton chemistry approaches: Opportunities and limitations. *Chemosphere*, *173*, 520–528.

Marco-Urrea, E., Radjenovic, J., Caminal, G., Petrovic, M., Vicent, T., & Barcelo, D. (2010). Oxidation of atenolol, propranolol, carbamazepine and clofibric acid by a biological Fenton-like system mediated by the white-rot fungus *Trametes versicolor*. *Water Research*, *44*(2), 521–532.

Martinez, D., Challacombe, J., Morgenstern, I., Hibbett, D., Schmoll, M., Kubicek, C. P., et al. (2009). Genome, transcriptome, and secretome analysis of wood decay fungus *Postia placenta* supports unique mechanisms of lignocellulose conversion. *Proceedings of the National Academy of Sciences*, *106*(6), 1954–1959.

Paszczynski, A., Crawford, R., Funk, D., & Goodell, B. (1999). De novo synthesis of 4,5-dimethoxycatechol and 2, 5-dimethoxyhydroquinone by the brown rot fungus *Gloeophyllum trabeum*. *Applied and Environmental Microbiology*, *65*(2), 674–679.

Rodgers, C. J., Blanford, C. F., Giddens, S. R., Skamnioti, P., Armstrong, F. A., & Gurr, S. J. (2010). Designer laccases: A vogue for high-potential fungal enzymes? *Trends in Biotechnology*, *28*(2), 63–72.

Schlosser, D., Fahr, K., Karl, W., & Wetzstein, H. G. (2000). Hydroxylated metabolites of 2,4-dichlorophenol imply a Fenton-type reaction in *Gloeophyllum striatum*. *Applied and Environmental Microbiology*, *66*(6), 2479–2483.

Schulze, A., Stoelzer, A., Striegler, K., Starke, S., & Prager, A. (2015). Biocatalytic self-cleaning polymer membranes. *Polymers*, 7(9), 1837–1849.

Sutzl, L., Laurent, C., Abrera, A. T., Schutz, G., Ludwig, R., & Haltrich, D. (2018). Multiplicity of enzymatic functions in the CAZy AA3 family. *Applied Microbiology and Biotechnology*, *102*(6), 2477–2492.

Suzuki, M. R., Hunt, C. G., Houtman, C. J., Dalebroux, Z. D., & Hammel, K. E. (2006). Fungal hydroquinones contribute to brown rot of wood. *Environmental Microbiology*, *8*(12), 2214–2223.

Theerachat, M., Guieysse, D., Morel, S., Remaud-Simeon, M., & Chulalaksananukul, W. (2019). Laccases from marine organisms and their applications in the biodegradation of toxic and environmental pollutants: A review. *Applied Biochemistry and Biotechnology*, *187*(2), 583–611.

Vasiliadou, I. A., Molina, R., Pariente, M. I., Christoforidis, K. C., Martinez, F., & Melero, J. A. (2019). Understanding the role of mediators in the efficiency of advanced oxidation processes using white-rot fungi. *Chemical Engineering Journal*, *359*, 1427–1435.

Wei, D., Houtman, C. J., Kapich, A. N., Hunt, C. G., Cullen, D., & Hammel, K. E. (2010). Laccase and its role in production of extracellular reactive oxygen species during wood decay by the brown rot basidiomycete *Postia placenta*. *Applied and Environmental Microbiology*, *76*(7), 2091–2097.

Wei, H., Xu, Q., Taylor, L. E., 2nd, Baker, J. O., Tucker, M. P., & Ding, S. Y. (2009). Natural paradigms of plant cell wall degradation. *Current Opinion in Biotechnology*, *20*(3), 330–338.

Wetzstein, H. G., Schmeer, N., & Karl, W. (1997). Degradation of the fluoroquinolone enrofloxacin by the brown rot fungus *Gloeophyllum striatum*: Identification of metabolites. *Applied and Environmental Microbiology*, *63*(11), 4272–4281.

Yelle, D. J., Ralph, J., Lu, F., & Hammel, K. E. (2008). Evidence for cleavage of lignin by a brown rot basidiomycete. *Environmental Microbiology*, *10*(7), 1844–1849.

CHAPTER FIVE

Characterization of biodegradation of plastics in insect larvae

Wei-Min Wu* and Craig S. Criddle
Department of Civil & Environmental Engineering, Stanford University, Stanford, CA, United States
*Corresponding author: e-mail address: billwu@stanford.edu

Contents

Abstract

Biodegradation of plastics has been observed at rapid turnover rate by some insect larvae, especially those of *Coleoptera*, in particular *Tenebrionidae*. *Tenebrio molitor* larva is well studied and capable of biodegrading polystyrene (PS), polyethylene (PE), polypropylene (PP), and polyvinyl chloride (PVC) in their digestive intestine in synergy with their gut microflora. This chapter includes the methods, protocols, and procedures used to characterize biodegradation of plastics in *T. molitor* larvae and their gut microbiomes with polystyrene as the model feedstock. The methods used can be expanded to enable investigation of other plastics and/or insects.

Methods in Enzymology, Volume 648
ISSN 0076-6879
https://doi.org/10.1016/bs.mie.2020.12.029

1. Introduction

Synthetic polymers are widely used in daily life and by many industries with an annual global production of over 359 million tons in 2018 (Europe, 2019). The six most frequently used plastic polymers, are polyethylene (PE, 29.7%, including low-density PE (LDPE), and high-density PE(HDPE)), polypropylene (PP, 19.3%), polyvinyl chloride (PVC,10%), polyurethane (PUR, 7.9%), polyethylene terephthalate (PET, 7.7%) and polystyrene (PS, 6.4%). Disposal and accumulation of synthetic plastic wastes, with PE, PP, PVC, PUR, PET and PS as major components, have been an environmental concern for decades.

Researchers have investigated biodegradation of plastic wastes since the 1960s. Due to their highly stable C—C and C—H covalent bonds, high molecular weights, hydrophobic properties, and the absence of readily oxidizable and/or hydrolyzable groups, most plastic polymers are extremely resistant to biodegradation. In general, biodegradation of plastics involves at least three reactions: (a) depolymerization or cleavage of polymer chains; (b) formation of oxidized intermediates, and (c) mineralization of intermediates to CO_2 and H_2O and to Cl^- when PVC is a feedstock (Peng, Li, et al., 2020; Yang, Yang, Wu, Zhao, & Jiang, 2014).

Reports dating to the 1950s have implicated insect larvae in the penetration and deterioration of plastics (Cline, 1978; Gerhardt & Lindgren, 1954). Over the last 10 years, research has focused on insect larvae-mediated biodegradation of PS, LDPE, PP and PVC, within the darkling beetle families (Coleoptera: Tenebrionidae), *i.e., Tenebrio molitor, Tenebrio obscurus, Zophobas atratus, Tribolium castaneum* and *Plesiophthalmus davidis* (Peng, Li, et al., 2020; Peng et al., 2019; Wang, Xin, Shi, & Zhang, 2020; Woo, Song, & Cha, 2020; Yang et al., 2015a); larvae of pest moths (Lepidoptera: Pyralidae), *i.e.,* Indian mealmoths (*Plodia interpunctella*) (Yang et al., 2014), greater waxworms (*Galleria mellonella*) (Bombelli, Howe, & Bertocchini, 2017), and lesser waxworms (*Achroia grisella*) (Kundungal, Gangarapu, Sarangapani, Patchaiyappan, & Devipriya, 2019). Plastic biodegradation in these invertebrates occurs within hours during passage through their digestive tract, a rate that far exceeds those of microorganisms in laboratory cultures or in the environment where degradation occurs on the scale of weeks, months, or even years.

Darkling beetles are the common name for a large family of beetles (Tenebrionidae) that is cosmopolitan in distribution with more than 20,000 estimated members. To date, five species (*T. molitor, T. obscurus,*

Z. atratus, *Tribolium castaneum* and *Plesiophthalmus davidis*) are confirmed for degradation of PS and LDPE (Peng, Chen, et al., 2020; Peng et al., 2019; Wang et al., 2020; Woo et al., 2020; Yang et al., 2015a). In their native environment, larvae of the genus *Tenebrio* inhabit the decaying wood of forests (Calmont & Soldati, 2008) where they consume dried leaves containing lignocellulosic materials. These larvae are also able to biodegrade PS, LDPE, PVC and PP as well as vulcanized styrene-butadiene rubber (SBR) and tire crumb (Peng, Li, et al., 2020; Yang, Ding, et al., 2021a; Yang, Gao, et al., 2021b, 2015a). *Tenebrio molitor* has four life stages: egg, larva, pupa, and beetle (Roberson, 2005). *T. molitor* larvae typically measure about 2.5 cm or more, whereas adults are generally between 1.25 and 1.8 cm in length. After mating with the male, the female lays eggs. Over their lifespan, a female will, on average, lay about 500 tiny white eggs. The eggs hatch within 1–3 weeks. During the larval stage, the larva feed on wheat bran, vegetation and dead insects and molt, depending upon larval stage, or instar (they complete development with 9–20 instars, 180–270 days). After the final molt, the larva become pupa for 7 days or more (up to 20 days at lower temperatures). The pupa then become adult beetles for 30–90 days. The new beetle is whitish and turns brown over time.

2. Biodegradation of plastics in *Tenebrio molitor*

2.1 Incubation conditions for biodegradation of plastics

Factors impacting and/or controlling plastic degradation in *T. molitor* larvae are directly associated with the properties of polymers and larval physiological conditions (growth, development stage), metabolic activities and structure of the gut microbiome as described below.

2.1.1 Plastic feedstock

All major plastics (PE, PE, PP, PVC, PET, PUR, *etc.*) as well polylactic acid (PLA) have been tested for biodegradation by *T. molitor* larvae in labs at Stanford University and collaborators. The rate of polymer degradation is influenced by chemical and physical structures, molecular weights, and degree of crystallinity. Polymer materials have higher molecular weights and a higher degree of crystallinity are more resistant to biodegradation. The larvae prefer plastic materials as foam, thin films or tiny particles (size less than their mouth). For large and rigid plastic materials, pretreatment (size reduction, conversion of crystalline to amorphous state) is needed.

Microplastic (MP) materials have a size <0.1 mm for *T. molitor*. A mixture of MPs plus wheat bran can be made with 1.6–2.0% jelly as adhesive to prepare co-diet.

2.1.2 Effect of co-diet

Composition of the diet or feedstock impacts plastic consumption rate and metabolic activities. When larvae are fed plastic polymers as sole diet, they can maintain plastic degradation and survive for 3–5 weeks, but with decreasing their body weight. This is because digestion of plastics can only provide energy and carbon for maintenance but nutrients necessary for growth are not provided by a plastic-only diet. Cannibalism is frequently observed among larvae, providing them with additional energy source and nutrients. Co-diets that include wheat bran, fruits, or vegetables facilitate plastic degradation, larval growth, and completion of life stages. A co-diet ratio of wheat bran *versus* plastics at 8 to 12 (w/w) is recommended (Yang, Brandon, et al., 2018).

2.1.3 Temperature and humidity

Temperature impacts mealworm survival and plastic consumption. *T. molitor* larvae from most countries (China, Europe and the USA) show optimal viability between 25 and 28 °C but the *T. molitor* larvae in South Asia can tolerate temperature >32 °C. Humidity is usually controlled at 60–80%.

2.1.4 Rearing density

For a plastic degradation test, a proper density of 2–3 larvae/cm^2 is recommended. *T. molitor* larvae exhibit competitive feeding behaviors.

2.2 Gut microbes

Biodegradation of plastics in insects is believed to be contributed by synergistic reactions of the digestive system (enzymes, digestive fluid, *etc.*) and gut microbes. The essential role of gut microbes in plastic degradation has been evaluated based on the changes in molecular weights with antibiotic suppression of gut microbes as opposed to controls without antibiotic treatment. To date, the results showed that when the gut microbes in *T. molitor* larvae were inhibited by gentamicin or mixed antibiotics, PS and PVC depolymerization basically or completely stopped; but LDPE and PP depolymerization still continued to a lesser degree or with a change in the depolymerization pattern (Yang, Brandon, et al., 2018; Yang, Ding, et al., 2021a; Yang, Gao, et al., 2021b). PS and PVC degradation are categorized as gut microbe-dependent, while LDPE and PP are less dependent upon gut microbes and can even be gut microbe-independent. To date, plastic (PS and LDPE)-degrading bacterial cultures have been isolated from *T. molitor* and other darkling beetles (Wang et al., 2020; Yang et al., 2015a; Yin, Xu, & Zhou, 2020).

Characterization of gut microbiomes is commonly applied to understand the community shift, dynamics, and predominance during plastics biodegradation. Studies indicate that gut microbial community structures within the gut of *T. molitor* larvae are diverse, and shift depending upon dietary changes (Brandon et al., 2018; Yang, Wu, et al., 2018). Feeding plastics such as LDPE and PS shifts the gut microbiome to a community that can biodegrade LDPE / PS (Brandon et al., 2018; Yang, Gao, et al., 2021b). Shifts in community structure are quantified using ecological metrics of richness - the number of species in the community- and by the evenness and dominance of species, as quantified by the Shannon and Simpson indices, metrics accounting for differences in the relative abundance of species: the fact that some species are common while others are rare. Principal coordinate analyses (PCoA) by diet (with and without plastic, for example) can be quantified using the Bray-Curtis dissimilarity index and used to determine how diet affects community structure at the levels of genera and species. Microorganism types most closely associated with plastic biodegradation can be identified at the level of genera and species.

Two sampling methods have been used to collect gut microbial samples, *i.e.*, rinsing of extracted gut tissue to collect gut microflora in phosphate buffered water (pH 7.0) for DNA extraction (Brandon et al., 2018) and collecting intestinal tissue for DNA extraction (Yang, Wu, et al., 2018). The latter method can collect all of the microorganisms including those attached to the intestinal wall but also collect insect DNA that can interfere with sequence analysis and data processing. For current sequencing, the Illumina MiSeq platform is recommended for 16S rRNA gene amplicon, PacBio and NanoPore for full-length 16S rRNA genes, and HiSeq, NextSeq, and NovaSeq platforms for shot-gun metagenome sequencing. Detailed methods for gut microbiome analysis are provided in the references (Brandon et al., 2018; Peng et al., 2019; Yang, Gao, et al., 2021b).

2.3 Enrichment and isolation of plastic-degrading gut microbes

Isolation and characterization of plastic-degrading microbial species from insect larvae are needed to understand plastic-degradation and to identify and characterize plastic-degrading enzymes. To date, plastic (*e.g.*, PS, LDPE)-degrading bacterial strains have been isolated from guts of darkling beetles (Yang et al., 2015b; Yin et al., 2020) and moth larvae (Yang et al., 2014). Reported bacterial cultures include LDPE-degrading *Bacillus* sp. YP1 and *Enterobacter asburiae* YT1 (Yang et al., 2014), LDPE-degrading co-culture of *Acinetobacter* sp. strain NyZ450 and *Bacillus* sp. strain NyZ451 (Yin et al., 2020), and PS-degrading *Exiguobacterium* sp. strain YT2 (Yang et al., 2015b) and *Acinetobacterium* sp. (Wang et al., 2020). Isolation methods are described in Section 4.

3. Analytical methods for plastic biodegradation

To characterize biodegradation of plastics by insects and their gut microorganisms, analytical methods are commonly used to determine mass loss, physical and chemical modification, changes in chemical structure, and biodegraded intermediates and products. Methods commonly used in published reports are summarized as below.

3.1 Gravimetric determination of weight loss

Weight loss is a widely used and essential indicator of biodegradation. To achieve accurate results, care is required in the selection of the proper mass of plastics material, clean-up of residual plastics, and recovery of residual polymer from frass *via*. solvent extraction.

3.2 Chemical and structure modification

3.2.1 Gel permeation chromatography (GPC)

Gel permeation chromatography (GPC) and high temperature gel permeation chromatography (HT-GPC) are essential tools used to study depolymerization and biodegradation, and provide information on the relative molecular weight of polymer samples and the distribution of molecular weights. If comparable standards are used (PS is a common standard), the molecular weights determined can be within $\pm 5\%$ accuracy. Outputs include: the number average molecular weight (M_n), the weight average molecular weight (M_w), the size average molecular weight (M_z), and the viscosity average molecular weight (M_v). In general, M_n provides information about the lowest molecular weight fraction, M_w is the average closest to the center of the distribution curve, and M_z represents the highest molecular weight portion. M_v represents the viscosity of the polymer sample based on the relationship using an empirical Mark-Houwink equation. In a typical distribution curve, the average values are related to each other as follows: $M_n < M_v < M_w < M_z$. The dispersity (also known as the polydispersity index or PDI) of a sample is defined as $PDI = M_n / M_w$ and gives an indication broadness of the polymer distribution. As results of biodegradation, significant changes in M_n, M_v, M_w and M_z will be observed; and the changes in MWD and PDI also occur. The method is not highly sensitive because the analysis is performed on the bulk polymer and cannot determine initial minor changes in biodegradation

3.2.2 Fourier transform infrared spectroscopy (FTIR)

FITR is commonly used to characterize chemical modifications of polymer structure and to monitor chemical changes on the polymer surface by detecting the formation of functional groups resulting from biodegradation or enzymatic attack. This method works well for the plastics with little or low additive content. To test biodegradation of plastics by insects, it is preferable to use the frass sample of the insect fed plastics as the sole diet. For tests with bacterial or fungal cultures, complete removal of biofilm or attached biomass on the surface of residual plastics is mandatory to avoid misleading results. Pretreatment with ultrasound plus 2% sodium dodecyl sulfate (SDS) is commonly used

3.2.3 Thermogravimetrical analysis (TGA)

TGA determines the thermal stability of a polymer. Biodegradation of plastics results in the decrease of the stability as an indicator. When frass of insects is tested using TGA, the samples containing co-diet such as wheat bran, starch, *etc.* could cause significant misinterpretation of results. A TG analyzer equipped with Fourier transform infrared spectroscopy (TG-FTIR) is commonly used to identify decomposition products from samples. Newly produced components from residual polymer sample are indicators of biodegradation

3.2.4 Proton nuclear magnetic resonance (^1H NMR)

^1H NMR is the application of nuclear magnetic resonance in NMR spectroscopy with respect to hydrogen-1 nuclei within the molecules of a substance, in order to determine the molecular structure. It has been used to study the chemical modifications of the polymers after biodegradation based on the formation of new peak(s) in the spectra of samples of biotreated polymers

3.2.5 Differential scanning calorimetric (DSC) analysis

DSC is useful for assessment of different thermal properties of materials and investigation of the thermal transition of synthetic polymers, such as glass transition (Tg). The shifts of Tg to lower temperatures indicates a decrease of stability of the polymer as a result of biodegradation. DSC can also be used for determination of M_v and crystallinity

3.3 Surface observation

3.3.1 Microscopic observations

This type of method is useful to observe the activities of gut microbes. Formation of biofilm or microbial attachment on plastic polymer surface can be observed using high magnification microscopes and fluorescence microscopes. Staining with the LIVE/DEAD BacLight Bacterial Viability Kit or other reagents can be used to observe or count live and dead cells. Biodegradation of plastic polymers by microorganisms has been evaluated using scanning electron microscopy (SEM) for years. SEM can be used for direct observation of microbial colonization on the plastic surface and formation of cracks and holes on polymer samples. However, colonization of polymer surfaces is not sufficient proof of the biodegradation given that microorganisms can grow on inert media as a support for biofilm formation and may survive by consuming biodegradable additives (such as plasticizers).

3.3.2 Water contact angle (WCA)

Determination of the WCA of the plastic polymer surface is commonly used to monitor the changes in hydrophobicity on polymer surfaces. Decrease in hydrophobicity can be an indicator of biodegradation activity and can be used to characterize the activities of gut microbes and gut digestive enzyme(s). Prior to the analysis, polymer samples must be cleansed to remove the biofilm and attached organics using SDS (2%, w/v) for at least 4 h prior to conducting a WCA measurement.

3.4 Isotopic labeling and other methods

3.4.1 Isotopic labeling

The biodegradation and mineralization of plastics can be assessed using labeled ^{14}C carbon in synthetic polymers. The production of $^{14}CO_2$ and $^{14}CH_4$ (if tested anaerobically) upon degradation is a direct indicator of mineralization. The release of soluble ^{14}C containing organics is also an indicator of biodegraded intermediates. ^{13}C labeled isotopic polymers are also used to examine biodegradation and mineralization by measuring $\delta^{13}C$ in the

CO_2 released and ^{13}C in biomass components such as fatty acids (Yang et al., 2015a). Biodegradation of ^{13}C labeled polymer will result in a significant increase in $\delta^{13}C$ in the released CO_2 and corresponding effects on biomass composition. The isotopic labeling is effective and powerful, especially employing ^{14}C but the application is limited due to difficulties and costs of preparing isotopic polymers and measures required for management and access to equipment.

3.4.2 CO_2 and methane production-based biodegradation methods

American Society for Testing and Materials (ASTM) and International Standards Organization (ISO) have developed methods for the evaluation of the biodegradation of plastics/polymers under aerobic or anaerobic conditions. According to ASTM, biodegradable means that the material can undergo decomposition into CO_2, CH_4, H_2O, inorganic compounds, or biomass. ISO definition requests only a chemical change of material (*e.g.*, oxidation). ASTM methods include tests for biodegradation under liquid, compost and solid conditions (see ASTM D5210–92(2007), D5338–11, D5511–2, D5526–12, D5988-12D6691–09). Although these methods were developed for examination of biodegradability of biodegradable polymers, they can also be used to evaluate enrichment-mediated or isolate-mediated biodegradation of plastic polymers and further degradation of plastic-containing frass.

3.4.3 Other methods

Other techniques have also been reported to evaluate biodegradation of plastics, including X-ray photoelectron spectroscopy (XPS), which was used to verify the changes in surface chemical components and functional groups (Yang et al., 2015b); solid-state ^{13}C cross-polarization/magic angle spinning nuclear magnetic resonance (CP/MAS NMR), which was used to identify newly appearing functional groups in solid samples (Yang et al., 2015a); attenuated total reflection Fourier transform infrared spectroscopy (ATR-FTIR), which identifies new functional groups in either liquid or sold samples during microbial biodegradation of plastics (Yin et al., 2020).

4. Protocols for the characterization of plastic degradation by *T. molitor* larvae

The following protocols focus on biodegradation of polystyrene (PS) in *T. molitor* larvae. They can also be employed for the degradation of other plastics as noted in the text using other members of darkling beetle larvae (*e.g., T. obscurus, Z. atratus, etc.*). These protocols include methods for the physiological characterization of the larvae responding to plastics, characterization of plastic degradation, sampling for gut microbiome, and isolation of plastic-degrading bacteria. To characterize biodegradation, analyses with GPC, FTIR, ^{1}H NMR and TGA can provide valuable results. SEM and WCA are used for gut microbial degradation studies.

4.1 Equipment, supplies and chemicals

4.1.1 Equipment and supplies

- Analytical balances.
- Centrifuge (10,000 rpm, 1.5 mL PP tubes).
- Environment incubator with temperature control (20–30 °C) and humidity displays.
- Freezer (−80 °C) to store DNA samples.
- Food grade polypropylene or polyethylene containers (300–500 mL in volume) suitable for the incubator.
- Glass vials with cap containing Teflon coated septum for DNA sampling.
- Glass vials with Teflon septum cover (5–10 mL) for polymer extraction.
- Norm-Ject 1 mL Polypropylene Luer Slip Tip Capsule Syringe for transfer THF extraction.
- PVDF (0.22 μm) sterile syringe filter (Thermo Fisher Scientific Inc., Dublin, Ireland).
- Refrigerator for frass samples and extracts.
- *Tenebrio molitor* larvae from online or local pet store with medium size (around 1.5 cm) or with an average weight of 60–70 mg each. The larvae should be fed on wheat bran without antibiotic supplements. The larvae should be healthy and contain no black dead larvae.

4.1.2 Chemicals

- Absolute ethanol (>99.5% purity, w/w).
- Plastic feedstocks in foam or MPs (<0.15 mm): polystyrene (PS), high density polyethylene (HDPE), low density polyethylene (LDPE), polypropylene (PP), polylactic acid (PLA), polyethylene terephthalate (PET), and polyvinyl chloride (PVC).
- Sodium dodecyl sulfate (SDS) used for the clean-up of plastic surface.
- Soy protein from nutrition store as nutrition-rich co-diet.
- Tetrahydrofuran (THF) used for extraction of PS and PVC.

- 1,2 Dichlorobenzene (DCB), >99.9% purity (w/w), used for the GPC analysis of LDPE and PP. DCB is considered hazardous by the 2012 OSHA Hazard Communication Standard (29 CFR 1910.1200).
- Dichloromethane (DCM) >99.9% (w/w) for LDPE, PP and PLA extraction. DCM is considered hazardous by the 2012 OSHA Hazard Communication Standard (29 CFR 1910.1200).
- Wheat bran from any pet store or farmer's market.
- Jelly powders or Jello (sugar free) from food store.
- Chemicals (analytical grade) for microbial media include: KH_2PO_4, K_2HPO_4, $MgSO_4$, NH_4NO_3, NaCl, $FeSO_4{\cdot}7H_2O$, $ZnSO_4$, $MnSO_4{\cdot}H_2O$; and agar, bacteriological tryptone, yeast extract.

4.2 Operational protocols

The following protocols are described for a test of biodegradation of PS foam or PS MPs in *T. molitor* larvae with and without wheat bran as co-diet. They can be modified for LDPE, HDPE, PP, PVC, PET, and PLA, and supplementing different mass ratios of co-diet. Other co-diet, *e.g.*, soy protein powders can also be considered.

4.2.1 Feedstock preparation

Plastic foam from manufacture can be used as feedstock directly. PS foams include expanded PS foam (EPS), which is used for building insulation and packing, and extracted PS foam (XPS), which is used as cups and food trays. Measure the density of the foam (g/cm^3) prior to use and contact manufacturer to understand the composition of additives in the foam. GPC analysis (see Section 3.2.1) is needed to determine the molecular weights and molecular weight distribution (MWD).

PS MPs can be used as feedstock. Jelly powders is used as adhesive to make PS-bran mixed diets. The procedures are:

1. The optimal mixture for the larval growth is at a ratio of bran/PS of 5:1–8:1 (Yang, Brandon, et al., 2018) but the ratio may be changed depending on tests.
2. Jelly powder (0.5 g) is dissolved into 30 mL boiled deionized water in a 200 mL glass beaker.
3. PS MPs (5 g) is added into the beaker.
4. After 10 min of stirring, the beaker is transferred to −4 °C refrigerator to mature the mixture for 1 h.
5. Finally, wheat bran (25 g) is added in the beaker and mixed completely with the solidified jell.

6. The mixture is then put in an aluminum pan and turned to air-drying for 2 days prior to feeding the larvae.
7. To prepare the control with wheat bran only, 30 g bran is mixed with 0.5 g Jelly.
8. Different mass ratios of wheat bran *versus* PS or other plastic MPs are prepared using the same procedures with a total 30 g weight of mixture plus 0.5 g jelly.

4.2.2 Set up of incubators with plastic foam as target feedstock

1. Add 300–350 larvae in each container at 25 °C, humidity of 70%.
2. Treatments include, *e.g.*, (i) unfed control; (ii) bran-fed control; (iii) PS only fed; (vi) PS plus bran-fed.
3. Duplicate each treatment.
4. Initial feedstock dosage: PS foam, 2 g; and wheat bran, 2 g.
5. Supplement wheat bran (2 g) once per week.
6. Supplement PS foam (2 g) when it is nearly consumed.
7. The incubation period for a short-term test is 5 weeks, and it can be up to 50 weeks for long-term test.
8. At the end of the test, collect all frass, weigh it, measure the dry weights, and determine content of residual PS (%, w/w) in frass to calculate PS mass balance.

4.2.3 Set up of incubators with microplastics (MPs) as feedstock

1. Add 300–350 larvae in each container at 25 °C, humidity of 70%.
2. Treatments include, *e.g.*, (i) unfed control; (ii) bran-fed control; (iii) PS MPs only fed; (vi) PS MPs plus bran-fed.
3. Duplicate each treatment.
4. Initial feed stock dosage: PS MPs-bran mixture, 30 g; control wheat bran, 30 g.
5. Supplement the mixed feedstock as it is consumed by observation.
6. The incubation period for a short-term test is 5 weeks, and it can be up to 50 weeks for a long-term test.
7. At the end of the test, collect all frass, weigh it, measure the dry weights, and determine content of residual PS (%, w/w) for biodegradation/digestion mass balance.

4.2.4 Larval counting and frass collection

1. Count number of live larvae, dead larvae, pupae and beetles and record their respective weights once per week. Discharge dead larvae.

2. Move the live larvae from each incubator, cleanse the larvae using compressed air flushing and then transfer them to a clean PP container. Keep them over night to allow them egest frass in it in order to collect frass.
3. Transfer the larvae back to the original incubator.
4. Collect frass into a pre-weighed PP centrifuge tube (1.5 mL), weigh the frass and keep the tubes with frass at −20 °C for GPC and other analysis (FTIR, TGA, ^{1}H NMR).

4.2.5 Sampling for gut microbiome analysis

1. Prior to the test, take 15–20 larvae, weigh them, and record the average weight. Then, transfer them into a glass vial (20 mL), fill the vial with absolute ethanol (purity >99.5%) to kill the larvae, and seal the vial with ethanol. The sample serves as background control.
2. For a short-term test (5 weeks), sampling with 15 larvae for each treatment is conducted at the end of the test using the same method as above.
3. For a long-term test, multiple sampling is conducted during the incubation period (weekly or monthly) and at the end of test.
4. All specimen in ethanol are preserved at −80 °C prior to DNA extraction.

4.2.6 Calculation of physiological parameters of the larvae

Survival rate (SR), cannibal rate (CR), pupation rate (PR) and the total mass of frass generated are determined periodically and must be calculated or measured at the end of test. The number of live larvae was counted to calculate SR as:

$$SR\ (\%) = (N_c)/N \times 100\%$$

where N is initial number of the larvae, and N_c is the number of live larvae (plus pupae and beetles if found) on day c. The numbers of pupae and disappeared larvae are counted to calculate PR and CR accordingly.

$$PR(\%) = N_p/N \times 100\%$$
$$\text{and } CR(\%) = (N - N_c - N_d)/N \times 100\%$$

where N_p is accumulated number of pupae, N_d is the number of dead larvae.

4.2.7 Residual fractions in frass

Determination of residual polymer content in frass egested from the larvae is a basic procedure to characterize biodegradation rate, extent, and mass balance, as part of the extraction of residual polymer for the GPC analysis.

The frass sample contains (i) water soluble fraction (soluble salts, dissolved organic products, *etc.*); (ii) ethanol extractable fraction (biodegraded intermediates, residues of co-diet, *etc.*); and (iii) solvent (THF, 1,2 DCB or DCM) extractable fraction. To conduct solvent extraction, THF is used to extract residual polymer of PS and PVC;1,2 DCB is used for LDPE and PP; and dichloromethane (DCM) is used for PLA, LDPE and PP.

For PS extraction, the following steps are used.

1. Frass sample (1.0 g) from each diet is first mixed with deionized water (20 mL) in a conical flask after ultrasonic treatment for 3 min and extracted in a shaker with 100 rpm for 3 h.
2. The water solution is filtered with a 0.22 μm poly(ether sulfone) (PES) membrane filter or fiber glass filter and then stored at 4 °C for any further analytical use.
3. After filtration, the residue is vacuum dried and weighed to calculate the mass of water-soluble fraction.
4. Subsequently, the residue is further mixed with ethanol (20 mL), collected ethanol-extracted fraction as above steps 1–3, in a 50 mL glass vial. Ethanol extracted solution is stored in glass vials capped with Teflon coated septum cover for any further use. The solid residue is air-dried and weighed.
5. Finally, the residue is mixed with solvent (THF, 20 mL) to obtain the THF-extractable fraction in a 50 mL glass vial capped with Teflon coated septum cover. THF extracted solution is stored in glass vials capped with Teflon coated septum cover for GPC analysis and/or further use.
6. After evaporation of THF, the residue solid is air-or vacuum dried, and weighed.
7. Calculation of water soluble-, ethanol extracted- and DCM-extractable fraction of frass as below:

$$C_w\ (g/g) = (W_f - W_w)/W_f \times 100\%$$

where C_w = water soluble fraction (%, w/w), W_f = weight of frass sample (g), W_w = weight of dried residual solid after water extraction (g).

$$C_e\ (\%) = (W_w - W_e)/W_f \times 100\%$$

where C_e = ethanol extracted fraction (%, w/w), W_e = weight of dried residual solid after ethanol extraction (g).

$$C_{THF}\ (\%) = (W_e - W_d)/W_f \times 100\%$$

where C_{THF} = THF extractable fraction (%, w/w), W_d = weight of vacuum dried residual solid after THF extraction (g).

The residual PS content in frass of the larvae fed with PS plus different ratios of wheat bran (x %, w/w) is calculated as.

$$C_{PS} = C_{dx} - C_{db} \times (100 - x)/100$$

where, C_{PS} is PS content in frass (%, w/w), C_{dx} is the THF extractable fraction of frass of the larvae consumed PS and bran at a ratio of x %. The x is the ratio calculated from mass loss of PS feedstock and bran (%, w/w), or x is the percentage of PS in the MPs-bran diet (%, w/w) when PS MPs are used to prepare PS-bran mixed feedstock. C_{db} is the THF extracted fraction of frass from the larvae fed on wheat bran only.

4.2.8 Polymer (PS) removal

Specific polymer (PS) consumption rate (SPCR) is calculated on the basis of the mass of PS consumed by 100 larvae over the test period as:

$$SPCR = W_{ps} \times 100/(N_a \times t)$$

where SPCR is specific polymer (PS) consumption rate (mg/100 larvae-d); Wps is the mass of total PS consumed (mg); Na is averaged number of the larvae over test period; t is the total days of test period.

At the end of the test, PS removal efficiency is calculated as.

PS removal (%, w/w) = $(W_{PS} - W_f \times C_{PLA})/W_{PS} \times 100\%$

where, W_{PS} is the total weight of PS consumed by larvae, W_f is the total weight of frass generated (g), C_{PS} is the PS content (%, w/w) in the frass.

4.2.9 GPC analysis

Gel permeation chromatography (GPC) is applied to analyze changes of molecular weight of the PS polymer (*e.g.*, employing Agilent 1260, Agilent Technologies Inc., USA).

1. Preparation of samples of frass are similar to the THF extraction test described previously (4.2.7). A sample of PS feedstock is dissolved directly in THF.
2. PS feedstock (0.2–1.0 g) and frass samples (0.5–1.0 g) from PS-fed larvae are commonly used.
3. THF extraction is conducted in a 50 mL glass vial capped with Teflon coated septum cover overnight.

4. After filtration with a 0.22 μm PVDF sterile syringe filter (Thermo Fisher Scientific Inc., Dublin, Ireland), the THF solution is transferred into a pre-weighed 50 mL glass vial, mixed on a magnetic stirrer with gentle heating (60 °C) or vacuum-dried overnight.
5. The vial with dried PS residue is weighed to measure the mass of residual PS.
6. The PS residue is re-dissolved with THF to achieve a concentration of 5 mg/mL.
7. Finally, the THF solution is transferred into a 1.0 mL GC vial and sealed for further GPC analysis.
8. The injection volume and eluent used depends on the model of GPC or HT-GPC used.

4.2.10 FTIR Analysis

Fourier transform infrared spectroscopy (FTIR) is used to detect changes of major functional groups of residual PS in frass (using, *e.g.*, Nicolet iS05 FTIR Spectrometer, Thermo Fisher Scientific, U.S.A.).

1. Prior to the analyses, the frass sample is lyophilized for at least 36 h to avoid deviation from water.
2. Then the sample is grinded with KBr to prepare a homogenous KBr pellet for scanning. The concentration of the sample in KBr should be in the range of 0.2% to 1%.
3. Depending on model of FTIR employed, for plastic samples, the peaks are identified in the range of 4000–500 cm^{-1} using, *e.g.*, OMNIC software (Thermo Fisher Scientific Inc., Pittsburgh, PA).

4.2.11 TGA analysis

Thermal gravimetric analysis (TGA) is conducted to characterize thermal changes from PS feedstock to the frass sample (using, *e.g.*, TA-Q500, TA Instruments, U.S.A.). The heating program included two different atmospheres in order to study pyrolysis of the sample under nitrogen and air ambience, respectively.

1. Frass sample (10 mg) and PS sample (10 mg) are first heated from ambient temperature to 800 °C at a heating rate of 20 °C/min under high-purity nitrogen ambience (99.999%, v/v).
2. Then cooled down to 500 °C.
3. Finally heated again to 800 °C under air ambience.
4. Differential scanning calorimetry was conducted to further study the crystallization temperature and melting process

4.2.12 ^{1}H NMR analysis

To characterize changes in the end groups of the egested polymer, liquid-state ^{1}H nuclear magnetic resonance (^{1}H NMR) analysis is conducted at ambient temperature.

1. The residual PS was extracted from frass with THF as described in Section 4.2.7, and air- or vacuum dried.
2. The dried residue from frass sample (50 mg) or PS feedstock is placed in a 10-mL glass vials and extracted for 2 h with 2 mL chloroform-D (purity 99.8%, Cambridge Isotope Laboratories, Inc., Tewksbury, MA).
3. Extracts are filtered through 0.22 μm PVDF filters and transferred to a clean 10-mL glass vial. Approximately 1 mL of extract is transferred into 5 mm NMR sample tube (Wilmad, LabGlass, Vineland, NJ, USA).
4. ^{1}H NMR spectra were measured on a 500-MHz NMR spectrometer (32 scans, delay time (d1) = 0.0 s). ^{1}H spectra are reported in parts per million (ppm) and referenced to a peak for residual deuterated chloroform (1H-7.26 ppm).

4.2.13 Water contact angle (WCA) analysis

1. Prior to the test, the plastic samples (pieces or films) are separated from microbial culture and rinsed with DI water.
2. The solid samples are mixed with 2% (w/v) sodium dodecyl sulfate (SDS) solution for 4 h and then rinsed with deionized water.
3. Check the sample carefully under a microscope and no visible biomass/biofilm should be observed. Otherwise, further clean-up is needed.
4. Measurement can be conducted using a contact angle measuring device (OCA40, DataPhysics, Filderstadt, Germany) or others.

4.2.14 Gut Microbial enrichment and bacterial isolation

1. Liquid carbon free basal medium (LCFBM) is used to enrich and characterize plastic-degrading gut bacteria. It contains (per 1000 mL deionized water) 0.7 g of KH_2PO_4, 0.7 g of K_2HPO_4, 0.7 g of $MgSO_4{\cdot}7H_2O$, 1.0 g of NH_4NO_3, 0.005 g of NaCl, 0.002 g of $FeSO_4{\cdot}7H_2O$, 0.002 g of $ZnSO_4{\cdot}7H_2O$, and 0.001 g of $MnSO_4{\cdot}H_2O$ according to the ASTM standard for determining the resistance of plastics to bacteria (G22–76, 1996).
2. After preparation, LCFBM (100 mL) is distributed in Erlenmeyer flask (250 mL), autoclaved at 121 °C for 40 min, then cooled prior to use.
3. The Luria-Bertaini (LB) medium contains (per 1000 mL deionized water) 10 g of bacteriological tryptone, 5 g of yeast extract, 10 g of NaCl and 15 g of agar. It is used for the isolation of single colonies.

4. LB is prepared in a 1000 flask with 500 mL Erlenmeyer flask according to the recipe, then add the desired amount of agar (normally about 1% w/v).
5. Autoclave LB for 25 min.
6. Cool the LB medium-agar mix to 55 °C.
7. Pour the plates. Use about 30 mL for each plate in a 100 mm diameter plate.
8. Allow the plates to set. Dry the plates in the laminar flow hood with the lid slightly off for 30 min (or in a 37 °C incubator for 2–3 h) prior to use.
9. If the plates are not used immediately seal them with Parafilm, or in the bag that the plates came in. Store the plates at 4 °C.
10. The whole guts of 10–15 larvae are extracted and pooled into a 10 mL centrifuge tube containing 5 mL saline solution.
11. After vibrating by a vortex for 5 min, the gut tissue is carefully removed, and the remaining suspension is moved into a 250 mL Erlenmeyer flask containing 100 mL LCFBM with 1 g PS micro powders (0.1–0.2 mm) or PS film pieces (1 cm × 1 cm) as the sole carbon source.
12. The mixture is incubated at 25 °C with shaking at 120–180 rpm for 15–60 days.
13. When turbidity occurs, the culture is serially diluted and spread on the LB agar plate to form single colonies at 25 °C.
14. Usually, emerged single colonies are observed within 24–72 h and then picked.
15. The picked colony is transferred in LCFBM with PS pieces or PS MPs for incubation.
16. Check morphotype under a microscope as cells grow (increase in turbidity).
17. Repeat steps 12–16 till single culture or same morphotype cells are observed.
18. Pure cultures can be obtained by repeatedly plate streaking. The abstained pure bacterial cultures are then further verified for PS degradation capacity in PS-containing LCFBM.

5. Results and analysis

T. molitor larvae actively chew, ingest and digest PS foam as well other major plastics (LDPE, PVC, PLA, PET, and PP) in either foam forms or MPs mixed with co-diet bran. The ingestion of plastics by the larvae impacts their physiology. Here we use the results of PS biodegradation as examples.

5.1 Larval activities and degradation rates

5.1.1 Survival rates and PS consumption of mealworms

T. molitor larvae chew and ingest PS foam regardless whether PS is supplied as sole feedstock or co-diets with wheat bran or soy protein or other nutrition-rich diet. At 25 °C, the larvae consumed PS (Fig. 1A) with much higher survival rates than unfed larvae in short term (Fig. 1B). One test showed that over a 32-day period, the SRs of the larvae are: unfed, 54.2 ± 2.5%; PS only, 86.7 ± 3.3%; PS plus bran, 90.0 ± 0.8%. During the test period, the larvae fed with PS (1.8 g) consumed 0.83 ± 0.04 g PS with average specific consumption rate of 11.8 mg ± 0.1 mg PS/100 larva per day. The unfed larvae lost 2.6 ± 0.2% of their average weight; the larvae fed with PS only maintained a stable weight; and bran-fed larvae experienced 32.0 ± 1.5% weight increase. The results indicated that ingestion of PS supports the life activities of the larvae but does not support their growth due to lack of necessary nutrition.

Another test with the co-diet with bran soy protein enhanced PS consumption significantly (Fig. 1C). Over a 32-day period, the larvae fed with PS only consumed PS at SPCR of 22.2 ± 1.8 mg/100 larvae per day but SCCR of the larvae fed co-diet bran and soy protein were 44.1 ± 4.8 and 49.1 ± 4.1 mg/100 larvae per day, respectively. The respective PS mass removal was 39.1 ± 1 3.6%, 67.6 ± 4.3% and 76.8 ± 2.8%.

A long-term test at 25 °C with 350 larvae for each treatment indicates that the diet conditions impact the survival rate (SR) pupariation rate (PR), and cannibal rate (CR). For the unfed group, the parameters were: SR, 4.0%; PR, 4.0%; CR, 96%. The larvae fed PS only had: SR, 9.7%; PR, 9.7%; CR,90%. The larvae fed PS plus bran showed: SR, 81%; PR, 75%; CR, 18%. The total mass of PS consumed was 1.13 g by the larvae fed on PS only *versus* 5.20 g by the larvae fed PS plus bran.

5.2 Characterization of biodegradation

5.2.1 THF extractable fraction in frass

THF extractable fraction from frass is used as indicator to estimate the content of residual PS in frass. Because PS (or other plastics) does not dissolve in water and ethanol, the previous two steps of extraction with DI water and ethanol can remove most organic contaminants, additives, and biodegraded intermediates from the frass. The THF extractable fraction is composed of mainly PS polymer with white color (Fig. 1D).

THF extractable fraction was measured on day 4, 8, 12, 16, 20,24, 28 and 32 during a 32-day test illustrated in Fig. 1B and D. The PS residue in frass (w/w, %) of the larvae fed PS only decreased from 66.2±2.3% on day 4 to 35.2±1.2% by day 24, stabilizing thereafter. The results indicate a progressively increased PS digestion capacity.

5.2.2 GPC analysis

GPC analysis provides information on three key indicators of depolymerization and degradation of plastic materials: M_n, M_w and MWD. Here we use the results of the test in Fig. 1C as example. The samples were obtained for PS residues extracted from frass collected on day 4, 8, 12, 16, 20, 24, 28, and 32. GPC analysis of the extracted PS residues (Fig. 1 D) reveals a progressive shift of MWD from higher to lower molecular weights over time compared to the PS feedstock (control), showing a typical broad depolymerization pattern (Fig. 1F), with about 16 days required to reach relatively stable lower levels of M_w and M_n (Fig. 1H). M_n and M_w values for PS residues in the frass after 32 days are significantly lower than M_n and M_w for PS in the feed. M_n decreased from 95,800±1300 to 77,000±4000 ($P=0.0214$); M_w decreased from 239,000±2400 to 212,000±4800 ($P=0.0206$). The results (Fig. 1E and F) suggest that the PS degradation activity increased gradually but stabilized after a 16- to 24-day adaptation period.

Depolymerization is the essential step of biodegradation of plastic polymers. Two different depolymerization patterns have been observed during biodegradation of plastics by insect larvae and microbial cultures., broad depolymerization or the decrease in both M_n and M_w of the plastic polymer (Fig. 1F); and limited extent depolymerization or increase in M_n and decrease in M_w or increase in both M_w and M_n (Peng, Chen, et al., 2020; Peng, Li, et al., 2020; Yang, Ding, et al., 2021a). However, decrease in M_z has been observed in both patterns.

5.2.3 FTIR analysis of frass

Here we present typical FTIR spectra to determine how PS polymers are modified using the frass samples of the larvae fed with PS only, bran plus PS, and bran only. FTIR is analyzed in comparison with FTIR spectra of PS feedstock (Fig. 2A). The larvae were incubated under four diet conditions at 25 °C for 32 days. The intensities of the peaks at 625–970 cm^{-1} (ring-bending vibration) are strong in PS feedstock but much weaker in frass samples. Characteristic peaks known to represent the PS benzene ring (C=C stretch, 1550–1610 and 1800–2000 cm^{-1}) almost disappear in frass

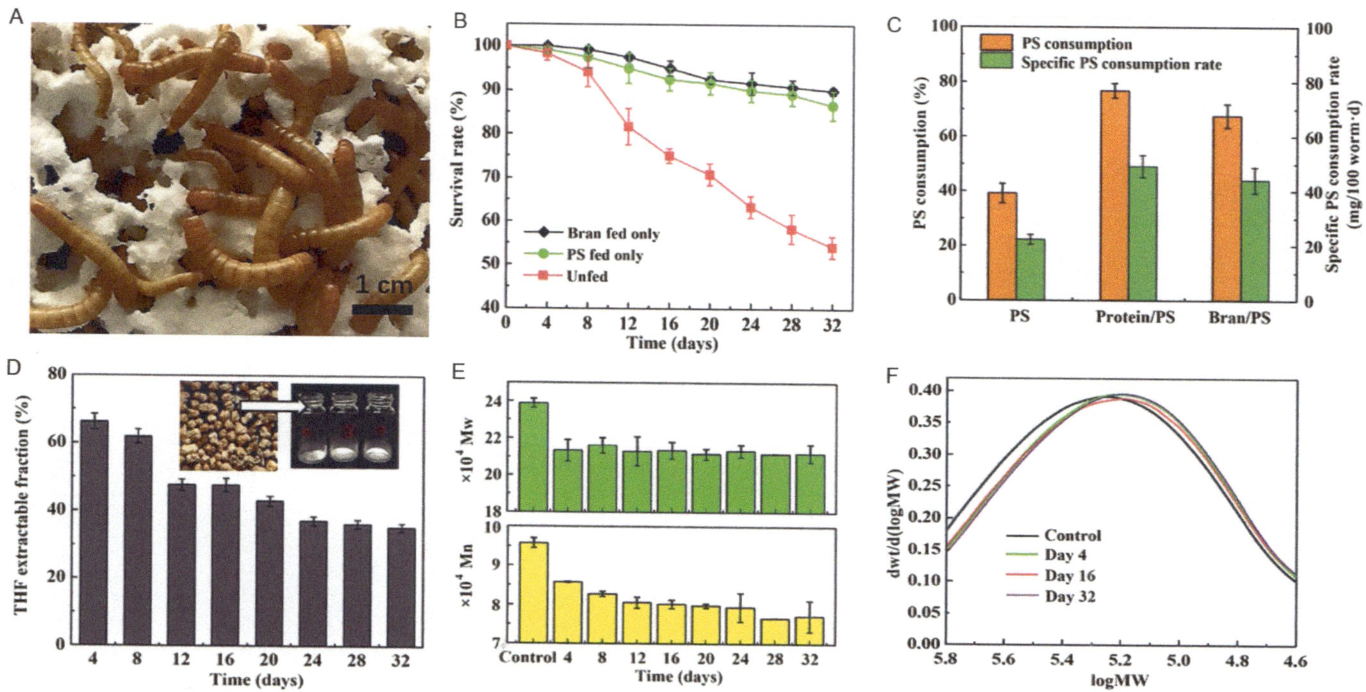

Fig. 1 (A) The larvae of *Tenebrio molitor* chew and ingest PS foam. (B) The survival rates of the larvae fed with different diets, *i.e.*, wheat bran plus PS, PS only and unfed. (C) PS consumption (%) and specific PS consumption rates with different co-diets over a 32-day period. (D) The decrease in HF extractable fraction (%) of frass indicating the enhanced digestion extent in larval guts. Insert photos: left is frass, right is the residual PS extracted from the frass with THF. (E) GPC analysis shows decrease in Mn and Mw of residual PS polymer extracted from the frass during the incubation period. (F) Shift in molecular weight distribution of residual PS polymer *versus* the original PS feedstock. *Data of panels (B–F) adapted from Yang, S.S., Brandon, A.M., Andrew Flanagan, J.C., Yang, J., Ning, D., Cai, S.Y. et al. (2018). Biodegradation of polystyrene wastes in yellow mealworms (larvae of* Tenebrio molitor *Linnaeus): Factors affecting biodegradation rates and the ability of polystyrene-fed larvae to complete their life cycle.* Chemosphere 191, *979–989.*

samples, providing evidence of ring cleavage. Further evidence of biodegradation is the observed decrease in intensities of peaks characteristic for PS and the appearance of carbonyl groups (C=O stretch, 1700 cm^{-1}). The broadening of peaks at 2500–3500 cm^{-1} in all FTIR spectra of frass samples is associated with the hydrogen bond of hydroxyl groups and/or carboxylic acid groups, suggesting a shift from hydrophobic to more hydrophilic surface properties.

5.2.4 ^1H NMR analysis

The same samples used for above FTIR analysis were extracted with chloroform-D for ^1H NMR analysis. The comparison of spectra for PS to the spectra of frass extracts reveals new peaks in the frass from the larvae fed PS only and PS plus bran (Fig. 2B). These peaks are detected in regions of chemical shift associated with -CH=CH-, $-CH_2-$, carbonyl ($H_2C=O$), and hydroxyl (-OH) groups. Their presence in PS residues of frass, but not control PS, is evidence of transformations within the larval gut, indicating new oxidized functional groups.

5.2.5 TGA analysis

TGA is used to detect the thermal modification of PS ingested by *T. molitor* larvae at the end of a test (Fig. 1C). PS foam had a 98.29% weight loss occurring at 360 °C ~ 480 °C, with the maximum decomposition rate temperature of about 450 °C (Fig. 2C). There are three decomposition stages of the frass fed PS: in the first stage, 6.62% weight loss occurred at 175 °C ~ 275 °C; in the second stage, 12.89% weight loss occurred at 275–360 °C; in the third stage, 53.12% weight loss occurred at 360–480 °C. The bran fed larvae frass has three decomposition stages: in the first stage, 14.46% weight loss occurred at 175–275 °C; in the second stage, 22.96% weight loss occurred at 275–360 °C; 18.11% weight loss occurred at 360–480 °C. Under the same heating program, PS fed frass has more decomposition stages than PS foam, indicating that the frass of *T. molitor* larvae contains new components other than PS foams. In the third stage, the weight loss rate of the frass is significantly lower than that of PS foams, indicating that the PS components in the frass are significantly reduced. Decompositions below 100 °C are classified as volatile organic compounds (intestinal secretions, carboxylic acid compounds from PS, *etc.*), while decomposition from 100 °C to 360 °C may be attributed to other organic waste and biodegradable residues. In the stage of 360–480 °C, the mass loss rate of PS is 98.29%, and the loss rate of the PS fed frass is 53.12%. The three-stage decomposition of PS fed larvae frass is similar to those of bran-fed larvae frass,

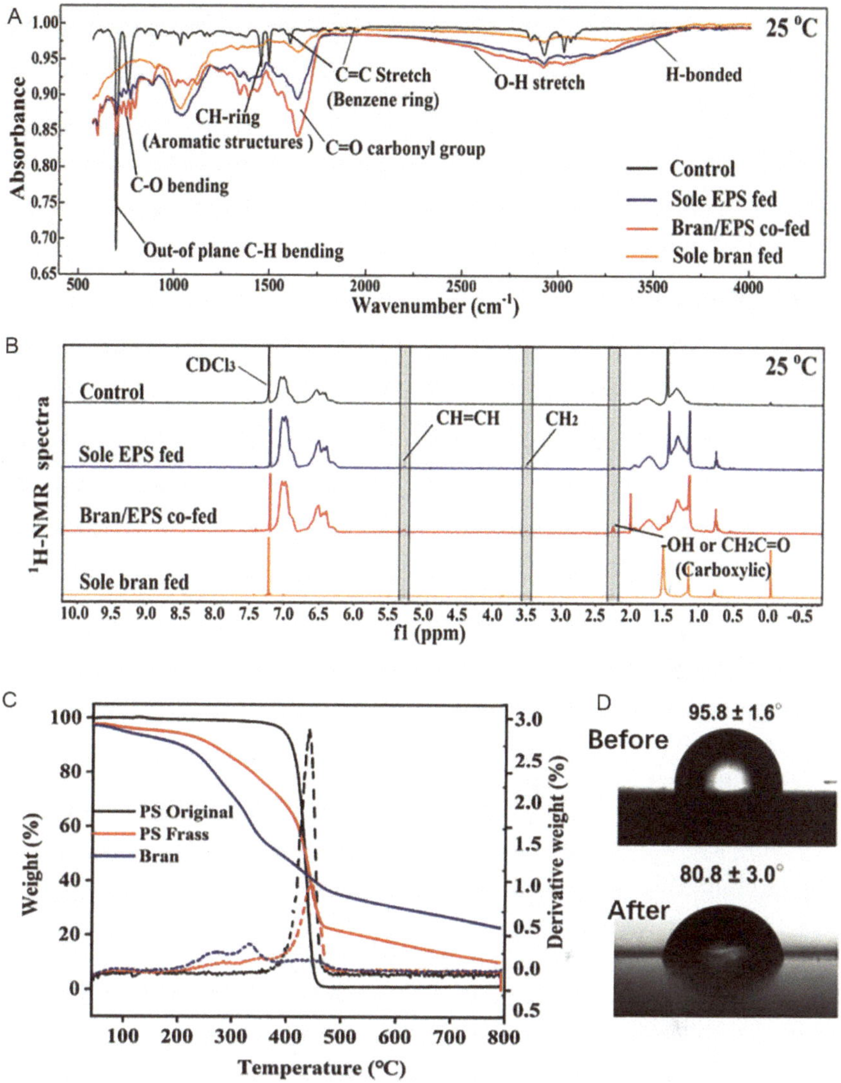

Fig. 2 (A) Spectra of FTIR and ^{1}H NMR of control (PS feedstock) *versus* frass samples of the larvae under different diet conditions, *i.e.*, fed with PS only, fed with PS plus bran and fed with bran only. The samples were collected after 32 days. (C) TGA analysis of PS feedstock *versus* the frass of the larva fed with PS only and bran only. (D) The change in WAC of PS pieces after incubation with strain YT1 for 60 days with 7.4% mass reduction. *Panels A–B adapted from Yang, S.S., Brandon, A.M., Andrew Flanagan, J.C., Yang, J., Ning, D., Cai, S.Y. et al. (2018). Biodegradation of polystyrene wastes in yellow mealworms (larvae of* Tenebrio molitor *Linnaeus): Factors affecting biodegradation rates and the ability of polystyrene-fed larvae to complete their life cycle.* Chemosphere, 191, *979–989. Panel C adapted from Yang, L., Gao, J., Liu, Y., Zhuang, G., Peng, X., Wu, W.-M., et al. (2021). Biodegradation of expanded polystyrene and low-density polyethylene foams in larvae of* Tenebrio molitor *Linnaeus (Coleoptera: Tenebrionidae): Broad versus limited extent depolymerization and microbe-dependence versus independence.* Chemosphere, 262, *127818. Panel D adapted from Yang, Y., Yang, J., Wu, W.M., Zhao, J., Song, Y.L., Gao, L.C. et al. (2015b). Biodegradation and mineralization of polystyrene by plastic-eating mealworms: Part 2. Role of gut microorganisms.* Environmental Science & Technology, 49, *12087–12093.*

indicating that PS fed larvae produced frass which has changed their composition to degraded products. The results indicate that PS is modified and oxidized after passage through *T. molitor* larvae. Similar TGA results were observed during biodegradation of LDPE and PVC in *T. molitor* larvae (Peng, Chen, et al., 2020; Peng, Li, et al., 2020; Yang, Gao, et al., 2021b).

5.3 Plastic-degraders

The presence of gut plastic-degraders provides evidence of plastic degradation in the larval guts. Plastic degrading bacteria were isolated using the methods described above by several researchers, *e.g.*, PS-degrading *Exiguobacterium* sp. strain YT2 (CGMCCC 10521) which grow on PS pieces as sole carbon source (Yang et al., 2015b). A total 13 single bacterial cultures were isolated by picking up colonies formed on LB agar plates and characterization of PS degrading capacity showed that *Exiguobacterium* sp. strain YT2 removed 7.4 ± 0.4% of mass of PS pieces over a 60 day period in LCFBM. Biodegradation was confirmed by decrease M_n from 40,400 to 37,500 Da and M_w from 124,200 to 110,100 Da using GPC analysis, formation of oxidized functional groups -C-O- by XPS, and decrease in WCA from 95 ± 1.6% to 80.8 ± 3.0%. (Fig. 2.D).

Using the same methods with LDPE powders as sole carbon source, PE-degrading bacterial strains were also isolated from *Plodia interpunctella*, *e.g.*, *Enterobacter absburiae* strain YT1 (GenBank KJ466896) and *Bacillus* sp. YP1 (KJ466897), growing on LDPE as sole carbon source (Yang et al., 2014); and *Acinetobacter* sp. strain NyZ450 (GenBank MT459299) and *Bacillus* sp. strain NyZ451 (MT459300), which can grow as co-culture on LDPE as sole carbon source (Yin et al., 2020).

6. Summary

Characterization of rapid biodegradation of plastics in insect larvae is described with a focus on *Tenebrio molitor* and its gut microbiome. The essential methods and protocols, and procedures are described to assess impacts on insect physiology, plastic mass balance, analysis of changes in polymer structure modification and oxidation, and isolation plastic-degrading bacterial species within the insect gut. The results of plastic degradation in *T. molitor* and isolation of plastic-degrading bacterial cultures are analyzed using examples. We envision that the methods used can be expanded to enable investigation of other synthetic polymer types and a broad assay of insects.

Acknowledgments

The authors appreciate support from the Woods Institute for Environment at Stanford University (Award 1197667-10-WTAZB) and acknowledge the use of data from publications by Dr. A.M. Brandon, Professor, R. M. Waymouth, Stanford University; Dr. S.S. Yang, Professor D.F. Xing, Harbin Institute of Technology; Mr. B.Y. Peng, Professor Y. L. Zhang, Tongji University; Professor J. Yang, Dr. Y. Yang, Beihang University; Professor M.E. Benbow, Mr. J. Receveur, Michigan State University; Professor D. H. Phillips, Queen's University of Belfast; Ms. L. Yang, Professor X.L. Zhuang, Chinese Academy of Sciences.

References

ASTM International. (1996). *ASTM G 22–76 (1996). Standard practice for determining resistance of plastics to bacteria*. West Conshohocken, PA: ASTM Interantional.

Bombelli, P., Howe, C. J., & Bertocchini, F. (2017). Polyethylene bio-degradation by caterpillars of the wax moth galleria mellonella. *Currunt Biology.*, *27*(8), R292–R293.

Brandon, A. M., Gao, S. H., Tian, R., Ning, D., Yang, S. S., Zhou, J., et al. (2018). Biodegradation of polyethylene and plastic mixtures in mealworms (larvae of *Tenebrio molitor*) and effects on the gut microbiome. *Environmental Science & Technology*, *52*, 6526–6533.

Calmont, B., & Soldati, F. (2008). Ecologie et biologie de *Tenebrio opacus* Duftschmid, 1812 Distribution et détermination des espèces françaises du genre *Tenebrio Linnaeus*, 1758 (Coleoptera, Tenebrionidae). *R.A.R.E.*, T. *27*, 81–87.

Cline, L. D. (1978). Penetration of seven common flexible packaging materials by larvae and adults of eleven species of stored-product insects. *Journal of Economic Entomology*, *71*, 726–729.

Gerhardt, P., & Lindgren, D. (1954). Penetration of packaging films: Film materials used for food packaging tested for resistance to some common stored-product insects. *California Agriculture*, (June), 3–4.

Kundungal, H., Gangarapu, M., Sarangapani, S., Patchaiyappan, A., & Devipriya, S. P. (2019). Efficient biodegradation of polyethylene (HDPE) waste by the plastic-eating lesser waxworm (*Achroia grisella*). *Environmental Science and Pollution Research*, *26*, 18509–18519.

Peng, B. Y., Chen, Z. B., Chen, J. B., Yu, H. R., Zhou, X. F., Criddle, C. S., et al. (2020). Biodegradation of polyvinyl chloride (PVC) in *Tenebrio molitor* (Coleoptera: Tenebrionidae) larvae. *Environment International*, *145*, 106106.

Peng, B.-Y., Li, Y., Fan, R., Chen, Z., Chen, J., Brandon, A. M., et al. (2020). Biodegradation of low-density polyethylene and polystyrene in superworms, larvae of *Zophobas atratus* (Coleoptera: Tenebrionidae): Broad and limited extent depolymerization. *Environmental Pollution*, *266*, 115206.

Peng, B.-Y., Su, Y., Chen, Z., Chen, J., Zhou, X., Benbow, M. E., et al. (2019). Biodegradation of polystyrene by dark (*Tenebrio obscurus*) and yellow (*Tenebrio molitor*) mealworms (Coleoptera: Tenebrionidae). *Environmental Science & Technology*, *53*(9), 5256–5265.

Plastics Europe (2019). Plastics-the facts. https://www.plasticseurope.org.

Roberson, W. H. (2005). *Urban insects and arachnids, a handbook of urban entomology* (pp. 126–127). Cambridge, UK: Cambridge University Press.

Wang, Z., Xin, X., Shi, X. F., & Zhang, Y. L. (2020). A polystyrene-degrading *Acinetobacterium* isolated from the larvae of *Tribolium castaneum*. *Science of the Total Environment*, *726*, 138564.

Woo, S., Song, I., & Cha, H. J. (2020). Polystyrene by the gut microbial flora of *Plesiophthalmus davidis* larvae. *Applied and Environmental Microbiology, 86*. https://doi.org/10.1128/AEM.01361-20. e01361–20.

Yang, S. S., Brandon, A. M., Andrew Flanagan, J. C., Yang, J., Ning, D., Cai, S. Y., et al. (2018). Biodegradation of polystyrene wastes in yellow mealworms (larvae of *Tenebrio molitor* Linnaeus): Factors affecting biodegradation rates and the ability of polystyrene-fed larvae to complete their life cycle. *Chemosphere, 191*, 979–989.

Yang, S. S., Ding, M. Q., He, L., Zhang, C. H., Li, Q. X., Xing, D. F., et al. (2021). Biodegradation of polypropylene by yellow mealworms (*Tenebrio molitor*) and superworms (*Zophobas atratus*) via gut-microbe-dependent depolymerization. *Science of the Total Environment, 756*, 144087. https://doi.org/10.1016/j.scitotenv.2020.144087. https://doi.org/10.1016/j.scitotenv.2020.144087.

Yang, L., Gao, J., Liu, Y., Zhuang, G., Peng, X., Wu, W.-M., et al. (2021). Biodegradation of expanded polystyrene and low-density polyethylene foams in larvae of *Tenebrio molitor* Linnaeus (Coleoptera: Tenebrionidae): Broad versus limited extent depolymerization and microbe-dependence versus independence. *Chemosphere, 262*, 127818.

Yang, S. S., Wu, W. M., Brandon, A. M., Fan, H. Q., Receveur, J. P., Li, Y., et al. (2018). Ubiquity of polystyrene digestion and biodegradation within yellow mealworms, larvae of *Tenebrio molitor* Linnaeus (Coleoptera: Tenebrionidae). *Chemosphere, 212*, 262–271.

Yang, J., Yang, Y., Wu, W. M., Zhao, J., & Jiang, L. (2014). Evidence of polyethylene biodegradation by bacterial strains from the guts of plastic-eating waxworms. *Environmental Science & Technology, 48*, 13776–13784.

Yang, Y., Yang, J., Wu, W. M., Zhao, J., Song, Y. L., Gao, L. C., et al. (2015a). Biodegradation and mineralization of polystyrene by plastic-eating mealworms: Part 1. Chemical and physical characterization and isotopic tests. *Environmental Science & Technology, 49*, 12080–12086.

Yang, Y., Yang, J., Wu, W. M., Zhao, J., Song, Y. L., Gao, L. C., et al. (2015b). Biodegradation and mineralization of polystyrene by plastic-eating mealworms: Part 2. Role of gut microorganisms. *Environmental Science & Technology, 49*, 12087–12093.

Yin, C. F., Xu, Y., & Zhou, N. Y. (2020). Biodegradation of polyethylene mulching films by a co-culture of *Acinetobacter* sp. strain NyZ450 and *Bacillus* sp. strain NYZ451 isolated from *Tenebrio molitor* larvae. *International Biodeterioration and Biodegradation, 155*, 105089.

CHAPTER SIX

Quantification of polystyrene plastics degradation using ^{14}C isotope tracer technique

Lili Tian[a,b], Yini Ma[a,*], and Rong Ji[a]

[a]State Key Laboratory of Pollution Control and Resource Reuse, School of the Environment, Nanjing University, Nanjing, China
[b]Institute of Agricultural Resources and Environment, Jiangsu Academy of Agricultural Sciences, Nanjing, China
*Corresponding author: e-mail address: mayini@nju.edu.cn

Contents

Abstract

^{14}C-isotope tracer technique is an effective tool for quantitative analysis of compounds. Based on its unique ^{14}C signal, ^{14}C-isotope tracer has been widely used in degradation of pollutants, especially in a complex environmental matrix. In this chapter, we introduce methods and examples for studying the degradation of polystyrene (PS) plastics using the ^{14}C-isotope tracer technique. ^{14}C-based gel permeation chromatography (GPC), high performance liquid chromatography (HPLC) as well as liquid scintillation counter (LSC) are introduced for analyzing changes in the molecular weight of plastic polymers, generation of hydrophilic products, and complete mineralization of plastics during microbial and UV-induced degradation. With the ^{14}C-isotope tracer technique, further studies on mechanisms for degradation of plastics/microplastics under complex natural environment conditions could be facilitated.

Methods in Enzymology, Volume 648
ISSN 0076-6879
https://doi.org/10.1016/bs.mie.2020.12.014

1. Introduction

Plastic pollution has developed into a global concern—large amounts of mismanaged solid waste release and low degradation rates lead to plastic debris accumulation in the ocean, fresh water and terrestrial ecosystems (Vethaak & Leslie, 2016). Those plastic debris could be ingested by various animals, such as seabirds, turtles, fish, causing serious injury or even death (Boerger, Lattin, Moore, & Moore, 2010). By ultraviolet (UV) irradiation, biological, physical and chemical effects, large plastics debris break down in the environment into fine plastics (FPs), including microplastics in the micrometer (1 μm–5 mm) (Andrady, 2011), or sub-micrometer range (100 nm − 1 μm) and even nanoplastics (1–100 nm) (Li, Luo, et al., 2020). These fine plastics are considered to be persistent in the environment (Cózar et al., 2014) and can spread widely throughout air, soil and water systems (Vethaak & Leslie, 2016). During degradation and fragmentation, chemical additives released from plastics have been found to be toxic for marine animals (Li et al., 2016).

Depending on the driving factors, plastics degradation can be subdivided into photodegradation, thermal degradation, ozone-induced degradation, catalytic degradation and biodegradation. Among these pathways, photodegradation is essential to plastics degradation in the environment, especially under UV irradiation. UV induces physical, chemical and optical changes of plastics, such as yellowing, embrittlement and surface oxidation as well as changes of molecular weight (Kaczmarek, Felczak, & Szalla, 2008; Shang, Chai, & Zhu, 2003; Ward, Armstrong, Walsh, Jackson, & Reddy, 2019). Under UV irradiation and other mechanical disturbance, micro-, even nano-sized plastics particles are generated (Lambert & Wagner, 2016; Song et al., 2017). However, the characterization of photodegradation of these FPs, especially nanoplastics, are missing, due to the limitation of current detection methods. Moreover, information about possible toxic products released during photodegradation, including small sized molecules and plastic additives, is urgently needed for a full risk assessment of FPs (Gewert, Plassmann, Sandblom, & MacLeod, 2018).

In addition to photodegradation, biodegradation of plastics has been considered to be an eco-friendly way to address disposed plastics waste, however, many widely used petrochemical plastics cannot be degraded in the environment without suitable enzymes and conditions (Wei et al., 2020). Using microscopic, spectroscopic as well as gel chromatographic methods, previous studies have investigated biodegradation of polystyrene (PS) and low-density polyethylene (LDPE) plastics (Li, Wei, et al., 2020; Yang et al., 2015). The methods used in previous studies turn out to have limitations at certain conditions, such as (1) gravimetry is not sensitive enough to detect slow biodegradation or biochemical alterations not leading to significant mass losses; (2) characterization methods such as FTIR, X-ray photoelectron spectroscopy (XPS) can detect the change in functional groups resulting from plastic degradation while they cannot quantify the biodegradation rate; (3) the generation of CO_2 during biodegradation of plastics is hard to discriminate from respiration in complex systems (Krueger, Harms, & Schlosser, 2015). Therefore, suitable quantitative methods for biodegradation of plastics are required.

^{14}C isotope tracer is a quantitative method for studying carbon-based chemicals with high sensitivity. With a unique ^{14}C signal, it was applied in studies on degradation and transformation of organic pollutants in complex environmental compartments such as soil, plants or even organisms (Gu et al., 2017; Liu et al., 2013; Sun et al., 2014). In this chapter, a quantitative method based on ^{14}C isotope tracer for characterizing the degradation processes of plastics is introduced, including the synthesis of ^{14}C-PS plastics films and nanoplastics and their application to study biodegradation by *Penicillium variabile* (*P. variabile*) and photodegradation under UV irradiation.

2. Protocols

2.1 Synthesis of [β-^{14}C]-styrene and [U-ring-^{14}C]-styrene

Methods to synthesize [β-^{14}C]-styrene and [U-ring-^{14}C]-styrene are introduced in the following parts. [β-^{14}C]-styrene is synthesized from [β-^{14}C]-malonic acid (Fig. 1, I to III) and [U-ring-^{14}C]-styrene is synthesized from [U-ring-^{14}C]-cinnamon acid (Fig. 1, IV to V), respectively.

2.1.1 Materials

- Malonic acid (solid, chemical purity 98%)
- Cinnamic acid (solid, chemical purity 98%)
- Piperidine (10.9 $mol\,L^{-1}$, chemical purity 99.5%)
- Benzaldehyde (9.8 $mol\,L^{-1}$, chemical purity 99.5%)
- Pyridine (12.4 $mol\,L^{-1}$, chemical purity 99.5%)
- Ethyl acetate (10 $mol\,L^{-1}$, chemical purity 99%)
- *n*-Hexane (7.7 $mol\,L^{-1}$, chemical purity 99%)

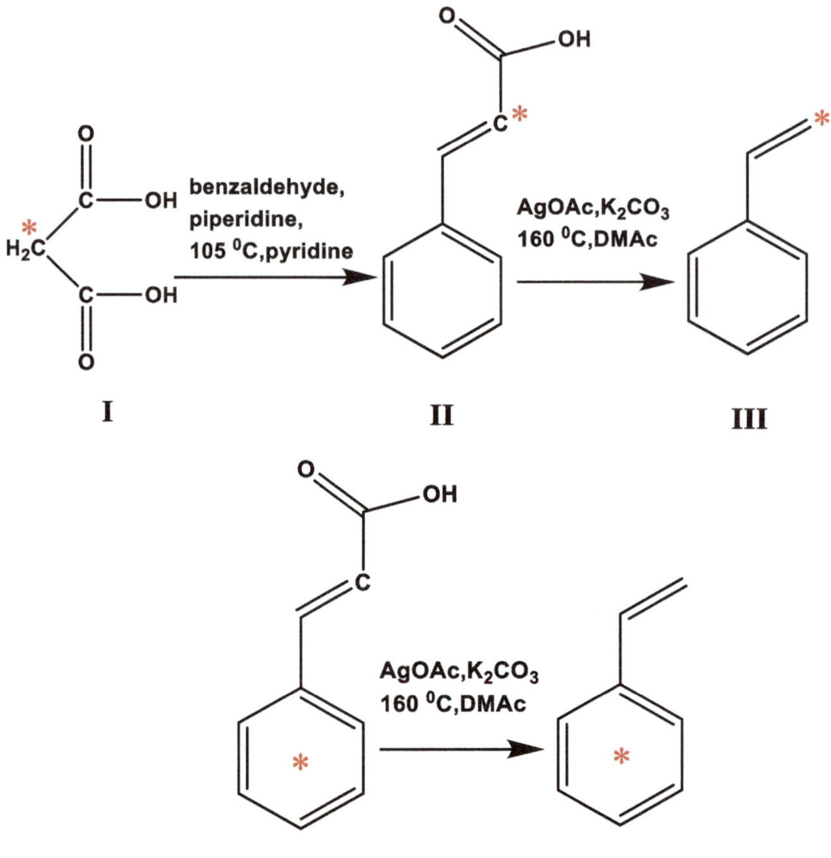

I II III

IV V

Fig. 1 The synthetic routes of [β-^{14}C]-styrene (**III**) (**I**: [β-^{14}C]-malonic acid; **II**: [β-^{14}C]-cinnamic acid) and [U-ring-^{14}C]-styrene (**V**) (**IV**: [U-ring-^{14}C]-cinnamic acid). AgOAc: silver acetate, K_2CO_3: potassium carbonate, DMAc: *N*, *N*-dimethylacetamide.

- Formic acid (26.5 mol L^{-1}, chemical purity 99.5%)
- Sodium sulfate (anhydrous Na_2SO_4, solid, chemical purity 99%)
- Silver acetate (AgOAc, solid, chemical purity 99%)
- Potassium carbonate (K_2CO_3, solid, chemical purity 98%)
- *N,N*-dimethylacetamide (DMAc, 10.8 mol L^{-1}, super dry solvent, chemical purity 99.5%)
- Developer (10 mL *n*-hexane, ethyl acetate 10 mL and formic acid 0.1 mL)
- Hydrochloric acid (HCl, chemical purity 99%, 1 mol L^{-1}, 43 mL 12 M concentrated HCl dissolved in 457 mL water)

- Sodium bicarbonate ($NaHCO_3$, chemical purity 99%, 0.6 mol L^{-1}, 5 g $NaHCO_3$ dissolved in 100 mL water)
- Sodium chloride (NaCl, chemical purity 99%, 6 mol L^{-1}, 36 g NaCl dissolved in 100 mL water)
- Sodium hydroxide (NaOH, chemical purity 99%, 1 mol L^{-1}, 4 g NaOH dissolved in 100 mL water)
- [β-^{14}C]-malonic acid (2100 MBq $mmol^{-1}$, Moravek Biochemicals, USA)
- [U-ring-^{14}C]-cinnamic acid (130 MBq $mmol^{-1}$, Moravek Biochemicals, USA)

2.1.2 Equipment

- Oil-bath pans (ETS-D5, IKA, Germany)
- Rotary evaporator (R210, Buchi, Switzerland)
- Thin layer chromatography (TLC, Qingdao Hailang, China)
- Typhoon Image Scanner (Typhoon Trio +, GE Healthcare, USA)
- Centrifuge (SL DR, Thermo, USA).

2.1.3 Method

2.1.3.1 Synthesis of [β-^{14}C]-styrene

(1) Mix [β-^{14}C]-malonic acid (15 MBq, 0.007 mmol) with non-labeled malonic acid 0.0355 g (0.348 mmol) in 1 mL pyridine and add to a 25 mL flask equipped with a magnetic stir bar and reflux condenser. Then add 177.5 μL of piperidine and 108.4 μL of benzaldehyde under nitrogen.

(2) Heat the mixture to 105 °C and react for 14 h under an oil bath and then cool down to room temperature.

(3) Acidify the mixture with 20 mL of 1 M HCl, extract with 10 mL of ethyl acetate by successively vortexing and precipitation for 3 times. Combine all organic extracts and concentrate by rotary evaporator (vacuum is 240 mbar, water bath is 40 °C) to 0.5 mL.

(4) Purify the concentrated samples on a preparative TLC plate. The preparative TLC plate with loaded samples is developed in developer and scanned by Typhoon Image Scanner. Scrape the bands of cinnamic acid (R_f=0.67) from the plate and extract with 15 mL of ethyl acetate for 7 times. Combine all organic extracts and concentrate by rotary evaporator (vacuum is 240 mbar, water bath is 40 °C).

(5) After concentration, [β-^{14}C]-cinnamic acid (12 MBq, 42 MBq $mmol^{-1}$) is obtained with a radiochemical yield of 82.8%.

(6) Mix [β-^{14}C]-cinnamic acid (12 MBq, 0.286 mmol) with 0.3807 g of non-labeled cinnamic acid (2.57 mmol) resulting in 0.4440 g ^{14}C-cinnamic acid (2.86 mmol) and add to a 25 mL flask. Then add 0.1011 g AgOAc (0.6 mmol) and 0.1254 g K_2CO_3 (0.9 mmol). Add 2 mL of *N,N*-dimethylacetamide (DMAc) dropwise with a syringe under nitrogen atmosphere and the resulting mixture is stirred at 160 °C for 8 h.

(7) Allow the mixture to cool down to room temperature. Add 2 mL *n*-hexane and 5 mL of 1 M NaOH. Extract with 10 mL *n*-hexane for 7 times. Combine the organic extracts and wash with 5 mL 0.6 M $NaHCO_3$ for 3 times and then wash with 5 mL 6 M NaCl. Dry the extracts with anhydrous Na_2SO_4 and concentrate with a rotary evaporator (vacuum is 330 mbar, water bath is 40 °C) to 0.5 mL.

(8) [β-^{14}C]-styrene (4.2 MBq, 4.2 MBq mmol^{-1}, 99.99% radiochemical purity) is obtained with a radiochemical yield of 33.7%.

2.1.3.2 Synthesis of [U-ring-^{14}C]-styrene

(1) Mix [U-ring-^{14}C]-cinnamic acid (11 MBq, 0.085 mmol) with 0.0791 g non-labeled cinnamic acid (0.761 mmol) and add the mixture to a 25 mL flask. Then add 0.0029 g of AgOAc (0.6 mmol) and 0.0036 g of K_2CO_3 (0.9 mmol). Add 1 mL of 10.8 M *N,N*-dimethylacetamide (DMAc) dropwise by syringe under nitrogen atmosphere and the resulting mixture is stirred at 160 °C for 8 h.

(2) Allow the mixture to cool down to room temperature. Add 2 mL *n*-hexane and 5 mL of 1 M NaOH. Extract with 10 mL *n*-hexane for seven times. Combine the organic extracts and wash with 5 mL 0.6 M $NaHCO_3$ for 3 times and then wash with 5 mL 6 M NaCl. Dry the extracts with anhydrous Na_2SO_4 and concentrate with a rotary evaporator (vacuum is 330 mbar, water bath is 40 °C) to 0.5 mL [U-ring-^{14}C]-styrene (1.0 MBq, 13 MBq mmol^{-1}, 99.99% radiochemical purity) is obtained with radiochemical yield 9.5%.

2.2 Synthesis of ^{14}C-PS film

A method to synthesize ^{14}C-PS film is introduced in the following parts. ^{14}C-PS film is synthesized from ^{14}C-styrene through bulk polymerization without catalyst. To eliminate the effect of the solvent *n*-hexane, an autoclave is selected as the reaction vessel. The high temperature and pressure in the autoclave can evaporate *n*-hexane.

2.2.1 Materials

- Styrene (8.7 mol L^{-1}, chemical purity 99%)
- Chloroform (12.4 mol L^{-1}, chemical purity 99%)
- Methanol (24.7 mol L^{-1}, chemical purity 99%, stored at 4 °C before use)
- [β-^{14}C]-styrene (4.2 MBq, 42 MBq $mmol^{-1}$)
- [U-ring-^{14}C]-styrene (1.3 MBq, 1.5 MBq $mmol^{-1}$)

2.2.2 Equipment

- Autoclave (D1–30, Faen Tech&Trade, China)

2.2.3 Method

(1) Dilute [U-ring-^{14}C]-styrene (1.3 MBq) with 938 μL of non-labeled styrene and [β-^{14}C]-styrene (4.2 MBq) with 881 μL of non-labeled styrene. Add diluted ^{14}C-styrene into a 10 mL glass tube and react in the autoclave at 121 °C for 8 h.

(2) Dissolve the obtained products in 10 mL chloroform and pour into 200 mL methanol to clean up the unreacted styrene or oligomers. Dissolve the precipitation in 10 mL chloroform again and repeat this purification procedure for three times.

(3) Dissolve the purified solid products in chloroform, pour into a glass petri plate and ^{14}C-PS films will form on the plate after the evaporation of chloroform. ^{14}C-PS film with different diameter can be prepared using different containers or cut into small pieces.

(4) [U-ring-^{14}C]-PS (0.1568 g, 542 Bq mg^{-1}, 85 kBq) is obtained with number average molecular weight (M_n) 3.3 kDa and polydispersity (PD) 2.2. [β-^{14}C]-PS (0.4072 g, 3200 Bq mg^{-1}, 1.3 MBq) is obtained with M_n of 19 kDa, and a Polydispersity (PD) of 1.9. The diameter of the ^{14}C-PS film is 3 cm.

2.3 Synthesis of ^{14}C-PS nanoplastics

The procedure of ^{14}C-PS nanoplastics synthesis is conducted according to the reference (Tian et al., 2019). ^{14}C-PS nanoplastics are synthesized from ^{14}C-styrene by suspension polymerization in water with $K_2S_2O_8$ as initiator.

2.3.1 Materials

- Styrene (8.7 mol L^{-1}, chemical purity 99%)
- [β-^{14}C]-styrene (2.5 MBq, 1.2 MBq $mmol^{-1}$)
- Potassium persulfate (0.085 mol L^{-1} $K_2S_2O_8$)
- Ethanol (17.1 mol L^{-1}, chemical purity 99%)

2.3.2 Equipment

- Oil bath pan (ETS-D5, IKA, Germany)
- Centrifuge (SL DR, Thermo, USA)

2.3.3 Method

(1) Add 70 mL of distilled water into a 250-mL three-necked bottle equipped with a stirring rod and a reflux condenser. Bubble with nitrogen for 30 min to remove oxygen.
(2) Add 0.2 g of ^{14}C-styrene (2.5 MBq, 1.2 MBq mmol^{-1}) dropwise to the water under nitrogen gas and disperse by stirring at 180 rpm for 30 min.
(3) Add 10 mL of 0.085 M $K_2S_2O_8$ dropwise to the bottle under stirring.
(4) Heat the resulting mixture to 70 °C and react for 24 h under nitrogen gas protection in the oil bath pan. Keep the stirring rate at 180 rpm during the whole reaction process.
(5) Separate the products from solvent by centrifuging at 9500 *g*, 20 °C for 30 min. Wash the precipitate sequentially with water (45 mL × 3) and ethanol (45 mL × 3) by re-suspension and centrifugation to remove unreacted reagents, respectively.
(6) ^{14}C-PS nanoplastics (0.0544 g, 0.18 MBq) are obtained after vacuum drying for 48 h.

2.4 Biodegradation of ^{14}C-PS film by *Penicillium variabile*

One case of the ^{14}C isotope technology applied in biodegradation of plastics is introduced here. ^{14}C-PS film is incubated with a strain of the fungus *P. variabile* in Czapek-Dox medium for 4 months, the mineralization rate is detected during the incubation period. It turns out ^{14}C-PS film can be mineralized by *P. variabile*.

2.4.1 Materials

- [β-^{14}C]-PS film (3200 Bq mg^{-1})
- [U-ring-^{14}C]-PS film (480 Bq mg^{-1})
- Sucrose (87.5 mmol L^{-1})
- Sodium hydrate (NaOH, 1 mol L^{-1}, 4 g NaOH dissolved in 100 mL water)
- Tetrahydrofuran (THF, 12.3 mol L^{-1}, chemical purity 99.9%)
- *Penicillium Variabile*
- Czapek-Dox medium (35 mmol L^{-1} $NaNO_3$, 5.7 mmol L^{-1} K_2HPO_4, 2 mmol L^{-1} $MgSO_4 \cdot 7H_2O$, 6.7 mmol L^{-1} KCl, 0.03 mmol L^{-1} $FeSO_4 \cdot 7H_2O$, pH 7.5)

2.4.2 Equipment

- Shaker (KS 4000i control, IKA Technology, Germany)
- Ozone generator (Ozonia LAB2B; Degrémont Technologies–Triogen, Switzerland)

2.4.3 Method

(1) Biodegradation of ^{14}C-PS film is conducted in the medium with *P. variabile* (Tian et al., 2017). [β-^{14}C]-PS and [U-ring-^{14}C]-PS film is prepared and cut into small pieces (0.5–2 cm).
(2) To investigate the effect of ozonation pre-treatment, a group of [β-^{14}C]-PS plastics are ozonated for 3 h in the ozone generator with a flow rate of 5 L min^{-1}, which is equivalent to 17 g of ozone during the whole ozonation process.
(3) The *P. variabile* strain is pre-cultivated in sterilized Czapek-Dox medium supplemented with 30 g of sucrose (final concentration is 87.6 mM) for 7 days (24 °C, 120 rpm). *P. variabile* is washed with sugar-free Czapek-Dox medium for three times before use.
(4) Add *P. variabile* cells to the glass bottle containing either [β-^{14}C]-PS or [U-ring-^{14}C]-PS films. The final concentration of dry biomass of *P. variabile* cells in the medium is typically 0.67 mg mL^{-1}.
(5) Hang vials containing 1 mL 1 M NaOH on the top of bottles to trap $^{14}CO_2$ during the incubation. Place the bottles on a rotary shaker and incubate for 112 d (24 °C, 120 rpm). All treatments are performed in triplicate. Control treatment without *P. variabile* is prepared with the same procedure.
(6) Measure the radioactivity in NaOH by liquid scintillation counter (LSC, see below) at 0, 7, 14, 28, 56, 84 and 112 d. More details are shown in Fig. 2.
(7) Measure the molecular weight of [β-^{14}C]-PS before and after incubation by gel permeation chromatography (GPC) coupled with LSC (^{14}C-GPC, see below) after dissolved in THF and the changes are shown in Fig. 3.

2.5 Photodegradation of [β-^{14}C]-PS nanoplastics

One case of ^{14}C isotope technology applied in photodegradation of plastics is introduced here. ^{14}C-PS nanoplastics are exposed to UV irradiation for 48 h. Mineralization, molecular weight changes of nanoplastics and the leachates generated during UV irradiation are measured.

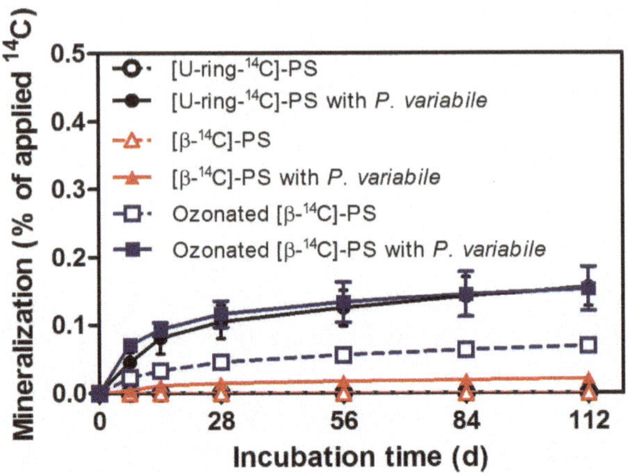

Fig. 2 Mineralization of [U-ring-^{14}C]-PS, [β-^{14}C]-PS and ozonated [β-^{14}C]-PS plastics during 112 d incubation in medium with and without *P. variabile* (Tian et al., 2017). The values are means of triplicates with standard deviations represented by error bars.

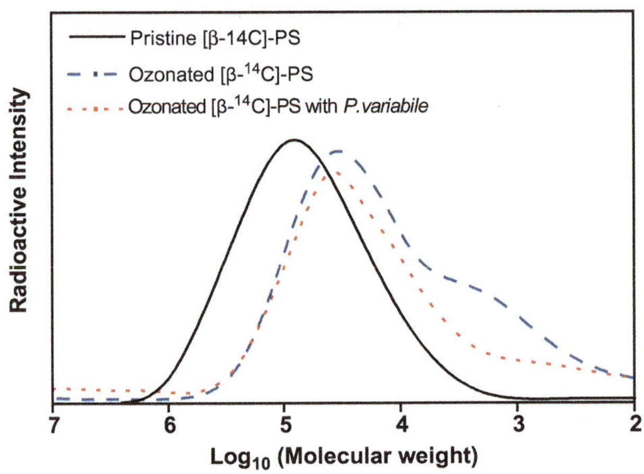

Fig. 3 Molecular weight change of ozonated [β-^{14}C]-PS plastics after 112 d incubation in medium with (red dot line) and without (blue dashed line) *P. variabile* compared to pristine PS (black solid line) characterized by ^{14}C-GPC (Tian et al., 2017). Data are normalized by the radioactive amount.

2.5.1 Materials

- [β-^{14}C]-PS nanoplastics (230 nm, 8000 Bq)
- Sodium hydroxide (NaOH, 1 mol L^{-1}, 4 g NaOH dissolved in 100 mL water)
- Tetrahydrofuran (THF, 12.3 mol L^{-1}, chemical purity is 99.9%)
- Cellulose filter (0.22 μm, choose filter with pore size smaller than the particles used in the experiment)

2.5.2 Equipment

- Quartz bottles (diameter 3 cm × height 12 cm, UV could penetrate quartz with little loss)
- UV254 light (GPH436T5L; Light Sources Inc., USA)
- Scintillation cocktail (Gold star, Meridian Biotechnologies Ltd., UK)
- Liquid scintillation counter (LSC, LS6500, Beckman Coulter, USA)

2.5.3 Method

Photodegradation of [β-^{14}C]-PS nanoplastics is conducted using UV irradiation (Tian et al., 2019).

(1) Place [β-^{14}C]-PS nanoplastics (1 mg) in quartz bottles and expose to UV254 light with an intensity of 0.7 mW cm^{-2} in a weathering chamber. Dark control without UV irradiation is conducted at the same time. All treatments are performed in triplicate.

(2) Seal the bottles with rubber stoppers and hang a vial containing 1 mL 1 M NaOH on each stopper to trap $^{14}CO_2$ generated during irradiation.

(3) Measure the radioactivity in NaOH after UV irradiation for 24 h and 48 h by LSC (see below) and the mineralization amount is shown in Fig. 4A.

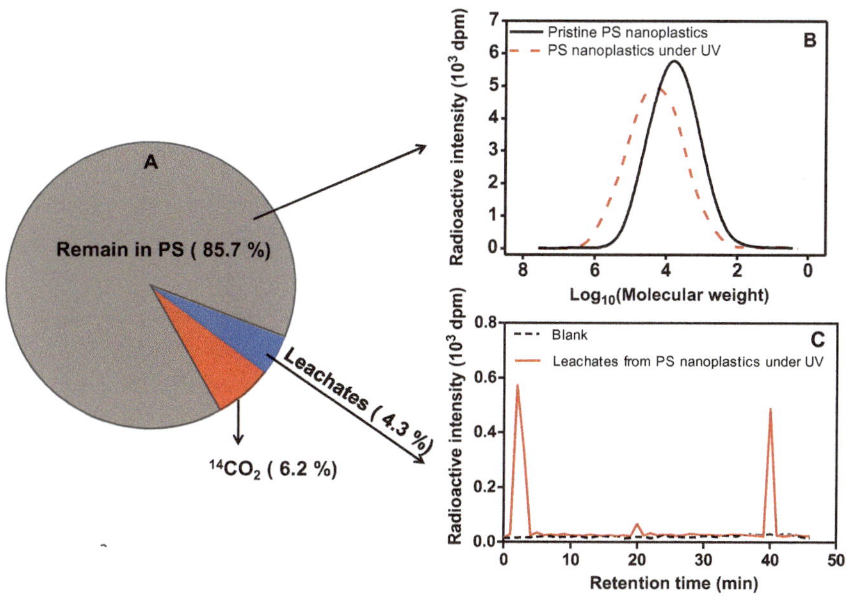

Fig. 4 Changes of ^{14}C-PS nanoplastics after 48 h UV irradiation (A), ^{14}C-GPC results of the molecular weight distribution of ^{14}C-PS nanoplastics (B) and characterization of water-soluble leachates by ^{14}C-HPLC (C) (Tian et al., 2019). Data are normalized by the radioactive amount.

(4) Wash the ^{14}C-PS nanoplastics with pure water after 48 h irradiation and filter them through a 0.22-μm cellulose filter membrane to separate the nanoplastics particles from possible water-soluble leachates.
(5) Collect ^{14}C-PS nanoplastics on the membrane and measure the molecular weight by ^{14}C-GPC after dissolved in THF. The changes of molecular weight are shown in Fig. 4B.
(6) Quantify the leachates in water by LSC and analyze by high performance liquid chromatography coupled with LSC (^{14}C-HPLC, see below) after concentrated. Different photodegradation products are found in water (Fig. 4C).

2.6 LSC analyses for ^{14}C in liquid phases

All the quantification of ^{14}C in liquid phase is performed by LSC, the instrument settings of LSC and detection procedure are introduced in these parts.

2.6.1 Materials

- Tetrahydrofuran (THF, 12.3 mol L^{-1}, chemical purity is 99.9%)
- Scintillation cocktail (Gold star, Meridian Biotechnologies Ltd., UK)

2.6.2 Equipment

- Liquid scintillation counter (LSC, LS6500, Beckman Coulter, USA)

2.6.3 Method

(1) The instrument settings of LSC are the following: the counting time is 2 min and the quench limit ranges from 20 to 337.1. The calibration curve is obtained using the radioactivity standards. The samples have a background of <0.5 Bq, and no chemiluminescence from the cocktail is observed.
(2) The radioactivity of ^{14}C-PS film or nanoplastics is detected after dissolving in THF and mixing with 3 mL scintillation cocktail.
(3) The amount of water-soluble products is measured by LSC after they are directly mixed with 3 mL scintillation cocktail.

2.7 ^{14}C-GPC analyses for molecular weight distribution of ^{14}C-plastics

The detection of molecular weight change of the ^{14}C plastics film or ^{14}C nanoplastics is performed by GPC coupled with LSC. The effluent is collected by time after the solvent passes through detector of the GPC and is detected by LSC after mixing with Scintillation cocktail.

2.7.1 Materials

- Tetrahydrofuran (THF, $12.3\,mol\,L^{-1}$, chemical purity is 99.9%)
- Scintillation cocktail (Gold star, Meridian Biotechnologies Ltd., UK)
- PL gel-MIXED B column ($10\,\mu m$, $300 \times 7.5\,mm$)
- PL gel-MIXED C column ($5\,\mu m$, $300 \times 7.5\,mm$)
- Polystyrene Calibration standards ($580–2.7 \times 10^6\,Da$)

2.7.2 Equipment

- GPC (PL-GPC 120, Agilent Technologies, USA)
- Refractive index detector (PL-GPC 120, Agilent Technologies, USA)

2.7.3 Method

(1) Dissolve ^{14}C-PS film or nanoplastics in THF and measure via a GPC chromatograph.

(2) Use a combination of two polystyrene gel columns of PL gel-MIXED B ($10\,\mu m$, $300 \times 7.5\,mm$) and PL gel-MIXED C ($5\,\mu m$, $300 \times 7.5\,mm$). The effluent is THF, flow rate is $1.0\,mL\,min^{-1}$ and temperature is $40\,°C$. The columns are calibrated using polystyrene standards with different molecular weight.

(3) Collect the effluent every minute in vials and measure the radioactivity after thorough mixing with $3\,mL$ cocktail.

2.8 ^{14}C-HPLC analyses for degradation products in liquid phase

The leachates generated from ^{14}C-PS nanoplastics during UV irradiation are analyzed on HPLC coupled with LSC. Collect the effluent by time after the solvent passes through the detector of the HPLC and detect by LSC after mixing with Scintillation cocktail.

2.8.1 Materials

- Methanol ($24.7\,mol\,L^{-1}$, chemical purity 99%)
- Acetonitrile ($19.1\,mol\,L^{-1}$, chemical purity 99%)
- $0.22\,\mu m$ PTFE filters
- HPLC vials

2.8.2 Equipment

- HPLC (1100 series, Agilent Technologies, USA)
- Reversed phase column Eclipse XDB-C18 ($250\,mm \times 4.6\,mm$, $5\,\mu m$; Agilent Technologies, USA)
- Diode array detector (Agilent Technologies, USA)

2.8.3 Method

(1) Concentrate the leachates after separation from nanoplastics particles and re-dissolve the leachates in methanol. Filter through 0.22 μm PTFE filter into HPLC vials.
(2) An Agilent HPLC 1100 system equipped with a vacuum degasser, a quaternary pump, an autosampler, a thermostatic column compartment, and a diode array detector is used. The column is Eclipse XDB-C18 and temperature set to 30°C.
(3) The mobile phase is a mixture of water (A) and acetonitrile (B) with a gradient elution program starting at 90% A for 2 min, decreasing to 40% A within 30 min, then kept at 40% A for 8 min, and finally increasing to 90% A within 5 min. The flow rate is 1 mL min^{-1}.
(4) Collect the effluent every minute in vials and measure the radioactivity after thorough mixing with 3 mL cocktail.

3. Summary and outlook

With ^{14}C isotope tracer, we developed a method to carry out the quantification of plastics degradation. Using the synthesized ^{14}C-PS plastics and nanoplastics, mineralization and decrease of molecular weight demonstrated that *P. variabile* can degrade PS at a low rate initially while ozonation pre-treatment can enhance the biodegradation significantly. PS nanoplastics can be photodegraded significantly by UV irradiation in short time and the transformation products with more hydrophilic groups may leach from nanoplastics to the surrounding water. Besides the application mentioned in this study, ^{14}C isotope tracer can be applied in research on degradation of plastics in more complex systems to mimic the environmental conditions with the help of the specific ^{14}C signal.

4. Safety notes

All the experiments with ^{14}C radioisotope should be done in a specific area with ^{14}C radioactive operating qualification. The operators should wear gloves, mask, shield, and protective suit to avoid the direct exposure of radiation. All the ^{14}C waste generated during experiments should be disposed by specific companies. During the ozonation experiment, the operators should wear gloves, mask, and protective suit to avoid ozone inhalation and skin contact.

Acknowledgments

This work was supported by the National Key Research and Development Program of China (2016YFC1402203) and National Natural Science Foundation of China (21876079).

References

Andrady, A. L. (2011). Microplastics in the marine environment. *Marine Pollution Bulletin, 62*(8), 1596–1605. https://doi.org/10.1016/j.marpolbul.2011.05.030.

Boerger, C. M., Lattin, G. L., Moore, S. L., & Moore, C. J. (2010). Plastic ingestion by planktivorous fishes in the North Pacific central gyre. *Marine Pollution Bulletin, 60*(12), 2275–2278. https://doi.org/10.1016/j.marpolbul.2010.08.007.

Cózar, A., Echevarría, F., González-Gordillo, J. I., Irigoien, X., Úbeda, B., Hernández-León, S., et al. (2014). Plastic debris in the open ocean. *Proceedings of the National Academy of Sciences, 111*(28), 10239–10244. https://doi.org/10.1073/pnas.1314705111.

Gewert, B., Plassmann, M., Sandblom, O., & MacLeod, M. (2018). Identification of chain scission products released to water by plastic exposed to ultraviolet light. *Environmental Science & Technology Letters, 5*(5), 272–276. https://doi.org/10.1021/acs.estlett.8b00119.

Gu, J., Jing, Y., Ma, Y., Sun, F., Wang, L., Chen, J., et al. (2017). Effects of the earthworm Metaphire guillelmi on the mineralization, metabolism, and bound-residue formation of tetrabromobisphenol a (TBBPA) in soil. *Science of the Total Environment, 595*, 528–536. https://doi.org/10.1016/j.scitotenv.2017.03.273.

Kaczmarek, H., Felczak, A., & Szalla, A. (2008). Studies of photochemical transformations in polystyrene and styrene–maleic anhydride copolymer. *Polymer Degradation and Stability, 93*(7), 1259–1266. https://doi.org/10.1016/j.polymdegradstab.2008.04.011.

Krueger, M. C., Harms, H., & Schlosser, D. (2015). Prospects for microbiological solutions to environmental pollution with plastics. *Applied Microbiology and Biotechnology, 99*(21), 8857–8874. https://doi.org/10.1007/s00253-015-6879-4.

Lambert, S., & Wagner, M. (2016). Characterisation of nanoplastics during the degradation of polystyrene. *Chemosphere, 145*, 265–268. https://doi.org/10.1016/j.chemosphere.2015.11.078.

Li, H.-X., Getzinger, G. J., Ferguson, P. L., Orihuela, B., Zhu, M., & Rittschof, D. (2016). Effects of toxic leachate from commercial plastics on larval survival and settlement of the barnacle *Amphibalanus amphitrite*. *Environmental Science & Technology, 50*(2), 924–931. https://doi.org/10.1021/acs.est.5b02781.

Li, L., Luo, Y., Li, R., Zhou, Q., Peijnenburg, W. J. G. M., Yin, N., et al. (2020). Effective uptake of submicrometre plastics by crop plants via a crack-entry mode. *Nature Sustainability, 3*(11), 929–937. https://doi.org/10.1038/s41893-020-0567-9.

Li, Z., Wei, R., Gao, M., Ren, Y., Yu, B., Nie, K., et al. (2020). Biodegradation of low-density polyethylene by Microbulbifer *hydrolyticus IRE-31*. *Journal of Environmental Management, 263*(110402). https://doi.org/10.1016/j.jenvman.2020.110402.

Liu, J., Wang, Y., Jiang, B., Wang, L., Chen, J., Guo, H., et al. (2013). Degradation, metabolism, and bound-residue formation and release of tetrabromobisphenol a in soil during sequential anoxic–oxic incubation. *Environmental Science & Technology, 47*(15), 8348–8354. https://doi.org/10.1021/es4014322.

Shang, J., Chai, M., & Zhu, Y. (2003). Photocatalytic degradation of polystyrene plastic under fluorescent light. *Environmental Science & Technology, 37*(19), 4494–4499. https://doi.org/10.1021/es0209464.

Song, Y. K., Hong, S. H., Jang, M., Han, G. M., Jung, S. W., & Shim, W. J. (2017). Combined effects of UV exposure duration and mechanical abrasion on microplastic fragmentation by polymer type. *Environmental Science & Technology, 51*, 4368–4376. https://doi.org/10.1021/acs.est.6b06155.

Sun, F., Kolvenbach, B. A., Nastold, P., Jiang, B., Ji, R., & Corvini, P. F.-X. (2014). Degradation and metabolism of tetrabromobisphenol a (TBBPA) in submerged soil and soil–plant systems. *Environmental Science & Technology*, *48*(24), 14291–14299. https://doi.org/10.1021/es503383h.
Tian, L., Chen, Q., Jiang, W., Wang, L., Xie, H., Kalogerakis, N., et al. (2019). A carbon-14 radiotracer-based study on the phototransformation of polystyrene nanoplastics in water versus in air. *Environmental Science: Nano*, *6*(9), 2907–2917. https://doi.org/10.1039/c9en00662a.
Tian, L., Kolvenbach, B., Corvini, N., Wang, S., Tavanaie, N., Wang, L., et al. (2017). Mineralisation of (14)C-labelled polystyrene plastics by *Penicillium variabile* after ozonation pre-treatment. *New Biotechnology*, *38*(Pt B), 101–105. https://doi.org/10.1016/j.nbt.2016.07.008.
Vethaak, A. D., & Leslie, H. A. (2016). Plastic debris is a human health issue. *Environmental Science & Technology*, *50*, 6825–6826. https://doi.org/10.1021/acs.est.6b02569.
Ward, C. P., Armstrong, C. J., Walsh, A. N., Jackson, J. H., & Reddy, C. M. (2019). Sunlight converts polystyrene to carbon dioxide and dissolved organic carbon. *Environmental Science & Technology Letters*, *6*(11), 669–674. https://doi.org/10.1021/acs.estlett.9b00532.
Wei, R., Tiso, T., Bertling, J., O'Connor, K., Blank, L. M., & Bornscheuer, U. T. (2020). Possibilities and limitations of biotechnological plastic degradation and recycling. *Nature Catalysis*, *3*(11), 867–871. https://doi.org/10.1038/s41929-020-00521-w.
Yang, Y., Yang, J., Wu, W.-M., Zhao, J., Song, Y., Gao, L., et al. (2015). Biodegradation and mineralization of polystyrene by plastic-eating mealworms: Part 2. Role of gut microorganisms. *Environmental Science & Technology*, *49*(20), 12087–12093. https://doi.org/10.1021/acs.est.5b02663.

CHAPTER SEVEN

Exploring the global metagenome for plastic-degrading enzymes

Pablo Pérez-García, Dominik Danso, Hongli Zhang, Jennifer Chow, and Wolfgang R. Streit*

Department of Microbiology and Biotechnology, University of Hamburg, Hamburg, Germany
*Corresponding author: e-mail address: wolfgang.streit@uni-hamburg.de

Contents

Abstract

Plastics are extensively used in our daily life, but they are also a major pollutant of our biosphere accumulating in both the ocean and the land. In the recent years, few enzymes and microorganisms have been discovered with the ability to degrade even fewer synthetic polymers. Nevertheless, more active species and enzymes need to be discovered and described in order to gain more knowledge about protein adaptation to the degradation of not-naturally-occurring polymers. Within this chapter, we focus on efficient methods to identify novel polyethylene terephthalate-degrading enzymes (PETases) from culturable and non-culturable microorganisms by a combination of sequence- and function-based screening. This protocol can be adapted to discover other plastic hydrolases and in general for other enzymes, for which not many characterized specimens are yet available.

Methods in Enzymology, Volume 648
ISSN 0076-6879
https://doi.org/10.1016/bs.mie.2020.12.022

1. Introduction

Plastics are synthetic polymers that serve us in our daily life as containers and packaging materials, clothing, working surfaces, machine parts, devices or surgery tools to name only a few examples. The main manufactured plastics in decreasing order of the yearly production amount are polyethylene (PE), polypropylene (PP), polyvinylchloride (PVC), polyethylene terephthalate (PET), polyurethane (PU), polystyrene (PS) and polyamide (PA) (Danso, Chow, & Streit, 2019). The overall plastic production amounted to more than 368 million tons in 2019 and has been growing since (PlasticsEurope, 2020). Most of the yearly production is used for packaging, which results in a short product lifetime (Geyer, Jambeck, & Law, 2017).

With 7% of the general demand for plastics in Europe and being used mainly for packaging (food containers and bottles), polyethylene terephthalate (PET) is one of the most relevant synthesized plastics. PET is a solid thermoplastic which is nearly insoluble in water and has a melting temperature of 250–260 °C (Lim, 2017). It was first discovered by John Rex Whinfield in 1941 (Rex & Tennant, 1949) by a polycondensation of its monomers terephthalic acid (TPA) and ethylene glycol (EG) (Fig. 1). Nowadays, numerous physically diverse variants of this polymer have been developed with different degrees of crystallinity and melting temperatures (Rieckmann & Völker, 2004). Because of its properties, PET is mostly turned into food/beverage containers (mainly bottles), but also foils or even synthetic textile fibers can be made out of it.

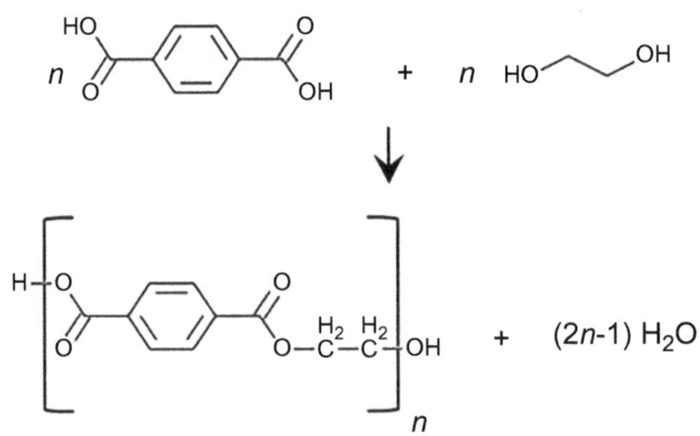

Fig. 1 Synthesis of polyethylene terephthalate (PET) by polycondensation of its monomers terephthalic acid (TPA) and ethylene glycol (EG).

In theory, plastics can be recycled and reused. In Europe most of them are incinerated to produce energy (PlasticsEurope, 2020) at the cost of generating CO_2 and other toxic substances. Yet, at a global scale, recycling or incineration rate is way below the level of production and so it is estimated that around 58% of it ends up in landfills or in the natural environment (Geyer et al., 2017).

1.1 Plastics as environmental pollutants

After disposal, plastics mount up in the environment and virtually do not degrade in nature due to their chemical properties. Earlier studies estimated already 100–500 years for the degradation of a single plastic bottle and a more recent study assumed 10–20 years for a plastic bag (Barnes, Galgani, Thompson, & Barlaz, 2009; Shaw & Day, 1994). Of special concern is the accumulation of plastic litter in aquatic systems, where plastics are carried by currents and mostly transported into so called "garbage patches." Some estimation suggested that approximately 250,000 tons of plastic are floating in the oceans (Eriksen et al., 2014), and these numbers might be higher in the present. More than 80% of the plastics in the ocean had its origin in land (Jambeck et al., 2015). In addition to accumulating in water, plastics are also ingested by aquatic animals and birds, filling their stomachs and leading to death by starvation (Derraik, 2002). Furthermore, abandoned fishing nets, bags and other products can also become lethal for marine animals due to entanglement. Severe health issues, also to humans, are suspected to be caused by the toxicity of degradation products and additives. Microplastics already dominate numerically marine human debris and microbial plastic ecology is an emerging research area (Amaral-Zettler, Zettler, & Mincer, 2020; Zettler, Mincer, & Amaral-Zettler, 2013).

1.2 Degradation processes of plastic

As mentioned above, the high stability and durability associated with petroleum-based plastics make them practically not degradable in nature (Webb, Arnott, Crawford, & Ivanova, 2013). Nevertheless, a very low rate of decomposition by both abiotic and biotic factors is assumed (Shaw & Day, 1994).

The first steps of plastic degradation in the environment include mechanical disruption caused by sea waves, wind and friction or crash against other solid materials (*e.g.*, rocks). These effectors break the plastics into smaller pieces (micro- and nanoplastics), resulting in a much larger surface area that can be further degraded in an abiotic or biotic manner (Cooper & Corcoran, 2010; Shaw & Day, 1994). For PET, "weathering" or "photodegradation" is also considered to be one of the main reasons for initial depletion. Terephthalic acid contains an aromatic ring that acts as a chromophore able to absorb UV light (Fig. 1). Long exposure of PET to ultraviolet radiation causes the clear polymer to undergo a yellowing process (Falkenstein et al., 2020; Shaw & Day, 1994). Warmer ambient and surface temperatures as well as humidity are other factors speeding up the degradation of plastics and can act in synergism with UV exposure (Fagerburg & Clauberg, 2004).

1.2.1 Biotic disruption of PET

In recent years, biotic degradation of some plastics and/or their oligomers (including PET, polyurethane, polyamide, polystyrene and polyethylene) has been described as reviewed recently (Danso et al., 2019; Wei & Zimmermann, 2017). Thereby bacteria and fungi have been observed to produce enzymes that are able to partially or even completely degrade few of the most important polymers and among those PET and PUR. Microorganisms able to degrade PET are affiliated with the genera *Thermobifida* and *Thermonospora, Bacillus, Pseudomonas, Vibrio, Ideonella* and others including few metagenome-derived enzymes (see references in Danso et al., 2019; Wei & Zimmermann, 2017). Furthermore, fungal PET depletion has been described for *Fusarium oxysporum* or *Humicola insolens* (Carniel, Valoni, Nicomedes, Gomes, & de Castro, 2017; Ronkvist, Xie, Lu, & Gross, 2009; Vertommen, Nierstrasz, van der Veer, & Warmoeskerken, 2005).

An important factor restricting the ease of enzymatic conversion is the degree of crystallinity. Most daily-used PET as packaging materials is semicrystalline. Highly crystalline fragments are difficult to attack. However, amorphous and low crystalline regions are susceptible to enzymatic hydrolysis (Webb et al., 2013; Wei et al., 2019). Due to its size and properties (Table 1), PET as a high-molecular weight polymer cannot be imported into the cell and enzymes can only erode its surface. Just as any other enzyme attacking large polymers (*e.g.,* cutinases, cellulases), promiscuous enzymes able to degrade PET are believed to be secreted into the environment. After hydrolysis, intake of smaller, water-soluble intermediates can take place (Mueller, 2006). This is similar to the degradative mechanisms known for other native polymers like cellulose.

In 2016, Yoshida et al. described the first full *in vivo* hydrolysis mechanism of PET. The system was observed in the bacterium *Ideonella sakaiensis* and comprised two enzymes: a PETase and a MHETase. The first enzyme

Table 1 Physical properties of PET. Values with ranges indicate properties which vary depending on crystallinity and degree of polymerization (Awaja & Pavel, 2005).

Property	Value
Average molecular weight	30,000–80,000 $g\,mol^{-1}$
Density	1.41 $g\,cm^{-3}$
Melting temperature	255–265 °C
Glass transition temperature	69–115 °C
Young's modulus	1700 MPa
Water absorption (24 h)	0.5%

is used to partially hydrolyze PET to its building blocks, including the main intermediate mono-(2-hydroxyethyl) terephthalate (MHET). The second protein breaks MHET into TPA and EG, which can be incorporated into the metabolism (Yoshida et al., 2016). Unfortunately, a recent work by Wallace et al. demonstrated that *I. sakaiensis* is not able to grow on the highly crystalline PET found in regular plastic water bottles (Wallace et al., 2020).

1.2.2 PETases: PET-degrading α/β-hydrolases

Carboxylic ester hydrolases act on ester bonds (EC 3.1.1.) and are often applied in several biotechnological processes. Triacylglycerol lipases (EC 3.1.1.3), belonging to this category, are used in feed and food production as well as in laundry powder or as catalysts for pharmaceutic and fine chemical production (Hasan, Shah, & Hameed, 2006; Houde, Kademi, & Leblanc, 2004). Other hydrolases like, *e.g.*, carboxylesterases (EC 3.1.1.1.), cutinases (EC 3.1.1.74) or phospholipases (EC 3.1.1.4) are also applied in similar processes (Borrelli & Trono, 2015; Panda & Gowrishankar, 2005; Pio & Macedo, 2009). All these enzymes share the typical α/β-fold described in the early 1990s, formed by an α/β sheet of eight β-sheets connected by α-helixes (Fig. 2). A nucleophile-His-acid catalytic triad (usually Ser-His-Asp) is located in loops that are the best-conserved regions of the fold (David & Paul, 2009; Ollis et al., 1992).

The EC number 3.1.1.101 was assigned to PET hydrolyzing enzymes, representing a novel group of hydrolases with currently just few representative entries in the BRENDA database (https://www.brenda-enzymes.org/enzyme.php?ecno=3.1.1.101). The crystal structure of those proteins revealed the typical α/β-hydrolase fold consisting of a central twisted β-sheet

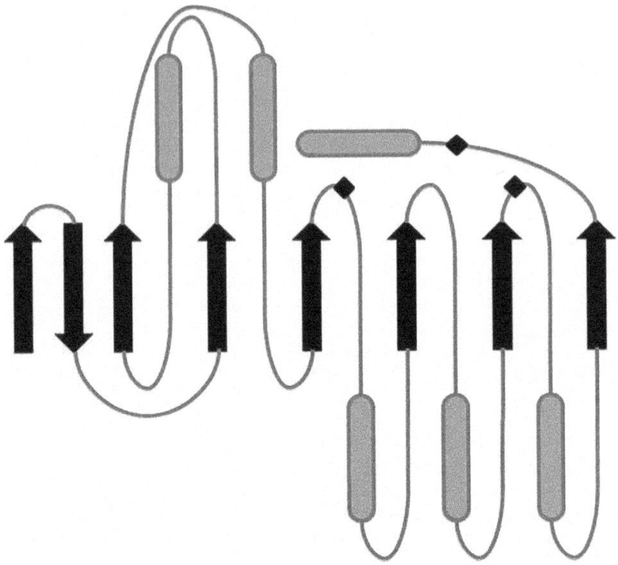

Fig. 2 The canonical structure of α/β-hydrolases. Black diamonds indicate catalytic residues (Ser, His, Asp). Adapted from Ollis et al., 1992 and David & Paul, 2009. The core of all α/β-hydrolases shares this typical fold consisting of eight β-sheets (black) connected by α-helices (gray).

flanked by α-helixes. The proteins appear as monomers in solution and are probably active as such on the substrate (Kitadokoro et al., 2012). No lid structure covering the active pocket is present, in contrast to the structure of true lipases where it promotes activity on esters with long acyl chains (Khan et al., 2017). True lipases display only poor or no activity at all against PET. A lid domain covering the catalytic pocket probably interferes with the bulky PET and prevents activity, as long as no compound is added which initiates interfacial activation of the protein (Eberl et al., 2009; Guebitz & Cavaco-Paulo, 2008).

A highly active PET hydrolase was found in *Humicola insolens*. The protein was able to almost completely degrade amorphous PET film at 70 °C over a period of 96 h (Ronkvist et al., 2009). Temperatures around 65 °C, close to the glass transition temperature (T_g) of PET, can result in an increment of the hydrolysis rate. Amorphous regions within the polymer become flexible and facilitate the accessibility to the catalytic pocket (Alves, Mano, Balaguer, Duenas, & Ribelles, 2002; Parikh, Gross, & Mccarthy, 1992). Thermostable enzymes are thus clearly beneficial for upcoming PET hydrolysis processes. Addition of divalent metal ions like Ca^{2+} or Mg^{2+} have been shown to yield positive effects on enzyme hydrolytic activity, promoting thermostability as shown for actinobacterial enzymes (Kawai et al., 2014; Miyakawa et al., 2015; Sulaiman, You, Kanaya, Koga, & Kanaya, 2014; Then et al., 2016; Thumarat, Nakamura, Kawabata, Suzuki, & Kawai, 2012). A similar effect was observed for phosphate on the metagenome derived LCC and the cutinase TfCut2 from *Thermobifida fusca* (Schmidt et al., 2016). Notably, Danso et al. identified *adhoc* in their study a number of thermostable enzymes implying that thermostability is a very common trait among PET esterases (Danso et al., 2018).

PETases seem to undergo product inhibition by the water soluble MHET and BHET. At concentrations of 0.5–2mM, these products are competitive inhibitors with similar binding constants (Barth et al., 2015). Hence, a thermostable PET hydrolase and parallel removal of released intermediates within a buffer system supplying phosphor and/or divalent metal ions would result in the elimination of the product inhibition and maximization of efficiency, pushing the reaction towards the product formation. The addition of the highly promiscuous CalB lipase from *Candida antarctica* (Carniel et al., 2017) or TfCa from *T. fusca* (Barth et al., 2016) has been shown to further convert the PET intermediate degradation product MHET to TPA, thus eliminating the inhibition effect. Other successful efforts to further increase the catalytic efficiency include the fusion to hydrophobins or non-catalytic binding domains (Espino-Rammer et al., 2013; Ribitsch et al., 2015; Ribitsch et al., 2013).

1.3 Metagenome screening for novel PETases

New biocatalysts, mainly enzymes from microorganisms, are used nowadays for a variety of applications. Factors such as cost reduction, eco-friendliness and simplicity of enzymatic reactions keep on raising the interest of the industry for these proteins, moving on from a chemistry-only-based catalysis (Li, Yang, Yang, Zhu, & Wang, 2012; Singh, Kumar, Mittal, & Mehta, 2016). However, the fact that 99.8% of the microorganisms present in many environments are not cultivable under laboratory conditions makes the discovery and characterization of new and better-performing candidates difficult. Functional metagenomic approaches are a possibility to screen environmental DNA and find new enzymes in the microbial "dark matter" without the need to isolate and cultivate the specific microbes (Streit & Schmitz, 2004). Numerous function-based screening protocols were developed, providing tools for the detection of biocatalysts with enhanced activity and/or other desired properties (Steele, Jaeger, Daniel, & Streit, 2009). Depending on the chemical properties and function of the enzyme, individual functional screening assays were developed (Ferrer et al., 2016). Still, functional metagenomics relies on heterologous expression in suitable hosts, *E. coli* being the preferred organism (Charles, Liles, & Sessitsch, 2017), which is the major bottleneck of this method. Additional advances are made by using *in vitro* expression technologies (Markel et al., 2020). Sequencing of environmental DNA and fast sequence alignment algorithms opened the door to sequence-based metagenomics, which allow for a faster, host-independent analysis of both the taxonomic composition and the metabolic potential of virtually any niche in the world (Hug et al., 2016; Methe et al., 2012; Pettersson, Lundeberg, & Ahmadian, 2009; Thompson et al., 2017; Venter et al., 2004). Furthermore, large amounts of sequencing projects are broadly available at public domain resources such as NCBI or IMG.

Considering PETases, traditional microbiological cultivation and functional metagenome screening methods did not result in a high taxonomic variety (Barth et al., 2015; Billig, Oeser, Birkemeyer, & Zimmermann, 2010; Danso et al., 2018; Kale, Deshmukh, Dudhare, & Patil, 2015; Kawai et al., 2014; Wilkes & Aristilde, 2017; Ribitsch et al., 2012; Roth et al., 2014; Sulaiman et al., 2014; Wei, Oeser, & Zimmermann, 2014; Yoshida et al., 2016). This bottleneck may be bypassed by means of sequence-based approaches (*e.g.*, profile Hidden-Markov Model searches), which can help us to better understand the actual variety of PET hydrolyzing organisms and enzymes.

2. Materials

For the bioinformatical analysis in Section 3.1, a GNU-like platform (any Linux distribution or Mac OS) is required. The commands used in the example hereunder have been tested on a system with the Linux distribution Ubuntu 18.04 LTS (4 CPU, 16 GB RAM) and the packages T-Coffee v.11.00 (Notredame, Higgins, & Heringa, 2000), HMMER v.3.1b2 (Finn, Clements, & Eddy, 2011) and SAMtools v.1.7 (H. Li et al., 2009) were used.

For the functional screening in Section 3.2, the following materials and reagents are needed:

- Polycaprolactone (PCL, Sigma-Aldrich, Germany, average MW ~14,000)
- Bis(2-hydroxyethyl) terephthalate (BHET, Sigma-Aldrich, Germany, MW: 254.24 Da)
- Amorphous PET film (ES301445, thickness: 250 μm, Goodfellow Ltd., Germany)
- Acetone (Carl Roth GmbH, Germany)
- Dimethyl sulfoxide (DMSO, Carl Roth GmbH, Germany)

- 1,1,1,3,3,3-Hexafluoro-2-propanol (Sigma-Aldrich, Germany)
- Paper punch
- Magnetic stirrer
- ULTRA TURRAX® T18 basic homogenizer (IKA WORKS Inc., USA)
- Glass beakers
- PYREX® Squibb Pear-shaped Separatory Funnel
- Rotary evaporator
- Petri dishes
- 10 mM PBS buffer pH 7.0
- Lysogeny broth (LB) medium (Bertani, 1951)
- Agar-agar Kobe I (Carl Roth GmbH, Germany)
- *E. coli* expression host [*e.g.*, BL21 (DE3), T7 SHuffle, NEB, Germany] transformed with putative PETase-coding genes in a suitable expression vector [*e.g.*, pET21a(+), pET28a(+)]

3. Methods

In order to identify novel PETases within large sequence datasets, we used a combined *in silico* and functional screening method (Fig. 3). A profile Hidden Markov Model [HMM (Eddy, 2011)] was constructed on the basis of already known and functionally tested enzymes (Table 2). After a sequence-based screening of a protein database, several putative PETase-coding genes were synthesized, and the respective proteins were tested for activity on indicator plates containing different (polymeric) substrates. The hereby described processes of *in silico* and functional screenings can be submitted to several rounds of iteration to refine the profile HMM and improve the results (Fig. 3). This methodology has been applied to discover and enlarge the knowledge on PET-degrading enzymes and their occurrence in both marine and terrestrial metagenomes (Danso et al., 2018). It could potentially be applied for other plastic degrading enzymes as well.

3.1 Construction of a profile hidden Markov model (HMM) and screening of a protein database for putative PETases

Construction of a profile HMM requires information on the amino acid sequences of already described proteins from the family of interest. Table 2 includes some of the firstly described PETases, which are used to construct the first model.

1. After retrieving the amino acid sequences from Table 2 to a file called "PETases.faa" (see "Retrieve/ID mapping" tool within the UniProt server, https://www.uniprot.org/), a multiple sequence alignment (MSA) is

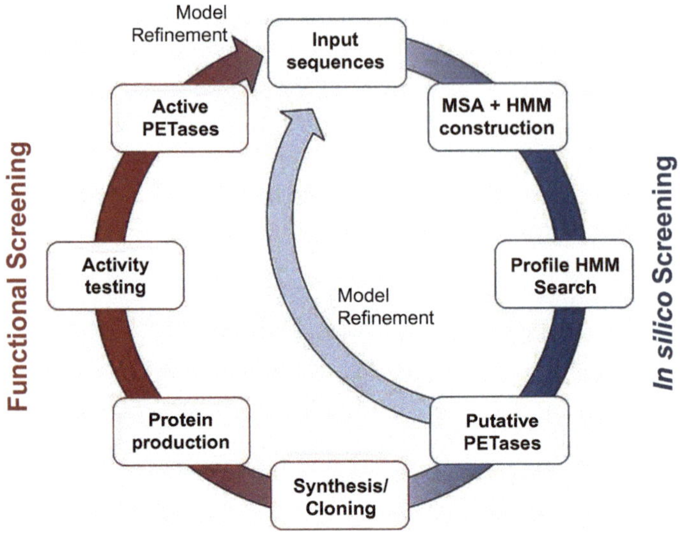

Fig. 3 Overall procedure of constructing the profile HMM and subsequent database search for PET hydrolases. The initial search rounds rely on previously described sequences as input. With subsequent functional testing of putative PET hydrolase sequences, the model can be refined and optimized. MSA: multiple sequence alignment.

Table 2 Protein sequences used for the initial construction of a PET hydrolase HMM.

No.	UniProt	Organism	Source
1.	W0TJ64	*Saccharomonospora viridis*	Kawai et al. (2014)
2.	E9LVI0	*Thermobifida fusca (Thermomonospora fusca)*	Dresler, van den Heuvel, Muller, and Deckwer (2006)
3.	E5BBQ3	*Thermobifida fusca (Thermomonospora fusca)*	Chen et al. (2008)
4.	D1A9G5	*Thermomonospora curvata*	Wei et al. (2014)
5.	E9LVH7	*Thermobifida alba*	Hu, Thumarat Zhang, Tang, and Kawai (2010)
6.	H6WX58	*Thermobifida halotolerans*	Ribitsch et al. (2012)
7.	E9LVH9	*Thermobifida celluloysilityca*	Acero et al. (2011)
8.	A0A0K8P6T7	*Ideonella sakaiensis*	Yoshida et al. (2016)
9.	G9BY57	uncultured bacterium	Sulaiman et al. (2012)

Note: The here included sequences belong to the first characterized enzymes. Meanwhile, novel PETases have been uncovered and their implementation has to be carefully evaluated with respect to activity, novelty of the sequences, phylogenetic distribution and other criteria (Danso et al., 2019; Danso et al., 2018).

performed with the software package T-Coffee. Its simplest usage is:
`$ t_coffee PETases.faa`

2. This command generates by default an alignment file ("PETases.aln") in Clustal format. A reformatting is needed to proceed with the next program in the pipeline. The T-Coffee package includes a tool for this purpose, which can be run as:
`$ t_coffee -other_pg seq_reformat -in PETases.aln -output fasta_aln > PETases_aln.fasta`
3. A profile HMM is generated with "hmmbuild," included in the HMMER software package:
`$ hmmbuild PETase.hmm PETases_aln.fasta`
4. A logo of the resulting "PETase.hmm" profile HMM can be visualized *via* the Skylign server (https://skylign.org/, Wheeler, Clements, & Finn, 2014, Fig. 4). It gives a fast, graphical overview of a profile HMM.
5. The program hmmsearch, also included in the HMMER software package, allows searching a protein profile HMM against a given protein sequence database in fasta format ("db.faa"; *e.g.*, the ORFs/CDSs of an organism/metagenome or the NCBI nr-database, see https://ftp.ncbi.nlm.nih.gov/). This can be achieved with the following command:
`$ hmmsearch -o PETase_hmm_out PETase.hmm db.faa`

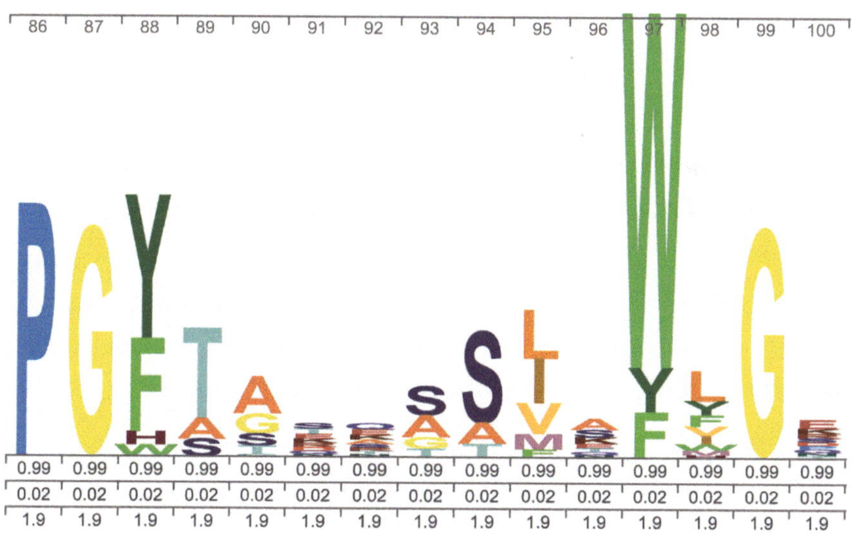

Fig. 4 PETase HMM logo (extract). Sequences used are those in Table 2. Stack height of each position is that position's information content. Positive-scoring (above background frequency) amino acids are shown, subdividing stack height according to estimated amino acid probability. Bottom numbers indicate occupancy, insert probability and insert length. Depicted are aligned amino acid positions 85–100 only.

6. The standard "hmmsearch" output ("PETase_hmm_out") includes a table with scores for complete sequences, domain annotation for each sequence (and alignments) and an internal pipeline statistics summary. A parseable table of per-sequence hits or per-domain hits can be created when setting the flags "--tblout <f>" or "--domtblout <f>", respectively. There are different tools to extract fasta sequences of desired hits from a protein database. Here we show how to extract (all) obtained hits with a combination of extended arguments (xargs) and the program faidx, included in the SAMtools software package. First, extract the first column of the parseable table (or the names of the wanted sequences only, according to a defined e-value or score threshold) to a file called "target_names.txt". Then, run the following command to extract the target sequences to a file called "hits_PETases.faa":

```
$ xargs samtools faidx db.faa < target_names.txt > hits_PETases.faa
```

The putative PETase sequences can be synthesized for the expression in a bacterial host (*e.g.*, *E. coli*). If available and for direct cloning purposes, the hit sequences can be also extracted from a fasta file containing the DNA sequences ("db.fna") coding for the proteins in "db.faa" with the abovementioned commands.

3.2 Functional screening on polymer-containing indicator plates

This part of the protocol requires several putative PETase-coding genes from Section 3.1 already synthesized/cloned in an expression vector and transformed into a suitable expression host like, *e.g.*, *E. coli*. The activity assay is based on the formation of a halo on indicator plates containing different dissolved polymers. The agar plates can be prepared either with PBS buffer or LB medium (supplemented with antibiotics and IPTG, 10–100 μg/mL, for induction of the gene of interest) and can contain polyethylene terephthalate (PET), polycaprolactone (PCL) or bis(2-hydroxyethyl) terephthalate (BHET). For the preparation of LB medium (10 g/L tryptone from casein, 5 g/L NaCl, 5 g/L yeast extract) and PBS buffer (137 mM NaCl, 2.7 mM KCl, 10 mM Na_2HPO_4, 1.8 mM KH_2PO_4) the required components are dissolved in H_2O. The pH of PBS buffer is adjusted with HCl to 7.0. Then, 1.5% (w/v, final concentration) agar is added and the solution is sterilized by autoclaving for 20 min at 121 °C. After autoclaving, the mixture is cooled down to 50–60 °C. The polymer solution described below is added as needed under constant stirring by using a magnetic stirrer. Finally, approximately 20 mL of solution per 92 mm petri dish are poured.

3.2.1 PET-containing plates

PET nanoparticles are prepared as previously published (Welzel, Müller, & Deckwer, 2002, Fig. 5):

1. Cut PET film into 5 mm circular pieces with the help of a paper punch
2. Dissolve 0.1 g of PET fragments in 10 mL 1,1,1,3,3,3-hexafluoro-2-propanol in a separatory funnel
3. Slowly drop the solution into a glass beaker filled with 100 mL pre-chilled H_2O under vigorous stirring
4. Remove the organic solvent from the mixture using a rotary evaporator
5. Add the complete solution to 200 mL 10 mM PBS buffer (pH 7.0) containing 1.5% (w/v) agar and mix by thorough stirring
6. Pour plates as described above

Nanoparticles can be used without size estimation. The final concentration of PET is approximately 0.3 mg/mL.

3.2.2 PCL-containing plates

1. Add 0.25 g PCL to 15 mL acetone and incubate at 60 °C until the polymer is completely dissolved
2. Transfer the mixture into 500 mL sterile LB medium or PBS buffer agar under constant stirring
3. Allow for complete evaporation of acetone under a fume hood
4. Add other supplements if needed

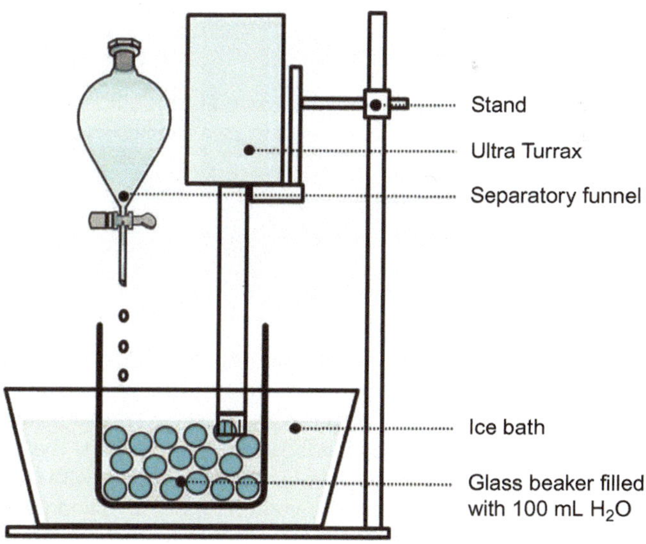

Fig. 5 Schematic diagram for the production of Polyethylene terephthalate (PET) nanoparticles. *Modified after Pütz, 2006.*

5. Pour the PCL-containing agar into petri dishes

Resulting PCL plates contain 500 mg/L of the substrate.

3.2.3 BHET-containing plates

1. Prepare a 1 M stock solution of BHET in DMSO
2. Heat the solution to 60 °C and mix by agitation or vortexing to dissolve the substrate
3. Add BHET solution to PBS/LB-agar to a final concentration of 5 mM, mix by stirring and pour plates directly

Figure 6 shows some results of the functional screening for PETases. The putative PETase-coding genes were synthesized in the expression vector

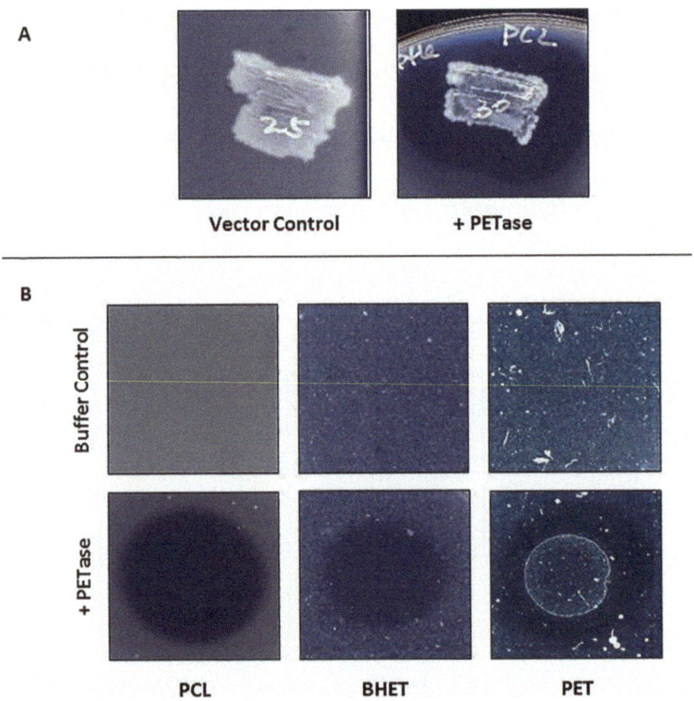

Fig. 6 Functional screenings of potential PETases. (A) Induction of potential PETase on the PCL indicator plates. A colony of *E. coli* BL21 (DE3) empty vector control (left) and a PETase expression strain (right) grown on the PCL containing LB agar plates. The formation of the clear halo indicates that the secreted protein can degrade the complex polymer PCL. LB plates contained IPTG (100 μg/mL). (B) Activity assay with His-tag purified PETases. Elution buffer was used as negative control (upper row) and the purified proteins (lower row) were spotted on a PCL, BHET, PET-nanoparticle containing PBS agar plate, respectively. Clear halo formation was observed.

pET21a(+) and transformed into the expression host *E. coli* BL21(DE3). For the expression, the colonies of an empty vector and a PETase-producing strain were streaked onto the PCL-containing LB plates (Fig. 6A). Additionally, the purified proteins were also tested on plates containing PCL, BHET and PET (Fig. 6B). After incubation of the plates over night at 37 °C formation of clear halos can be observed, which indicates hydrolytic activity. When assaying the activity of bacteria expressing putative PETases, allow for some of them to lysate in order to release the enzyme-containing cytosol by incubating the plates for longer time.

4. Notes

The example on how to perform a HMM-search includes only the basic commands. For more information on advanced features and customization of the default parameters and inclusion thresholds, run "package_name -h" for additional help or "man package_name" for displaying its manual. We highly recommend running the t_coffee MSA with the expresso mode (insert flag "-mode expresso"). This mode fetches the templates automatically by BLAST (requires internet connection and additional packages, see manual) and uses structural information of PDB structures whose similarity to the original sequence is higher than a specified cutoff (default is 30%). Note that "hmmsearch" also accepts a protein alignment as input. However, there are some advantages to building a profile HMM (see "hmmbuild -h"). We also highly recommend making use of multi-threading options when possible (see "package_name -h"), especially when handling large files. The program "esl-sfetch" included in the HMMER software package can also be used to extract fasta sequences from a database. You may want to check for the presence of signal peptides in your hits and remove them prior to gene synthesis. This can be performed with SignalP 5.0 either as a locally installed program or through their server version (Almagro Armenteros et al., 2019).

In the above mentioned examples in Section 3.2, the production of putative PETases was performed with the expression vector pET21a(+) in the host *E. coli* BL21 (DE3). Nevertheless, other expression vectors and hosts (*e.g.*, pET28a(+), *E. coli* T7 SHuffle) can also be used. In the example on how to prepare the indicator plates either PBS buffer or LB medium were used. We recommend using the PBS indicator plates to assay the activity of purified proteins and the LB medium indicator plates for the induction of putative genes in host cells.

5. Outlook and summary

In summary, our outlined method allows a quick and reliable identification of enzyme candidate genes in metagenome and in general in genome sequences. One should of course keep in mind that these candidates need to be verified experimentally using biochemical, structural and other sophisticated methods. Notably, the above-described protocol is certainly suitable to not only identifying enzymes linked to plastic degradation, but it will also be very useful tool for the identification of other enzymes of interest for biotechnology, biomedical applications and other fields of research. Using large sequence data sets and mining these is in general a very tedious and time-consuming approach. Within this framework the method described in this chapter offers perhaps a useful tool to speed up this approach and narrow the searches down to get faster access to the most promising genes and enzymes. HMM-motif-based searches—as outlined here—will further contribute toward developing a completely cell free screening pipeline by combining the *in-silico* searches with in *in vitro* technologies (Markel et al., 2020). Combining *in silico* screenings with high-throughput *in vitro* screening technologies will thereby effectively shorten the timeline between the sequence accumulation and identification and delivery of the most promising enzyme candidates.

References

Acero, E. H., Ribitsch, D., Steinkellner, G., Gruber, K., Greimel, K., Eiteljoerg, I., et al. (2011). Enzymatic surface hydrolysis of PET: Effect of structural diversity on kinetic properties of Cutinases from Thermobifida. *Macromolecules*, *44*(12), 4632–4640. https://doi.org/10.1021/ma200949p.

Almagro Armenteros, J. J., Tsirigos, K. D., Sønderby, C. K., Petersen, T. N., Winther, O., Brunak, S., et al. (2019). SignalP 5.0 improves signal peptide predictions using deep neural networks. *Nature Biotechnology*, *37*(4), 420–423. https://doi.org/10.1038/s41587-019-0036-z.

Alves, N. M., Mano, J. F., Balaguer, E., Duenas, J. M. M., & Ribelles, J. L. G. (2002). Glass transition and structural relaxation in semi-crystalline poly(ethylene terephthalate): a DSC study. *Polymer*, *43*(15), 4111–4122. https://doi.org/10.1016/S0032-3861(02)00236-7.

Amaral-Zettler, L. A., Zettler, E. R., & Mincer, T. J. (2020). Ecology of the plastisphere. *Nature Reviews. Microbiology*, *18*(3), 139–151. https://doi.org/10.1038/s41579-019-0308-0.

Awaja, F., & Pavel, D. (2005). Recycling of PET. *European Polymer Journal*, *41*(7), 1453–1477. https://doi.org/10.1016/j.eurpolymj.2005.02.005.

Barnes, D. K. A., Galgani, F., Thompson, R. C., & Barlaz, M. (2009). Accumulation and fragmentation of plastic debris in global environments. *Philosophical Transactions of the Royal Society B: Biological Sciences*, *364*(1526), 1985–1998. https://doi.org/10.1098/rstb.2008.0205.

Barth, M., Honak, A., Oeser, T., Wei, R., Belisario-Ferrari, M. R., Then, J., et al. (2016). A dual enzyme system composed of a polyester hydrolase and a carboxylesterase enhances the biocatalytic degradation of polyethylene terephthalate films. *Biotechnology Journal, 11*(8), 1082–1087. https://doi.org/10.1002/biot.201600008.

Barth, M., Oeser, T., Wei, R., Then, J., Schmidt, J., & Zimmermann, W. (2015). Effect of hydrolysis products on the enzymatic degradation of polyethylene terephthalate nanoparticles by a polyester hydrolase from Thermobifida fusca. *Biochemical Engineering Journal, 93*, 222–228. https://doi.org/10.1016/j.bej.2014.10.012.

Bertani, G. (1951). Studies on lysogenesis. 1. The mode of phage liberation by lysogenic Escherichia-Coli. *Journal of Bacteriology, 62*(3), 293–300. Retrieved from <Go to ISI>://WOS:A1951UK07700004 http://jb.asm.org/content/62/3/293.full.pdf.

Billig, S., Oeser, T., Birkemeyer, C., & Zimmermann, W. (2010). Hydrolysis of cyclic poly(ethylene terephthalate) trimers by a carboxylesterase from Thermobifida fusca KW3. *Applied Microbiology and Biotechnology, 87*(5), 1753–1764. https://doi.org/10.1007/s00253-010-2635-y.

Borrelli, G. M., & Trono, D. (2015). Recombinant lipases and phospholipases and their use as biocatalysts for industrial applications. *International Journal of Molecular Sciences, 16*(9), 20774–20840. https://doi.org/10.3390/ijms160920774.

Carniel, A., Valoni, E., Nicomedes, J., Gomes, A. D., & de Castro, A. M. (2017). Lipase from Candida antarctica (CALB) and cutinase from Humicola insolens act synergistically for PET hydrolysis to terephthalic acid. *Process Biochemistry, 59*, 84–90. https://doi.org/10.1016/j.procbio.2016.07.023.

Charles, T. C., Liles, M. R., & Sessitsch, A. (2017). *Functional metagenomics: Tools and applications*. Springer.

Chen, S., Tong, X., Woodard, R. W., Du, G. C., Wu, J., & Chen, J. (2008). Identification and characterization of bacterial cutinase. *Journal of Biological Chemistry, 283*(38), 25854–25862. https://doi.org/10.1074/jbc.M800848200.

Cooper, D. A., & Corcoran, P. L. (2010). Effects of mechanical and chemical processes on the degradation of plastic beach debris on the island of Kauai, Hawaii. *Marine Pollution Bulletin, 60*(5), 650–654. https://doi.org/10.1016/j.marpolbul.2009.12.026.

Danso, D., Chow, J., & Streit, W. R. (2019). Plastics: Microbial degradation, environmental and biotechnological perspectives. *Applied and Environmental Microbiology*. https://doi.org/10.1128/AEM.01095-19.

Danso, D., Schmeisser, C., Chow, J., Zimmermann, W., Wei, R., Leggewie, C., et al. (2018). New insights into the function and global distribution of polyethylene terephthalate (PET)-degrading Bacteria and enzymes in marine and terrestrial metagenomes. *Applied and Environmental Microbiology, 84*(8). https://doi.org/10.1128/aem.02773-17.

David, L. O., & Paul, D. C. (2009). a/b-hydrolase fold: An update. *Protein Peptide Lett., 16*(10), 1137–1148. https://doi.org/10.2174/092986609789071298.

Derraik, J. G. (2002). The pollution of the marine environment by plastic debris: A review. *Marine Pollution Bulletin, 44*(9), 842–852. Retrieved from http://www.ncbi.nlm.nih.gov/pubmed/12405208. https://www.sciencedirect.com/science/article/pii/S0025326X02002205?via%3Dihub.

Dresler, K., van den Heuvel, J., Muller, R. J., & Deckwer, W. D. (2006). Production of a recombinant polyester-cleaving hydrolase from Thermobifida fusca in Escherichia coli. *Bioprocess and Biosystems Engineering, 29*(3), 169–183. https://doi.org/10.1007/s00449-006-0069-9.

Eberl, A., Heumann, S., Bruckner, T., Araujo, R., Cavaco-Paulo, A., Kaufmann, F., et al. (2009). Enzymatic surface hydrolysis of poly(ethylene terephthalate) and bis(benzoyloxyethyl) terephthalate by lipase and cutinase in the presence of surface active molecules. *Journal of Biotechnology, 143*(3), 207–212. https://doi.org/10.1016/j.jbiotec.2009.07.008.

Eddy, S. R. (2011). Accelerated profile HMM searches. *PLoS Computational Biology*, 7(10), e1002195. https://doi.org/10.1371/journal.pcbi.1002195.

Eriksen, M., Lebreton, L. C. M., Carson, H. S., Thiel, M., Moore, C. J., Borerro, J. C., et al. (2014). Plastic pollution in the World's oceans: More than 5 trillion plastic pieces weighing over 250,000 tons afloat at sea. *PLoS One*, *9*(12). doi:ARTN e111913.

Espino-Rammer, L., Ribitsch, D., Przylucka, A., Marold, A., Greimel, K. J., Acero, E. H., et al. (2013). Two novel class II Hydrophobins from Trichoderma spp. stimulate enzymatic hydrolysis of poly(ethylene terephthalate) when expressed as fusion proteins. *Applied and Environmental Microbiology*, *79*(14), 4230–4238. https://doi.org/10.1128/Aem.01132-13.

Fagerburg, D. R., & Clauberg, H. (2004). Photodegradation of Poly(Ethylene Terephthalate) and Poly(Ethylene/1,4-Cyclohexylenedimethylene Terephthalate). In *Modern Polyesters: Chemistry and Technology of Polyesters and Copolyesters* (pp. 609–641). John Wiley & Sons, Ltd.

Falkenstein, P., Gräsing, D., Bielytskyi, P., Zimmermann, W., Matysik, J., Wei, R., et al. (2020). UV Pretreatment impairs the enzymatic degradation of polyethylene terephthalate. *Frontiers in Microbiology*, *11*(689). https://doi.org/10.3389/fmicb.2020.00689.

Ferrer, M., Martínez-Martínez, M., Bargiela, R., Streit, W. R., Golyshina, O. V., & Golyshin, P. N. (2016). Estimating the success of enzyme bioprospecting through metagenomics: Current status and future trends. *Microbial Biotechnology*, *9*(1), 22–34.

Finn, R. D., Clements, J., & Eddy, S. R. (2011). HMMER web server: Interactive sequence similarity searching. *Nucleic Acids Research*, *39*, W29–W37. https://doi.org/10.1093/nar/gkr367.

Geyer, R., Jambeck, J. R., & Law, K. L. (2017). Production, use, and fate of all plastics ever made. *Science Advances*, *3*(7). https://doi.org/10.1126/sciadv.1700782. ARTN e1700782.

Guebitz, G. M., & Cavaco-Paulo, A. (2008). Enzymes go big: Surface hydrolysis and functionalisation of synthetic polymers. *Trends in Biotechnology*, *26*(1), 32–38. https://doi.org/10.1016/j.tibtech.2007.10.003.

Hasan, F., Shah, A. A., & Hameed, A. (2006). Industrial applications of microbial lipases. *Enzyme and Microbial Technology*, *39*(2), 235–251. https://doi.org/10.1016/j.enzmictec.2005.10.016.

Houde, A., Kademi, A., & Leblanc, D. (2004). Lipases and their industrial applications - An overview. *Applied Biochemistry and Biotechnology*, *118*(1–3), 155–170. https://doi.org/10.1385/Abab:118:1-3:155.

Hu, X., Thumarat, U., Zhang, X., Tang, M., & Kawai, F. (2010). Diversity of polyester-degrading bacteria in compost and molecular analysis of a thermoactive esterase from Thermobifida alba AHK119. *Applied Microbiology and Biotechnology*, *87*(2), 771–779. https://doi.org/10.1007/s00253-010-2555-x.

Hug, L. A., Baker, B. J., Anantharaman, K., Brown, C. T., Probst, A. J., Castelle, C. J., et al. (2016). A new view of the tree of life. *Nature Microbiology*, *1*(5), 16048. https://doi.org/10.1038/Nmicrobiol.2016.48.

Jambeck, J. R., Geyer, R., Wilcox, C., Siegler, T. R., Perryman, M., Andrady, A., et al. (2015). Plastic waste inputs from land into the ocean. *Science*, *347*(6223), 768–771. https://doi.org/10.1126/science.1260352.

Kale, S. K., Deshmukh, A. G., Dudhare, M. S., & Patil, V. B. (2015). Microbial degradation of plastic: a review. *Journal of Biochemical Technology*, *6*(2), 952–961. Retrieved from < Go to ISI >://WOS:000382960000004.

Kawai, F., Oda, M., Tamashiro, T., Waku, T., Tanaka, N., Yamamoto, M., et al. (2014). A novel Ca2 +−activated, thermostabilized polyesterase capable of hydrolyzing polyethylene terephthalate from *Saccharomonospora viridis* AHK190. *Applied Microbiology and Biotechnology*, *98*(24), 10053–10064. https://doi.org/10.1007/s00253-014-5860-y.

Khan, F. I., Lan, D., Durrani, R., Huan, W., Zhao, Z., & Wang, Y. (2017). The lid domain in lipases: Structural and functional determinant of enzymatic properties. *Frontiers in Bioengineering and Biotechnology*, *5*, 16. https://doi.org/10.3389/fbioe.2017.00016.

Kitadokoro, K., Thumarat, U., Nakamura, R., Nishimura, K., Karatani, H., Suzuki, H., et al. (2012). Crystal structure of cutinase Est119 from *Thermobifida alba* AHK119 that can degrade modified polyethylene terephthalate at 1.76 angstrom resolution. *Polymer Degradation and Stability*, *97*(5), 771–775. https://doi.org/10.1016/j.polymdegradstab.2012.02.003.

Li, H., Handsaker, B., Wysoker, A., Fennell, T., Ruan, J., Homer, N., et al. (2009). The sequence alignment/map format and SAMtools. *Bioinformatics*, *25*(16), 2078–2079. https://doi.org/10.1093/bioinformatics/btp352.

Li, S., Yang, X., Yang, S., Zhu, M., & Wang, X. (2012). Technology prospecting on enzymes: Application, marketing and engineering. *Computational and Structural Biotechnology Journal*, *2*, e201209017. https://doi.org/10.5936/csbj.201209017.

Lim, H. C. A. (2017). Chapter 20 - Thermoplastic polyesters A2 - Gilbert, marianne. In *Brydson's Plastics Materials (Eighth Edition)* (pp. 527–543). Butterworth-Heinemann.

Markel, U., Essani, K. D., Besirlioglu, V., Schiffels, J., Streit, W. R., & Schwaneberg, U. (2020). Advances in ultrahigh-throughput screening for directed enzyme evolution. *Chemical Society Reviews*, *49*(1), 233–262. https://doi.org/10.1039/c8cs00981c.

Methe, B. A., Nelson, K. E., Pop, M., Creasy, H. H., Giglio, M. G., Huttenhower, C., et al. (2012). A framework for human microbiome research. *Nature*, *486*(7402), 215–221. https://doi.org/10.1038/nature11209.

Miyakawa, T., Mizushima, H., Ohtsuka, J., Oda, M., Kawai, F., & Tanokura, M. (2015). Structural basis for the Ca2+−enhanced thermostability and activity of PET-degrading cutinase-like enzyme from *Saccharomonospora viridis* AHK190. *Applied Microbiology and Biotechnology*, *99*(10), 4297–4307. https://doi.org/10.1007/s00253-014-6272-8.

Mueller, R. J. (2006). Biological degradation of synthetic polyesters - enzymes as potential catalysts for polyester recycling. *Process Biochemistry*, *41*(10), 2124–2128. https://doi.org/10.1016/j.procbio.2006.05.018.

Notredame, C., Higgins, D. G., & Heringa, J. (2000). T-coffee: A novel method for fast and accurate multiple sequence alignment. *Journal of Molecular Biology*, *302*(1), 205–217. https://doi.org/10.1006/jmbi.2000.4042.

Ollis, D. L., Cheah, E., Cygler, M., Dijkstra, B., Frolow, F., Franken, S. M., et al. (1992). The alpha/beta hydrolase fold. *Protein Engineering*, *5*(3), 197–211. Retrieved from http://www.ncbi.nlm.nih.gov/pubmed/1409539. http://peds.oxfordjournals.org/content/5/3/197.full.pdf.

Panda, T., & Gowrishankar, B. S. (2005). Production and applications of esterases. *Applied Microbiology and Biotechnology*, *67*(2), 160–169. https://doi.org/10.1007/s00253-004-1840-y.

Parikh, M., Gross, R. A., & Mccarthy, S. P. (1992). The effect of crystalline morphology on enzymatic degradation kinetics. *Abstracts of Papers of the American Chemical Society*, *203*, 205. Retrieved from <Go to ISI>://WOS:A1992HK16300827.

Pettersson, E., Lundeberg, J., & Ahmadian, A. (2009). Generations of sequencing technologies. *Genomics*, *93*(2), 105–111. https://doi.org/10.1016/j.ygeno.2008.10.003.

Pio, T. F., & Macedo, G. A. (2009). Cutinases: Properties and industrial applications. *Advances in Applied Microbiology*, *66*(66), 77–95. https://doi.org/10.1016/S0065-2164(08)00804-6.

PlasticsEurope. (2020). *Plastics – The Facts 2020*. https://www.plasticseurope.org/de/resources/publications/4312-plastics-facts-2020.

Pütz, A. (2006). Isolierung, Identifizierung und biochemische Charakterisierung Dialkylphthalat spaltender Esterasen. In *PhD PhD thesis*. Düsseldorf: Heinrich-Heine-Universität Düsseldorf.

Rex, W. J., & Tennant, D. J. (1949). *Polymeric linear terephthalic esters*. Google Patents.
Ribitsch, D., Acero, E. H., Greimel, K., Dellacher, A., Zitzenbacher, S., Marold, A., et al. (2012). A new esterase from *Thermobifida halotolerans* hydrolyses polyethylene terephthalate (PET) and Polylactic acid (PLA). *Polymers*, *4*(1), 617–629. https://doi.org/10.3390/polym4010617.
Ribitsch, D., Acero, E. H., Przylucka, A., Zitzenbacher, S., Marold, A., Gamerith, C., et al. (2015). Enhanced Cutinase-catalyzed hydrolysis of polyethylene terephthalate by covalent fusion to Hydrophobins. *Applied and Environmental Microbiology*, *81*(11), 3586–3592. https://doi.org/10.1128/Aem.04111-14.
Ribitsch, D., Yebra, A. O., Zitzenbacher, S., Wu, J., Nowitsch, S., Steinkellner, G., et al. (2013). Fusion of binding domains to *Thermobifida cellulosilytica* Cutinase to tune sorption characteristics and enhancing PET hydrolysis. *Biomacromolecules*, *14*(6), 1769–1776. https://doi.org/10.1021/bm400140u.
Rieckmann, T., & Völker, S. (2004). Poly(Ethylene Terephthalate) polymerization – Mechanism, catalysis, kinetics, mass transfer and reactor design. In *Modern Polyesters: Chemistry and Technology of Polyesters and Copolyesters* (pp. 29–115). John Wiley & Sons, Ltd.
Ronkvist, A. M., Xie, W. C., Lu, W. H., & Gross, R. A. (2009). Cutinase-catalyzed hydrolysis of poly(ethylene terephthalate). *Macromolecules*, *42*(14), 5128–5138. https://doi.org/10.1021/ma9005318.
Roth, C., Wei, R., Oeser, T., Then, J., Follner, C., Zimmermann, W., et al. (2014). Structural and functional studies on a thermostable polyethylene terephthalate degrading hydrolase from *Thermobifida fusca*. *Applied Microbiology and Biotechnology*, *98*(18), 7815–7823. https://doi.org/10.1007/s00253-014-5672-0.
Schmidt, J., Wei, R., Oeser, T., Belisario-Ferrari, M. R., Barth, M., Then, J., et al. (2016). Effect of Tris, MOPS, and phosphate buffers on the hydrolysis of polyethylene terephthalate films by polyester hydrolases. *FEBS Open Bio*, *6*(9), 919–927. https://doi.org/10.1002/2211-5463.12097.
Shaw, D. G., & Day, R. H. (1994). Colour- and form-dependent loss of plastic micro-debris from the North Pacific Ocean. *Marine Pollution Bulletin*, *28*(1), 39–43. https://doi.org/10.1016/0025-326X(94)90184-8.
Singh, R., Kumar, M., Mittal, A., & Mehta, P. K. (2016). Microbial enzymes: Industrial progress in 21st century. *3 Biotech*, *6*(2), 174. https://doi.org/10.1007/s13205-016-0485-8.
Steele, H. L., Jaeger, K.-E., Daniel, R., & Streit, W. R. (2009). Advances in recovery of novel biocatalysts from metagenomes. *Journal of Molecular Microbiology and Biotechnology*, *16*(1–2), 25–37. Retrieved from https://www.karger.com/Article/Pdf/142892.
Streit, W. R., & Schmitz, R. A. (2004). Metagenomics - the key to the uncultured microbes. *Current Opinion in Microbiology*, 7(5), 492–498. https://doi.org/10.1016/j.mib.2004.08.002.
Sulaiman, S., Yamato, S., Kanaya, E., Kim, J. J., Koga, Y., Takano, K., et al. (2012). Isolation of a novel cutinase homolog with polyethylene terephthalate-degrading activity from leaf-branch compost by using a metagenomic approach. *Applied and Environmental Microbiology*, *78*(5), 1556–1562. https://doi.org/10.1128/Aem.06725-11.
Sulaiman, S., You, D. J., Kanaya, E., Koga, Y., & Kanaya, S. (2014). Crystal structure and thermodynamic and kinetic stability of metagenome-derived LC-Cutinase. *Biochemistry*, *53*(11), 1858–1869. https://doi.org/10.1021/bi401561p.
Then, J., Wei, R., Oeser, T., Gerdts, A., Schmidt, J., Barth, M., et al. (2016). A disulfide bridge in the calcium binding site of a polyester hydrolase increases its thermal stability and activity against polyethylene terephthalate. *FEBS Open Bio*, *6*(5), 425–432. https://doi.org/10.1002/2211-5463.12053.
Thompson, L. R., Sanders, J. G., McDonald, D., Amir, A., Ladau, J., Locey, K. J., et al. (2017). A communal catalogue reveals Earth's multiscale microbial diversity. *Nature*, *551*(7681), 457. https://doi.org/10.1038/nature24621.

Thumarat, U., Nakamura, R., Kawabata, T., Suzuki, H., & Kawai, F. (2012). Biochemical and genetic analysis of a cutinase-type polyesterase from a thermophilic Thermobifida alba AHK119. *Applied Microbiology and Biotechnology*, *95*(2), 419–430. https://doi.org/10.1007/s00253-011-3781-6.

Venter, J. C., Remington, K., Heidelberg, J. F., Halpern, A. L., Rusch, D., Eisen, J. A., et al. (2004). Environmental genome shotgun sequencing of the Sargasso Sea. *Science*, *304*(5667), 66–74. https://doi.org/10.1126/science.1093857.

Vertommen, M. A. M. E., Nierstrasz, V. A., van der Veer, M., & Warmoeskerken, M. M. C. G. (2005). Enzymatic surface modification of poly(ethylene terephthalate). *Journal of Biotechnology*, *120*(4), 376–386. https://doi.org/10.1016/j.jbiotec.2005.06.015.

Wallace, N. E., Adams, M. C., Chafin, A. C., Jones, D. D., Tsui, C. L., & Gruber, T. D. (2020). The highly crystalline PET found in plastic water bottles does not support the growth of the PETase-producing bacterium Ideonella sakaiensis. *Environmental microbiology Reports*, *12*(5), 578–582.

Webb, H. K., Arnott, J., Crawford, R. J., & Ivanova, E. P. (2013). Plastic degradation and its environmental implications with special reference to Poly(ethylene terephthalate). *Polymers*, *5*(1), 1–18. https://doi.org/10.3390/Polym5010001.

Wei, R., Breite, D., Song, C., Gräsing, D., Ploss, T., Hille, P., et al. (2019). Biocatalytic degradation efficiency of postconsumer polyethylene terephthalate packaging determined by their polymer microstructures. *Advanced Science (Weinheim, Baden-Wurttemberg, Germany)*, *6*(14), 1900491. https://doi.org/10.1002/advs.201900491.

Wei, R., Oeser, T., Then, J., Kuhn, N., Barth, M., Schmidt, J., et al. (2014). Functional characterization and structural modeling of synthetic polyester-degrading hydrolases from *Thermomonospora curvata*. *AMB Express*, *4*, 44. https://doi.org/10.1186/s13568-014-0044-9.

Wei, R., Oeser, T., & Zimmermann, W. (2014). Synthetic polyester-Hydrolyzing enzymes from thermophilic Actinomycetes. *Advances in Applied Microbiology*, *89*(89), 267–305. https://doi.org/10.1016/B978-0-12-800259-9.00007-X.

Wei, R., & Zimmermann, W. (2017). Biocatalysis as a green route for recycling the recalcitrant plastic polyethylene terephthalate. *Microbial Biotechnology*, *10*(6), 1302–1307.

Welzel, K., Müller, R. J., & Deckwer, W. D. (2002). Enzymatischer Abbau von Polyester-Nanopartikeln. *Chemie Ingenieur Technik*, *74*(10), 1496–1500. https://doi.org/10.1002/1522-2640(20021015)74:10<1496::AID-CITE1496>3.0.CO;2-P.

Wheeler, T. J., Clements, J., & Finn, R. D. (2014). Skylign: A tool for creating informative, interactive logos representing sequence alignments and profile hidden Markov models. *BMC Bioinformatics*, *15*(1), 7.

Wilkes, R. A., & Aristilde, L. (2017). Degradation and metabolism of synthetic plastics and associated products by Pseudomonas sp.: Capabilities and challenges. *Journal of Applied Microbiology*, *123*(3), 582–593. https://doi.org/10.1111/jam.13472.

Yoshida, S., Hiraga, K., Takehana, T., Taniguchi, I., Yamaji, H., Maeda, Y., et al. (2016). A bacterium that degrades and assimilates poly(ethylene terephthalate). *Science*, *351*(6278), 1196–1199. https://doi.org/10.1126/science.aad6359.

Zettler, E. R., Mincer, T. J., & Amaral-Zettler, L. A. (2013). Life in the "plastisphere": Microbial communities on plastic marine debris. *Environmental Science & Technology*, *47*(13), 7137–7146. https://doi.org/10.1021/es401288x.

CHAPTER EIGHT

Cutinases from thermophilic bacteria (actinomycetes): From identification to functional and structural characterization

Masayuki Oda[a], Nobutaka Numoto[b], Gert-Jan Bekker[c], Narutoshi Kamiya[d], and Fusako Kawai[a,*]

[a]Graduate School of Life and Environmental Sciences, Kyoto Prefectural University, Kyoto, Japan
[b]Medical Research Institute, Tokyo Medical and Dental University (TMDU), Bunkyo-ku, Tokyo, Japan
[c]Institute for Protein Research, Osaka University, Suita, Osaka, Japan
[d]Graduate School of Simulation Studies, University of Hyogo, Kobe, Japan
*Corresponding author: e-mail address: fkawai@kit.ac.jp

Contents

Methods in Enzymology, Volume 648
ISSN 0076-6879
https://doi.org/10.1016/bs.mie.2020.12.031

Abstract

Thermophilic cutinases are mainly obtained from thermophilic actinomycetes, and are categorized into two groups, i.e., those with higher (>70 °C) or lower (<70 °C) thermostabilities. The thermostabilities of cutinases are highly relevant to their ability to degrade polyethylene terephthalate (PET). Many crystal structures of thermophilic cutinases have been solved, showing that their overall backbone structures are identical, irrespective of their ability to hydrolyze PET. One of the unique properties of cutinases is that metal ion-binding on the enzyme's surface both elevates their melting temperatures and activates the enzyme. In this chapter, we introduce the methodology for the identification and cloning of thermophilic cutinases from actinomycetes. For detailed characterization of cutinases, we describe the approach to analyze the intricate dynamics of the enzyme, based on its crystal structures complexed with metal ions and model substrates using a combination of experimental and computational techniques.

1. Introduction

The global plastic production reached 359 million tons in 2018, among which the largest production is in Asia (51%; 30% in China, 4% in Japan and 17% in the rest of Asia) (PlasticsEurope, 2019). Currently, while most of the collected plastic waste is still either burned or deposited in landfills, non-collected waste ends up in nature, due to littering and unauthorized dumping, causing serious environmental issues, e.g., marine pollution and microplastics. The improvement of current recycling methods is a key issue for a circular economy in the future.

Enzymatic recycling is a branch of chemical recycling and suitable for hydrolysable plastics, such as polyesters, among which polyethylene terephthalate (PET) occupies the majority. The annual demand of PET amounted to approximately 5% of the total plastic production, which was mainly used for packaging including PET bottles, and fibers.

To depolymerize PET polymer to yield the building blocks, two important features of PET hydrolases are required. First, a high substrate specificity towards the bulky aromatic ring structure of PET and second, a temperature optimum and thermostability of over 70 °C, which corresponds to the glass transition temperature of PET, are required to access the PET chains with increased flexibility, thereby enabling highly efficient PET hydrolysis. Given these requirements, four enzymes belonging to the cutinase group (EC 3.1.1.74) have thus far been identified as suitable candidates to be used as PET hydrolases (Kawai, Kawabata, & Oda, 2019, 2020; Wei & Zimmermann, 2017).

2. Identification of thermophilic cutinases

2.1 Identification and cloning of thermophilic cutinases from actinomycetes

PET hydrolase was first cloned from *Thermobifida fusca* (Müller, Schrader, Profe, Dresler, & Deckwer, 2005). Since then, extensive research on polyesterases from *T. fusca* and other actinomycetes has revealed that all the enzymes cloned from these actinomycetes belong to the cutinase group (EC 3.1.1.74), which is a subgroup of carboxyl ester hydrolases (EC 3.1.1). Cutinases were originally identified as enzymes able to degrade cutin that is part of the cuticular layer in leaves, or the suberin part of tree bark, shown by research on cutinases that began with phytopathogenic fungi such as *Fusarium solani pisi* (Heredia, 2003; Kollattukudy, 1981). Later, it was found that cutinases are able to hydrolyze polyesters (Baker, Poultney, Liu, Gross, & Montclare, 2012; Murphy, Cameron, Huang, & Vinopal, 1996).

To date, the majority of esterases with promiscuous polymer degrading activity have been identified as cutinases. Carboxyl ester hydrolases (EC 3.1.1) are members of the serine hydrolase group that share the same catalytic triad consisting of Ser-His-Asp. Lipases (EC 3.1.1.3) form the largest subgroup under EC 3.1.1 and possess a lid that covers their active site, which is a prerequisite for the interfacial activation caused by interaction with lipids, but such a lid would limit the access of a macromolecule to the active site. In contrast, cutinases (EC 3.1.1.74) have no lids and share a common structure with an open and shallow active site, which enables the accommodation of a bulky polyester chain segment, such as PET, and facilitating consequently the access of the active serine to the target ester bond. Thermophilic actinomycetes have been isolated as habitants in compost, such as several *Thermobifida* species, which are sources for various thermophilic cutinases.

All actinomycete cutinases identified so far are highly homologous enzymes with approximately the same molecular mass in the range of 28–30 kDa (Kawai et al., 2019, 2020). As described above, cutinases are members of the serine hydrolase family that share the catalytic triad Ser-His-Asp with the catalytic serine residue in the consensus sequence motif (G-X-S-X-G). PCR amplification of a cutinase gene can thus be performed using primer pairs designed for this consensus sequence. An expression vector that enables the incorporation of a His-tag either at the N- or C-terminus is generally used to allow the purification of the recombinant enzyme via immobilized metal affinity chromatography (IMAC) (Kawai et al., 2014; Roth et al., 2014;

Thumarat, Nakamura, Kawabata, Suzuki, & Kawai, 2012). The plasmid harboring the cutinase gene is then transformed into a host strain such as *Escherichia coli* BL21, although recombinant expression of actinomycete cutinases has been also reported in other hosts than *E. coli*. Table 1 shows several published examples of vectors and hosts that can be used for this purpose.

2.2 Purification and characterization of polyester-hydrolyzing cutinases

Actinomycete cutinases are in general recombinantly expressed with a His-tag and initially purified from a cell-free extract with an IMAC column, where the purity of protein elution is then confirmed by SDS-PAGE. When carefully purified on an IMAC column using cell lysate containing over-expressed cutinases, more than 90% purity can be obtained via a single step of purification, but if necessary, further purification such as ion exchange chromatography and size exclusion chromatography can be performed. For characterization of a purified cutinase, activity measurements with different substates are essential as detailedly described below in Protocol 4.3.

3. Structural and thermodynamic analysis of PET-hydrolyzing cutinases

3.1 Crystal structures of thermophilic cutinases

Many crystal structures of thermophilic cutinases have been solved to date (Table 2), which can be categorized into two groups based on their temperature optima and thermostabilities, but they share the same approximate molecular mass (28–30 kDa) and the classical α/β hydrolase fold, which is comprised of nine centrally located twisted β-strands and 8 α-helices that are located on both sides of the β-sheet structure.

Differences in their temperature optima and thermostabilities are relevant to their catalytic efficiency in PET hydrolysis. For several cutinases, the beneficial effects of divalent ions such as Ca^{2+} and Mg^{2+} on their thermostability and activity against PET have been reported (Sulaiman et al., 2014; Then et al., 2015; Thumarat et al., 2012, 2015). Est119 from *T. alba* retains the same overall structure with or without Ca^{2+} (PDB IDs: 3WYN and 3VIS). Also, cutinase 1 or 2 from *T. cellulosilytica* (Thc_Cut1 and Thc_Cut2) and Thc_Cut2 mutants retain the same overall structures independent of the binding of either Mg^{2+}/Ca^{2+} (PDB IDs: 5LUI, 5LUJ, 5LUK, and 5LUL). Exceptionally Cut190S226P from *Saccharomonospora viridis* AHK190 revealed uniquely either an active or inactive form depending on the binding of Ca^{2+}

Table 1 Vectors and hosts used for cloning and recombinant expression of thermophilic cutinases.

Gene	Origin	Vector	Host	References
Expression in *E. coli*				
Tfu_882 Tfu_883	*Thermobifida fusca* (laboratory stock)	pET20b(+)	*E. coli* BL21 (DE3)	Chen et al. (2008) Su, Woodard, Chen, and Wu (2013)
Cut1 Cut2	*T. fusca* NRRL B-8184	pET22b(+)	*E. coli* BL21 (DE3)	Hedge and Veeranki (2013)
TfCut2	*T. fusca* KW3 (DSM6013)	pET20b(+)	*E. coli* BL21 (DE3)	Roth et al. (2014)
Tha_Cut1	*T. alba* DSM43185	pET26b(+)	*E. coli* BL21-Gold (DE3)	Ribitsch et al. (2012)
Est119 Est1	*T. alba* AHK119	pQE80L	*E. coli* Rosseta-gami B (DE3)	Thumarat et al. (2012) and Thumarat et al. (2015)
Thc_cut1, Thc_cut2	*T. cellulosilytica* DSM44535	pET26b(+)	*E. coli* BL21-Gold (DE3)	Herrero Aceto et al. (2011)
Tcur1278, Tcur0390	*T. curvata* DSM43138	pBAD TOPO	*E. coli* Top10	Wei et al. (2014)
LC-cutinase	Metagenome from leaf-branch compost	pET25b	*E. coli* BL21-CodonPlus (DE3)-RP	Sulaiman et al. (2012)
		pET20b(+)	*E. coli* BL21 (DE3)	Schmidt et al. (2016)
		pET26b	*E. coli* BL21 (DE3)	Tournier et al. (2020)
Cut190	*Saccharomonospora viridis* AHK190	pQE80L	*E. coli* Rosseta-gami B (DE3)	Kawai et al. (2014)
Expression in other than *E. coli*				
BTA-1	*T. fusca* DSM43793	pMM1525	*Bacillus megaterium* MS941 and WH323	Yang, Malten, Grote, Jahn, and Deckwer (2007)
BTA-1	*Thermobifida* sp. BCC23166	pIJ8600	*Streptomyces rimosus* R7	Sinsereekul et al. (2010)
LC-cutinase	Metagenome from leaf-branch compost	pPJ912	*P. pastoris* (ATCC76273)	Shirke et al. (2018)
TfCut2	*T. fusca* KW3 (DSM6013)	pDAM	*B. subtilis* strain RH11496	Wei et al. (2019)

Table 2 Crystallographic structures of actinomycete cutinases.

Enzyme	Source	PDB ID	References
Lipase[a]	*Streptomyces exfoliatus*	1JFR	Wei et al. (1998)
TfCut2	*Thermobifida fusca* KW3	4CG1, 4CG2, 4CG3 (complexed with a serine inhibitor)	Roth et al. (2014)
Est119	*Thermobifida alba* AHK119	3VIS, 3WYN (complexed with Ca^{2+}) 6AID (complexed with Ca^{2+}, ethyl lactate and lactate)	Kitadokoro et al. (2012) Kitadokoro et al. (2019)
Thc_Cut1 and Thc_Cut2	*Thermobifida cellusilytica* DSM44535	5LUI, 5LUJ, 5LUK, 5LUL (complexed with Ca^{2+}/Mg^{2+})	Ribitsch et al. (2017)
LC-cutinase	Metagenome from leaf-compost	4EB0	Sulaiman, You, Kanaya, Koga, and Kanaya (2014)
LC-cutinase mutants	Metagenome from leaf-compost	6THT, 6THS	Tournier et al. (2020)
Cut190 mutant, Cut190^{S226P}	*Saccharomonospora viridis* AHK190	4WFI, 4WFJ, 4WFK	Miyakawa et al. (2015)
Cut190 mutants, Cut190$^{S226P/R228S}$S176A (Cut190*S176A)	*Saccharomonospora viridis* AHK190	5ZNO, 5ZRQ, 5ZRR, 5ZRS (complexed with Ca^{2+}/Zn^{2+} and ET-succinate/ Et-adipate)	Numoto et al. (2018)
Cut190* mutants[b], Cut190** and Cut190**S176A	*Saccharomonospora viridis* AHK190	7CEF, 7CEH	Senga et al. (2021)

[a]Apparently cutinase type, as the enzyme has no lid.
[b]Cut190** and Cut190**S176A, in which three C-terminal amino acids (Lys-Leu-Asn) are deleted from Cut190*, and its inactive form (Cut190*S176A).

ions (PDB IDs: 4WFI, 4WFJ, and 4WFK) or its absence, respectively (Miyakawa et al., 2015). Nevertheless, the active form of Cut190S226P corresponds to those of the other cutinases listed in Table 2. Due to their sequence homology and structural similarity, we hypothesized that the dependence and advantageous effects of divalent cations on their thermostability and hydrolytic activity against PET is most probably shared by known actinomycete cutinases (Table 2).

3.2 Structural dynamics driven by Ca^{2+}-binding and complexation with model substrates

To date, the crystal structures of cutinases in complex with divalent cations and model substrates are mostly restricted to Cut190 mutants. The crystal structure of Cut190S226P with an N-terminal His-tag was solved both in the absence and presence of Ca^{2+} at 1.45 Å and 1.75 Å resolution, respectively (Miyakawa et al., 2015), forming a common cutinase structure, as shown in Fig. 1A. Here, a Ca^{2+} is located at the β1-β2 loop, which is called Site 1 with Ser76, Ala78, and Phe81 bound to the Ca^{2+}, in addition to Asn133. Upon Ca^{2+} binding, a large conformational change is observed in three loops: the β1-β2, β3-α2, and β4-α3 loops, as shown in Fig. 1B.

In addition, using the inactive mutant of S226P/R228S (Cut190*), Cut190*S176A, the crystal structures in complex with Ca^{2+} [1], Ca^{2+} and Zn^{2+} [2], Zn^{2+} and mono-ethyl succinate (Et-succinate) [3], and Ca^{2+}, Zn^{2+} and mono-ethyl adipate (Et-adipate) [4] were determined at 1.60, 1.12, 1.34, and 1.40 Å resolution, respectively (PDB IDs: 5ZNO, 5ZRQ, 5ZRR, and 5ZRS) (Numoto et al., 2018). The overall structure of Cut190*S176A in complex with Ca^{2+} [1] is like that of Cut190S226P in complex with Ca^{2+}. In addition to Site 1, two additional Ca^{2+}-binding sites, Sites 2 and 3, were observed. Site 3 is considered as a common Ca^{2+} or Mg^{2+} binding site found also in other thermophilic cutinases. The overall structure of the Zn^{2+}-bound form [2] is almost identical to that of the Ca^{2+}-bound form. The structure in complex with Et-succinate [3] corresponds to that of the pre-reaction state of the enzyme and was designated as the engaged form. It should also be noted that no electron density of Ca^{2+} was observed at Site 1 in [3]. The crystal structure of Cut190*S176A in complex with Et-adipate [4] provided additional information about the enzymatic reaction, in particular, about the post-reaction state. The overall structure is comparable with that of the open form, but unlike [3], an electron density corresponding to a Ca^{2+} ion at Site 1 was observed.

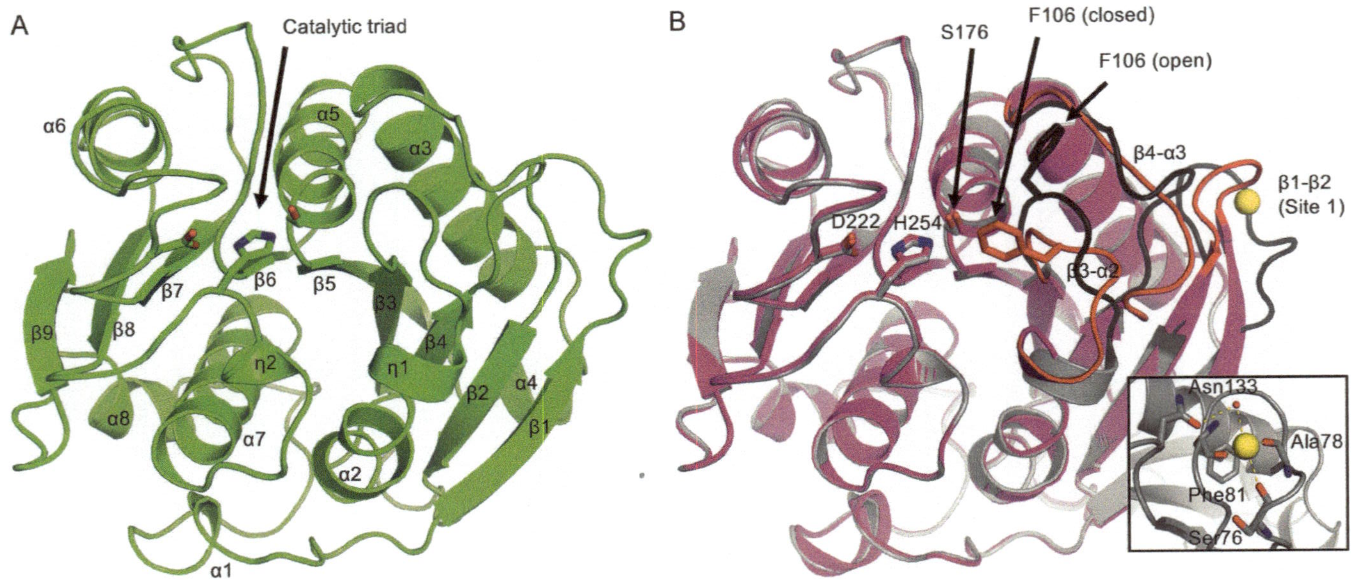

Fig. 1 (A) Overall structure of cutinase, represented by that of Cut190 (PDB ID: 5ZNO). The secondary structure elements are indicated, with the residues part of the catalytic triad drawn as a stick model. (B) Structural changes of Cut190 induced by Ca^{2+} binding. The Ca^{2+}-bound form (gray and black for cutinase, and a yellow sphere for Ca^{2+}) has been superposed onto the Ca^{2+} free form (magenta and red for cutinase). The most significant structural changes are observed at the three loops (black and red for Ca^{2+}-bound and Ca^{2+} free forms, respectively) adjacent to the active site. Phe106 of the β3-α2 loop is responsible for opening/closing the active site to interact with the ligands. The inset shows a closeup view of the Ca^{2+} binding site (Site 1). Interaction between cutinase to Ca^{2+} is indicated by dashed yellow lines. Colors are visible in the electronic version.

Molecular dynamics (MD) simulations were used to analyze the local dynamics, stabilities, and effects of Ca^{2+} binding. Integrating experimental and computational results, the enzymatic reaction process can be demonstrated via the obtained coarse snapshots (Fig. 2).

Cutinase can naturally transit between the open and closed forms, where Ca^{2+} binding at Site 1 shifts the balance towards the open form, so that the

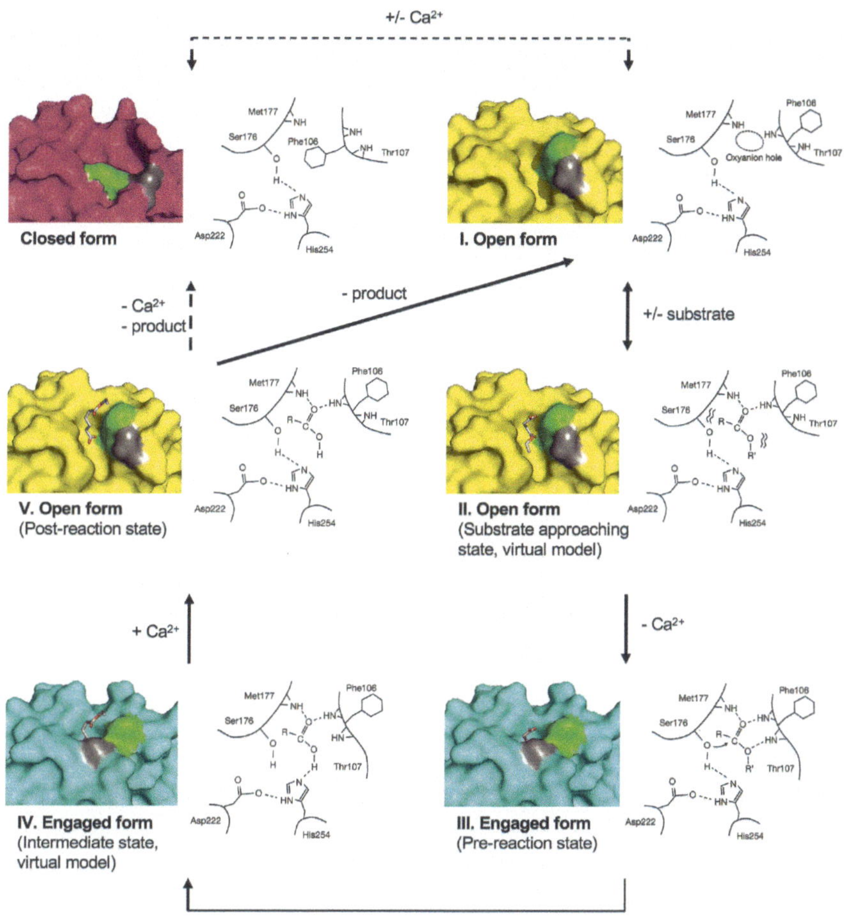

Fig. 2 Schematic diagram of the reaction cycle of Cut190. Structural changes around the active site including Phe106 (green) and Thr107 (gray) are shown as surface models. Colors are visible in the electronic version. *Reproduced from Numoto, N., Kamiya, N., Bekker, G. -J., Yamagami, Y., Inaba, S., Ishii, K., et al. (2018). Structural dynamics of the PET-degrading cutinase-like enzyme from Saccharomonospora viridis AHK190 in substrate-bound states elucidates the Ca^{2+}-driven catalytic cycle.* Biochemistry, 57, 5289–5300. *Copyright 2018 (American Chemical Society).*

enzyme enters the substrate awaiting state (I). The substrate initially approaches the active site of the open form (II), but the open form would not be able to stably accommodate the substrate, which is demonstrated by MD simulations (Numoto et al., 2018). Once Ca^{2+} is released from Site 1, it induces a conformational change to the engaged form (III), where the enzyme tightly binds to the substrate and initiates the hydrolysis reaction. Once the cleavage has been completed, the product needs to quickly leave the active site. MD simulations suggest that the engaged form could still strongly bind the product (IV), suggesting a prolonged dissociation time and reduced the enzymatic throughput. Ca^{2+} re-binding at Site 1 accelerates the switch to the open form again and promotes an efficient exchange of the product for a new substrate (V) (Supplementary video in the online version at https://doi.org/10.1016/bs.mie.2020.12.031). Further research on this reaction mechanism would be a way to elucidate the reaction mechanisms of metal ion-regulated enzymatic hydrolysis also for other cutinases.

Since substrate-bound structures of Cut190 are obtained using small molecules of Et-succinate and Et-adipate, the engaged form might be a little different to better accommodate the larger PET polymer. However, there is still some rationale behind the assumption that the structural change to the engaged form is required for efficient activity. For example, it might prevent water molecules from entering the active site or distort the conformation of the substrate PET chains to destabilize them, thereby promoting the cleavage reaction. Interestingly, there is a short tunnel near the active site of the engaged form. It should be verified whether bound PET chains go through this tunnel in a processive manner, as is the case for cellobiohydrolase, for which a long active site tunnel has been reported (Divne et al., 1994).

3.3 Thermodynamic analysis of metal binding and its effects on stability

Metal binding to cutinases can be analyzed using isothermal titration calorimetry (ITC). The binding parameters of Mn^{2+}, Zn^{2+}, Mg^{2+}, and Ca^{2+} to Cut190*S176A were compared (Senga et al., 2019). While an endothermic response was observed upon adding Mn^{2+}, Zn^{2+} and Mg^{2+}, only a limited response was observed when Ca^{2+} was added. It should be noted that the binding stoichiometry, i.e., the binding ratio of an ion to Cut190*S176A, corresponds 3 for Zn^{2+}, while it is only 1 for Mn^{2+} and Mg^{2+}, although three Ca^{2+}-binding sites were identified, as described above. The rationale behind these observations could be elucidated by assessing their different binding affinities.

Cut190 and other cutinases have unique features that enable Ca^{2+} binding to increase their thermal stability, which can be confirmed by circular dichroism (CD) and differential scanning calorimetry (DSC) (Inaba, Kamiya, Bekker, Kawai, & Oda, 2019; Senga et al., 2019; Then et al., 2015). The increased thermal stability upon Ca^{2+} binding can be endorsed using MD. Based on the analyses, only the Ca^{2+} bound structure showed perturbation of the loop regions that bind to Ca^{2+}. Combined analyses by experimental measurements using ITC, CD and DSC, and computational MD are indispensable for comprehensive understanding of the subtle regulation mechanism by metal ions (possibly a common effect for all cutinases).

4. Protocols

4.1 Cloning of cutinases

Here we describe the representative cloning protocols of actinomycetes cutinases, using Cut190 from *S. viridis* AHK190 as an example.

4.1.1 Equipment

- PCR cycler
- Constant temperature incubation shaker
- High speed refrigerated centrifuge
- Ultrasonic cell disrupter
- Autoclave

4.1.2 Materials

- Total DNA extraction kit, such as Wizard® (Promega)
- DNA polymerase, such as KOD Plus (Toyobo)
- LB-agar plate for blue-white selection: Autoclave LB-agar medium and cool down to 50 °C, to which add 10 μL 5-bromo-4-chloro-3-indolyl-β-D-galactopyranoside dissolved in dimethylformamide (20 mg/mL), 10 μL of 100 mM isopropyl-β-D-1-thiogalactopyranoside (IPTG) sterilized by passing an ultrafiltration membrane (0.2 μm pore size) and an appropriate antibiotic, mixed well and poured into a Petri dish
- SOB medium (New England BioLabs) or attached to competent cells
- Plasmid isolation (miniprep) kit from any supplier

4.1.3 Procedure

1. Design primers, based on the amino acid sequences conserved in *Saccharomonospora viridis* DSM 43017 (CP001683.1), for example,

Cut190F (5′-GACCAGTTGCCTGGACTCTC-3′) and Cut190R (5′-GCGGACTCCGATATTCAGC-3′), targeting the whole sequence of a cutinase gene.

2. Using designed primers, amplify a DNA fragment from total DNA from *S. viridis* AHK190 (FERM P-21507). PCR mixtures (25 μL) contain 2.5 μL 10 × KOD Plus buffer, 0.3 pmol/μL of each primer, 1 mM $MgCl_2$, 0.2 mM of each deoxyribonucleotide, 1 μL dimethyl sulfoxide, 0.5 μL total DNA, 0.5 U KOD Plus, and MilliQ water.
3. PCR is performed for 1 cycle with 4 min at 94 °C; 30 cycles with 30 s at 94 °C; 30 s at 58 °C and 5 min at 68 °C; 1 cycle with 10 min at 68 °C.
4. Purify DNA fragments from agarose gels using MagExtractor DNA purification kit (TOYOBO) for sequencing through a commercial company.
5. Ligate into the pGEM-T easy vector (Promega) following the protocols of the provider.
6. Add the plasmid solution (3–4 μL) to 50 μL competent cells of *E. coli* DH5α (Takara Bio) on ice, keep the cells for 20 min on ice, heat-shock them at 42 °C for 45 s. and then keep them for 5 min on ice. Add SOB medium (150 μL) and incubate at 37 °C for 20 min.
7. Spread each 100 μL transformant solution on a blue-white selection agar plate and incubate at 37 °C overnight until blue/white colonies appear.
8. Extract plasmids of white colonies and sequence both strands of them.
9. For sequence assembly to obtain the open reading frame (ORF) of the enzyme gene (*cut190*), use the GENETYX software. Use the SignalP Server (http://www.cbs.dtu.dk/services/SignalP/) to predict the presence and length of a signal sequence in ORF.
10. Select appropriate cloning sites, for example *Bam*HI and *Hin*dIII, of an expression vector that are not present in the sequenced gene, and amplify a gene fragment, using Cut190F2 (5′-GCCGGATCCCAGGACAACCCTTACGAACG-3′) and Cut190R2 ((5′-GCGAAGCTTGTACGGGCAGGTCGAGCGG-3′)) primers (the *Bam*HI and *Hin*dIII restriction sites are underlined, respectively).
11. Digest the amplified fragment with *Bam*HI and *Hin*dIII, purify with a miniprep kit and ligate into the *Bam*HI and *Hin*dIII sites of pQE80L to clone the *cut190* as a 6 × his-tagged fusion protein, using a ligation kit (Takara Bio).
12. To remove the polyhistidine tag, a DNA sequence coding for a PreScission site (LEVLFQGP) should be introduced between a

polyhistidine tag and the mature protein sequence, by designing appropriate primers and performing PCR, as described above

13. Transform the resultant plasmid (pQE80L-*cut190*) into a host strain such as *E. coli* BL21 or Rosetta-gami B(DE3). If necessary, select another host such as that shown in Table 1. For the transformation, follow the processes described below.
    - **13.1** Thaw competent cells (50 μL) on ice for 10 min.
    - **13.2** Add the plasmid (4 μL), tap gently and kept on ice for 20 min.
    - **13.3** Place in a heat block heated at 42 °C, keep for 35 s and immediately transfer to an ice bath for 2 min.
    - **13.4** Add SOB medium (150 μL) and incubate at 37 °C for 1 h.
    - **13.5** Apply all the volume to a LB-agar plate supplemented with 100 μg/mL ampicillin and spread on the entire surface with a spreader.
    - **13.6** Incubate a plate at 37 °C overnight.
    - **13.7** Pick up colonies.

4.2 Purification of cutinases

Recombinant cutinase with a His-tag can be purified by IMAC and further on by, e.g., anion exchange chromatography, if necessary.

4.2.1 Equipment

- Constant temperature incubation shaker
- High speed refrigerated centrifuge
- Ultrasonic cell disrupter
- Spectrophotometer
- SDS-PAGE
- Water bath shaker
- High pressure liquid chromatography (HPLC)

4.2.2 Materials

- IMAC & anion exchange columns
- IPTG
- Antibiotic
- LB broth
- Lysis buffer: 20 mM NaH_2PO_4 (pH 7.4) containing 500 mM NaCl and 20 mM imidazole
- Equilibration buffer: 20 mM Tris-HCl (pH 8.0) containing 300 mM NaCl and 10 mM imidazole

- Wash buffer: 20 mM Tris-HCl (pH 8.0) containing 300 mM NaCl and 40 mM imidazole
- Elution buffer: 50 mM Tris-HCl buffer (pH 8.0) containing 50 mM NaCl and 200 mM imidazole
- Desalting column, such as PD-10 (Cytiva)
- Equilibration buffer for PD-10: 25 mM Tris-HCl (pH 8.0) containing 150 mM NaCl
- PreScission protease, such as Turbo 3C (Acceragen)
- Ultrafiltration Centrifugal Filter Unit, such as Amicon Ultra (MWCO of 10,000; Merck)
- Protein assay kit (Bio-Rad)
- Bovine serum albumin
- Polymer emulsion, such as poly(butylene glycol succinate-*co*-adipate) (PBSA) (Bionolle™ EM-301; weight average MW = 1.0×10^5) (Showa Denko K. K.)

4.2.3 Procedure

1. Cultivate an expression host, for example *E. coli* Rosetta-gami B(DE3) carrying pQE 80L-*cut190*, in 100 mL of LB broth containing 50 μg/mL ampicillin at 37 °C to an approximate OD_{600} of 0.4–0.5.
2. Add 100 mM IPTG at a final concentration of 0.1 mM (if expression is insufficient, optimize) to the culture and further cultivate the cells at 28 °C for 12 h.
3. Collect the cells by centrifugation at 4000*g* for 15 min at 4 °C and wash the pellets twice with saline (0.9% NaCl).
4. Resuspend the pellets in lysis buffer, disrupt by sonication, and centrifuge at 8000*g* at 4 °C to remove cell debris.
5. Put the resulting supernatant (soluble fraction) on an IMAC column, for example Ni-nitrilotriacetic column (Qiagen) equilibrated with equilibration buffer, wash with 30 mL of wash buffer and elute the enzyme using 15 mL of elution buffer.
6. Apply the elution (2.5 mL) on a PD-10 column equilibrated with equilibration buffer and wash with 8 mL of equilibration buffer, fractionated in 1 mL steps.
7. When a polyhistidine tag needs to be removed, add PreScission protease, for example, Turbo 3C, 1 U for 300 μg protein purified by IMAC and incubate at 4 °C for 16 h.
8. Apply the protease-treated solution (10–12 mL) on a IMAC column (4 mL) to remove Turbo 3C and polyhistidine tag, wash with 8 mL of equilibration buffer for PD-10 and collect all the effluent.

9. For further purification, load the elution onto an anion exchange column such as HiTrap Q FF 2.1, elute with 20 mM Tris-HCl buffer (pH 8.0) (according to the isoelectric point, optimize) and apply a 1–500 mM NaCl gradient, collect the enzyme peak.
10. Apply the enzyme solution (5 mL) on an Amicon Ultra410K (Merck) to lower NaCl concentration to less than 1 mM and exchange elution buffer with an appropriate buffer, by repeating centrifugation at 3000*g* for 5–10 min at 4 °C to approximately half volume and addition of 50 mM Tris-HCl (pH 7.0) to the concentrated protein solution 10–15 times and finally add an aliquot of the same buffer to adjust the concentration of protein.
11. The protein concentrations should be determined using a protein assay kit with bovine serum albumin as a standard.

4.3 Activity assay of cutinases

The classical definition of cutinase originates from the hydrolysis of cutin existing in the cuticular layer of plants, but activity of cutinases related to polyester hydrolysis is measurable by three methods described here.

4.3.1 Equipment

- Photometer equipped with a constant temperature cell holder
- Deep freezer
- Constant temperature incubator
- Water bath shaker
- HPLC

4.3.2 Chemicals

- *para*-nitrophenol butyrate (*p*NPB): Solubilize *p*NPB in isopropanol at 50 mM and store at −20 °C until use.
- Potassium phosphate (pH 7.0) or Tris-HCl buffer (pH 7.0)
- Polyesters used for agar plates, for example, polycaprolactone, poly(butylene succinate), and PBSA.
- Polymer emulsion such as PBSA (an absorbance of 1 at 600 nm corresponds to 0.25-mg dry weight PBSA per milliliter and a decrease of one OD_{600} of PBSA suspension corresponds to 1.5 nkat.)
- Bis(hydroxyethyl)terephthalate (BHET), mono(hydroxyethyl)terephthalate (MHET; monomer unit) and terephthalate (TPA)

4.3.3 Procedure

Activity measurements can be conducted by three possible methods.

1. *pNPB activity*:
   1.1. The assay mixture contains 50 μL of enzyme (appropriately diluted), 930 μL of 50 mM sodium phosphate buffer (pH 6.0). Start the reaction by adding 20 μL *p*NPB solution at 50 °C.
   1.2. Record the absorbance of *p*NP at 346 nm for 1 min against the control without enzyme. The molar extinction coefficient of *p*-nitrophenol (*p*NP) is 5.48 $mmol^{-1} cm^{-1}$. One enzyme unit (U) is defined as the amount of enzyme liberating 1 μmol of *p*NP per min.
2. *Polyester-hydrolyzing activity*:
   2.1. The qualitative assay of polyester degradation activity is confirmed by detecting halo formation on an LB-agar plate containing 0.1% polyester particle suspension.
      2.1.1. Dissolve polyester in dichlomethane (0.2 g per 5 mL), add the solution to 30 mL of potassium phosphate buffer or Tris-HCl buffer (pH 7.0), and sonicate for 7–8 min.
      2.1.2. Finally, incubate the sonicated solution at 50 °C for 10 min to evaporate the dichloromethane.
      2.1.3. Mix the polymer suspension with autoclaved LB-agar medium at a final concentration of 0.1% and solidify on Petri dishes.
      2.1.4. Make an appropriate hole by a sterile pipette tip on a LB-polymer agar plate and place 30–50 μL enzyme solution therein, followed by incubation at 50 °C overnight to 2 days.
      2.1.5. Relative activities are evaluated by the size of a clear zone formed around the hole.
   2.2. The polyester degradation is also quantitatively measured as follows:
      2.2.1. A reaction mixture including polymer suspension such as PBSA (500 μL, corresponding to a final OD_{600} of 1.0–1.5); 100 mM Tris-HCl buffer (pH 8.2), 1000 μL; 100 mM $CaCl_2$, 50 μL; and ultrapure (e.g., obtained with a MilliQ system) water, 400 μL (for enzyme reaction) or 450 μL(for control reaction) is prepared in a total volume of 2 mL and kept on ice
      2.2.2. Mix control reaction mixtures gently by turning upside down repeatedly and measure OD_{600} (Measure 1). Add cutinase (50 μL) to the enzyme reaction mixture and mix

well. Incubate control and enzyme reaction mixtures at 37 °C for 30 min by gentle shaking on a water bath shaker. Then put all the reaction mixtures on ice to stop the reaction.

2.2.3. Measure OD_{600} of all the reaction mixtures. Measure 2 is defined as OD_{600} for control mixtures and Measure 3 is defined as OD_{600} for enzyme mixtures. Value 1 (Measure 1-Measure 2) indicates non-enzymatic hydrolysis. Value 2 (Measure 3-Value 1) indicates the OD_{600} change as a result of the net enzymatic hydrolysis. A decrease of 1 OD_{600} corresponds to 1.5 nkat

3. *PET-hydrolyzing activity*:

3.1. Hydrolysis products of PET, such as BHET, MHET and TPA is measured by HPLC or by measuring the absorbance at 240 nm, based on the molar absorption coefficient of TPA (13.8×10^4).

3.2. HPLC conditions are as follows: a reversed phase column; elution with a mobile phase (acetonitrile/water/formic acid (15/85/0.1) at a flow rate of 1.0 mL/min. The amounts of BHET, MHET and TPA are calculated from respective peak areas obtained by standard solutions

4.4 Crystallization and crystal structure analysis

This sub-section introduces the typical protocols of the crystallization, ligand soaking, and X-ray crystal structure analysis to obtain cutinase structures.

4.4.1 Equipment

- High speed refrigerated centrifuge
- Stereo microscope (×10 to ×50)
- Incubator
- Crystal mounts and loops
- Sample pin storage and shipping container (Universal V1-Puck) and the handling tools

4.4.2 Materials

- Centrifugal filters with low protein-binding membranes
- 24 or 96 well crystallization plates
- Siliconized glass cover slides
- Cutinase

- Protein crystallization screening kits
- Ligands
- Liquid nitrogen
- XDS (Kabsch, 2010)
- CCP4 (Winn et al., 2011)
- PHASER (McCoy et al., 2007)
- MOLREP (Vagin & Teplyakov, 1997)
- COOT (Emsley, Lohkamp, Scott, & Cowtan, 2010)
- REFMAC (Murshudov et al., 2011)
- PHENIX (Adams et al., 2010)
- MOLPROBITY (Chen et al., 2010)

4.4.3 Procedure

1. Concentrate protein solution up to 5–10 mg/mL using a centrifugal filter such as Amicon Ultra 10K MWCO (Merck) at 4 °C.
2. Screen crystallization conditions using commercial screening kits (e.g., Hampton Research) with the hanging drop or sitting drop vapor diffusion method, by mixing equal volume of protein and reservoir solutions (typically, 1–2 μL of each) on the pedestal.
3. Seal crystallization chamber immediately by sealing tape or glass cover slides.
4. Incubate the crystallization plate at 20 °C. Monitor every 1 or 2 days by a stereo microscope until crystals appear. Put the plate back into the incubator immediately after observation. It will take a week to a month for sufficient crystal growth.
5. The next five steps are for the crystal freezing and ligand soaking. When sufficiently large sizes (more than 50 μm at least one side) of crystals (Fig. 3A) are obtained, prepare the cryoprotectant solution containing 20% (v/v) glycerol along with their respective reservoir buffers
6. Pre-cool Universal V1-Puck in liquid nitrogen at least 10 min

   When handling liquid nitrogen, use appropriate goggles and gloves to protect your eyes and body. It is mandatory to perform the procedure in a well-ventilated room and use an oxygen monitor.
7. Pick the crystals using crystal mounts and loops. A slightly larger size of the loop than the target crystal should be selected (Fig. 3B). Soak the crystals in 1–2 μL of the cryo-protectant solutions on a glass cover slide. Soaking time should be several seconds to a minute.
8. In case a ligand complexed structure is targeted, add 10–100 mM of the ligand to the cryoprotectant solutions or soak the crystals in step 4.

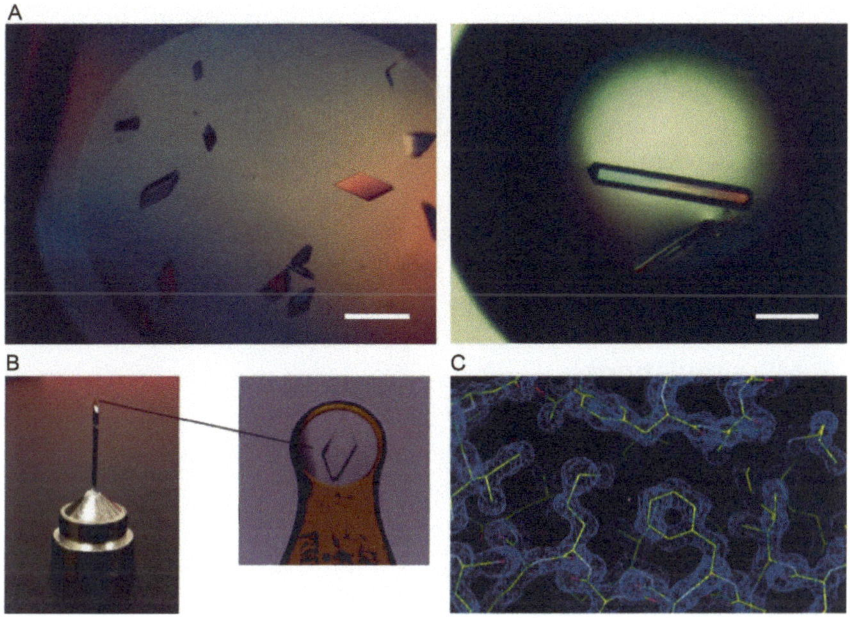

Fig. 3 (A) Crystals of Cut190. The scale bar indicates 0.2 mm. (B) Crystals picked up via crystal mounts and loops. (C) Refined model and electron density map at 1.4 Å resolution.

A cryoprotectant with a larger molecular weight, such as polyethylene glycol (PEG) 400, may be useful because glycerol could potentially inhibit the ligand binding of cutinases.

9. Pick the crystal from the soaking solution, and flash-freeze in liquid nitrogen and directly load into the cassette. You can store the frozen crystals over months in the liquid nitrogen dewars and ship to the synchrotron radiation facilitates by using a dry shipper.
10. Perform X-ray diffraction experiments at a home source or synchrotron. To avoid serious radiation damage, keep the total X-ray dose under 20 MGy. For more information, contact beamline scientists.
11. The next four steps are for the structure determination and model refinement. X-ray diffraction images are processed using the XDS program package. Integrated and scaled data were then truncated using the CCP4 program suite. For further refinement, 5% of the independent, randomly selected reflections are set apart as a test set to calculate the R_{free} values.
12. Determine the initial phases by the molecular replacement method using PHASER or MOLREP. Since the cutinase structures are usually

similar, most known structures such as the structures of the Cut190S226P mutant (PDB ID: 4WFI or 4WFJ) can be used as a search model.

13. Repeat several cycles of manual model rebuilding using COOT and refinement using REFMAC or PHENIX (Fig. 3C), until the R_{free} value converges. Typical R_{free} values range around one tenth of the resolution in angstrom (Å). If the R_{free} value is significantly higher, a serious problem of the crystal symmetry, space group assignment or alike may be present.

14. Validate the refined models by MOLPROBITY (Fig. 3).

4.5 MD simulations

MD simulations are used to analyze both the effect of the Ca^{2+} concentration on cutinase and the structural landscape and dynamics of cutinase in the different states and with different bound ligands.

4.5.1 Materials

- Protein Data Bank Japan (https://pdbj.org/)
- Gromacs MD software (Abraham et al., 2015) with Zero Dipole Method support (Bekker et al., 2017; Kamiya, Fukuda, & Nakamura, 2013)
- Custom preparation tools (https://gitlab.com/gjbekker/gromacs/-/blob/bekker/bekker-scripts/autoBuild.py) and analysis scripts
- Amber ff99SB-ILDN protein force field parameters (Lindorff-Larsen et al., 2010)
- GAFF force field parameters (Wang, Wolf, Caldwell, Kollman, & Case, 2004)
- Monovalent (Joung & Cheatham, 2008) and divalent (Li, Roberts, Chakravorty, & Merz, 2013) ion force field parameters
- Biological Structure Model Archive (Bekker, Kawabata, & Kurisu, 2020)
- Q-value analysis (Best, Hummer, & Eaton, 2013)
- Molmil (https://pdbj.org/molmil2/; Bekker, Nakamura, & Kinjo, 2016)

4.5.2 Procedure

1. Obtain the structures either via the experimental method described above (Section 4.4) or otherwise from the Protein Data Bank Japan or the RCSB Protein Data Bank. Then, for each system, the cutinase molecule (with co-crystalized ions and ligand, if present) should be placed inside an adequately sized dodecahedron periodic box (to optimize the size-to-performance ratio), and then filled with explicit water molecules and the desired $CaCl_2$ concentration.

2. An initial energy minimization is used to minimize the energy of the system, without perturbing the experimental structure (primarily to optimize the position of generated water molecules, ions, and hydrogens).
3. A second energy minimization stage is used to minimize the system energy as a whole (after any issues involving the generated particles have been solved).
4. To prepare for the production MD simulation, create 10 parallel trajectories for each system, each with a different random seed for the initial velocity, and execute a 100 ps NVT MD simulation (using the Bussi thermostat) to bring the system to the target temperature (e.g., 300 K).
5. Next, execute a 100 ps NPT MD simulation (using Parrinello-Rahman barostat) for each trajectory to achieve the correct pressure (1 bar).
6. For the production MD simulations, perform a 100 ns NVT MD simulations per trajectory, with a snapshot structure saved in the compressed XTC format at 10 ps intervals. Detailed settings used for the preparation and simulations, as well as the raw data produced and used for the subsequent analysis can be found in the Biological Structure Model Archive under BSM-00003 (https://bsma.pdbj.org/entry/3/path/data/raw) and BSM-00004 (https://bsma.pdbj.org/entry/4/path/data/raw).
7. Analyze the dependency on Ca^{2+} concentration for the stability of cutinase (BSM-00003) by using the Q-value to measure the stability of cutinase at different Ca^{2+} concentrations (0, 25, 100 and 250 mM). The Q-value is defined as

$$Q = \frac{1}{N} \sum_{(ij)} \frac{1}{1 + \exp\left(\beta\left(r_{ij}(X) - \Lambda r_{ij}^{0}\right)\right)}$$

where N is the number of native contact pairs with a distance less than 4.5 Å, $r_{ij}(X)$ the distance of the pair (i, j) in configuration X, r_{ij}^{0} the distance in the native configuration (i.e., the experimental structure), β a smoothing parameter and Λ a parameter to account for the fluctuations formed by the contact, set to 5 Å^{-1} and 1.8 Å, respectively. Calculate the average Q-values (with standard deviations) using the final 70 ns of the trajectories for each system. Higher averages correspond to more stable systems, while lower averages to less stable ones.
8. To analyze the dynamics of the cutinase molecule in different states (BSM-00004), including the bound compounds such as adipate and succinate, use Principal Component Analysis (PCA) to project the structural ensembles onto a 2-dimensional space. To do so, however, molecules

must first be first superposed onto each other based on their Cα-atoms, followed by PCA on the xyz coordinates. Representative structures along the first principal component are to be taken and can then be analyzed using our molecular viewer Molmil to show the structural changes that occur during the conformational change between the open to the closed state.

For steps 2 and 4, position restraints on the heavy solute atoms must be used. For steps 2–5, a short-range cutoff of 12 Å is used, with the Zero Dipole Method to treat the electrostatic interactions.

For steps 1–5, our custom preparation script, which makes use of the MD program Gromacs and the force field parameters (Amber ff99SB-ILDN, GAFF, ions) listed above can be used to automate the procedures.

4.6 ITC analysis

Metal-binding thermodynamics to cutinase are analyzed using ITC.

4.6.1 Equipment

- iTC200 or PEAQ-ITC calorimeter (Malvern Panalytical)

4.6.2 Chemicals

- 500 mM $CaCl_2$
- 500 mM $ZnCl_2$
- 500 mM $MnCl_2$
- 500 mM $MgCl_2$
- 10 mM Tris-HCl buffer (pH 8.0)

4.6.3 Procedure

1. A solution of a metal (1–2 mM) is titrated into a solution of cutinase (0.03–0.06 mM) at 25 °C. Each titration consists of a preliminary 0.4 μL injection followed by 19 subsequent 2 μL additions.
2. The heat for each injection is subtracted from the heat of dilution of the injectant, which is measured by injecting the metal solution into the experimental buffer, 10 mM Tris-HCl buffer (pH 8.0).
3. Each corrected heat is divided by the moles of metal injected and is analyzed with the Microcal Origin software supplied by the manufacturer. Binding stoichiometry, equilibrium association constant (K_a), and binding enthalpy change (ΔH) are directly determined from the fitting procedure. Gibbs free energy change (ΔG) is calculated from the equation, $\Delta G = -RT \ln K_a$, and binding entropy change (ΔS) is calculated from the equation, $\Delta G = \Delta H - T\Delta S$.

4.7 CD analysis

Secondary structures and thermal stabilities of cutinase are analyzed using CD.

4.7.1 Equipment

- J-725 or J-810 spectropolarimeter (JASCO)

4.7.2 Chemicals

- 500 mM $CaCl_2$
- 500 mM $ZnCl_2$
- 500 mM $MnCl_2$
- 500 mM $MgCl_2$
- 10 mM Tris-HCl buffer (pH 8.0)

4.7.3 Procedure

1. CD spectra are measured at 20 °C for protein concentrations of 0.04 mg mL^{-1} in the absence or presence of metal ions placed in quartz cells with a path length of 10 mm.
2. Spectra for CD between 200 and 250 nm are obtained using a scanning speed of 20 nm min^{-1}, a time response of 1 s, a bandwidth of 1 nm, and an average over eight scans.
3. Thermal denaturation curves are recorded in the temperature mode at 222 nm from 20 °C to 80–95 °C with a heating rate of 1.0 °C min^{-1}.
4. The analysis of the transition curves is performed on the basis of two-state transition model. The molar ellipticity data of temperature-scanning CD are fitted with the weighted-mean combination of the two linear functions for the folded and the unfolded states.

4.8 DSC analysis

Folding thermodynamics of cutinase are analyzed using DSC.

4.8.1 Equipment

- VP-capillary DSC calorimeter (Malvern Panalytical)

4.8.2 Chemicals

- 500 mM $CaCl_2$
- 10 mM Tris-HCl buffer (pH 8.0)

4.8.3 Procedure

1. The data are collected for protein concentrations of 1.0 mg mL^{-1} in the absence or presence of metal ions by heating the solution from 20 °C to 80–95 °C with a heating rate of 1.0 °C min^{-1}.
2. The sample is cooled down and reheated without exchanging the solution in the cells to check the reversibility of denaturation.
3. Data are analyzed using the Microcal Origin software supplied by the manufacturer. The ΔC_p is calculated from the difference in the C_p values between unfolded and folded states at the denaturation temperature. The calorimetric enthalpy change is calculated by integrating the area in each heat capacity curve.

5. Summary

This chapter describes for the sources of thermophilic cutinases, the experimental procedures for cloning, recombinant expression and purification of them in an appropriate host as his-tagged proteins. Recombinant cutinases can be used not only as PET hydrolases, but also in the degradation of other aliphatic polyester (e.g., PCL) or aliphatic-aromatic co-polyester (e.g., PBSA). Here we briefly summarized methodologies to determine the hxdrolytic activities and thermostability of selected cutinases relevant to Ca^{2+}-binding, based on their X-ray structures, using experimental measurements (ITC, CD and DSC) and computational MD simulations. This chapter provides a broad spectrum of options for characterizing thermophilic cutinases possessing PET hydrolysis ability.

Acknowledgments

We thank Xiaoping Hu, Uschara Thumarat, Akiko Matsubara, Aya Iwaki, Fumihiko Takeuchi, Toshihiro Matsunami, Shinji Iijima, Takaaki Takeda, Yohei Tamiaki, Yuma Kobayashi, Satoshi Suzua, Yuki Kimura, Satomi Inaba, Yuri Yamagami, Akane Senga, and Miho Emori for their help in setting up basics of the herein described techniques. This work was supported by funding from the Institute for Fermentation, Osaka (Japan).

References

Abraham, M. J., Murtola, T., Schulz, R., Páll, S., Smith, J. C., Hess, B., et al. (2015). GROMACS: High performance molecular simulations through multi-level parallelism from laptops to supercomputers. *SoftwareX*, *1–2*, 19–25.

Adams, P. D., Afonine, P. V., Bunkóczi, G., Chen, V. B., Davis, I. W., Echols, N., et al. (2010). *PHENIX*: A comprehensive python-based system for macromolecular structure solution. *Acta Crystallographica Section D*, *66*, 213–221.

Baker, P. J., Poultney, C., Liu, Z., Gross, R., & Montclare, J. K. (2012). Identification and comparison of cutinases for synthetic polyester degradation. *Applied Microbiology and Biotechnology*, *93*, 229–240.

Bekker, G.-J., Kamiya, N., Araki, M., Fukuda, I., Okuno, Y., & Nakamura, H. (2017). Accurate prediction of complex structure and affinity for a flexible protein receptor and its inhibitor. *Journal of Chemical Theory and Computation, 13*, 2389–2399.

Bekker, G.-J., Kawabata, T., & Kurisu, G. (2020). The biological structure model archive (BSM-arc): An archive for in silico models and simulations. *Biophysical Reviews, 12*, 371–375.

Bekker, G.-J., Nakamura, H., & Kinjo, A. R. (2016). Molmil: A molecular viewer for the PDB and beyond. *Journal of Cheminformatics, 8*, 42.

Best, R. B., Hummer, G., & Eaton, W. A. (2013). Native contacts determine protein folding mechanisms in atomistic simulations. *Proceedings of the National Academy of Science of the United States of America, 110*, 17874–17879.

Chen, V. B., Arendall, W. B., 3rd, Headd, J. J., Keedy, D. A., Immormino, R. M., Kapral, G. J., et al. (2010). MolProbity: All-atom structure validation for macromolecular crystallography. *Acta Crystallographica Section D, 66*, 12–21.

Chen, S., Tong, X., Woodard, R. W., Du, G., Wu, J., & Chen, J. (2008). Identification and characterization of bacterial cutinase. *The Journal of Biological Chemistry, 283*, 25854–25862.

Divne, C., Ståhlberg, J., Reinikainen, T., Ruohonen, L., Pettersson, G., Knowles, J. K., et al. (1994). The three-dimensional crystal structure of the catalytic core of cellobiohydrolase I from *Trichoderma reesei*. *Science, 265*, 524–528.

Emsley, P., Lohkamp, B., Scott, W. G., & Cowtan, K. (2010). Features and development of Coot. *Acta Crystallographica Section D, 66*, 486–501.

Hegde, K., & Veeranki, V. D. (2013). Production optimization and characterization of recombinant cutinases from *Thermobifida fusca* sp. NRRL B-8184. *Applied Biochemistry and Biotechnology, 170*, 644–675.

Heredia, A. (2003). Biochemical and biophysical characteristics of cutin, a plant barrier biopolymer. *Biochimica et Biophysica Acta, 1620*, 1–7.

Herrero Acero, E., Ribitsch, D., Steinkellner, G., Gruber, K., Greimel, K., Eiteljoerg, I., et al. (2011). Enzymatic surface hydrolysis of PET: Effect of structural diversity on kinetic properties of cutinases from *Thermobifida*. *Macromolecules, 44*, 4632–4640.

Inaba, S., Kamiya, N., Bekker, G.-J., Kawai, F., & Oda, M. (2019). Folding thermodynamics of PET-hydrolyzing enzyme Cut190 depending on Ca^{2+} concentration. *Journal of Thermal Analysis and Calorimetry, 135*, 2655–2663.

Joung, I. S., & Cheatham, T. E. (2008). Determination of alkali and halide monovalent ion parameters for use in explicitly solvated biomolecular simulations. *The Journal of Physical Chemistry B, 112*, 9020–9041.

Kabsch, W. (2010). XDS. *Acta Crystallographica Section D, 66*, 125–132.

Kamiya, N., Fukuda, I., & Nakamura, H. (2013). Application of zero-dipole summation method to molecular dynamics simulations of a membrane protein system. *Chemical Physics Letters, 568–569*, 26–32.

Kawai, F., Kawabata, T., & Oda, M. (2019). Current knowledge on enzymatic PET degradation and its possible application to waste stream management and other fields. *Applied Microbiology and Biotechnology, 103*, 4253–4268.

Kawai, F., Kawabata, T., & Oda, M. (2020). Current state and perspectives related to the PET hydrolases available for biorecycling. *ACS Sustainable Chemistry & Engineering, 8*, 8894–8908.

Kawai, F., Oda, M., Tamashiro, T., Waku, T., Tanaka, N., Yamamoto, M., et al. (2014). A novel Ca^{2+}-activated, thermostabilized polyesterase capable of hydrolyzing polyethylene terephthalate from *Saccharomonospora viridis* AHK190. *Applied Microbiology and Biotechnology, 98*, 10053–10064.

Kitadokoro, K., Kakara, M., Matsui, S., Osokoshi, R., Thumarat, U., Kawai, F., et al. (2019). Structural insights into the unique polylactate-degrading mechanism of *Thermobifida alba* cutinase. *The FEBS Journal, 286*, 2087–2098.

Kitadokoro, K., Thumarat, U., Nakamura, R., Nishimura, K., Karatani, H., Suzuki, H., et al. (2012). Crystal structure of cutinase Est119 from *Thermobifida alba* AHK119 that can degrade modified polyethylene terephthalate at 1.76 Å resolution. *Polymer Degradation and Stability*, *97*, 771–775.

Kollattukudy, P. E. (1981). Structure, biosynthesis, and biodegradation of cutin and suberin. *Annual Review of Plant Physiology*, *32*, 539–567.

Li, P., Roberts, B. P., Chakravorty, D. K., & Merz, K. M. (2013). Rational design of particle mesh Ewald compatible Lennard-Jones parameters for +2 metal cations in explicit solvent. *Journal of Chemical Theory and Computation*, *9*, 2733–2748.

Lindorff-Larsen, K., Piana, S., Palmo, K., Maragakis, P., Klepeis, J. L., Dror, R. O., et al. (2010). Improved side-chain torsion potentials for the Amber ff99SB protein force field. *Proteins*, *78*, 1950–1958.

McCoy, A. J., Grosse-Kunstleve, R. W., Adams, P. D., Winn, M. D., Storoni, L. C., & Read, R. J. (2007). Phaser crystallographic software. *Journal of Applied Crystallography*, *40*, 658–674.

Miyakawa, T., Mizushima, H., Ohtsuka, J., Oda, M., Kawai, F., & Tanokura, M. (2015). Structural basis for the Ca^{2+}-enhanced thermostability and activity of PET-degrading cutinase from *Saccharomonospora viridis* AHK190. *Applied Microbiology and Biotechnology*, *99*, 4297–4307.

Müller, R.-J., Schrader, H., Profe, J., Dresler, K., & Deckwer, W.-D. (2005). Enzymatic degradation of poly(ethylene terephthalate): Rapid hydrolysis using a hydrolase from *T. fusca*. *Macromolecular Rapid Communications*, *26*, 1400–1405.

Murphy, C. A., Cameron, J. A., Huang, S. J., & Vinopal, R. T. (1996). *Fusarium* polycaprolactone depolymerase is cutinase. *Applied and Environmental Microbiology*, *62*, 456–460.

Murshudov, G. N., Skubák, P., Lebedev, A. A., Pannu, N. S., Steiner, R. A., Nicholls, R. A., et al. (2011). REFMAC5 for the refinement of macromolecular crystal structures. *Acta Crystallographica Section D*, *67*, 355–367.

Numoto, N., Kamiya, N., Bekker, G.-J., Yamagami, Y., Inaba, S., Ishii, K., et al. (2018). Structural dynamics of the PET-degrading cutinase-like enzyme from *Saccharomonospora viridis* AHK190 in substrate-bound states elucidates the Ca^{2+}-driven catalytic cycle. *Biochemistry*, *57*, 5289–5300.

PlasticEurope. (2019). *Plastics-the Facts 2019*. Retrieved from https://www.plasticseurope.org/application/files/9715/7129/9584/FINAL_web_version_Plastics_the_facts2019_14102019.pdf.

Ribitsch, D., Acero, E. H., Greimel, K., Eiteljoerg, I., Trotscha, E., Freddi, G., et al. (2012). Characterization of a new cutinase from *Thermobifida alba* for PET-surface hydrolysis. *Biocatalysis and Biotransformation*, *30*, 2–9.

Ribitsch, D., Hromic, A., Zitzenbacher, S., Zartl, B., Gamerith, C., Pellis, A., et al. (2017). Small cause, large effect: Structural characterization of cutinases from *Thermobifida cellulosilytica*. *Biotechnology and Bioengineering*, *114*, 2481–2488.

Roth, C., Wei, R., Oeser, T., Then, J., Föllner, C., Zimmermann, W., et al. (2014). Structural and functional studies on a thermostable polyethylene terephthalate degrading hydrolase from *Thermobifida fusca*. *Applied Microbiology and Biotechnology*, *98*, 7815–7823.

Schmidt, J., Wei, R., Oeser, T., Belisário-Ferrari, M. R., Barth, M., Then, J., et al. (2016). Effect of Tris, MOPS, and phosphate buffers on the hydrolysis of polyethylene terephthalate films by polyester hydrolases. *FEBS Open Bio*, *6*, 919–927.

Senga, A., Hantani, Y., Bekker, G.-J., Kamiya, N., Kimura, Y., Kawai, F., et al. (2019). Metal binding to cutinase-like enzyme from *Saccharomonospora viridis* AHK190 and its effects on enzyme activity and stability. *The Journal of Biochemistry*, *166*, 149–156.

Senga, A., Numoto, N., Yamashita, M., Iida, A., Ito, N., Kawai, F., et al. (2021). Multiple structural states of Ca^{2+} regulated PET hydrolase, Cut190, and its correlation with activity and stability. *The Journal of Biochemistry*. in press.

Shirke, A., White, C., Englaender, J. A., Zwarycz, A., Butterfoss, G. L., Linhardt, R. J., et al. (2018). Stabilizing leaf and branch compost cutinase (LCC) with glycosylation; mechanism and effect on PET hydrolysis. *Biochemistry*, *57*, 1190–1200.

Sinsereekul, N., Wangkam, T., Thamchaipenet, A., Srikhirin, T., Eurwilaichitr, L., & Champreda, V. (2010). Recombinant expression of BTA hydrolase in *Streptomyces rimosus* and catalytic analysis on polyesters by surface plasmon resonance. *Applied and Environmental Microbiology*, *86*, 1775–1784.

Su, L., Woodard, R. W., Chen, J., & Wu, J. (2013). Extracellular location of *Thermobifida fusca* cutinase expressed in *E. coli* BL21(DE3) without mediation of a signal peptide. *Applied and Environmental Microbiology*, *79*, 4192–4198.

Sulaiman, S., Yamato, S., Kanaya, E., Kim, J.-J., Koga, Y., Takano, K., et al. (2012). Isolation of a novel cutinase homolog with polyethylene terephthalate-degrading activity from leaf-branch compost by using a metagenomic approach. *Applied Environmental Microbiology*, *78*, 1556–1562.

Sulaiman, S., You, D. J., Kanaya, E., Koga, Y., & Kanaya, S. (2014). Crystal structure and thermodynamic and kinetic stability of metagenome-derived LC-cutinase. *Biochemistry*, *53*, 1858–1869.

Then, J., Wei, R., Oeser, T., Barth, M., Belisario-Ferrari, M. R., Schmidt, J., et al. (2015). Ca^{2+} and Mg^{2+} binding site engineering increases the degradation of polyethylene terephthalate films by polyester hydrolase from *Thermobifida fusca*. *Biotechnology Journal*, *10*, 592–598.

Thumarat, U., Kawabata, T., Nakajima, M., Nakajima, H., Sugiyama, A., Yazaki, K., et al. (2015). Comparison of genetic structures and biochemical properties of tandem cutinase-type polyesterases from *Thermobifida alba* AHK119. *Journal of Bioscience and Bioengineering*, *120*, 491–497.

Thumarat, U., Nakamura, R., Kawabata, T., Suzuki, H., & Kawai, F. (2012). Biochemical and genetic analysis of a cutinase-type polyesterase from a thermophilic *Thermobifida alba* AHK119. *Applied Microbiology and Biotechnology*, *95*, 419–430.

Tournier, V., Topham, C. M., Gilles, A., David, B., Folgoas, C., Moya-Leclair, E., et al. (2020). An engineered PET depolymerase to break down and recycle plastic bottles. *Nature*, *580*, 216–219.

Vagin, A., & Teplyakov, A. (1997). MOLREP: An automated program for molecular replacement. *Journal of Applied Crystallography*, *30*, 1022–1025.

Wang, J. M., Wolf, R. M., Caldwell, J. W., Kollman, P. A., & Case, D. A. (2004). Development and testing of a general amber force field. *Journal of Computational Chemistry*, *25*, 1157–1174.

Wei, R., Breite, D., Song, C., Gräsing, D., Ploss, T., Hille, P., et al. (2019). Biocatalytic efficiency of postconsumer polyethylene terephthalate packaging determined by their polymer microstructures. *Advanced Science*, *6*, 190049.

Wei, R., Oeser, T., Then, J., Kühn, N., Barth, M., Schmidt, J., et al. (2014). Functional characterization and structural modeling of synthetic-polyester-degrading hydrolases from *Thermomonospora curvata*. *AMB Express*, *4*, 44.

Wei, Y., Swenson, L., Casdtro, C., Derewenda, U., Minor, W., Arai, H., et al. (1998). Structure of a microbial homologue of mammalian platelet-activating factor acetylhydrolases: *Streptomyces exfoliates* lipase at 1.9 Å resolution. *Structure*, *6*, 511–519.

Wei, R., & Zimmermann, W. (2017). Biocatalysis as a green route for recycling the recalcitrant plastic polyethylene terephthalate. *Microbial Biotechnology*, *10*, 1302–1307.

Winn, M. D., Ballard, C. C., Cowtan, K. D., Dodson, E. J., Emsley, P., Evans, P. R., et al. (2011). Overview of the CCP4 suite and current developments. *Acta Crystallographica Section D*, *67*, 235–242.

Yang, Y., Malten, M., Grote, A., Jahn, D., & Deckwer, W.-D. (2007). Codon optimized *Thermobifida fusca* hydrolase secreted by *Bacillus megaterium*. *Biotechnology and Bioengineering*, *96*, 780–794.

CHAPTER NINE

Ideonella sakaiensis, PETase, and MHETase: From identification of microbial PET degradation to enzyme characterization

Shosuke Yoshida[a,†], Kazumi Hiraga[b,†], Ikuo Taniguchi[c], and Kohei Oda[b,*]

[a]Institute for Research Initiatives, Nara Institute of Science and Technology & Division of Biological Science, Nara Institute of Science and Technology, Ikoma, Nara, Japan
[b]Kyoto Institute of Technology, Kyoto, Japan
[c]International Institute for Carbon-Neutral Energy Research, Kyushu University, Fukuoka, Japan
*Corresponding author: e-mail address: bika@kit.ac.jp

Contents

† Equal contribution.

Methods in Enzymology, Volume 648
ISSN 0076-6879
https://doi.org/10.1016/bs.mie.2020.12.007

Abstract

Few reports have described the biological degradation or utilization of poly(ethylene terephthalate) (PET) to support microbial growth. We screened environmental samples from a PET bottle recycling site and identified the microbial consortium no. 46, which degraded amorphous PET at ambient temperature; thereafter, we isolated the resident *Ideonella sakaiensis* 201-F6 strain responsible for the degradation. We further identified two hydrolytic enzymes from *I. sakaiensis*, PET hydrolase (PETase) and mono(2-hydroxyethyl) terephthalate hydrolase (MHETase), which synergistically converted PET into its monomeric building blocks. Here, we provide original methods of microbial screening and isolation of PET degrading microbe(s). These novel approaches can be adapted for exploring microorganisms that degrade PET and other plastics. Furthermore, our enzyme assay protocols to characterize PETase and MHETase can be applied to evaluate new enzymes that target PET and its hydrolysates.

1. Introduction

Poly(ethylene terephthalate) (PET) is a commonly used plastic and is nonbiodegradable in the natural environment (Devi et al., 2015; Tokiwa, Calabia, Ugwu, & Aiba, 2009). Potential factors restricting microbial or enzymatic degradation of PET are low chain mobility, high crystallinity, and high surface hydrophobicity (Shah, Hasan, Hameed, & Ahmed, 2008; Tokiwa et al., 2009; Webb, Arnott, Crawford, & Ivanova, 2013). The lack of enzymes that can cleave highly stable C—C bonds in nature deters the biodegradation of vinyl polymers. Abiotic scission activated via ultraviolet irradiation and oxidation to form carbonyl groups in polymers are needed to aid enzymatic reactions (Wei & Zimmermann, 2017b). Compared to these recalcitrant C—C bonds, ester bonds in the polymer backbone are theoretically more susceptible to biodegradation. However, a high ratio of aromatic terephthalate units in PET reduces chain mobility and markedly decreases accessibility of the hydrolysable ester bonds in their main chains. Moreover, PET is semicrystalline, comprising both crystalline and amorphous domains. Because enzymes can generally degrade the flexible amorphous domain, the biodegradation rate of plastics decreases with increasing crystallinity (Marten, Müller, & Deckwer, 2003, 2005; Tokiwa et al., 2009; Wei et al., 2019).

Despite the physiochemically recalcitrant nature of PET, numerous PET hydrolytic enzymes (PHEs) have been identified and recently reviewed (Taniguchi, Yoshida, Hiraga, Miyamoto, & Kimura, 2019). Most PHEs are cutinases (EC 3.1.1.74) that can hydrolyze cutin, an insoluble aliphatic polyester contained in plant cuticle. Cutinases are produced by plant pathogenic fungi and actinomycetes, also found in the unknown organisms included in the certain microbiome (Danso et al., 2018; Sulaiman et al., 2012; Taniguchi et al., 2019; Wei, Oeser, & Zimmermann, 2014). Cutinases are ubiquitous serine esterases that catalyze reactions with a wide range of substrates, such as short-chain soluble esters, water-insoluble medium- and long-chain triacylglycerols, and polyesters, including PET.

Although numerous active PHEs have been reported, only a few limited studies have described biological PET degradation by specific microorganisms (Nimchua, Eveleigh, Sangwatanaroj, & Punnapayak, 2008; Nimchua, Punnapayak, & Zimmermann, 2007; Yoshida et al., 2016a). In general, microorganisms in nature adhere to a surface where they develop consortiums with various microbial species. We used amorphous, low-crystallinity (Crystallinity 1.9% determined by differential scanning calorimetry) PET films as the major carbon source for screening microbes to achieve efficient PET degradation. We screened over 250 environmental samples of microbially contaminated PET debris from a PET bottle recycling site in Sakai, Osaka, Japan. We isolated microbial consortium no. 46, which completely degraded PET into CO_2 and water at ambient temperature (Yoshida et al., 2016a, 2016b). Cultured microbial consortium no. 46 adhered to and caused significant morphological changes in the PET film with the size of $20 \times 15 \times 0.15$–0.2 mm. As degradation only occurred at the film surface, the rate is displayed as 0.13 mg/cm^2 of film surface area/day at 30°C. Separate experiments showed that 75% of the carbon in the degraded PET film was converted to CO_2 at 28°C. X-ray photoelectron spectroscopy (XPS) revealed a significant increase in surface functional groups, indicating the occurrence of random chain scission or end-type degradation by PET hydrolytic enzymes. The microorganisms responsible for PET degradation were isolated by limiting dilutions of no. 46, followed by enrichment culture on PET films. The isolated strain that utilized PET as a major carbon and energy source for its growth was a new species of the genus *Ideonella*, leading to the proposed name, *Ideonella sakaiensis* 201-F6 (Tanasupawat, Takehana, Yoshida, Hiraga, & Oda, 2016).

Sequencing of the *I. sakaiensis* genome revealed a single open reading frame encoding a putative lipase sharing 51% amino acid sequence identity with a cutinase homolog from *Thermobifida fusca* (TfH), which has PET-hydrolytic activity (Müller, Schrader, Profe, Dresler, & Deckwer, 2005). Incubation of the corresponding recombinant proteins with PET film produced crater-like pitting on the film surface along with the hydrolysis products such as mono(2-hydroxyethyl) terephthalic acid (MHET) and terephthalic acid (TPA). The ratio of the activity towards PET to aliphatic esters in this enzyme was dramatically higher compared to the other enzymes, such as TfH, a cutinase homolog from leaf-branch compost metagenome (LC cutinase, or LCC) (Sulaiman et al., 2012), and *Fusarium solani* cutinase (FsC) from a fungus (Silva et al., 2005). Thus, this enzyme was designated as a PET hydrolase (PETase) and was assigned the new EC number 3.1.1.101. Another unique feature of PETase is that its thermolability, with hydrolytic activity peaking at 40°C, is distinct from general PHEs that exhibit higher thermostability (Kawai, Kawabata, & Oda, 2020; Taniguchi et al., 2019; Wei & Zimmermann, 2017a). Phylogenetic tree analysis of PHEs showed that both eukaryotes (e.g., fungal cutinases) and bacteria express these enzymes. The bacterial domain expressing PHEs is further divided into two major branches belonging to the phyla *Actinobacteria*, and *Proteobacteria* including *I. sakaiensis*.

Structure determination (Austin et al., 2018; Fecker et al., 2018; Han et al., 2017; Joo et al., 2018; Liu et al., 2018) and comparison of PETase from *I. sakaiensis* (proteobacterial PHE) with the *T. fusca* cutinase (actinobacterial PHE, PDB ID: 4CG1) revealed several unique features of PETase. The PET binding cleft is broader and shallower in PETase than in the *T. fusca* cutinase, which is largely brought by serine residue (Ser238 in PETase) instead of bulky phenylalanine (Phe209 in *T. fusca* cutinase), providing more space to accommodate PET as a substrate. Two unique cysteine residues in PETase form a disulfide bond at the vicinity of the active site, which regulates the active site flexibility of mesophilic PETase. These characteristics potentially contribute to efficient PET hydrolysis.

The rational structure-based design of PETase has overcome its low catalytic activities and low structural stability. Son et al., described a highly functional mutant harboring substitutions in three residues (S121E/D186H/R280A variant), displaying an increased Tm of 8.8°C and 4.3-fold enhanced PET degradation activity at 30°C and 9.1-fold at 40°C for 24h compared with the wild-type enzyme (Son et al., 2019). Cui et al., recently computationally redesigned PETase to generate a thermostable mutant S214H/I168R/W159H/S188Q/R280A/A180I/G165A/Q119Y/L117F/T140D (DuraPETase) that displayed an increased Tm of 31°C and 38-fold enhanced PET degradation activity at 37°C for 24h compared with the original enzyme (Cui et al., 2019). The reduction of activation energy by DuraPETase was identical to that by PETase, indicating that PETase rapidly loses its activity due to instability even at 37°C.

We also found that incubating PET film with 0.005% sodium tetradecyl sulfate for 1h before enzyme addition dramatically improved the initial

3-h catalytic activity of PETase at 30°C from 0.07 to 9.0 nmol min^{-1} cm^{-2}. This improvement in PET hydrolytic activity at 30°C was comparable to that of thermostable PHE (TfH, 2.2 nmol min^{-1} cm^{-2} at 60°C; LCC, 15 nmol min^{-1} cm^{-2} at 60°C). The cationic region formed by R53, R90, and K95 in PETase was crucial for the efficient acceleration of activity by anionic surfactants (Furukawa, Kawakami, Oda, & Miyamoto, 2018).

The major product of PET hydrolysis by PETase is MHET. A tannase family protein identified from *I. sakaiensis* efficiently hydrolyzes MHET to TPA and ethylene glycol (EG) with a k_{cat}/K_m of $4200 \pm 370\,s^{-1}\,mM^{-1}$, although it has no or very low activity with other ester compounds (Yoshida et al., 2016a). The enzyme was, therefore, designated as a MHET hydrolase (MHETase) and assigned the new EC number 3.1.1.102. The crystal structure of MHETase revealed that it is composed of an α/β-hydrolase domain and a lid domain (Palm et al., 2019). Structural and biochemical analyses of MHETase revealed that MHET binding to a catalytic pocket induces a structural change that closes the active site, and that six amino acid residues in the lid domain directly interact with the substrate, resulting in more stringent substrate recognition and more robust interaction.

The biochemical properties of PETase and MHETase, along with their localization predicted from signal peptides, indicated a PET metabolic model in *I. sakaiensis*. Both enzymes played specific roles, with PETase converting PET into oligomers including MHET as their primary component, and MHETase further hydrolyzing MHET to TPA and EG. This combination and arrangement of enzymes for metabolizing PET has not yet been identified in any other organism.

Regarding issues associated with PET pollution in the environment, microbial consortium no. 46, *I. sakaiensis* 201-F6, or new engineered microorganisms (Chen et al., 2020; Gong et al., 2018; Kim et al., 2020; Moog et al., 2019; Yan, Wei, Cui, Bornscheuer, & Liu, 2020) could be used to digest waste PET collected from oceans and coastlines into CO_2, and water in waste treatment facilities. Alternatively, PET-degrading microorganisms and their enzymes could be used to break down waste PET into its building blocks such as TPA and MHET for recovery and "bio-recycling." A potentially different application is the surface treatment of PET fibers to texturizing and staining textiles. PETase would also be applicable to the biodegradation of poly(ethylene furanoate) (PEF), which is a PET alternative derived from biomass (Austin et al., 2018; Hiraga, Taniguchi, Yoshida, Kimura, & Oda, 2019; Taniguchi et al., 2019).

2. Screening microorganisms that degrade PET and isolation of microbial consortium no. 46

The microbial screening procedures used to identify microorganisms that can grow on PET (such as *I. sakaiensis*) are described in this section. These methods include distinguishing biodegradation, quantifying the degree of biodegradation, and tracing the conversion of degraded PET. We also characterized PETase and MHETase to understand the mechanism underlying PET degradation. As PET is a polyester composed of the two building blocks (TPA and EG), estimation of hydrolytic products is easier than the degradation products of polyolefins. However, the hydrolysis products are quite heterogeneous in size, and includes both small water-soluble molecules (such as TPA, EG, MHET, and BHET) and larger water-insoluble molecules. By analyzing plastic surface properties and detecting the products of enzymatic PET hydrolysis, the degradation of plastic material can be evaluated.

2.1 Materials

1. Medium without main carbon source:
   0.05% yeast extract,
   0.2% ammonium sulfate,
   1% trace elements* (0.1% $FeSO_4 \cdot 7H_2O$, 0.1% $MgSO_4 \cdot 7H_2O$, 0.01% $CuSO_4 \cdot 5H_2O$, 0.01% $MnSO_4 \cdot 5H_2O$, and 0.01% $ZnSO_4 \cdot 7H_2O$),
   10 mM phosphate buffer (pH 7.0).
2. Modified lettuce and egg (MLE) medium:
   0.15% dried lettuce extract,
   0.15% egg extract,
   0.2% ammonium sulfate,
   1% of the trace elements,
   10 mM phosphate buffer (pH 7.0).
3. Carbon source:
   Low-crystallinity (1.9%) thin PET film (20 × 15 × 0.15–0.2 mm 60–65 mg, a kind gift from Teijin Co., Ltd).
   Amorphous PET is commercially available at Goodfellow Ltd., as Gf-PET.

2.2 Procedure

1. Collect PET-contaminated environmental samples such as sediment, soil, waste water, and activated sludge.
2. Autoclave 10 mL of medium in test tubes (Ø18 × 180 mm).

3. Sterilize PET film by immersion in 70% ethanol, then dry in sterile air at room temperature (RT) for 1 h.
4. Aseptically place PET film into medium in test tubes.
5. Add ~1 g of environmental sample to test tubes.
6. Shake test tubes at 250 strokes/min at 30°C (Fig. 1).
7. Refresh culture medium biweekly to maintain PET-degrading microorganisms.
8. Visually examine PET film and microbial growth.
9. Freeze microbial samples in 12.5% glycerol at −80°C for storage.
10. Remove PET film from test tubes.
11. Wash PET film with water, then air dry. If a microbial consortium is attached (for example, a biofilm) to the PET film, recover the attached material using a pipette and mildly disperse in water by sonication.
12. Examine PET film by eye or microscopy.

Consortium no. 46 was stably cultured with PET film in MLE medium.

3. Isolation of *I. sakaiensis* 201-F6 from microbial consortium no. 46

When target microorganisms are difficult to grow on agar plates, those responsible for PET degradation can be enriched in liquid cultures using PET as a sole carbon source. After concentrating microorganisms, agar plates were useful for single colony isolation of *I. sakaiensis*.

3.1 Materials

1. Yeast extract–sodium carbonate–vitamin (YSV) medium:
   0.01% yeast extract,
   0.02% sodium hydrogen carbonate,
   0.1% ammonium sulfate,
   0.01% calcium carbonate,
   0.1% vitamin mixture (0.05% thiamine-HCl, 0.05% niacin, 0.03% *p*-aminobenzoic acid, 0.01% pyridoxal-HCl, 0.01% pantothenate, 0.005% biotin, and 0.05% vitamin B12),
   1% trace elements*.
   10 mM phosphate buffer (pH 7.0).

3.2 Procedure

1. Serially dilute the microbial consortium with YSV medium in 10-fold steps.
2. Place diluted samples with PET film (Ø6 mm) in 96-well plates (Fig. 1).

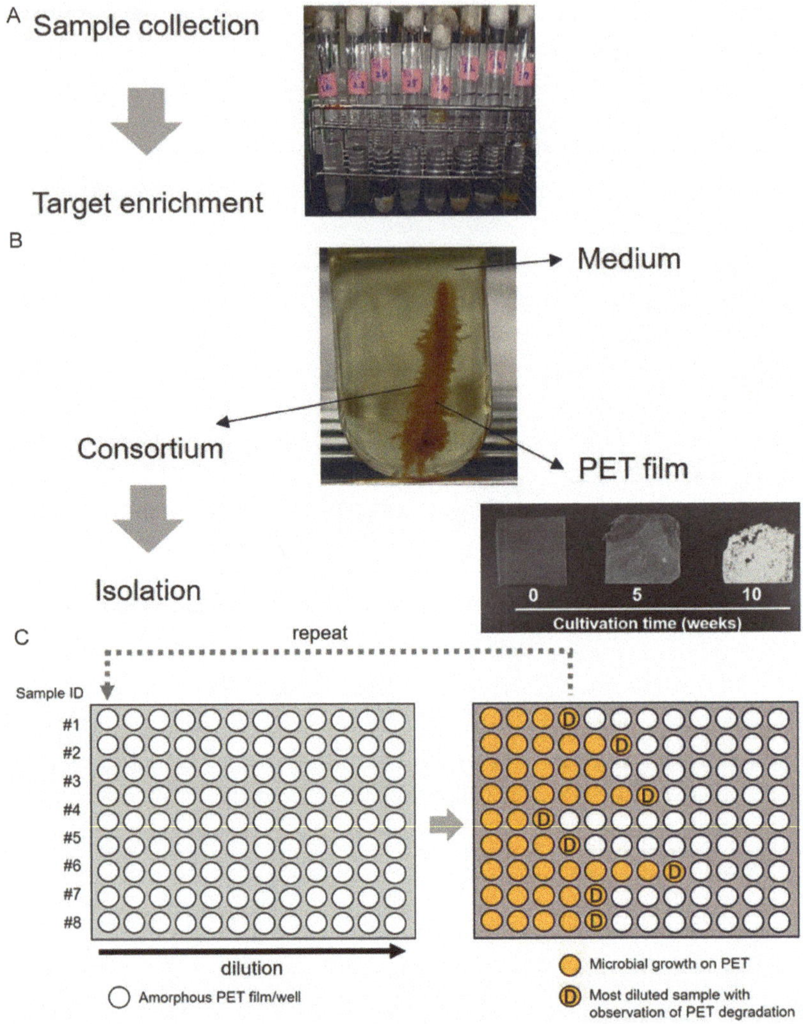

Fig. 1 Microbial screening for PET degradation. (A) Enrichment culture of PET-degrading and assimilating microorganisms. We used environmental samples contaminated with PET debris, and individually placed them in minimal medium with amorphous PET film as the major carbon source. Samples were cultured by shaking at 30°C. (B) Microbial consortium after 20 days of incubation. Degradation of PET is confirmed by weight loss and conversion into CO_2 and biomass. Biodegradation is evidenced by changes in film surface morphology and functional groups. (C) Isolation of PET-degrading and assimilating microorganism(s). Limiting dilution and enrichment culture on PET film both decrease the heterogeneity of the microbial consortium. We combined these two methods to develop a pipeline that can concentrate target microorganisms in 96-well plates.

3. Culture statically at 30°C.
4. Repeat the above three steps to concentrate PET-degrading microorganisms.
5. Monitor cultured microbes by PCR amplification of 16S rDNA V3–V5 region of genomic DNA extracted from cultures using a bacterial 16S rDNA-specific primer pair: forward, 5′-CGCCCGCCGCGCCCCGCGCCCGTCCCGCCGCCCCCGCCCGCCTACGGGAGGCAGCAG-3′; reverse, 5′-CCCCGTCAATTCCTTTGAGTTT-3′, (GC clamp is underlined), followed by denaturing gradient gel electrophoresis (Muyzer, de Waal, & Uitterlinden, 1993).
6. Spread microbes on YSV agar (0.5% w/v) plates for single colony isolation.

4. Detection of microbial PET degradation

The degradation of PET film can be evaluated as weight loss, CO_2 generation via PET catabolism, surface morphological changes in films determined by scanning electron microscopy (SEM), and as increase in numbers of surface functional groups determined by X-ray photoelectron spectroscopy (XPS). However, degradation in the initial state is difficult to study the mechanism using NMR and GPC because it usually proceeds from the surface and the bulk remains intact. In addition, rapid metabolism of intermediates limits detection of PET metabolites by high-performance liquid chromatography (HPLC).

4.1 Weight loss

4.1.1 Procedure

1. Remove PET films from cultures.
2. Immerse films in 1% SDS and gently rotate for 5 min.
3. Repeat step 2.
4. Place in distilled water and gently rotate for 5 min.
5. Repeat step 2 twice.
6. Immerse in ethanol.
7. Air-dry at RT for 1 h (vacuum drying is best).
8. Weigh film and calculate weight loss.

4.2 CO_2 generation

4.2.1 Equipment

1. CO_2 trap: Constructed according to the Japanese Standards Association (JSA) JIS K6951 method to entrap CO_2 generated in Ascarite (sodium hydroxide coated silica, Sigma-Aldrich).
2. Total organic carbon (TOC) analyzer.

4.2.2 Procedure

1. After cultivation, weigh Ascarite to calculate absorbed CO_2.
2. Separate biomass and a treated PET film in culture medium by centrifugation at 8000 rpm for 20 min.
3. Remove the PET film from the supernatant.
4. Analyze TOC contents of supernatant and precipitate.
5. Calculate the conversion rate from PET to CO_2 (R) as follows:

$$R(\%) = \frac{CO_2(PET+) - CO_2(PET-)}{\text{carbon weight of degraded PET film}} \times 100$$

where CO_2 (PET+) and CO_2 (PET−) indicate the carbon weight of CO_2 generated in the presence and absence of PET, respectively.

4.3 SEM

4.3.1 Equipment

1. Plasma ion coater (OPC-80; Nippon Laser & Electronics Lab., Nagoya, Japan).
2. FE-SEM (JEOL, Tokyo, Japan).

4.3.2 Procedure

1. Coat dried film with osmium tetroxide (highly toxic, but safety is ensured both in the glass ampule and in OPC-80) using a plasma ion coater.
2. Examine the film surface using FE-SEM operating at an electron beam intensity of 5 kV.

4.4 XPS

Surface atomic composition and chemical bonding modes are analyzed by detecting photoelectrons released from the substrate surface via X-ray irradiation. Peaks in the narrow scan of X-ray photoelectron spectroscopy at 284, 286, and 288 eV in the C_{1s} region are assigned to photoelectrons from *C*—C, *C*—O, and *C*=O. These three peaks overlap, and the contribution of each peak can be calculated by a deconvolution process on an analysis software equipped with the photoelectron spectrometer. However, for precise quantification, the resulting hydroxyl and carboxyl groups upon hydrolysis are separately labeled with F and N, respectively. The PET substrates were sequentially reacted with heptafluorobutyryl chloride followed by 1,1′-carbonyldiimidazole in anhydrous tetrahydrofuran to introduce F and N atoms to the hydroxyl and carboxyl groups, respectively. Photoelectrons from C_{1s}, F_{1s}, and N_{1s} were collected by the narrow scan to determine a surface atomic composition of the PET films. The surface functional density (groups per cm^2) was calculated with a PET density of $1.375\,g/cm^3$ (Yoshida et al., 2016b).

4.4.1 Equipment

1. JPS-9010MC/SP photoelectron spectrometer (JEOL Ltd., Tokyo, Japan) with a 100W MgKα X-ray source (λ 9.889 Å, 1253.5 eV) and take-off angle of 45°, which presumably collects photoelectrons from the surface to a depth of ~3 nm. Narrow scan spectra were obtained by accumulating 15 scans at intervals of 0.1 eV.

4.4.2 Chemicals

1. Highly corrosive heptafluorobutyryl chloride and 1,1'-carbonyldiimidazole (Sigma-Aldrich, St. Louis, MO, USA).

4.4.3 Procedure

1. Place the PET films ($20 \times 15 \times 0.15$–$0.2\,mm^3$) in 100 mL of heptafluorobutyryl chloride/tetrahydrofuran (0.27 mol/L) for 2 h under N_2 to label –OH groups with F.
2. Wash the films with tetrahydrofuran, methanol and hexane, and then dry.
3. Place the films in 100 mL of 1,1′-carbonyldiimidazole/tetrahydrofuran (0.12 mol/L) for 4 h under N_2 to label –COOH groups with N.
4. Wash the films with tetrahydrofuran, methanol and hexane, and then dry.
5. Collect photoelectrons of F_{1s} at 685 eV, N_{1s} at 400 eV, and C_{1s} at 285 eV in the narrow scan.

5. Characterization of PETase and MHETase

5.1 Cloning, expression, and purification of PETase and MHETase

The genes encoding PETase and MHETase, without nucleotide sequences corresponding to the signal peptide predicted by the LipoP 1.0 server (http://www.cbs.dtu.dk/services/LipoP) (Juncker et al., 2003), were codon-optimized for expression in *Escherichia coli* cells. The gene encoding PETase was cloned into the pET-21b vector (Novagen, San Diego, CA, USA). The gene was expressed in *E. coli* BL21 (DE3) CodonPlus RIPL competent cells (Agilent Technologies GmbH., Waldbronn, Germany) via 0.1 mM IPTG induction at 16°C. The gene encoding MHETase was cloned into the pCold II vector (Takara Bio, Otsu, Japan). Genes were expressed in

E. coli Rosetta-gami B(DE3) (Merck Millipore, Darmstadt, Germany) via 0.1 mM IPTG induction at 15°C. *E. coli* was harvested by centrifugation (5000 × *g*, 5 min, 4°C) and resuspended in lysis buffer (50 mM Tris–HCl, pH 7.5, 300 mM NaCl, 20 mM imidazole-HCl). Cells were sonicated on ice and the lysate was clarified by centrifugation (15,000 × *g*, 20 min, 4°C). The supernatant was then applied to a Ni-NTA column. Free proteins were removed by washing with lysis buffer, then bound proteins were eluted with 50 mM Tris–HCl, pH 7.5, 300 mM NaCl, 250 mM imidazole-HCl (elution buffer). The buffer was exchanged for 50 mM Na_2HPO_4-HCl, pH 7.0, then the proteins were purified by passage through a PD-10 desalting column (GE Healthcare, Piscataway, NJ, USA). Purified protein concentrations were determined from the calculated molar extinction coefficient at 280 nm.

5.2 Characterization of PETase and other PHEs

The hydrolytic activity of PETase was estimated by detecting and quantifying products released from PET into water using reverse-phase HPLC. The compounds detected at an absorbance of 240 nm comprised TPA, MHET, and bis(2-hydroxyethyl)terephthalic acid (BHET). PET degradation was also estimated by monitoring morphological changes in the PET surface by SEM.

Furthermore, we measured the activities against PET film and *p*-nitrophenol–linked aliphatic esters (*p*NP-aliphatic esters), widely used to characterize lipases and cutinases, at 30°C and pH 7.0, and calculated the ratio of the hydrolytic activities for PET to those for *p*NP-aliphatic esters (Yoshida et al., 2016a). Wherein PETase displayed substantially higher values of the ratio compared with other PHEs (TfH, LCC, and FsC), accounting for 380, 48, and 400 times, respectively. The catalytic preference of PET to aliphatic esters, compared with the other enzymes, led to its designation as a PETase. Therefore, this method could allow us to discriminate PETase enzymes, having higher substrate specificity for PET, from other PHEs.

In addition, the chain mobility of amorphous PET increased upon approaching the glass transition temperature at >65°C, which resulted in increased PET hydrolysis. Therefore, the temperature-dependent activity and thermostability of PHEs is important to determine when considering their practical industrial applications.

5.2.1 PET (solid material) degradation assay

5.2.1.1 Equipment

1. High performance liquid chromatography (HPLC): HPLC system (LC-2010A HT, Shimadzu Corporation, Kyoto, Japan) equipped with

a Cosmosil 5C18-AR-II guard column and Cosmosil 5C18-AR-II column (Nacalai Tesque Inc., Kyoto, Japan). Mobile phase: methanol/20 mM phosphate buffer (pH 2.5); flow rate, 1.0 mL/min. Monitoring wavelength: 240 nm. Elution: 0 to 15 min, 25% (v/v) methanol; 15 to 25 min, 25%–100% linear methanol gradient.

5.2.1.2 Procedure

1. Prepare 293 μL of 50 mM Na_2HPO_4-HCl (pH 6.0–8.0), Bicine-NaOH (pH 8.0–9.0), or 50 mM glycine-NaOH (pH 9.0–10 buffer) in 1.5-mL tubes.
2. Place low-crystallinity (1.9%) thin PET film (Ø6 mm) in the buffer and quickly spin down to immerse the film in the buffer.
3. Add 7.5 μL of 4 μM (final: 100 nM) purified PETase into the mixture, gently invert the tubes several times, and quickly spin down.
4. Incubate statically at 30°C (or temperature[s] of interest) for hours to days.
5. Terminate the reaction by diluting the aqueous solution with 90% 20 mM phosphate buffer (pH 2.5)/10% DMSO (v/v), followed by heating at 85°C for 10 min.
6. Centrifuge the tube (15,000 × *g*, 10 min), then decant and analyze the supernatant using HPLC (Fig. 2).
7. Wash the PET film with 1% SDS, distilled water, and ethanol, then air-dry it for 1 h at RT for SEM analysis (Fig. 2).

Total amounts of enzymatically released compounds from PET will provide consistent values in degradation performance with those derived from the resultant decrease in thickness of PET film (Fig. 5b in (Furukawa et al., 2018).

5.2.2 Assay of enzymatic BHET hydrolysis

5.2.2.1 Procedure

1. Prepare 233 μL of 50 mM Na_2HPO_4-HCl (pH 7.0) in 1.5-mL tubes.
2. Add 60 μL of 5 mM BHET in DMSO to the tubes and mix using a pipette.
3. Add 7.5 μL of 4 μM (final: 100 nM) of purified PETase to the mixture, gently invert the tube several times, and quickly spin down.
4. Incubate statically at 30°C for hours to days.
5. Terminate the reaction by dilution with 80% 20 mM phosphate buffer (pH 2.5)/20% DMSO (v/v), followed by heating at 85°C for 10 min.
6. Centrifuge the tubes (15,000 × *g*, 10 min).
7. Analyze the supernatant by HPLC.

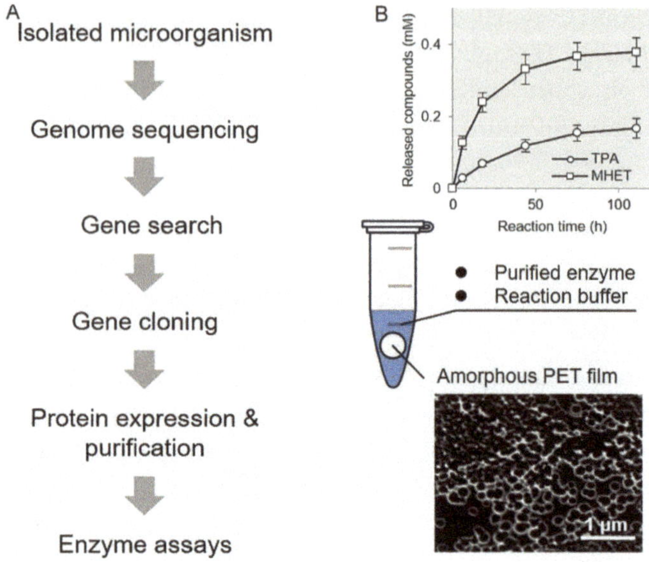

Fig. 2 Genome-based enzyme identification and assay of PET hydrolytic activity. (A) Schematic of genome-based identification of enzymes isolated from microorganisms using next-generation sequencing technology. This inexpensive method can provide genome information about unique microorganisms isolated in the laboratory. Genome-based technologies such as proteomics, transcriptomics, metabolomics, and in silico homology studies are powerful tools that can be used to search for target enzymes. (B) Simple assays of PET hydrolytic enzymes. *Upper panel* shows compounds released from PET by PETase. *Lower panel* shows surface morphological changes in PET film determined by SEM.

5.2.3 Plate reader assays for para-nitrophenol (pNP)-aliphatic esters

5.2.3.1 Chemicals and reagents

1. Substrate stock-solution (SS):
   10 mM *p*NP-acetate (Sigma-Aldrich),
   10 mM *p*NP-butyrate (Sigma-Aldrich),
   10 mM *p*NP-octanoate (Sigma-Aldrich),
   10 mM *p*NP-hexanoate (Tokyo Chemical Industry Co. Ltd., Tokyo, Japan). These are dissolved in acetonitrile.
2. Enzyme stock-solution (ES):
   250 nM PETase in 50 mM Na_2HPO_4-HCl (pH 7.0).
3. Reaction buffer (RB):
   50 mM Na_2HPO_4-HCl (pH 7.0), 0.3% Triton X-100.

5.2.3.2 Procedure

1. Add 20 μL of ES to each well of 96-well plates.
2. Mix 25 μL of SS and 200 μL of RB in other wells.

3. Incubate plates for 3 min at 30°C.
4. Transfer 180 μL of SS/RB to wells containing ES.
5. Continuously monitor *p*NP production at 415 nm using a plate reader.
6. Identify the initial catalytic rate with linearity.
7. Calculate specific activity using a *p*NP standard curve in 10% acetonitrile/90% RB (v/v) at pH 7.0.

5.2.4 Temperature-dependence assay

The PET hydrolytic activity of PETase and other PHEs (TfH, LCC, and FsC) was optimal at alkaline pH (commonly around pH 9.0) (Yoshida et al., 2016a). To identify the initial PET hydrolytic rate with good measurement sensitivity, the detection of soluble compounds released from PET during a short reaction period at alkaline pH is critical.

5.2.4.1 Procedure

1. Prepare 293 μL of Bicine-NaOH (pH 9.0) in 1.5-mL tubes.
2. Add amorphous PET film (Ø6 mm) into the buffer and quickly spin down to immerse the film in the buffer.
3. Incubate at an appropriate temperature for 3 min.
4. Add 7.5 μL of 4 μM (final: 100 nM) purified PETase into the mixture, gently invert the tube several times, and immediately centrifuge.
5. Incubate the mixture for 1 h.
6. Terminate the reaction by adding 100 μL of an equal amount of 90% 200 mM phosphate buffer (pH 2.5)/10% DMSO (v/v), followed by heating at 85°C for 10 min and centrifuge the tubes (15,000 × *g*, 10 min).
7. Analyze the supernatant using HPLC.

5.3 Characterization of MHETase

5.3.1 Enzymatic preparation of MHET crude solution

1. Mix 200 μL of 20 mM BHET in DMSO with 780 μL of 50 mM Na_2HPO_4-HCl (pH 7.0).
2. Add 20 μL of 4 μM (final: 80 nM) PETase in 50 mM Na_2HPO_4-HCl (pH 7.0).
3. Fill up to 1000 μL with 50 mM Na_2HPO_4-HCl (pH 7.0).
4. Incubate statically at 30°C for 24 h.
5. Confirm complete hydrolysis of BHET to MHET using HPLC.
6. Eliminate protein from reaction mixtures using Amicon Ultra 10 kDa (Merck Millipore), resulting in a 4 mM MHET crude solution containing EG (MHET stock).

5.3.2 Kinetic analysis of MHETase

1. Dilute MHET stock solution with 80% 50 mM Na_2HPO_4-HCl (pH 7.0)/20% DMSO (v/v) from 12.5 to 312.5 μM.
2. Place 640 μL of the diluted MHET stock solution in 1.5-mL tubes at 30°C for 3 min.
3. Place 10 nM MHETase protein in the 1.5-mL tubes at 30°C for 3 min.
4. Initiate reactions by adding 160 μL of MHETase (final concentration, 2 nM) to tubes containing diluted MHET stock solution and incubate at 30°C for 10, 20 and 30 min.
5. Sample 200 μL of reaction mixtures.
6. Terminate reactions with an equal volume of 80% 200 mM phosphate buffer (pH 2.5)/20% DMSO (v/v), followed by heating at 85°C for 10 min, and centrifuge at 15,000 × *g* for 10 min.
7. Analyze the supernatant using HPLC.
8. Detect the TPA product.
9. Identify the initial catalytic rate with linearity and plot the initial rates against MHET concentrations.
10. Determine the kinetic parameters according to the Michaelis–Menten equation using Graph Pad Prism (GraphPad Software, San Diego, CA, USA).

6. Summary

Accumulated plastic waste causes environmental pollution and thus threatens ecosystems. This chapter describes the identification of the consortium no. 46, *I. sakaiensis* 201-F6, and its enzymes that are involved in PET degradation (PETase and MHETase), and generalizes the methods to isolate PET-degrading microorganisms and to characterize the key hydrolytic enzymes. The discovery of *I. sakaiensis* shows that PET-degrading bacteria exist in nature. *I. sakaiensis* cannot grow on the medium including 3% (w/v) NaCl (Tanasupawat et al., 2016). Therefore, isolation of other PET-degrading microorganisms from high-salt environments might help to address problems caused by the spread of microplastics in oceans. Thermostable PETase and MHETase could be designed based on their 3D structures. The ability of *I. sakaiensis* to degrade PET could be potentially enhanced by synthetic biology approaches. These possibilities will inspire a new sustainable bioindustry along with novel environmental technologies, heralding a new "bio-economy" based on recycled and renewable resources instead of fossil resources. *I. sakaiensis* and its enzymes have great potential for PET degradation in a circular economy.

Acknowledgments

We thank A. Wlodawer for critical comments regarding this manuscript. This study was supported by grants-in-aid for scientific research (17H03794 to S.Y.). The nucleotide sequence data are available in the DNA Data Bank of Japan, European Molecular Biology Laboratory, and GenBank databases under the accession numbers BBYR01000001 to BBYR01000227. *Ideonella sakaiensis* 201-F6^T has been deposited as strains NBRC 110686^T and TISTR 2288^T.

Declarations of interest

None.

References

Austin, H. P., Allen, M. D., Donohoe, B. S., Rorrer, N. A., Kearns, F. L., Silveira, R. L., et al. (2018). Characterization and engineering of a plastic-degrading aromatic polyesterase. *Proceedings of the National Acadamy of Sciences of the United States of America, 115*, E4350–E4357.

Chen, Z. Z., Wang, Y. Y., Cheng, Y. Y., Wang, X., Tong, S. W., Yang, H. T., et al. (2020). Efficient biodegradation of highly crystallized polyethylene terephthalate through cell surface display of bacterial PETase. *Science of the Total Environment, 709*, 136138.

Cui, Y., Chen, Y., Liu, X., Dong, S., Tian, Y., Qiao, Y., et al. (2019). Computational redesign of a PETase for plastic biodegradation by the GRAPE strategy. *bioRxiv*. https://doi.org/10.1101/787069 (preprint).

Danso, D., Schmeisser, C., Chow, J., Zimmermann, W., Wei, R., Leggewie, C., et al. (2018). New insights into the function and global distribution of polyethylene terephthalate (PET)-degrading bacteria and enzymes in marine and terrestrial metagenomes. *Applied and Environmental Microbiology, 84*, e02773-17.

Devi, R. S., Kannan, V. R., Natarajan, K., Nivas, D., Kannan, K., Chandra, S., et al. (2015). The role of microbes in plastic degradation. In R. Chandra (Ed.), *Environmental waste management* (pp. 341–370). Florida: CRC Press.

Fecker, T., Galaz-Davison, P., Engelberger, F., Narui, Y., Sotomayor, M., Parra, L. P., et al. (2018). Active site flexibility as a hallmark for efficient PET degradation by *I. sakaiensis* PETase. *Biophysical Journal, 114*, 1302–1312.

Furukawa, M., Kawakami, N., Oda, K., & Miyamoto, K. (2018). Acceleration of enzymatic degradation of poly(ethylene terephthalate) by surface coating with anionic surfactants. *ChemSusChem, 11*, 4018–4025.

Gong, J. X., Kong, T. T., Li, Y. Q., Li, Q. J., Li, Z., & Zhang, J. F. (2018). Biodegradation of microplastic derived from poly(ethylene terephthalate) with bacterial whole-cell biocatalysts. *Polymers, 10*.

Han, X., Liu, W., Huang, J. W., Ma, J., Zheng, Y., Ko, T. P., et al. (2017). Structural insight into catalytic mechanism of PET hydrolase. *Nature Communications, 8*, 2106.

Hiraga, K., Taniguchi, I., Yoshida, S., Kimura, Y., & Oda, K. (2019). Biodegradation of waste PET. *EMBO Reports, 20*, e49365.

Joo, S., Cho, I. J., Seo, H., Son, H. F., Sagong, H. Y., Shin, T. J., et al. (2018). Structural insight into molecular mechanism of poly(ethylene terephthalate) degradation. *Nature Communications, 9*, 382.

Juncker, A. S., Willenbrock, H., Von Heijne, G., Brunak, S., Nielsen, H., & Krogh, A. (2003). Prediction of lipoprotein signal peptides in Gram-negative bacteria. *Protein Science, 12*, 1652–1662.

Kawai, F., Kawabata, T., & Oda, M. (2020). Current state and perspectives related to the polyethylene terephthalate hydrolases available for biorecycling. *ACS Sustainable Chemistry & Engineering*, *8*, 8894–8908.

Kim, J. W., Park, S. B., Tran, Q. G., Cho, D. H., Choi, D. Y., Lee, Y. J., et al. (2020). Functional expression of polyethylene terephthalate-degrading enzyme (PETase) in green microalgae. *Microbial Cell Factories*, *19*, 97.

Liu, B., He, L., Wang, L., Li, T., Li, C., Liu, H., et al. (2018). Protein crystallography and site-direct mutagenesis analysis of the poly(ethylene terephthalate) hydrolase PETase from *Ideonella sakaiensis*. *Chembiochem*, *19*, 1471–1475.

Marten, E., Müller, R. J., & Deckwer, W. D. (2003). Studies on the enzymatic hydrolysis of polyesters I. low molecular mass model esters and aliphatic polyesters. *Polymer Degradation and Stabilility*, *80*, 485–501.

Marten, E., Müller, R. J., & Deckwer, W. D. (2005). Studies on the enzymatic hydrolysis of polyesters. II. Aliphatic-aromatic copolyesters. *Polymer Degradation and Stabilility*, *88*, 371–381.

Moog, D., Schmitt, J., Senger, J., Zarzycki, J., Rexer, K. H., Linne, U., et al. (2019). Using a marine microalga as a chassis for polyethylene terephthalate (PET) degradation. *Microbial Cell Factories*, *18*, 171.

Müller, R. J., Schrader, H., Profe, J., Dresler, K., & Deckwer, W. D. (2005). Enzymatic degradation of poly(ethylene terephthalate): Rapid hydrolyse using a hydrolase from *T. fusca*. *Macromolecular Rapid Communications*, *26*, 1400–1405.

Muyzer, G., de Waal, E. C., & Uitterlinden, A. G. (1993). Profiling of complex microbial populations by denaturing gradient gel electrophoresis analysis of polymerase chain reaction-amplified genes coding for 16S rRNA. *Applied Environmental Microbiology*, *59*, 695–700.

Nimchua, T., Eveleigh, D. E., Sangwatanaroj, U., & Punnapayak, H. (2008). Screening of tropical fungi producing polyethylene terephthalate-hydrolyzing enzyme for fabric modification. *Journal of Industrial Microbiology & Biotechnology*, *35*, 843–850.

Nimchua, T., Punnapayak, H., & Zimmermann, W. (2007). Comparison of the hydrolysis of polyethylene terephthalate fibers by a hydrolase from *Fusarium oxysporum* LCH I and *Fusarium solani* f sp. *pisi*. *Biotechnology Journal*, *2*, 361–364.

Palm, G. J., Reisky, L., Bottcher, D., Muller, H., Michels, E. A. P., Walczak, M. C., et al. (2019). Structure of the plastic-degrading Ideonella sakaiensis MHETase bound to a substrate. *Nature Communications*, *10*, 1717.

Shah, A. A., Hasan, F., Hameed, A., & Ahmed, S. (2008). Biological degradation of plastics: A comprehensive review. *Biotechnology Advances*, *26*, 246–265.

Silva, C. M., Carneiro, F., O'Neill, A., Fonseca, L. P., Cabral, J. S. M., Guebitz, G., et al. (2005). Cutinase—A new tool for biomodification of synthetic fibers. *Journal of Polymer Science Part A: Polymer Chemistry*, *43*, 2448–2450.

Son, H. F., Cho, I. J., Joo, S., Seo, H., Sagong, H. Y., Choi, S. Y., et al. (2019). Rational protein engineering of thermo-stable PETase from *Ideonella sakaiensis* for highly efficient PET degradation. *ACS Catalysis*, *9*, 3519–3526.

Sulaiman, S., Yamato, S., Kanaya, E., Kim, J. J., Koga, Y., Takano, K., et al. (2012). Isolation of a novel cutinase homolog with polyethylene terephthalate-degrading activity from leaf-branch compost by using a metagenomic approach. *Applied Environmental Microbiology*, *78*, 1556–1562.

Tanasupawat, S., Takehana, T., Yoshida, S., Hiraga, K., & Oda, K. (2016). *Ideonella sakaiensis* sp. nov., isolated from a microbial consortium that degrades PET. *International Journal of Systematic and Evolutionary Microbiology*, *66*, 2813–2818.

Taniguchi, I., Yoshida, S., Hiraga, K., Miyamoto, K., Kimura, Y., & Oda, K. (2019). Biodegradation of PET: Current status and application aspects. *ACS Catalysis*, *9*, 4089–4105.

Tokiwa, Y., Calabia, B. P., Ugwu, C. U., & Aiba, S. (2009). Biodegradability of plastics. *International Journal of Molecular Science*, *10*, 3722–3742.

Webb, H. K., Arnott, J., Crawford, R. J., & Ivanova, E. P. (2013). Plastic degradation and its environmental implications with special reference to poly(ethylene terephthalate). *Polymers*, *5*, 1–18.

Wei, R., Breite, D., Song, C., Grasing, D., Ploss, T., Hille, P., et al. (2019). Biocatalytic degradation efficiency of postconsumer polyethylene terephthalate packaging determined by their polymer microstructures. *Advanced Science (Weinh)*, *6*, 1900491.

Wei, R., Oeser, T., & Zimmermann, W. (2014). Synthetic polyester-hydrolyzing enzymes from thermophilic actinomycetes. *Advances in Applied Microbiology*, *89*, 267–305.

Wei, R., & Zimmermann, W. (2017a). Biocatalysis as a green route for recycling the recalcitrant plastic polyethylene terephthalate. *Microbial Biotechnology*, *10*, 1302–1307.

Wei, R., & Zimmermann, W. (2017b). Microbial enzymes for the recycling of recalcitrant petroleum-based plastics: How far are we? *Microbial Biotechnology*, *10*, 1308–1322.

Yan, F., Wei, R., Cui, Q., Bornscheuer, U. T., & Liu, Y. J. (2020). Thermophilic whole-cell degradation of polyethylene terephthalate using engineered *Clostridium thermocellum*. *Microbial Biotechnology*. https://doi.org/10.1111/1751-7915.13580.

Yoshida, S., Hiraga, K., Takehana, T., Taniguchi, I., Yamaji, H., Maeda, Y., et al. (2016a). A bacterium that degrades and assimilates poly(ethylene terephthalate). *Science*, *351*, 1196–1199.

Yoshida, S., Hiraga, K., Takehana, T., Taniguchi, I., Yamaji, H., Maeda, Y., et al. (2016b). Response to Comment on "A bacterium that degrades and assimilates poly(ethylene terephthalate)". *Science*, *353*, 759.

CHAPTER TEN

GRAPE, a greedy accumulated strategy for computational protein engineering

Jinyuan Sun[a,b], Yinglu Cui[a], and Bian Wu[a,*]
[a]CAS Key Laboratory of Microbial Physiological and Metabolic Engineering, State Key Laboratory of Microbial Resources, Institute of Microbiology, Chinese Academy of Sciences, Beijing, China
[b]University of Chinese Academy of Sciences, Beijing, China
*Corresponding author: e-mail address: wub@im.ac.cn

Contents

Abstract

Nature harbors fascinating enzymatic catalysts with high efficiency, chemo-, regio- and stereoselectivity. However, the insufficient stability of the enzymes often prevents their widespread utilization for industrial processes. Not content with the finite repertoire of naturally occurring enzymes, protein engineering holds promises to extend the applications of the improved enzymes with desired physical and catalytic properties. Herein, we devised a computational strategy (greedy accumulated strategy for protein engineering, GRAPE) to enhance the thermostability of enzymes. Through scanning of

Methods in Enzymology, Volume 648
ISSN 0076-6879
https://doi.org/10.1016/bs.mie.2020.12.026

all point mutations of the structural and evolutionary consensus analysis, a library containing fewer than 100 mutations was established for characterization. After preliminary experimental verification, effective mutations are clustered in a multidimensional physical property space and then accumulated via the greedy algorithm to produce the final designed enzyme. Using the recently reported *Is*PETase from *Ideonella sakaiensis* that decomposes PET under ambient temperatures as a starting point, we adopted the GRAPE strategy to come up with a DuraPETase (T_M=77 °C, raised by 31 °C) which showed drastically enhanced degradation performance (300-fold) on semicrystalline PET films at 40 °C.

1. Introduction

As natural biocatalysts, enzymes have high catalytic efficiency and substrate specificity, and therefore are with great potential for industrial application. However, the low robustness of wild-type enzymes generally restricts their application under harsh industrial conditions. Many protein engineering methods were developed to tackle the bottleneck of stability. Of particular interest for protein engineering is directed evolution, which simulates the natural evolution pathway to screen for variants with desired properties in a randomly generated library round by round (Tracewell & Arnold, 2009). However, laboratory evolution usually requires high throughput expression and screening methods compatible with large mutant libraries. With the development of bioinformatic methods, computational enzyme design has significantly advanced from fairly simple to increasingly accurate and versatile methods. Several computational strategies have been developed to tackle the bottleneck of stability, leading to a group of enzymes with improved thermostability by significantly smaller wet-lab experiment efforts. Nevertheless, when the desired function is beyond a single or double mutation, not only the accuracy of computational methods is falling, but also the number of possible paths is growing exponentially, which is NP (nondeterministic polynomial time)-hard (Pierce & Winfree, 2002). An NP problem requires an exhaustive search to prove the solution's correctness by examining every possible answer if it cannot be reduced to a P problem ($P \neq NP$). For example, for a protein with 300 amino acids, the number of ways to make a four mutation variant is more than 50 trillion.

To tackle the NP-hard enzyme design problem, we introduce a computational strategy for protein engineering, termed GRAPE, to reduce the library scale and reduce the risk of combining mutations with antagonistic effects. The GRAPE strategy (Fig. 1) addresses the evolutionary hard

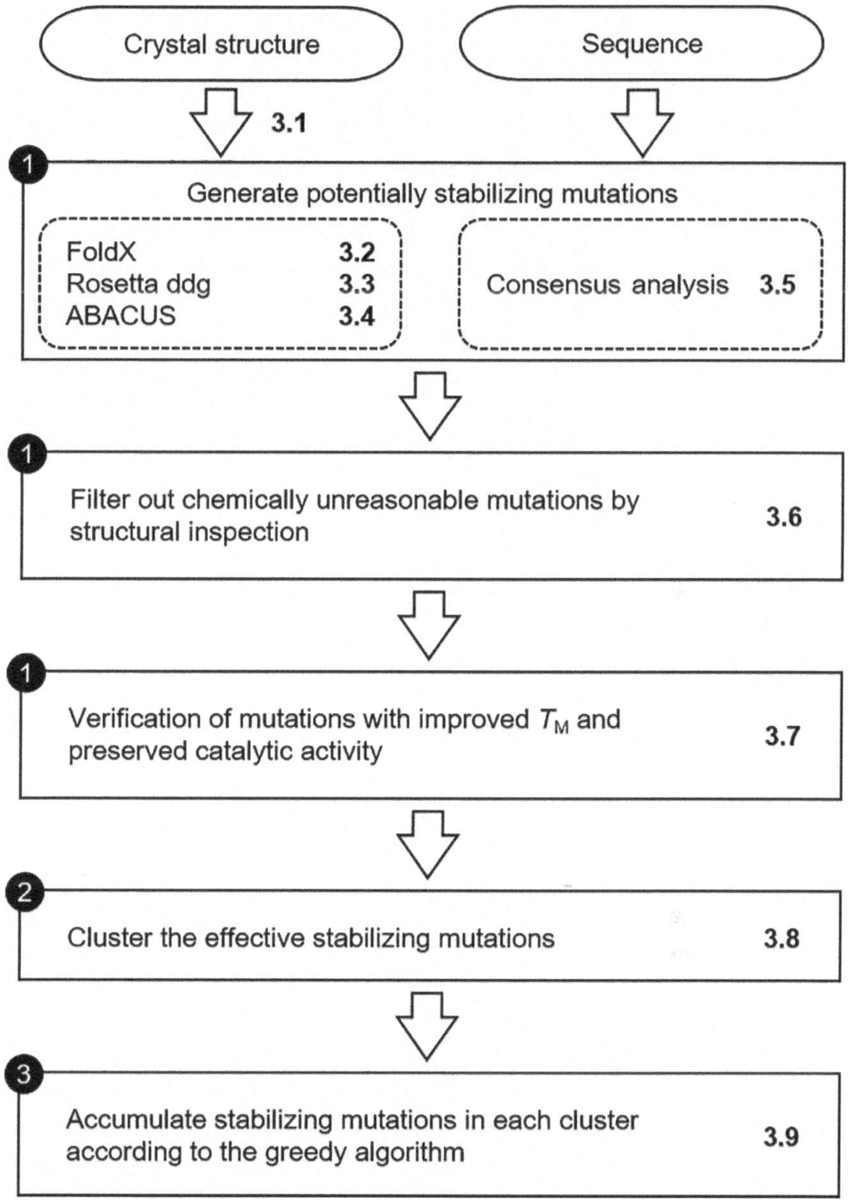

Fig. 1 The flowchart of greedy accumulated strategy for protein engineering (GRAPE). The left numbers 1 refers to the single mutation identification part, 2 to the clustering and 3 for the accumulating. Right numbers refer to sections in this protocol.

problem by in silico screening, clustering, and accumulation. It combines the clustering analysis with a greedy accumulation of computationally designed effective mutations, enhancing the probability of discovering the shortest adaptive route to improve the fitness of the multidimensional sequence-function space. This way, a relatively small library will be established, which does not require high throughput characterization methods.

To demonstrate the practicality of the GRAPE strategy, we applied it to engineer an enzyme to improve the protein thermostability. Biocatalytic degradation is considered an option of plastic waste disposal of e.g. polyethylene terephthalate (PET) in a more eco-friendly sustainable way. Most of the reported PET degrading enzymes are thermophilic cutinases that exhibit their highest activity at elevated temperatures >50 °C (Kawai, Kawabata, & Oda, 2020; Wei & Zimmermann, 2017). Recently, *Is*PETase from *Ideonella sakaiensis* was reported to degrade amorphous PET plastics at ambient temperature <40 °C (Yoshida et al., 2016). However, the low durability hampers the long-time application of this enzyme. We applied the GRAPE strategy to redesign *Is*PETase, and finally achieved the DuraPETase through 87 single point mutation tests and 67 accumulated multi-mutation tests. The apparent melting temperature (T_M) of DuraPETase (77 °C) was 31 °C higher than that of wild-type *Is*PETase. Meanwhile, the long-time degrading ability toward the semicrystalline (30%) PET films at 40 °C was increased over 300-fold as well.

2. Materials

The GRAPE workflow consists of executables and scripts to apply the four software packages to identify thermostabilizing single point mutations. It is tested under a user-friendly macOS system and Linux distribution Ubuntu 16.04 LTS (see *Note* 1). FoldX (Delgado, Radusky, Cianferoni, & Serrano, 2019), Rosetta (Leaver-Fay et al., 2011), and ABACUS (Xiong et al., 2014) are employed to predict energy changes upon point mutations. HMMER (Johnson, Eddy, & Portugaly, 2010), Muscle (Edgar, 2004), and Skylign (Wheeler, Clements, & Finn, 2014) servers are used for consensus analysis. Molecular graphics is performed using Pymol (DeLano, 2002). Rosetta and FoldX can be downloaded from https://groups.google.com/forum/#!forum/fresco-stabilization-of-proteins. ABACUS can be downloaded from http://biocomp.ustc.edu.cn/servers/abacus-design.php. Other scripts and files used in this protocol are available as Supplementary materials in the online version at https://doi.org/10.1016/bs.mie.2020.12.026.

Hardware requirements—The GRAPE workflow does not require a large amount of computation power. The calculation time to predict the ΔΔG of a single mutation is not very time-consuming, even within minutes using a MacBook computer (2.3 GHz Dual-Core Intel Core i5). However, for thousands of single point mutations to deal with, it is recommended to calculate on a workstation, which could be done in a few hours (see *Note* 2).

3. Methods

3.1 Setting up working directories and preparing target structure

The crystal structure (in .pdb file) requires preparations before ΔΔG calculations employing structure-based software. Software used in this protocol are not parameterized for most non-protein ligands which must be removed prior to the calculations.

3.1.1 Make a directory and several subdirectories for the GRAPE design job

```
mkdir ~/GRAPEsoft/
```

The FoldX software can be extract to the GRAPEsoft fdirectory from an archive downloaded from http://foldxsuite.crg.eu/products#foldx. Also the DistributeFoldX, DistributeRosettaddg and FLAGrow3 can be downloaded from https://groups.google.com/forum/#!forum/fresco-stabilization-of-proteins.

```
mkdir ~/GRAPE/
cd ~/GRAPE/
mkdir foldx rosetta abacus consensus_analysis
```

3.1.2 Get the crystal structure file of the protein of interest, which is usually in pdb format. Pdb files can be downloaded from the RCSB database (https://www.rcsb.org/). Here, the crystal structure of *Is*PETase (PDB ID: 5XH3) is used:

```
wget https://files.rcsb.org/download/5XH3.pdb
```

Use Pymol to open the 5XH3.pdb file. Remove ligands and other non-protein co-crystallized solvent molecules. Then, save the structure as 5XH3_protein.pdb (see *Note* 3). Also, the pdb file can be edited with a text editor or by command line:

```
grep ^ATOM 5XH3.pdb > 5XH3_protein.pdb
```

An edited pdb file can be found in Supplementary materials in the online version at https://doi.org/10.1016/bs.mie.2020.12.026.

3.2 ΔΔG calculation with FoldX

This part uses the FoldX software, which predicts the ΔΔG values for individual mutations. Point mutations meeting the cutoff (ΔΔG < −5 kcal/mol) will be selected for further visual inspections.

3.2.1 Copy the prepared pdb file into the Foldx directory and enter the directory

```
cp 5XH3_protein.pdb ~/GRAPE/foldx/
cd ~/GRAPE/foldx/
```

3.2.2 Create a table file that lists the protein residues according to the one-letter residue code, chain name and the serial number in a tab-delimited format. An example of 5XH3.tab is shown in Fig. 2A. Copy the FoldX executable file and rotabase.txt file into current directory

```
cp ~/GRAPEsoft/rotabase.txt ./
cp ~/GRAPEsoft/DistributeFoldX ./
```

3.2.3 Set up the FoldX calculations:

```
./DistributeFoldX Phasel 5XH3_protein.pdb 1 A 5XH3.tab 1000
~/GRAPEsoft/foldx5/FoldX.linux64
```

3.2.4 Start the calculations by running the todolist file which is the outcome of previous step:

```
./todolist&.
```

You can use the `top` command (exit top with 'q') to check whether the job has finished.

3.2.5 Once the job has finished, select desired variants by setting a cutoff of calculated ΔΔG. Here, point mutations with ΔΔG lower than −5 kcal/mol will be selected for inspections. An example output file of FoldX is shown in Fig. 2B

3.3 ΔΔG calculation using Rosetta

The very nature of this part is similar to the previous procedure. ΔΔG of point mutations are calculated and filtered with −5 kcal/mol as cutoff.

3.3.1 Go to the ~/GRAPE/rosetta directory and copy the pdb file into it:

```
cp ~/GRAPE/5XH3_protein.pdb ~/GRAPE/rosetta/
cd ~/GRAPE/rosetta/
```

3.3.2 Use the Rosetta_ddg script to distribute a ddg prediction job

```
cp ~/GRAPEsoft/DistributeRosettaddg ./
cp ~/GRAPEsoft/FLAGrow3 ./
./DistributeRosettaddg Phasel 5XH3.tab 1 A 1 5XH3_protein.pdb
1000 FLAGrow3 ~/rosetta_src_2016.32.58837_bundle/main/source/bin/
ddg_monomer.linuxgccrelease
```

(A)

```
N    A    1
P    A    2
Y    A    3
A    A    4
R    A    5
...
```

(D)

```
site: 1 native type: ASN
ALA   delta S1:   -1.730   delta S2:    3.572   delta SEF    1.842
CYS   delta S1:    0.741   delta S2:    1.511   delta SEF    2.252
ASP   delta S1:   -0.293   delta S2:   -1.514   delta SEF   -1.807
GLU   delta S1:   -0.770   delta S2:    2.274   delta SEF    1.504
...
```

(B)

```
Bellow are the mutations and the change in stability and SD from 5 calculations, all in KJ mol -1
T84P        -16.067     0.905
E202M       -15.857     2.003
Q90D        -15.648     0.351
S131F       -14.061     1.228
...
```

(C)

```
Bellow are the mutations and the change in stability and SD from 5 calculations, all in KJ mol -1
P181K       -55.501     0.000
R71W        -24.238     0.000
S146H       -21.184     0.000
D234H       -20.715     0.000
...
```

Fig. 2 (A) Example of the 5XH3.tab file. (B) Example of the FoldX output file. (C) Example of the Rosetta output file. (D) Example of the ABACUS output file. Only the first five lines are shown in the figure.

3.3.3 Start the calculations by running the todolist file:

```
./todolist&.
```

It will take 20 min to serval hours to run the job depending on the hardware of the used computer. Again, point mutations with ΔΔG lower than −5 kcal/mol will be selected for inspections. An example output file of Rosetta is shown in Fig. 2C. Details about the ΔΔG calculations using FoldX and Rosetta have been described previously.

3.4 ABACUS energy calculation by ABACUS and filtering

3.4.1 Go to the ~/GRAPE/abacus directory and copy the pdb file into it:

```
cp ~/GRAPE/5XH3_protein.pdb ~/GRAPE/abacus/
cd ~/GRAPE/abacus/
```

3.4.2 Preprocess the input PDB file with the shell script ABACUS_prepare which is part of the ABACUS software. Note that all residues must have the same chain ID even if the chain is actually discontinuous or there are multiple chains

```
ABACUS_prepare 5XH3_protein.pdb
```

3.4.3 Generate SEF(statistical energy function) energy tables with the shell script ABACUS_S1S2 which is part of the ABACUS software:

```
ABACUS_S1S2 5XH3_protein.pdb
```

3.4.4 Calculate the change of the various SEF energy terms associated with all possible single mutations

```
ABACUS_singleMutationScan 5XH3_protein.pdb abacus_output.txt
```

Mutations with an ABACUS energy < -3 A.e.u. were selected. An example output file of ABACUS is shown in Fig. 2D.

3.5 Consensus analysis

To further identify mutations improving stability, evolutionary information-based consensus design is employed in the GRAPE strategy.

3.5.1 Enter the ~/GRAPE/consensus_analysis/ directory:

```
cd ~/GRAPE/consensus_analysis/
```

3.5.2 Get the protein sequence corresponding to the crystal structure. The Uniprot (https://www.uniprot.org/) accession of *Is*PETase is A0A0K8P6T7, use wget command to fetch the sequence:

```
wget https://www.uniprot.org/uniprot/A0A0K8P6T7.fasta
```

3.5.3 Download the GRAPE_consensus_analysis.py and the fasta file of homologous thermophilic protein sequences (also in Supplementary materials in the online version at https://doi.org/10.1016/bs.mie.2020.12.026) to the current directory and run the consensus analysis:

```
wget https://wulab.cf/grape/online/scripts/GRAPE_consensus_analysis.py
wget https://wulab.cf/grape/online/target_refdb.fasta
```

```
python3 GRAPE_consensus_analysis.py -s A0A0K8P6T7.fasta -ds target_
refdb.fasta -m analysis
```

Dependencies of the script and how to select thermophilic homology sequences are described in the notes section (see *Notes* 4, 5).

3.5.4 Use the cat command to merge both fasta files of thermophilic homologies (target_refdb.fasta in Supplementary materials in the online version at https://doi.org/10.1016/bs.mie.2020.12.026) and the target sequence (A0A0K8P6T7.fasta):

```
cat target_refdb.fasta A0A0K8P6T7.fasta > petdataset.fasta
```

3.5.5 Use muscle to build the multiple sequence alignment (MSA) file in fasta format:

```
muscle -in petdataset.fasta -out petdataset.aln -maxiter 99
```

3.5.6 Use hmmbuild to build the Hidden Markov model (HMM) from the MSA file:

```
hmmbuild petdataset.hmm petdataset.aln
```

3.5.7 Then, petdataset.hmm can be uploaded to the Skylign server (http://skylign.org/) for further visualization and scoring (see *Note* 6). The MSA and HMM logo of selected conserved sites are shown in Fig. 3

3.6 Selecting computationally designed and consensus analysis yielded variants for wet-lab characterization

The potential stabilizing point mutations often have one or more identifiable biophysical errors due to simplifications in the energy functions and incomplete conformational sampling. With visual inspection, such variants are eliminated. This further improves the quality of the library that will be screened experimentally, and thus reducing the required screening effort. This step requires expertise and vision in protein physics.

3.6.1 Go to the ~/GRAPE/selected_pdb/ where stored pre-selected variants are found:

```
cd ~/GRAPE/selected_pdb/
```

3.6.2 Download the script inspectmutation.py (see Supplementary materials in the online version at https://doi.org/10.1016/bs.mie.2020.12.026) to the current directory:

```
wget https://wulab.cf/grape/online/scripts/inspectmutation.py
```

3.6.3 Inspect variants and the wild-type structure in Pymol using the inspectmutation.py script. The following parameters are wild-type structure, variant structure, position of mutation and chain number:

```
pymol inspectmutation.py 5XH3_protein.pdb GLU77_5XH3_protein_Repair.
pdb 77 A
```

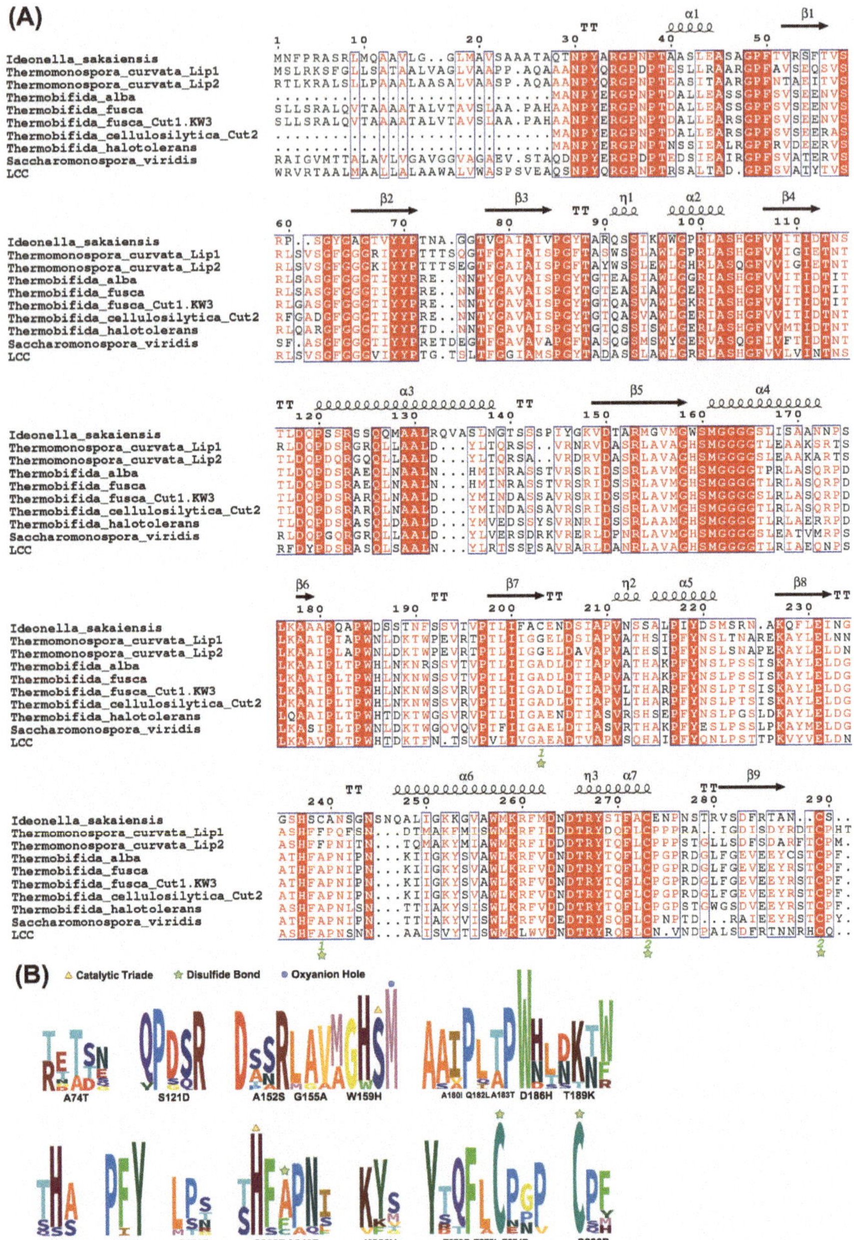

Fig. 3 Amino acid sequence alignment of PET hydrolases. (A) An alignment of the PET hydrolase sequences used in the Consensus analysis. Aligned sequences with Uniprot accession numbers are Ideonella_sakaiensis (*Is*PETase, *Ideonella sakaiensis*, A0A0K8P6T7), Thermomonospora_curvata_Lip1 (lipase, *Thermomonospora curvata*, D1A9G5, referred as Tcur1278 (Wei et al., 2014) in the original paper), Thermomonospora_curvata_Lip2 (lipase, *Thermomonospora curvata*, D1A2H1, referred as Tcur0390 (Wei et al., 2014) in the original paper), Thermobifida_alba (cutinase (Ribitsch et al., 2012), *Thermobifida alba*, E9LVH7),

More examples are shown in Fig. 4. The variant residue is colored purple and the wild-type residue is colored blue. Polar contacts are shown in yellow dashed lines (*see Note* 7).

3.6.4 The following criteria were used for visual inspection of the predicted mutant structures:

[1] Introduced residues are not allowed to break hydrogen-bonding or salt-bridge interactions, especially not inside a hydrophobic part of the protein

[2] Introduced residues are not allowed to form unsaturated hydrogen-bonding interactions, especially not inside a hydrophobic part of the protein

[3] Hydrophobic amino acids that are not located at the active site are not allowed on the protein surface if they are surrounded by the solvent

[4] The introduced residue is not allowed to make steric clashes with the remaining structure

[5] The introduced residue is not allowed to result in large cavities inside the protein

[6] The introduced residue is not allowed to destabilize the secondary structure of the protein

3.7 Experimental verification of the filtered mutations

Mutations passed through visual inspection need to be verified by experiments to measure the ΔT_M and enzymatic activity. Only variants that are more thermostable ($\Delta T_M \geq 1.5\,°C$) were selected for further clustering steps. The genetic manipulation follows standard protocols and is thus only outlined here for completeness.

Thermobifida_fusca (lipase (Chen et al., 2010), *Thermobifida fusca*, Q47RJ6), Thermobifida_fusca_Cut1.KW3 (cutinase (Herrero Acero et al., 2011), *Thermobifida fusca*, E5BBQ2), Thermobifida_cellulosilytica_Cut2 (cutinase 2 (Herrero Acero et al., 2011), *Thermobifida cellulosilytica*, E9LVH9), Thermobifida_halotolerans (serine hydrolase (Ribitsch et al., 2012), *Thermobifida halotolerans*, H6WX58), Saccharomonospora_viridis (lipase (Kawai et al., 2014), *Saccharomonospora viridis*, C7MVE8), LCC (cutinase (Sulaiman et al., 2012), uncultured bacteria isolated from metagenomics analysis, G9BY57). (B) Hidden Markov model (HMM) of mutations with consensus score > 2.0. The amino acid alignment from panel A was used to calculate an HMM profile. The HMM was consequently visualized as a logo with information content above the background (Skylign; http://skylign.org). *Reprinted (adapted) with permission from Cui, Y. L., Chen, Y. C., Liu, X. Y., Dong, S. J., Tian, Y. e., Qiao, Y. X., et al. (2021). Computational redesign of a PETase for plastic biodegradation under ambient condition by the GRAPE strategy.* ACS Catalysis, 11, *1340–1350. doi:10.1021/acscatal.0c05126.*

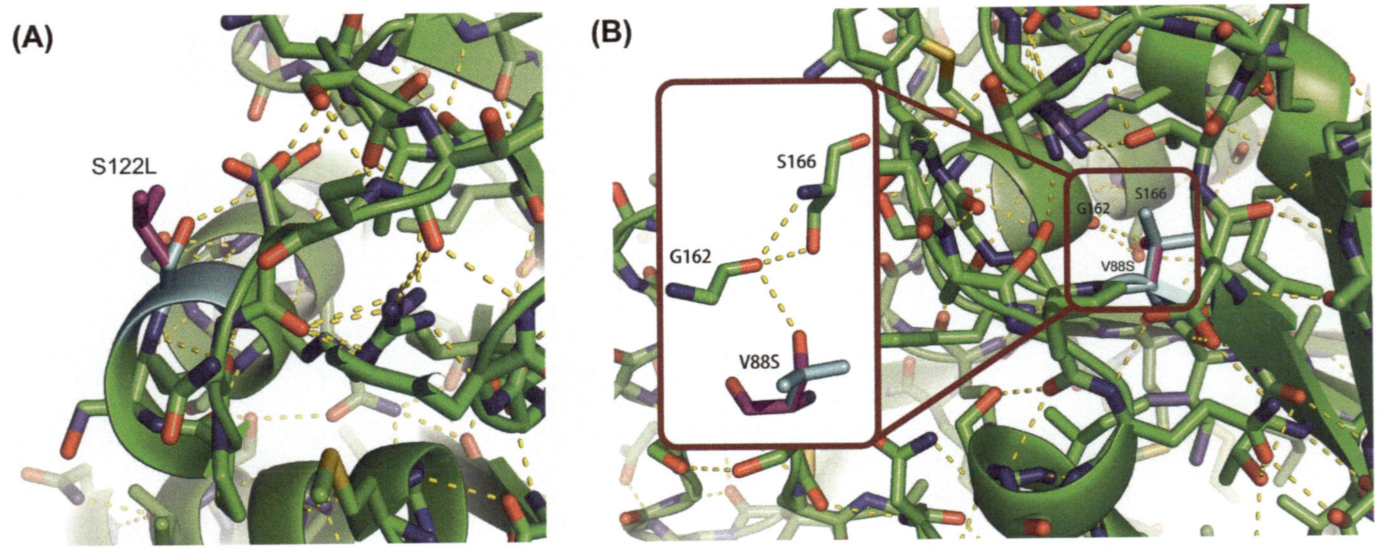

Fig. 4 Two examples of the most common mistakes occurring in prediction. The wild-type residue at the mutation position is colored light blue and the mutated residue is in purple. (A) The hydrophobic Leu was introduced to the surface and it replaced a Ser that formed a hydrogen bond with a water molecule. (B) The V88S buried in the hydrophobic core was replaced by a hydrophilic Ser. Although new hydrogen bond was created by Ser88 and the backbone oxygen atom of Gly162, the new hydrogen bond is not very stable and the transfer energy of Ser to a hydrophobic environment is high. Furthermore, the smaller sidechain of Ser will lead to a cavity in the hydrophobic core and destabilize the structure. Pymol is used for visualization and production of the figure.

3.7.1 The gene encoding *Is*PETase from *Ideonella sakaiensis* strain 201-F6 (GenBank accession number: GAP38373.1) should be synthesized following codon optimization for heterologous expression in the desired host cells and include a terminally fused His-tag for purification.

3.7.2 Single point mutations can be constructed using the QuickChange site-directed mutagenesis kit (Agilent Technologies, Santa Clara, CA, USA). This method is efficient and usually requires sequencing using only one single clone to verify the success of mutagenesis. Variants should be expressed, purified, and characterized for both stability (T_M) and activity data. Detailed information of wet-lab experiments are described elsewhere (Cui et al., 2021).

3.7.3 Out of 86 selected variants expressed for experimental characterization, 21 variants exhibiting ΔT_M more than 1.5 °C without significant loss in their activity (retaining >70% activity) are considered as effective and selected for further accumulation.

3.8 *K*-means clustering of mutations with significant improvement of protein melting temperature

K-means clustering is one of the most commonly used algorithms that has been successfully used in market segmentation, computer vision, and astronomy among many other domains (*see Notes* 8, 9).

3.8.1 A set of parameters, including T_M improvements, potential effects of the mutations, the interactions between the variants, the location of the variants and the distances between the Cα atoms of the mutated residue and the catalytic triad are characterized for clustering. Save these parameters in the 21.csv file (can be downloaded from https://wulab.cf/grape/online/21.csv, also can be found in Supplementary materials in the online version at https://doi.org/10.1016/bs.mie.2020.12.026) for clustering in the GRAPEscripts directory.

3.8.2 Use WEKA3.0 to run the *k*-means clustering. For the programme, the data set should be in the format .CSV or .ARFF. Click Open file button and open the cluster_input.csv file.

3.8.3 Click on the cluster button to perform clustering. For cluster parameters, we choose SimpleKMeans and set the number of clusters to 3 (Fig. 5A).

3.8.4 Visualize cluster assignments (Fig. 5B). Input data and the results of clustering are shown in Table 1.

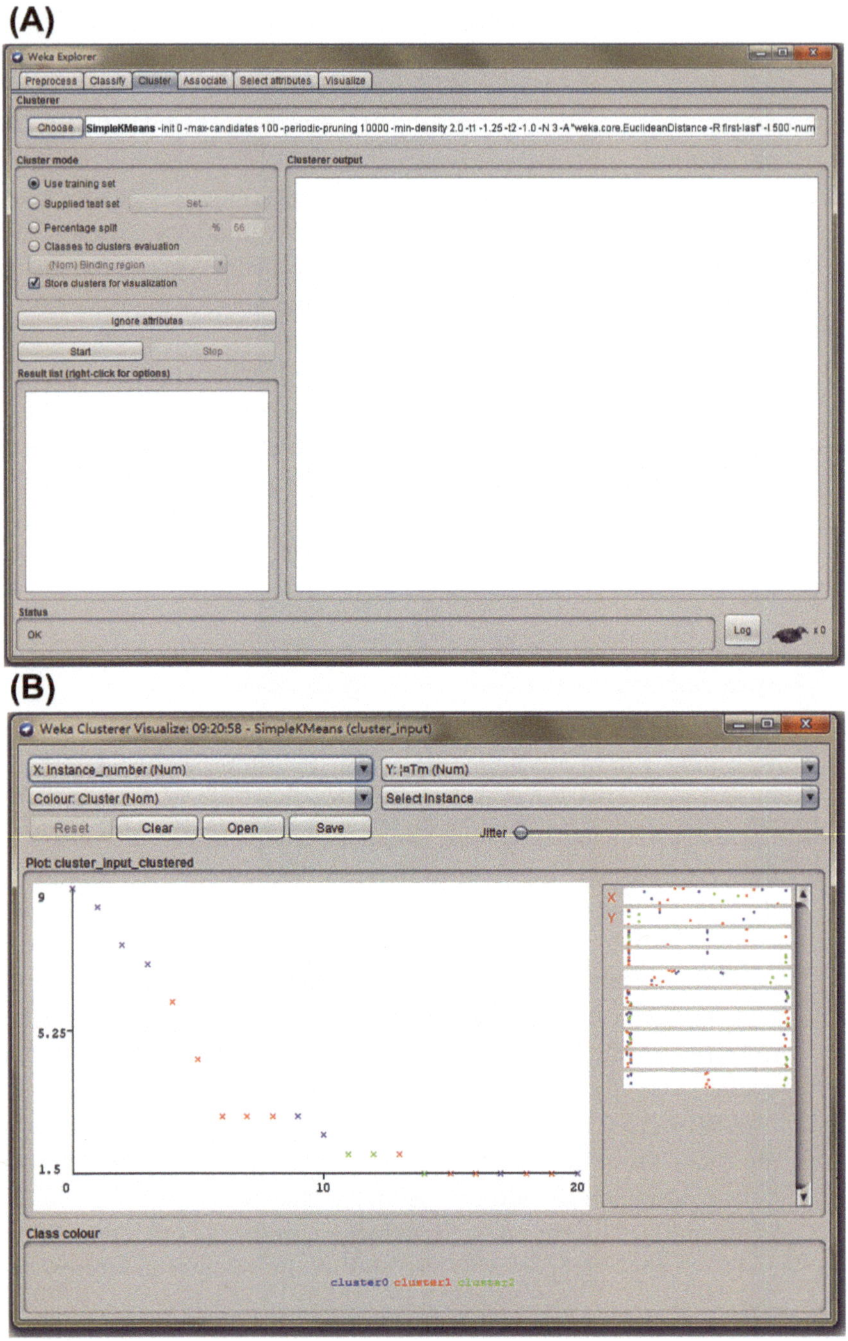

Fig. 5 Clustering uses Weka 3.8 (https://www.cs.waikato.ac.nz/ml/weka/). (A) *K*-means clustering algorithms. (B) Result of the *k*-means clustering.

Table 1 Clustering of the stabilizing point mutations based on their positions, efficacies and presumed effects.

Mutation	ΔT_M	Interaction cluster[a]	Region[b]	Distance[c]	Hbond[d]	Hydrophobic interactions[d]	Entropy change[d]	Binding region[e]	Cluster
S214H	9	0	1	9.62	FALSE	FALSE	FALSE	TRUE	1
W159H	8.5	1	1	3.81	TRUE	FALSE	FALSE	TRUE	1
I168R	7.5	2	1	12.55	TRUE	FALSE	FALSE	TRUE	1
D186H	7	2	1	8.00	TRUE	FALSE	FALSE	TRUE	1
P181A	6	0	1	7.98	FALSE	TRUE	TRUE	FALSE	2
Q119Y	4.5	3	1	10.12	FALSE	TRUE	FALSE	TRUE	2
L117F	3	3	1	12.12	FALSE	TRUE	FALSE	TRUE	2
S187W	3	3	1	11.22	FALSE	TRUE	FALSE	TRUE	2
N212R	3	0	1	14.24	FALSE	TRUE	FALSE	FALSE	2
A248P	3	0	2	21.09	FALSE	FALSE	TRUE	TRUE	1
K95A	2.5	0	1	14.79	FALSE	FALSE	FALSE	TRUE	1
T140D	2	0	3	32.49	TRUE	FALSE	FALSE	FALSE	3
K148W	2	0	3	29.81	FALSE	TRUE	FALSE	FALSE	3
S166T	2	4	1	9.65	FALSE	TRUE	FALSE	TRUE	2
T77E	1.5	0	3	30.22	TRUE	FALSE	FALSE	FALSE	3
G165A	1.5	4	1	8.55	FALSE	TRUE	TRUE	TRUE	2

Continued

Table 1 Clustering of the stabilizing point mutations based on their positions, efficacies and presumed effects.—cont'd

Mutation	ΔT_M	Interaction cluster[a]	Region[b]	Distance[c]	Hbond[d]	Hydrophobic interactions[d]	Entropy change[d]	Binding region[e]	Cluster
A180I	1.5	0	1	11.79	FALSE	TRUE	FALSE	TRUE	2
S188Q	1.5	2	1	12.98	TRUE	FALSE	FALSE	TRUE	1
I208V	1.5	0	1	8.59	FALSE	TRUE	FALSE	TRUE	2
S238F	1.5	1	1	9.05	FALSE	TRUE	FALSE	TRUE	2
R280A	1.5	0	2	20.62	FALSE	FALSE	FALSE	TRUE	1

[a]Mutations may have interactions with each other in each interaction cluster.
[b]Locations of each mutation. Each number represents a separate region.
[c]Distance between the Cα atoms of the mutation and catalytic Ser160.
[d]Possible mechanism of stabilization.
[e]Whether the mutation is located on the PET-binding region. The PET-binding region is according to the predicted model by molecular docking.

3.9 Greedy accumulation of beneficial mutations to an ultimate hyper-stable mutant

A greedy algorithm follows the problem-solving heuristic of making the locally optimal choice at each stage with the intent of finding a global optimum. For disambiguation, the term **round** is used to represent serval trails of combining mutations to a certain variant trying to find a best hit, and the term **stage** is the serial number of a round (Fig. 6). In this part, every step is based on the variant selected from the previous step.

3.9.1 Use the mutation with the highest T_M is considered as the starting point for the initial round of greedy accumulation.

3.9.2 Combine the remaining mutations in the cluster one by one to the achieved variant from the previous step and have its stability and activity tested for the best variant with improved stability without affecting the activity.

3.9.3 Repeat the last procedure until all variants combined to variants achieved by the previous round lead to a decrease of thermostability and activity.

3.9.4 Pick the best mutation in the remaining clusters and combine it to the latest variant and repeat the two previous steps until all clusters have been screened.

See *Note* 10 for example of *Is*PETase.

3.10 Further characterization of the mutants

3.10.1 Test the ability of a final variant to degrade a semicrystalline PET film (30% crystallinity). The degradation profiles at various temperatures demonstrated the robustness of the designed enzyme (referred to as DuraPETase). PET granules were purchased from Sigma-Aldrich Co. (St. Louis, MO, USA). All other chemicals were purchased from Sigma-Aldrich Co. in the highest available purity and were used without further purification

3.10.2 Soak PET film (ø8 mm) with 10 μL of enzyme (stock concentration 0.5 mg/mL) in 490 μL buffer containing 50 mM glycine-NaOH (pH 9.0) at different temperatures (37, 40, 50 and 60 °C) for 10 days.

3.10.3 Terminate reactions by heat treatment (100 °C, 10 min).

3.10.4 Analyze the supernatant obtained by centrifugation (18,000*g*, 5 min) by HPLC.

3.10.5 *Is*PETase completely lost its activity within 24 h at 37 °C, and the concentration of the degradation products finally reached 12 μM. As expected, DuraPETase bearing enhanced thermostability enabled long-term catalytic activity, which withstood for 10 days at 37 °C.

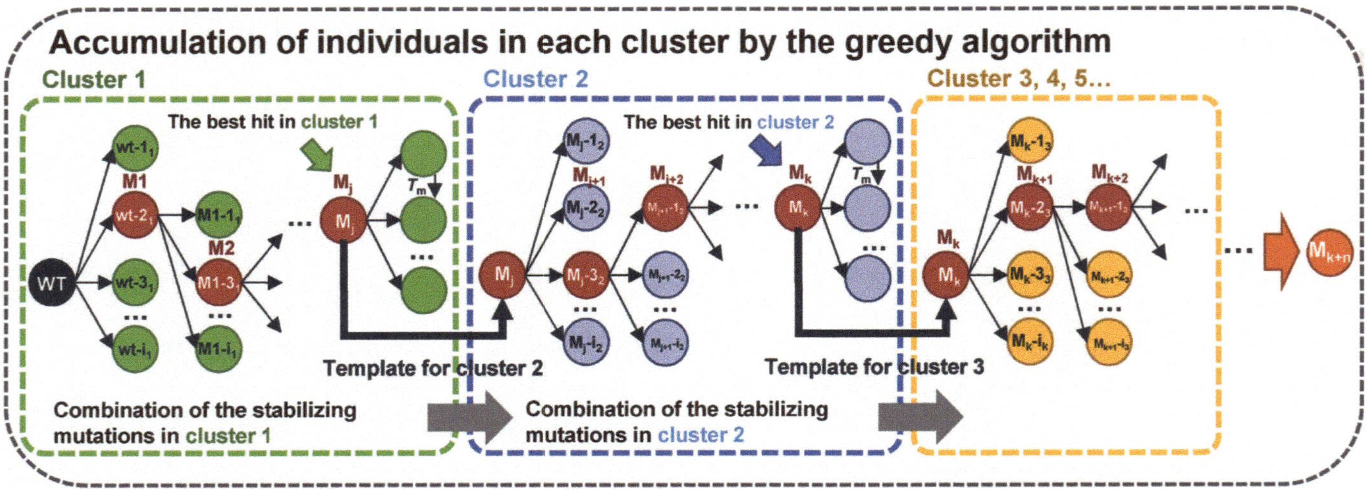

Fig. 6 Schematic representation of the accumulation step of the mutations in each cluster according to the greedy algorithm. Beneficial mutations in each cluster were crossed with the best hit of the current population until the remaining mutations in the cluster have been traversed or the T_M values of the combined variants decrease. If the variant showed high thermostability but a degradation reduction of >50%, the combined variant was not adopted. For example, the best hit in cluster 1 (Mj) served as the template for further cycles of accumulation in cluster 2, whereas the best hit in cluster 2 (Mk) was regarded as the parent for cycles in the next cluster. The best hit in each stage was sequentially referred to as M1, M2, ..., Mj, Mj+1, ..., Mk, Mk+1, ..., Mk+n. The red circles represent the mutants chosen from a combination stage that were used as the template for further accumulation (black arrows). *Reprinted (adapted) with permission from Cui, Y. L., Chen, Y. C., Liu, X. Y., Dong, S. J., Tian, Y. e., Qiao, Y. X., et al. (2021). Computational redesign of a PETase for plastic biodegradation under ambient condition by the GRAPE strategy.* ACS Catalysis, 11, *1340–1350. doi:10.1021/acscatal.0c05126. Copyright (2021) American Chemical Society.*

3.10.6 The concentration of total released compounds reached up to 3.1 mM, which was a 300-fold increase over those obtained with *Is*PETase.

3.10.7 Polymer hydrolysis was confirmed by scanning electron microscopy (SEM) images of the film surfaces following enzymatic treatment for 10 days (Fig. 7). Remarkably, severe erosion occurred with the formation of highly porous foam structures on the film surface after 10 days incubation by DuraPETase.

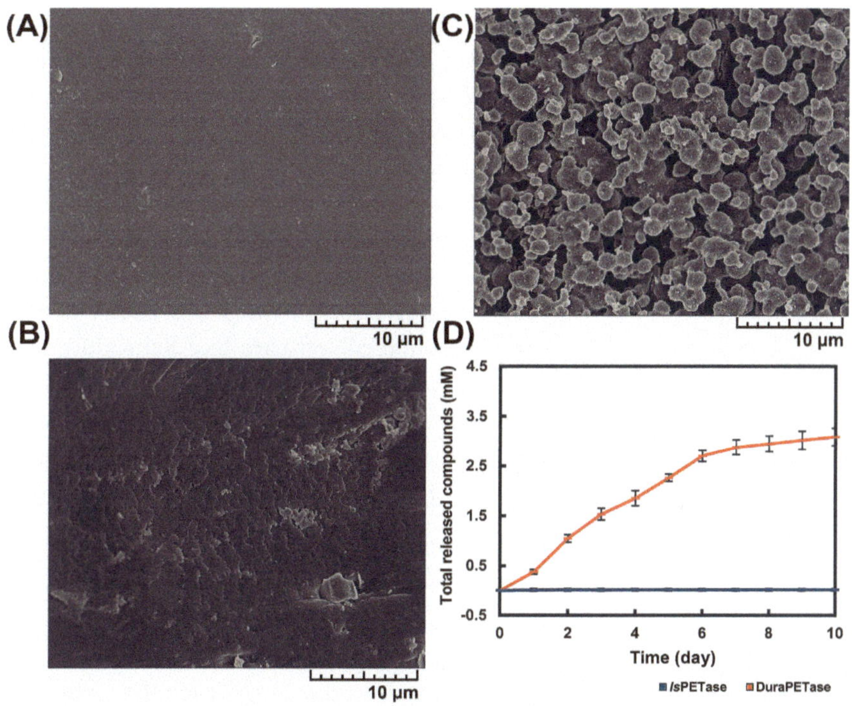

Fig. 7 SEM images for PET films after incubation with (A) DuraPETase, (B) *Is*PETase, and (C) buffer-only PET film control. (D) The total product release after incubation with *Is*PETase and DuraPETase for 10 days at 37 °C. All SEM images were taken after 10 days of incubation at an enzyme loading of 0.01 mg/mL in 50 mM glycine-NaOH buffer (pH 9.0) or with a buffer-only control. *Reprinted (adapted) with permission from Cui, Y. L., Chen, Y. C., Liu, X. Y., Dong, S. J., Tian, Y. e., Qiao, Y. X., et al. (2021). Computational redesign of a PETase for plastic biodegradation under ambient condition by the GRAPE strategy.* ACS Catalysis, 11, *1340–1350. doi:10.1021/acscatal.0c05126. Copyright (2021) American Chemical Society.*

4. Notes

1. This protocol should work appropriately on all Unix-like systems. This protocol assumes that the reader has only basic knowledge about shell command lines. However, challenges may still exist for installing the software. For readers not familiar with typical Linux distributions like Ubuntu or Debian, it is recommended to do the visual inspections on Windows or macOS system.
2. The increase of $\Delta\Delta G$ calculation time is in linear relation to the protein sequence length because the mutation library size is $19 \times n$ (n is the length of the protein). For a typical enzyme, the library size is within 10^3–10^4 mutations, which is feasible even for a home-use desktop to run the job on an acceptable time scale.
3. It is very important to select a structure of high resolution (smaller than 2.5 Å) and low R_{free} (smaller than 0.25). Models obtained by homology modeling are not good enough in most cases. It is also very important to correct mistakes in the model or they will seriously affect predictions. For beginners, public webserver like PROSS (https://pross.weizmann.ac.il/step/pross-terms/) and commercial software like Schrödinger are good options for structure preparation.
4. The *Kullback–Leibler* divergence, also called *relative entropy*, is a measure of how one probability distribution is different from a second reference probability distribution. The *relative entropy* of one alignment position is defined as:

$$D(p_i \,||\, q_i) = \sum_{1}^{L} p_i \ln\left(\frac{p_i}{q_i}\right) \tag{1}$$

The *Consensus Score* which is the probability adjusted by the relative entropy of the particular column in alignment Eq. (2) is defined as:

$$Score = p_i D(p_i \,||\, q_i) \tag{2}$$

The *Consensus Score* ranging from 0.045 (Leu) to 4.50 (Trp) is calculated in the 1% Leu and 99% Trp probability combination. The background probability q_i of each amino acid is taken from the amino acid frequency in the Uniprot database. The observed probability p_i is the probability of the alignment of the target and homologous sequences.

5. The GRAPE_consensus_analysis.py script depends on the software Muscle for multiple sequence alignment. The target_refdb.fasta file is a manually collected thermostable homologous protein sequences containing nine experimentally characterized thermostable homologs. When only few homologs have been characterized as thermostable, the DETECT consensus analysis online server (https://wulab.cf/depect/depect_conanalysis.html) can find more homologs from genomes of thermophilic organisms and preform the consensus analysis.
6. The *consensus score* calculated by GRPAE_consensus_analysis.py employs e as the base, which is different from the Skylign server that uses 2 as the base. Also, residue numbers returned by GRPAE_consensus_analysis.py are the original positions in the target sequence and the positions shown in the Skylign output are column numbers in the alignment, which needs to be translated carefully.
7. Visual inspection is a common practice for protein design projects. It requires knowledge in protein physics and experience. With access to enough computation power, a short molecular dynamic simulation is helpful.
8. *K*-means clustering uses k prototype vectors (i.e., centers or centroids of k clusters) to characterize data and minimizes a sum-of-squares cost function to find these prototypes with a coordinate descent optimization method. The algorithm uses Euclidean distance measures to compute distances between instances and clusters. Given a set of individuals (x_1, x_2, ..., x_n), *k*-means clustering aims to partition the individuals into k ($\leq$n) sets, which is known as the function given as follows:

$$\arg\min \sum_{i=1}^{k} \sum_{x \in S_i}^{k_i} \|x - \mu_i\|^2 \tag{3}$$

where $\|x - \mu_i\|$ is the Euclidean distance between x_i and μ_i, k_i is the number of individuals in the ith cluster and k is the number of clusters.

9. In the *Is*PETase example, 21 mutations enabling improvement of ΔT_M more than 1.5 °C were identified by experiments. It is impossible to test all combinations because the number of all possible combinations is 2,097,130. The rational way is to follow the stepwise path, select a starting point and add other mutations one by one as the first round, then use the best double mutation as the starting point for the second round scan. By using the greedy algorithm, the library scale was reduced to only 210 variants. Furthermore, dividing 21 mutations into

different clusters can reduce the testing space to 81 variants. In this case, the DuraPETase with 10 residue substitutions was achieved by 65 trials. Accordingly, 144 trials are required to obtain the same variant without clustering.

10. Herein, the starting point for the initial round of greedy accumulation is described. Then, other mutations in the first cluster (K95A, W159H, I168R, D186H, S188Q, A248P and R280A) were combined to the S214H mutation and resulted in a best hit (M2) in the second stage, which was selected as the new template for next round accumulation. After six rounds of the accumulations, a total of 25 combined mutants were explored experimentally. The best hit (S214H-I168R-W159H-S188Q-R280A, M5) obtained at the fifth accumulation stage in the first cluster served as the template for the next crossover of mutations in the second cluster. In the exploitation process of mutations in the second cluster, five rounds of accumulation were performed and a total of 35 combined mutants were explored. The best hit variant in the S214H-I168R-W159H-S188Q-R280A-A180I-G165A-Q119Y-L117F (*Is*PETase-M9) with a T_M value of 75 °C was considered the template for the accumulation of mutations in the third cluster. Finally, the most thermostable variant, S214H-I168R-W159H-S188Q-R280A-A180I-G165A-Q119Y-L117F-T140D (*Is*PETase-M10, referred to as DuraPETase) originating from the third cluster, exhibited an apparent melting temperature of 77 °C, which was drastically elevated by 31 °C. Detailed information are described elsewhere (Cui et al., 2021).

5. Summary

Here, we present the GRAPE strategy to enhance the thermostability of enzymes. Using the recently reported *Is*PETase from *Ideonella sakaiensis* that decomposes PET under ambient temperatures as a starting point, we adopted the GRAPE strategy to come up with a DuraPETase (T_M = 77 °C, raised by 31 °C) which showed 300-fold enhanced degradation performance on semicrystalline PET films at 37 °C. With improved stability and activity, the DuraPETase may be useful in further applications like enzyme-catalyzed plastic degradation or recycling to help address the challenge of a sustainable PET recycling. Also by adopting the GRAPE strategy, it is possible to achieve highly improved designed enzymes by designing and screening a relatively small library and efficiently accumulating beneficial mutations. The case study of *Is*PETase and this strategy may be employed to pave the way to design robust plastic degrading enzymes and other enzymes required by various industries.

References

Chen, S., Su, L. Q., Billig, S. S., Zimmermann, W., Chen, J., & Wu, J. (2010). Biochemical characterization of the cutinases from *Thermobifida fusca*. *Journal of Molecular Catalysis B: Enzymatic*, *63*(3–4), 121–127. https://doi.org/10.1016/j.molcatb.2010.01.001.

Cui, Y. L., Chen, Y. C., Liu, X. Y., Dong, S. J., Tian, Y.e., Qiao, Y. X., et al. (2021). Computational redesign of a PETase for plastic biodegradation under ambient condition by the GRAPE strategy. *ACS Catalysis*, *11*, 1340–1350. https://doi.org/10.1021/acscatal.0c05126.

DeLano, W. L. (2002). Pymol: An open-source molecular graphics tool. *CCP4 Newsletter on Protein crystallography*, *40*(1), 82–92.

Delgado, J., Radusky, L. G., Cianferoni, D., & Serrano, L. (2019). FoldX 5.0: Working with RNA, small molecules and a new graphical interface. *Bioinformatics (Oxford, England)*, *35*(20), 4168–4169. https://doi.org/10.1093/bioinformatics/btz184.

Edgar, R. C. (2004). MUSCLE: Multiple sequence alignment with high accuracy and high throughput. *Nucleic Acids Research*, *32*(5), 1792–1797. https://doi.org/10.1093/nar/gkh340.

Herrero Acero, E., Ribitsch, D., Steinkellner, G., Gruber, K., Greimel, K., Eiteljoerg, I., et al. (2011). Enzymatic surface hydrolysis of PET: Effect of structural diversity on kinetic properties of cutinases from *Thermobifida*. *Macromolecules*, *44*(12), 4632–4640. https://doi.org/10.1021/ma200949p.

Johnson, L. S., Eddy, S. R., & Portugaly, E. (2010). Hidden Markov model speed heuristic and iterative HMM search procedure. *BMC Bioinformatics*, *11*, 431. https://doi.org/10.1186/1471-2105-11-431.

Kawai, F., Kawabata, T., & Oda, M. (2020). Current state and perspectives related to the polyethylene terephthalate hydrolases available for biorecycling. *ACS Sustainable Chemistry & Engineering*, *8*(24), 8894–8908. https://doi.org/10.1021/acssuschemeng.0c01638.

Kawai, F., Oda, M., Tamashiro, T., Waku, T., Tanaka, N., Yamamoto, M., et al. (2014). A novel Ca^{2+}-activated, thermostabilized polyesterase capable of hydrolyzing polyethylene terephthalate from *Saccharomonospora viridis* AHK190. *Applied Microbiology and Biotechnology*, *98*(24), 10053–10064. https://doi.org/10.1007/s00253-014-5860-y.

Leaver-Fay, A., Tyka, M., Lewis, S. M., Lange, O. F., Thompson, J., Jacak, R., et al. (2011). ROSETTA3: an object-oriented software suite for the simulation and design of macromolecules. *Methods in enzymology*, *487*, 545–574. https://doi.org/10.1016/B978-0-12-381270-4.00019-6.

Pierce, N. A., & Winfree, E. (2002). Protein design is NP-hard. *Protein Engineering, Design and Selection*, *15*(10), 779–782. https://doi.org/10.1093/protein/15.10.779.

Ribitsch, D., Herrero Acero, E., Greimel, K., Eiteljoerg, I., Trotscha, E., Freddi, G., et al. (2012). Characterization of a new cutinase from *Thermobifida alba* for PET-surface hydrolysis. *Biocatalysis and Biotransformation*, *30*(1), 2–9. https://doi.org/10.3109/10242422.2012.644435.

Ribitsch, D., Herrero Acero, E., Greimel, K., Dellacher, A., Zitzenbacher, S., Marold, A., et al. (2012). A new esterase from *Thermobifida halotolerans* hydrolyses polyethylene terephthalate (PET) and polylactic acid (PLA). *Polymers*, *4*(1), 617–629. https://doi.org/10.3390/polym4010617.

Sulaiman, S., Yamato, S., Kanaya, E., Kim, J. J., Koga, Y., Takano, K., et al. (2012). Isolation of a novel cutinase homolog with polyethylene terephthalate-degrading activity from leaf-branch compost by using a metagenomic approach. *Applied and Environmental Microbiology*, *78*(5), 1556–1562. https://doi.org/10.1128/aem.06725-11.

Tracewell, C. A., & Arnold, F. H. (2009). Directed enzyme evolution: Climbing fitness peaks one amino acid at a time. *Current Opinion in Chemical Biology*, *13*(1), 3–9. https://doi.org/10.1016/j/cbpa.2009.01.017.

Wei, R., Oeser, T., Then, J., Kühn, N., Barth, M., Schmidt, J., et al. (2014). Functional characterization and structural modeling of synthetic polyester-degrading hydrolases from *Thermomonospora curvata*. *AMB Express*, 44. https://doi.org/10.1186/s13568-014-0044-9.

Wei, R., & Zimmermann, W. (2017). Biocatalysis as a green route for recycling the recalcitrant plastic polyethylene terephthalate. *Microbial Biotechnology*, *10*(6), 1302–1307. https://doi.org/10.1111/1751-7915.12714.

Wheeler, T. J., Clements, J., & Finn, R. D. (2014). Skylign: A tool for creating informative, interactive logos representing sequence alignments and profile hidden Markov models. *BMC Bioinformatics*, *15*, 7. https://doi.org/10.1186/1471-2105-15-7.

Xiong, P., Wang, M., Zhou, X. Q., Zhang, T. C., Zhang, J. H., Chen, Q., et al. (2014). Protein design with a comprehensive statistical energy function and boosted by experimental selection for foldability. *Nature Communications*, *5*, 5330. https://doi.org/10.1038/ncomms6330.

Yoshida, S., Hiraga, K., Takehana, T., Taniguchi, I., Yamaji, H., Maeda, Y., et al. (2016). A bacterium that degrades and assimilates poly(ethylene terephthalate). *Science*, *351*(6278), 1196–1199. https://doi.org/10.1126/science.aad6359.

CHAPTER ELEVEN

Mechanistic investigation of enzymatic degradation of polyethylene terephthalate by nuclear magnetic resonance

Patricia Falkenstein[a], Ren Wei[b,c], Jörg Matysik[a], and Chen Song[a,*]
[a]Institut für Analytische Chemie, Universität Leipzig, Leipzig, Germany
[b]Department of Biotechnology & Enzyme Catalysis, Institute of Biochemistry, University of Greifswald, Greifswald, Germany
[c]Junior Research Group Plastic Biodegradation, Institute of Biochemistry, University of Greifswald, Greifswald, Germany
*Corresponding author: e-mail address: chen.song@uni-leipzig.de

Contents

Abstract

The biocatalytic degradation of polyethylene terephthalate (PET) by thermophilic microbial enzymes has recently emerged as an option for a future eco-friendly recycling process for plastic waste, as it occurs under mild conditions and requires no harmful additives. In this chapter, we present a brief overview of solution and solid-state nuclear magnetic resonance (NMR) spectroscopic methods for the characterization of composition and chemical microstructure of PET and also associated chain dynamics over multiple time scales. Such detailed information provides an understanding of the enzymatic PET degradation mechanism by polyester hydrolases at the molecular level.

Methods in Enzymology, Volume 648
ISSN 0076-6879
https://doi.org/10.1016/bs.mie.2020.11.002

1. Introduction

Thermophilic microbial cutinases, such as TfCut2 from *Thermobifida fusca* (Roth et al., 2014) and LC-cutinase derived from leaf-branch compost (Sulaiman et al., 2012) are particularly suitable for the biocatalytic degradation of PET because the polymer chains of the amorphous fraction will get more mobile at elevated temperatures near the glass transition temperature (T_g) of about 75 °C which leads to higher accessibility of the PET ester bonds (Ronkvist, Xie, Lu, & Gross, 2009; Wei & Zimmermann, 2017). A variety of solution and solid-state NMR spectroscopic approaches have been applied to study the polymer properties of PET in association with enzymatic degradation processes in a wide range of temperatures (Falkenstein et al., 2020; Wei, Breite, et al., 2019; Wei, Song, et al., 2019). NMR spectroscopy could show, in particular, site-specific/segmental structural and dynamic information on the polymer chains at atomic resolution, making it possible to add molecular details to the degradation mechanism currently under debate (Austin et al., 2018; Joo et al., 2018; Wei, Song, et al., 2019).

The versatility of solution NMR has been proven in providing high-resolution structural information for small and hydrophobic systems (Günther, 2013) and in precise chemical characterization of polymer chains (de Ilarduya & Muñoz-Guerra, 2014). However, solution NMR is limited by the molecular size of the target biomolecule. PET can be completely dissolved in certain fluorinated solvents without the formation of high amounts of polymeric by-products (Berkowitz, 1984; Fox, Moad, van Diepen, Willing, & Cook, 1997). Accordingly, solution NMR analyses of PET have mostly been carried out using, e.g., 1,1,1,3,3,3-hexafluoro-2-propanol (HFIP) or trifluoroacetic acid (TFA) in combination with D-chloroform ($CDCl_3$, de Ilarduya & Muñoz-Guerra, 2014; Ma, Agarwal, Vekemans, & Sikkema, 2003; Peez, Janiska, & Imhof, 2019). To quantify the end groups of PET by solution NMR, a number of NMR methods have been proposed such as (i) carbodiimide-mediated esterification of the carboxylic acid end groups with HFIP at room temperature (Ma et al., 2003); (ii) quantitative reaction of hydroxyl and carboxyl end groups with trichloroacetyl isocyanate (Donovan & Moad, 2005), and (iii) the analysis in HFIP/$CDCl_3$ by ^{13}C NMR (Fox et al., 1997).

Unlike in solution, high-resolution solid-state NMR under magic-angle spinning (MAS) is applicable to molecules that undergo slow molecular tumbling or are fully immobilized, and the spectral resolution is immune to size of molecules. In this technique, the solid sample is spun rapidly in a cylindrical rotor around an axis oriented at a so-called "magic angle" of 54.74° to the applied magnetic field which leads to the suppression of anisotropic interactions, and thus produces solution-like spectra with narrow lines. However, in many circumstances the overall averaging of these interactions by MAS is undesirable, for example, dipolar interactions between low-γ nuclei encode information about nuclear proximity. Therefore, a wide variety of recoupling techniques have been introduced for restoring specific anisotropic interactions under MAS conditions (Duer, 2002). Compared with solution NMR, MAS NMR is advantageous for studying PET in its intrinsic solid state similarly as that used in the enzymatic degradation (Choudhury, Lee, Kriegel, Koros, & Beckham, 2012; Schmidt-Rohr, Hu, & Zumbulyadis, 1998; Sefcik, Schaefer, Stejskal, & McKay, 1980; Wilhelm & Spiess, 1996). Furthermore, solid-state NMR is unique in determining the degree of crystallinity and the mobility of the amorphous fraction of a certain PET sample which are known to be closely associated with the enzymatic degradation performance (Ronkvist et al., 2009; Vertommen, Nierstrasz, van der Veer, & Warmoeskerken, 2005; Wei, Breite, et al., 2019). More specifically, solid-state NMR studies on PET allow investigation of intriguing polymer chain dynamics over a broad time scale (Choudhury et al., 2012). In this chapter, we provide experimental procedures of selected NMR methods that are particularly applicable to the study of PET polymers in the context of biocatalytic degradation.

2. ^{1}H solution NMR analysis to quantify PET chain scissions

^{1}H solution NMR analysis allows the determination of the average chain length of PET, thereby evaluating the extent of decomposition as a result of, e.g., enzymatic degradation (Wei, Breite, et al., 2019) or ultraviolet (UV) irradiation (Falkenstein et al., 2020). For the acquisition of ^{1}H NMR spectra of PET dissolved in HFIP/$CDCl_3$, it is crucial to accumulate a sufficient number of scans to obtain a signal-to-noise ratio of at least 30:1 for the weakest signal at about 4.0 ppm (Fig. 1) which represents the methylene protons in α-position to the hydroxyl end group of the PET chain (Kenwright et al., 1999). Using the relative integration ratios of this signal and those at about 4.7 and 8.1 ppm, representing the protons of the oxyethylene unit and the terephthalate ring in the PET main chain, respectively (de Ilarduya & Muñoz-Guerra, 2014; Kenwright et al., 1999), the degree of polymerization (DP) can be calculated

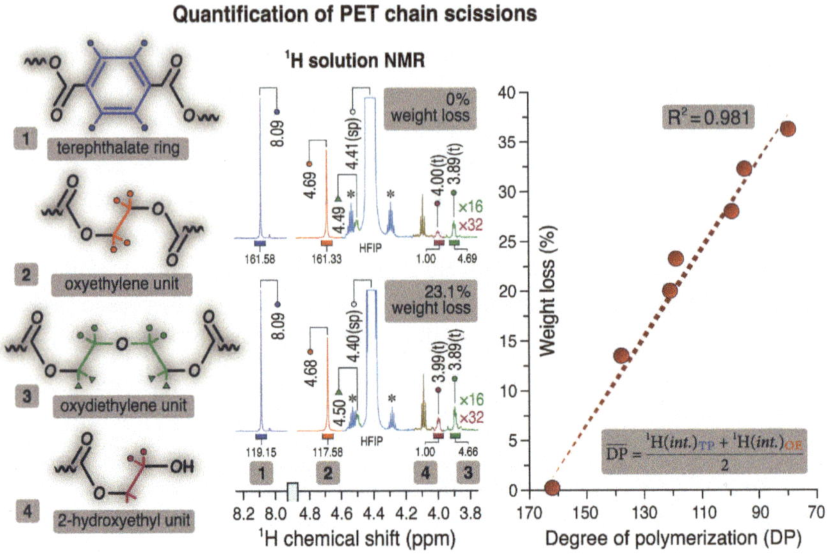

Fig. 1 (*left*) Structural elements occurring in a PET polymer chain and identifiable by NMR spectroscopy (de Ilarduya & Muñoz-Guerra, 2014; Kenwright, Peace, Richards, Bunn, & MacDonald, 1999). (*middle*) ^{1}H solution NMR spectra of (*top*) untreated and (*bottom*) enzyme-treated CP-PET. The characteristic spectral regions of 3.8–4.8 and 8.0–8.2 ppm are shown. Integration ratios are given below the respective signals relative to the CH_2 signal at about 4.0 ppm. For clarity, the intensities of the ^{1}H signals at about 3.9 and 4.0 ppm were increased by a factor of 16 and 32, respectively. ^{13}C satellites are marked with asterisks. (*right*) Relative weight loss of enzyme-treated CP-PET as a function of the DP including linear fit.

(Wei, Breite, et al., 2019). As shown in Fig. 1 (*right*), a linear relationship was found between the gravimetrically determined weight loss and the DP for PET from post-consumer food packaging containers (Carton Pack Srl, Rutigliano, Italy; termed as CP-PET) incubated with the purified PET degrading enzyme TfCut2 at 70 °C for different durations (Wei, Breite, et al., 2019).

2.1 Protocol

2.1.1 Equipment

- NMR spectrometer: Bruker DRX-600 NB spectrometer operating at a Larmor frequency of 600.13 MHz for ^{1}H using a 5-mm TXI probe (Rheinstetten, Germany)
- 5-mm NMR tubes (Wilmad, Merck KGaA, Darmstadt, Germany)

2.1.2 Chemicals

- Amorphous PET film (Goodfellow Ltd., Bad Nauheim, Germany)
- HFIP (≥99%, Merck KGaA, Darmstadt, Germany)
- $CDCl_3$ (99.5% D, Cambridge Isotope Laboratories, Tewksbury, MA) containing 0.03% (v/v) tetramethylsilane (TMS) as an internal standard

Caution: HFIP causes severe skin burns and eye damage. It may damage the organs by prolonged or repeated exposure and may probably affect fertility and damage the unborn child. Therefore, avoid inhalation of vapors or mist as well as contact with eyes and skin. Wear protective gloves, protective clothing, eye and face protection. Use this chemical only in a chemical fume hood.

$CDCl_3$ causes skin irritation, damage to organs through prolonged or repeated exposure, serious eye irritation, is harmful if swallowed and toxic if inhaled. It is suspected of causing cancer and damaging fertility as well as damaging the unborn child. Avoid eye and skin contact and do not breathe vapor or mist. Wear protective gloves, protective clothing, eye and face protection. Use this chemical only in a chemical fume hood.

2.1.3 Preparation

- Dissolve PET in HFIP and store at room temperature for ≥5 days with a final concentration of about 7.14 μg/μL.
- Mix with $CDCl_3$ containing 0.03% (v/v) TMS, e.g., at ratio 1:16 (v/v).
- Pipette the solution into a 5-mm NMR tube with a final volume of 500–700 μL.
- Seal the NMR tube with Parafilm to prevent loss of HFIP.
- Measure a ^1H NMR spectrum with 1k scans and a long recycle delay of 15 s at 20.0 ± 0.1 °C.

2.1.4 Data evaluation

- Correct the baseline of the ^1H spectra manually.
- Fit the ^1H signals at 4.0, 4.7 and 8.1 ppm using the program MestReNova (Mestrelab Research S.L., Santiago de Compostela, Spain).
- Create a new spectrum with the fitted peaks and integrate them. Set the signal at 4.0 ppm to unity and use the same integral regions for comparison of different spectra.
- Calculate the DP value from the relative integration ratios (including 1% for ^{13}C satellites).
- To calculate the number average molecular weight ($\overline{M}_n$), use the molar mass of a PET repeating unit which is 192.17 g/mol.

3. Solid-state NMR analysis to determine PET chain conformation and dynamics

Solid-state NMR spectroscopy is a powerful tool to determine the local structure and chain dynamics (over a broad time scale) in PET. In this chapter, a series of solid-state NMR methods are introduced for investigating the relationships between chain conformation, dynamics and crystallization behavior of PET at the atomic level. These properties are understood to have a considerable impact on the enzymatic degradation performance. Solid-state NMR measurements can thus provide a detailed picture of molecular and mechanistic aspects of the biocatalytic degradation of PET.

3.1 Protocol

3.1.1 Equipment

- NMR spectrometer: Bruker AVANCE III 400 WB spectrometer operating at Larmor frequencies of 400.15 MHz for ^{1}H and 100.62 MHz for ^{13}C using a 4-mm double-resonance MAS probe (Rheinstetten, Germany)
- 4-mm MAS rotor (CortecNet, Voisins-Le-Bretonneux, France)

3.1.2 Chemicals

- Amorphous PET powder with a particle size (Ø) of <0.25 mm or 0.25–0.5 mm (amorphous PET film purchased from Goodfellow Ltd., Bad Nauheim, Germany, and further cryogenically ground)
- $Pb(NO_3)_2$ (≥99%, Merck KGaA, Darmstadt, Germany)
- L-Tyrosine·HCl (≥98%, Merck KGaA, Darmstadt, Germany)

Caution: $Pb(NO_3)_2$ causes severe eye damage and is harmful if swallowed or inhaled. It causes damage to organs (blood, central nervous system, immune system, kidney) through prolonged or repeated exposure and may affect fertility or damage the unborn child. It may also cause an allergic skin reaction. Therefore, do not inhale this substance and avoid contact with eyes and skin. Wear protective gloves, protective clothing, eye and face protection. Use this chemical only in a chemical fume hood.

3.1.3 Preparation

- Fully load a 4-mm MAS rotor with the cryogenically ground PET powder sample.
- To have the real and precise value of the sample temperature (against thermocouple readout of the exit gas temperature), a calibration procedure

using ^{207}Pb chemical shift of $Pb(NO_3)_2$ must be performed prior to measurements (Guan & Stark, 2010).
- ^{13}C chemical shifts were externally referenced to the $C(O)O^-$ signal of solid L-tyrosine·HCl at 172.1 ppm.
- The sample should be incubated in the magnet for at least 12 h prior to the measurements to avoid spectral changes caused by thermodynamic nonequilibrium during data acquisition (Hutchinson, 1995; Qian, Shen, Sun, & Wu, 1996; Wei, Breite, et al., 2019).

3.1.4 Data evaluation

- Data processing of the various spectra can be found in detail either in the literature about the specific pulse sequence or in the TopSpin (Bruker Biospin) user manuals.
- Spectral fitting (e.g., Figs. 2 and 7) was done with the program MestReNova.
- Curve fitting was performed with the program Origin (OriginLab Corporation, Northampton, MA, USA) (e.g., Figs. 4 and 7) or with SIMPSON (Bak, Rasmussen, & Nielsen, 2000) (e.g., Fig. 6) and is described in the following sections.

3.2 Cross polarization for determination of PET chain conformation

Polymer chains of amorphous PET coexist in two distinct conformations, *gauche* (*g*) and *trans* (*t*), with different OC–CO torsion angles Ψ (Fig. 2), where the *g* content clearly exceeds the *t* content below the glass transition temperature (e.g., at 30 °C). Crystalline PET, by contrast, consists only of polymer chains in *t* conformation (Schmidt-Rohr et al., 1998). The ^{13}C cross polarization (CP)/MAS experiment is a useful tool to determine the *g*/*t* ratio of PET chains which correlates with the degree of crystallinity. This can be achieved by deconvolution of the ethylene carbon signal which consists of two components representing the *g* and *t* conformation and, accordingly, the amorphous and crystalline phase of PET (Choudhury et al., 2012; Gabriëlse, Angad Gaur, Feyen, & Veeman, 1994). To this end, the ethylene carbon signal was fitted using the Line Fitting feature of MestReNova 12.0.0. An example is shown in Fig. 2, where a smaller *g*/*t* ratio was found for the amorphous PET sample at 70 °C relative to that at 30 °C, indicating that the sample becomes more crystalline at higher temperatures. This can be attributed to the physical aging of the sample which leads to more ordered regions in

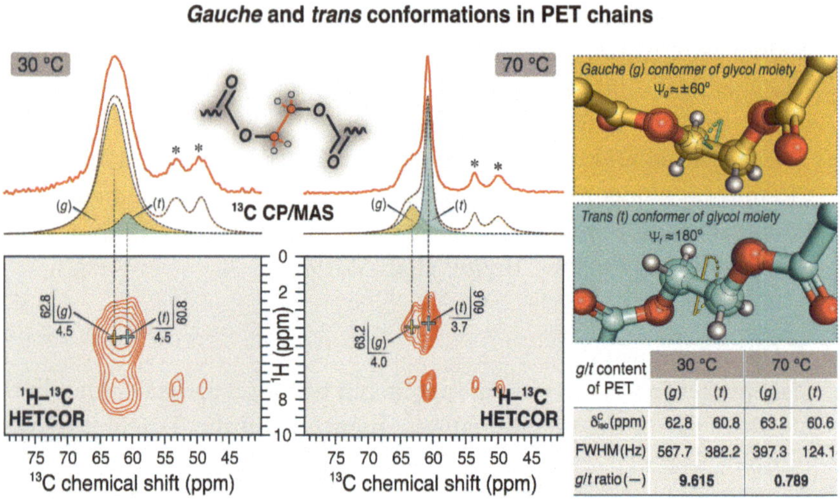

Fig. 2 (*top left*) ^{13}C CP/MAS and (*bottom left*) ^{1}H–^{13}C HETCOR spectra of amorphous PET (see Section 3.1.2), measured at 30 and 70 °C using a MAS frequency of 8 and 8.176 kHz, respectively. Quantification of *g* and *t* content was done by fitting of a Voigt function with a Lorentzian-to-Gaussian ratio of approximately 1:1 to the experimental data. Spinning sidebands are marked with asterisks. (*top right*) Schematic of the oxyethylene unit of PET in *g* and *t* conformation, corresponding to an average OC–CO torsion angle of $\Psi_g \approx \pm 60°$ and $\Psi_t \approx 180°$, respectively. (*bottom right*) The embedded table summarizes the spectral features of *g*/*t* content of amorphous PET at 30 and 70 °C.

PET (Hutchinson, 1995; Tournier et al., 2020; Wei, Breite, et al., 2019; Wei, Song, et al., 2019).

The precise ^{13}C chemical shift values of the two conformations can also be determined from ^{1}H–^{13}C heteronuclear correlation (HETCOR) experiments (Roberts, Vega, & Griffin, 1984) with supercycled phase-modulated Lee–Goldburg decoupling (PMLG5-S2) (Paul et al., 2009; Vinogradov, Madhu, & Vega, 2001) (Fig. 2, *bottom left*).

3.2.1 Data acquisition of the CP experiments

- ^{1}H–^{13}C magnetization transfer was achieved by using linear 70–100% ^{1}H-ramped CP (Metz, Wu, & Smith, 1994; Schaefer & Stejskal, 1976) with a contact time of 2 ms and a ^{13}C radio frequency (r.f.) lock field of 66 kHz.
- Spin-lock pulses should be optimized to fulfill the Hartmann–Hahn matching condition.
- For heteronuclear decoupling during acquisition swept-frequency two-pulse phase modulation (SW_f-TPPM) (Thakur, Kurur, & Madhu, 2006;

Vinod Chandran, Madhu, Kurur, & Bräuniger, 2008) at a ^{1}H r.f. field of 100 kHz was used (also for all of the other solid-state NMR experiments discussed in this chapter).

3.2.2 Data acquisition and processing of the HETCOR experiments

- ^{1}H–^{1}H homonuclear dipolar decoupling was accomplished by using the PMLG5-S2 scheme.
- Each PMLG5 block consists of 10 pulses, taking the following phases: 339.22°, 297.65°, 256.08°, 214.51°, 172.94°, 352.94°, 34.51°, 76.08°, 117.65°, and 159.22° (m5m shape in TopSpin 3.2 library).
- To complete the S2 supercycle, the PMLG5 block was then repeated with 180° phase shift and the optimized PMLG5 pulse was 1.30 μs with an r.f. field of 91 kHz.
- The scaling factor was set to be 0.32, calculated by dividing the observed difference between the centers of the CH_2 and NH_3 signals of solid α-glycine in the indirect dimension by their difference of 4.84 ppm. Dividing the spectral width by this scaling factor appropriately scaled the ^{1}H indirect dimension which was referenced by assigning the midpoint of two CH_2 proton peaks to 3.52 ppm.

3.3 Dynamics investigations

A number of solid-state NMR methods were applied for investigation of chain dynamics in PET on multiple time scales. In Fig. 3, an overview of the selected methods and corresponding dynamics across the time scales examined is given. The only pulse sequence mentioned in Fig. 3 for that we show no experimental data in this chapter is recoupling of chemical shift anisotropy (ROCSA) (Chan & Tycko, 2003). With this experiment recoupled powder patterns are obtained, similar to the separation of undistorted powder patterns by effortless recoupling (SUPER) experiment (Liu, Mao, & Schmidt-Rohr, 2002) introduced here. In contrast to ROCSA, the SUPER experiment can be carried out at low MAS frequencies, e.g., below 5 kHz (Liu et al., 2002). For the other methods in the figure, their applications and the information provided are described in more detail in the following sections.

3.3.1 Rotating-frame spin-lattice relaxation to investigate PET chain dynamics

Rotating-frame spin-lattice relaxation ($T_{1\rho}$) experiments provide information about molecular dynamics on a time scale from microseconds to

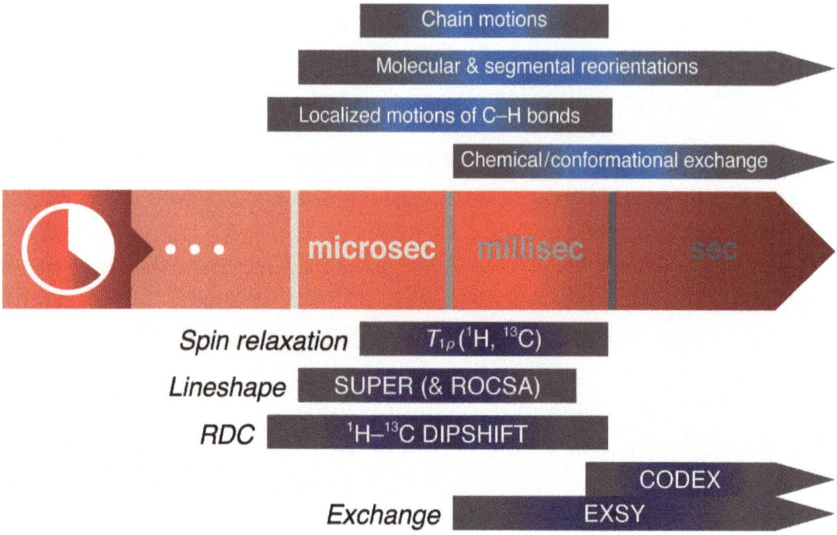

Fig. 3 Overview of NMR methods for determination of chain dynamics in PET over microseconds to seconds (and longer) time scales.

milliseconds (Schmidt-Rohr & Spiess, 1994). Proton spin-lattice relaxation times in the rotating frame, $T_{1\rho}(^1H)$, suffer from ^1H spin diffusion, thus leading to the averaging of relaxation times. Therefore, they provide more segmental information about PET chain motions (Gabriëlse et al., 1994; VanderHart & Garroway, 1979). During the $T_{1\rho}(^1H)$ experiment the proton spin-lock time τ is varied and followed by CP transfer which means that detection takes place via the neighboring carbons. The normalized signal intensities $M(\tau)/M(0)$ can be fitted to a single exponential to yield $T_{1\rho}(^1H)$ relaxation times for each carbon (Choudhury et al., 2012). In Fig. 4 (*left*), the normalized ^1H magnetization decay in the rotating frame is shown exemplarily for the protonated aromatic and the ethylene glycol (EG) carbons of untreated and UV-treated PET including the resulting $T_{1\rho}(^1H)$ values. For the UV-treated PET sample larger $T_{1\rho}(^1H)$ values were determined, indicating increased average motional correlation times and thus slower motions (Falkenstein et al., 2020).

Unlike $T_{1\rho}(^1H)$ relaxation times, carbon spin-lattice relaxation times in the rotating frame, $T_{1\rho}(^{13}C)$, are not averaged by spin diffusion and hence provide site-specific information on PET chain dynamics (Gabriëlse et al., 1994; Schaefer, Stejskal, Steger, Sefcik, & McKay, 1980). The $T_{1\rho}(^{13}C)$

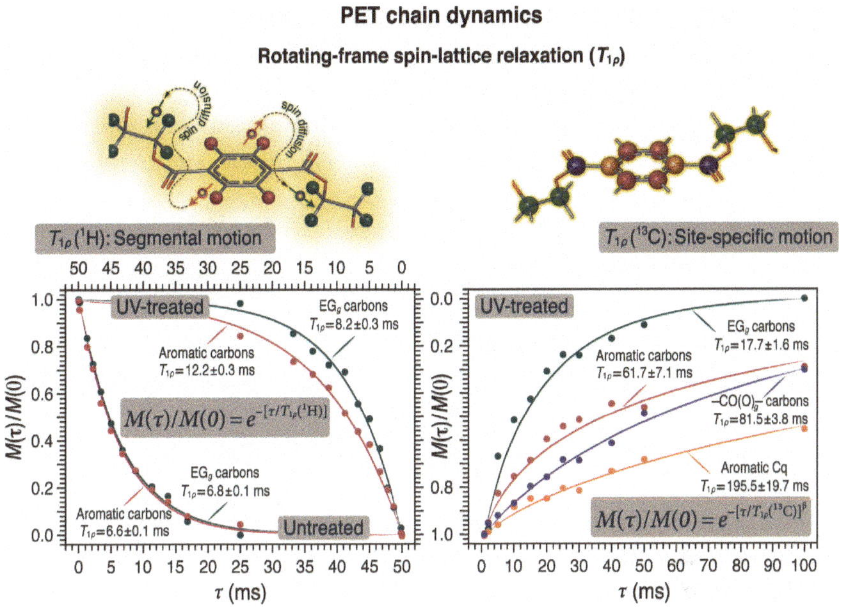

Fig. 4 (*left*) Normalized ^{1}H magnetization decay in the rotating frame for the protonated aromatic and the EG carbons of untreated (see Section 3.1.2) and UV-treated PET (amorphous PET powder treated with UV light over 14 days, for details see Falkenstein et al., 2020) at 30 °C. (*right*) Normalized ^{13}C magnetization decay in the rotating frame for the carbonyl and EG carbons in *g* conformation as well as for the protonated aromatic and quaternary aromatic carbons of UV-treated PET, measured at 30 °C.

pulse sequence consists of CP followed by a spin-lock time τ in the carbon channel which is varied during the experiment. To determine the $T_{1\rho}(^{13}C)$ relaxation times for each carbon, the normalized signal intensities $M(\tau)/M(0)$ can be fitted to a stretched exponential (Choudhury et al., 2012). The ^{13}C magnetization decay in the rotating frame is shown in Fig. 4 (*right*) for the protonated and quaternary aromatic as well as for the carbonyl and the EG carbons in *g* conformation of the UV-treated PET sample. Clearly, the $T_{1\rho}(^{13}C)$ values differ more significantly from each other than the corresponding $T_{1\rho}(^1H)$ values. The quaternary aromatic and the carbonyl carbons show larger $T_{1\rho}(^{13}C)$ relaxation times because no protons are attached. By comparing $T_{1\rho}(^{13}C)$ values at different temperatures or of different samples, as for the $T_{1\rho}(^1H)$ measurements, information about dynamic changes can be obtained (Choudhury et al., 2012).

3.3.1.1 Data acquisition and evaluation of the $T_{1\rho}$ experiments

- A 66kHz spin-lock field was used for ^{1}H and ^{13}C for $T_{1\rho}(^1H)$ and $T_{1\rho}(^{13}C)$ measurements, respectively.
- The length of the spin-lock pulse was incremented in 13 steps for $T_{1\rho}(^1H)$ measurements ranging from 0.01 to 50ms and in 11 steps for $T_{1\rho}(^{13}C)$ from 1 to 100ms.
- Both $T_{1\rho}(^1H)$ and $T_{1\rho}(^{13}C)$ experiments were performed at a MAS frequency of 8kHz.
- Experimental data were transferred to MestReNova and PET carbon signals were fitted in the stacked mode using the Line Fitting feature.
- The normalized signal intensities $M(\tau)/M(0)$ were plotted against τ and fitted to a single exponential for $T_{1\rho}(^1H)$ and to a stretched exponential for $T_{1\rho}(^{13}C)$ experiments, respectively, using OriginPro 8G.

3.3.2 Separation of undistorted powder patterns by effortless recoupling to study reorientations of the carbonyl group in PET

The SUPER experiment is a 2D MAS experiment which provides quasi-static chemical shift anisotropy (CSA) powder patterns, composed of one peak at the middle principal value of the chemical shift tensor δ_{22} and steps at the two other principal values δ_{11} and δ_{33}. The principal values can thus be read directly from the spectrum (Liu et al., 2002; Schmidt-Rohr & Spiess, 1994). The orientation of the principal axes of the PET carbonyl carbon was determined by density functional theory (DFT)-based calculations using bis (2-hydroxyethyl) terephthalate (BHET) as model compound (Fig. 5, *left*). Molecular dynamics in PET would have an effect on the shape and the principal values of the CSA tensors (Liu et al., 2002). The CSA powder patterns of the PET carbonyl carbon atom exemplified in Fig. 5 (*right*) have similar shapes and principal values to those obtained by Liu et al. (2002) which is also the typical lineshape for an ester group (Liu et al., 2002) although the simulated static spectra show less distortions and sharper steps. However, no significant changes were observed for the CSA tensors of the UV-treated PET or the samples measured at 70°C (Fig. 5).

3.3.2.1 Procedure to determine the orientation of principal axes for PET carbonyl carbon by DFT-based calculations

- All calculations on BHET were realized with the Gaussian 16 software suite.
- An initial guess of the molecular structure was generated with TmoleX (Steffen et al., 2010).

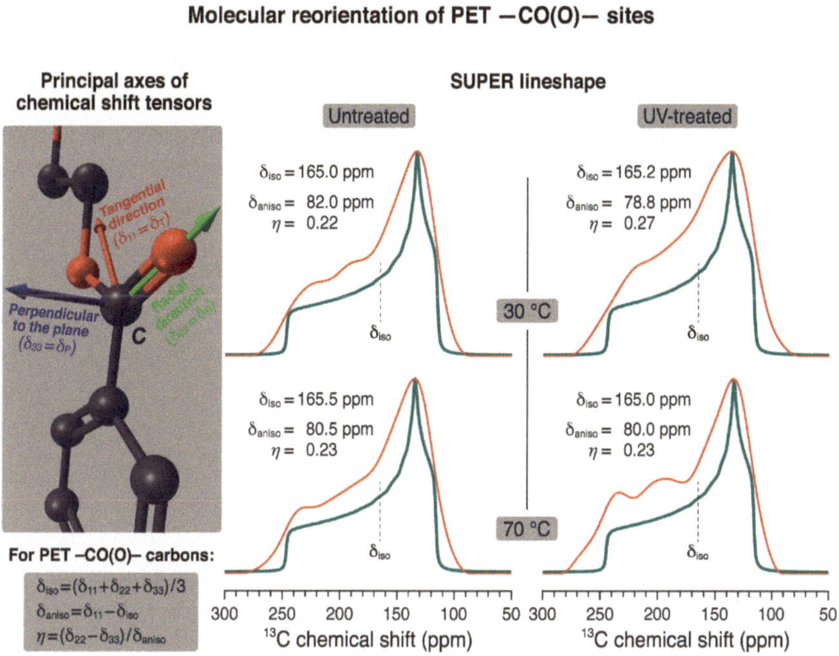

Fig. 5 (*left*) Orientation of the principal axes of the PET carbonyl carbon. (*right*) Carbonyl powder spectra obtained as cross sections from 2D SUPER (*thin lines*) and simulated static chemical-shift powder patterns (*thick lines*) before and after UV irradiation at 30 and 70 °C, respectively. The SUPER experiments were performed at a MAS frequency of 5 kHz.

- The BHET structure was optimized using the hybrid functional B3LYP (Becke, 1988, 1993; Lee, Yang, & Parr, 1988) in combination with Grimme dispersion correction with Becke/Johnson damping (D3BJ) (Grimme, Antony, Ehrlich, & Krieg, 2010; Grimme, Ehrlich, & Goerigk, 2011) and the cc-pVDZ basis set (Dunning Jr., 1989).
- For determination of the nuclear magnetic shielding tensor, the same functional was employed along with the pcSseg-3 basis set (Jensen, 2015) from the basis set exchange repository (Pritchard, Altarawy, Didier, Gibson, & Windus, 2019) for the gauge-including atomic orbital (GIAO) calculations (Cheeseman, Trucks, Keith, & Frisch, 1996).

3.3.2.2 Data acquisition and evaluation of the SUPER experiments

- An offset of 164 ppm was used for the ^{13}C channel since data analysis of this experiment becomes easier if the sites of interest are measured near resonance (Liu et al., 2002).

- To obtain the experimental powder patterns, cross sections were extracted from the experimental 2D SUPER spectra at the ppm values of the signals to be analyzed.
- Static chemical-shift powder patterns were obtained by SIMPSON simulations.

3.3.3 Dipolar chemical shift correlation to quantify localized dynamics

Dipolar chemical shift correlation (DIPSHIFT) (Hong, Gross, & Griffin, 1997) is a pseudo-2D experiment which allows the investigation of localized dynamic processes of molecules on the sub-microsecond time scale (deAzevedo et al., 2008; Huster, Xiao, & Hong, 2001; Ivanir-Dabora, Nimerovsky, Madhu, & Goldbourt, 2015). In Fig. 6, DIPSHIFT curves are shown for the EG and the aromatic C–H groups of untreated and UV-treated PET. For each curve, ^{1}H–^{13}C dipolar coupling order parameters S were calculated, ranging from 0 (isotropic motion) to 1 (rigid limit) (deAzevedo et al., 2008; Ivanir-Dabora et al., 2015). In order to calculate these order parameters, rigid-limit values for the CH and CH_2 groups were first determined which in the case of PET can be achieved by performing DIPSHIFT experiments on crystalline BHET (Merck KGaA, Darmstadt, Germany) and fitting the resulting curves site-specifically (Wei, Song, et al., 2019). In general, the higher

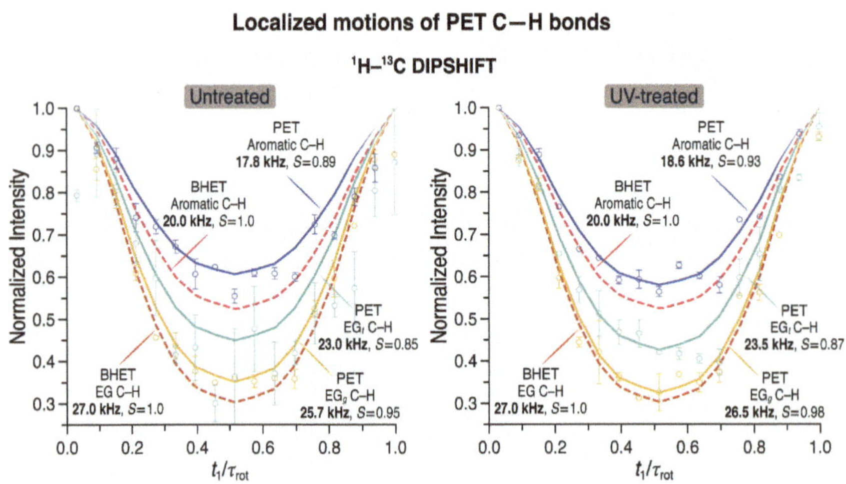

Fig. 6 ^{1}H–^{13}C dephasing curves for the phenylene and EG unit of (*left*) untreated and (*right*) UV-treated PET obtained from DIPSHIFT measurements at 30 °C under 8 kHz MAS. The rigid-limit values for the CH and CH_2 group (*dashed lines*) were determined using BHET.

order parameters calculated for the UV-treated PET sample (Fig. 6) indicate a reduced mobility of the phenylene and EG units in PET after UV irradiation (Falkenstein et al., 2020).

3.3.3.1 Data acquisition and evaluation of the DIPSHIFT experiments

- ^{1}H–^{1}H homonuclear dipolar decoupling was accomplished by using the PMLG5 scheme (Vinogradov et al., 2001).
- The optimized PMLG5 pulse was 2.08 μs with 80 kHz r.f. amplitude.
- All experimental DIPSHIFT curves were fitted with SIMPSON simulations using a PMLG scaling factor of 0.5. The scaling factor was determined before based on the *J*-splitting of the CH and CH_2 groups of adamantane.

3.3.4 Centerband-only detection of exchange to characterize reorientations of the PET phenylene unit

It has been argued that the phenylene unit of PET performs 180° ring flip motions about its 1,4-axis (Choudhury et al., 2012; Wilhelm & Spiess, 1996). Slow segmental reorientations on a time scale of milliseconds to seconds, like the phenylene ring flip, can be detected with a 1D exchange experiment called centerband-only detection of exchange (CODEX) (Fig. 7). This site-specific information is accessible by recording a series of CODEX spectra with different mixing times (t_m). In general, two spectra are required per mixing time, namely the normal CODEX spectrum and a reference spectrum, where the mixing time and the *z*-filter time (t_z) are exchanged (deAzevedo, Hu, Bonagamba, & Schmidt-Rohr, 1999). The CODEX spectrum must be subtracted from the reference spectrum in order to remove dominant signals from immobile sites as well as to correct losses due to T_1 relaxation, thus obtaining the so-called pure-exchange spectrum (deAzevedo et al., 1999; Hodgkinson, 2018) (Fig. 7, *bottom left*). Fitting of the normalized pure-exchange CODEX intensities $\Delta S/S_0$, meaning the ratio of pure-exchange CODEX to reference intensity, to a stretched exponential yields the correlation time of the ring flip motion, or more specifically the center of the correlation time distribution τ_c (Choudhury et al., 2012; deAzevedo et al., 1999). From the final exchange intensity E_∞ the number M of equivalent orientational sites accessed by a specific carbon during the motional process and the fraction f_m of mobile segments can be determined. The normalized pure-exchange CODEX intensities of the protonated aromatic carbons in amorphous PET (see Section 3.1.2)

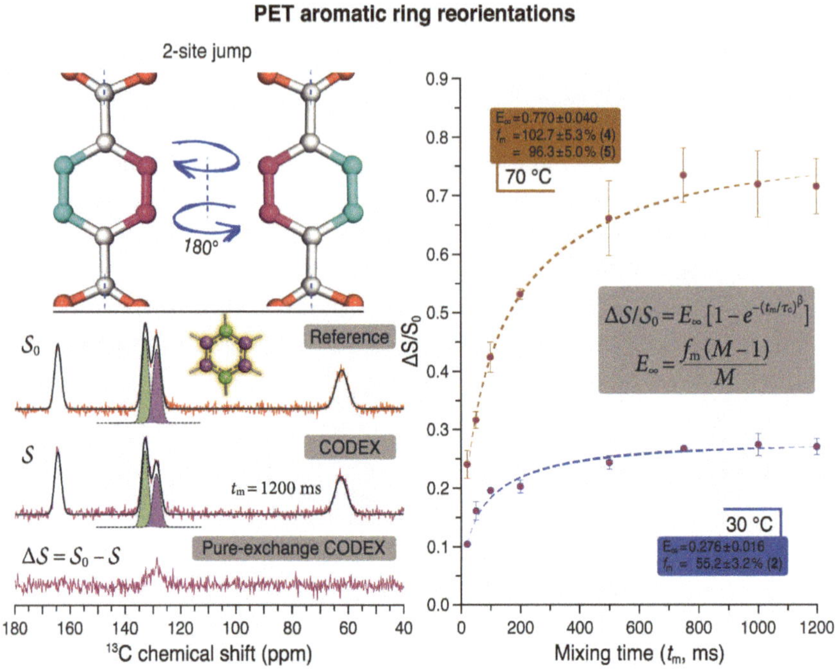

Fig. 7 (*top left*) Schematic of the 180° phenylene ring flip of PET. (*bottom left*) CODEX, reference and pure-exchange CODEX spectra of PET measured at 30 °C. (*right*) Normalized pure-exchange CODEX signal intensities of the protonated aromatic carbons in PET as a function of the mixing time, measured at 30 and 70 °C under 12.5 kHz MAS. The experimental data were fitted to a stretched exponential (*dashed lines*). For both curves the final exchange intensity E_∞, the fraction f_m of mobile segments as well as the number M of equivalent orientational sites accessed by one protonated aromatic carbon during the motional process (*bold; in brackets*) are given.

measured at 30 and 70 °C are shown in Fig. 7 (*right*) including the calculated values for E_∞ and f_m. At 70 °C both E_∞ and f_m are increased. The number of equivalent orientational sites accessed by one protonated aromatic carbon is $M=2$ at 30 °C that indicates the 180° phenylene ring flip mentioned above, whereby at 70 °C M is assumed to be 4 or 5, suggesting more complex motions of the phenylene unit. The correlation time was found to be very sensitive to the shape of CODEX curves and to single data points, leading to large errors of the τ_c values. Therefore, one should be careful when interpreting the results in the context of dynamic changes of different samples (Choudhury et al., 2012).

3.3.4.1 Data acquisition and evaluation of the CODEX experiments

- Mixing times from 0.02 to 1.2 s and a z-filter time of 1.28 ms were used.
- The sum of the preparation and refocusing time was 10 t_r ($N=10$).
- An extended phase cycle was employed for proper cancelation of unwanted transverse components during t_m and t_z under fast MAS (Reichert, Bonagamba, & Schmidt-Rohr, 2001).
- The experimental 1D spectra were fitted and deconvoluted using the Line Fitting feature of MestReNova.
- The normalized pure-exchange CODEX intensities of the protonated aromatic carbons were then plotted against t_m and fitted to the stretched exponential function using OriginPro 8G.

3.3.5 Exchange spectroscopy for investigation of gauche/trans conformational exchange in PET

Multidimensional exchange NMR is applied to characterize molecular motions on a time scale of milliseconds to several seconds. During the 2D ^{13}C exchange spectroscopy (EXSY) experiment, the exchange process takes place during the mixing time t_m leading to off-diagonal signals in the spectrum. Without slow dynamics during t_m, frequencies remain unchanged and only diagonal signals are observed. By repeating the experiment with different mixing times dynamics across the time scales can be examined (Jeener, Meier, Bachmann, & Ernst, 1979; Perrin & Dwyer, 1990; Schmidt-Rohr & Spiess, 1994). For example, Zemke, Chmelka, Schmidt-Rohr, and Spiess (1991) reported conformational exchange near the glass transition of atactic polypropylene by using 2D exchange NMR. 2D ^{13}C EXSY spectra of PET shown in Fig. 8 were acquired at 30 and 80 °C to investigate the potential exchange between the g and t conformations of PET below and above glass transition ($T_g \sim 75$ °C, Ronkvist et al., 2009).

For PET, no unambiguous evidence of g/t exchange was found at either temperature. Although the diagonal signal detected at 80 °C was significantly broadened compared to that at 30 °C, no fully resolved cross peaks were detected. It should be noted that in the case of PET, the chemical shifts of the EG carbons in g and t conformations are very similar and therefore the close-lying non-diagonal elements arising from chemical/conformational exchange can be masked by the strong diagonal elements. In this case, spin-echo based diagonal peak suppression would be employed (Banerjee & Chandrakumar, 2015; Wang et al., 2018).

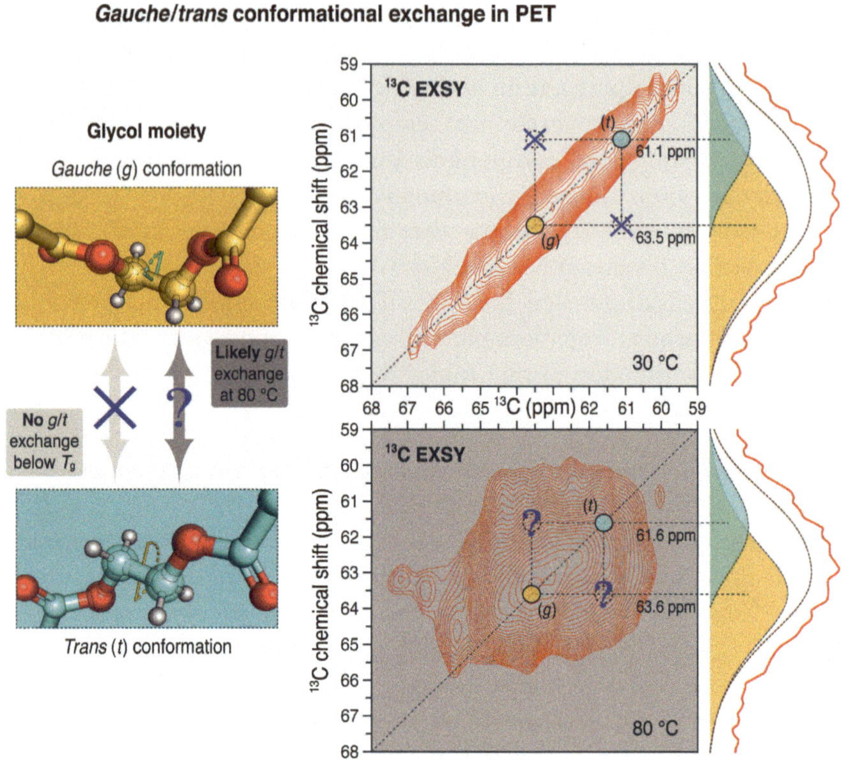

Fig. 8 (*left*) Schematic of the possible *g/t* conformational exchange in PET. (*right*) ^{13}C EXSY spectra of PET (see Section 3.1.2) measured at 30 and 80 °C under 12.5 kHz MAS and with a mixing time of 500 ms.

4. Summary

This chapter demonstrates how NMR spectroscopy can be applied as a tool to characterize PET polymers under the effect of enzymatic degradation or a desired degradation temperature. A variety of advanced NMR experiments have been overviewed which may contribute to clarify further questions about the biocatalytic degradation of plastics and to elucidate the underlying mechanism. It is in particular the combination of this "toolbox" that brings into light the various parameters that determine the accessibility of PET to directed and controlled degradation. The above-mentioned NMR spectroscopic methods thus open up possibilities for future development of recycling approaches for plastic waste, including but not limited to PET.

Acknowledgments

The authors would like to thank their colleagues Dr. C. Wiebeler for DFT-based calculations, Dr. P. Bielytskyi and Dr. D. Gräsing for technical assistance during various stages of the experiments, and Prof. W. Gärtner for valuable discussions and critical reading of the manuscript. Computations for this work were realized with resources of Leipzig University Computing Center.

Funding

P.F. gratefully acknowledges the Deutsche Bundesstiftung Umwelt (DBU, grant no. 20018/565) for financial support.

References

Austin, H. P., Allen, M. D., Donohoe, B. S., Rorrer, N. A., Kearns, F. L., Silveira, R. L., et al. (2018). Characterization and engineering of a plastic-degrading aromatic polyesterase. *Proceedings of the National Academy of Sciences of the United States of America, 115*, E4350–E4357.

Bak, M., Rasmussen, J. T., & Nielsen, N. C. (2000). SIMPSON: A general simulation program for solid-state NMR spectroscopy. *Journal of Magnetic Resonance, 147*, 296–330.

Banerjee, A., & Chandrakumar, N. (2015). Two-dimensional nuclear magnetic resonance: Exploiting spin echoes to maximize information content by suppression of diagonal peaks in homonuclear experiments. *The Journal of Physical Chemistry A, 119*, 482–487.

Becke, A. D. (1988). Density-functional exchange-energy approximation with correct asymptotic behavior. *Physical Review A, 38*, 3098–3100.

Becke, A. D. (1993). Density-functional thermochemistry. III. The role of exact exchange. *The Journal of Chemical Physics, 98*, 5648–5652.

Berkowitz, S. (1984). Viscosity–molecular weight relationships for poly(ethylene terephthalate) in hexafluoroisopropanol–pentafluorophenol using SEC–LALLS. *Journal of Applied Polymer Science, 29*, 4353–4361.

Chan, J. C. C., & Tycko, R. (2003). Recoupling of chemical shift anisotropies in solid-state NMR under high-speed magic-angle spinning and in uniformly ^{13}C-labeled systems. *Journal of Chemical Physics, 118*, 8378–8389.

Cheeseman, J. R., Trucks, G. W., Keith, T. A., & Frisch, M. J. (1996). A comparison of models for calculating nuclear magnetic resonance shielding tensors. *The Journal of Chemical Physics, 104*, 5497–5509.

Choudhury, R. P., Lee, J. S., Kriegel, R. M., Koros, W. J., & Beckham, H. W. (2012). Chain dynamics in antiplasticized and annealed poly(ethylene terephthalate) determined by solid-state NMR and correlated with enhanced barrier properties. *Macromolecules, 45*, 879–887.

de Ilarduya, A. M., & Muñoz-Guerra, S. (2014). Chemical structure and microstructure of poly(alkylene terephthalate)s, their copolyesters, and their blends as studied by NMR. *Macromolecular Chemistry and Physics, 215*, 2138–2160.

deAzevedo, E. R., Hu, W.-G., Bonagamba, T. J., & Schmidt-Rohr, K. (1999). Centerband-only detection of exchange: Efficient analysis of dynamics in solids by NMR. *Journal of the American Chemical Society, 121*, 8411–8412.

deAzevedo, E. R., Saalwächter, K., Pascui, O., de Souza, A. A., Bonagamba, T. J., & Reichert, D. (2008). Intermediate motions as studied by solid-state separated local field NMR experiments. *The Journal of Chemical Physics, 128*, 104505.

Donovan, A. R., & Moad, G. (2005). A novel method for determination of polyester end-groups by NMR spectroscopy. *Polymer, 46*, 5005–5011.

Duer, M. J. (Ed.). (2002). *Solid-state NMR spectroscopy: Principles and applications* Blackwell Science Ltd.
Dunning, T. H., Jr. (1989). Gaussian basis sets for use in correlated molecular calculations. I. The atoms boron through neon and hydrogen. *The Journal of Chemical Physics, 90*, 1007–1023.
Falkenstein, P., Gräsing, D., Bielytskyi, P., Zimmermann, W., Matysik, J., Wei, R., et al. (2020). UV pretreatment impairs the enzymatic degradation of polyethylene terephthalate. *Frontiers in Microbiology, 11*, 689.
Fox, B., Moad, G., van Diepen, G., Willing, I., & Cook, W. D. (1997). Characterization of poly(ethylene terephthalate) and poly(ethylene terephthalate) blends. *Polymer, 38*, 3035–3043.
Gabriëlse, W., Angad Gaur, H., Feyen, F. C., & Veeman, W. S. (1994). ^{13}C solid-state NMR study of differently processed poly(ethylene terephthalate) yarns. *Macromolecules, 27*, 5811–5820.
Grimme, S., Antony, J., Ehrlich, S., & Krieg, H. (2010). A consistent and accurate *ab initio* parametrization of density functional dispersion correction (DFT-D) for the 94 elements H-Pu. *The Journal of Chemical Physics, 132*, 154104.
Grimme, S., Ehrlich, S., & Goerigk, L. (2011). Effect of the damping function in dispersion corrected density functional theory. *Journal of Computational Chemistry, 32*, 1456–1465.
Guan, X., & Stark, R. E. (2010). A general protocol for temperature calibration of MAS NMR probes at arbitrary spinning speeds. *Solid State Nuclear Magnetic Resonance, 38*, 74–76.
Günther, H. (2013). *NMR spectroscopy: Basic principles, concepts and applications in chemistry* (3rd ed.). Wiley-VCH Verlag GmbH & Co. KGaA.
Hodgkinson, P. (Ed.). (2018). *Modern methods in solid-state NMR: A practitioner's guide* The Royal Society of Chemistry.
Hong, M., Gross, J. D., & Griffin, R. G. (1997). Site-resolved determination of peptide torsion angle ϕ from the relative orientations of backbone N–H and C–H bonds by solid-state NMR. *The Journal of Physical Chemistry B, 101*, 5869–5874.
Huster, D., Xiao, L., & Hong, M. (2001). Solid-state NMR investigation of the dynamics of the soluble and membrane-bound colicin Ia channel-forming domain. *Biochemistry, 40*, 7662–7674.
Hutchinson, J. M. (1995). Physical aging of polymers. *Progress in Polymer Science, 20*, 703–760.
Ivanir-Dabora, H., Nimerovsky, E., Madhu, P. K., & Goldbourt, A. (2015). Site-resolved backbone and side-chain intermediate dynamics in a carbohydrate-binding module protein studied by magic-angle spinning NMR spectroscopy. *Chemistry—A European Journal, 21*, 10778–10785.
Jeener, J., Meier, B. H., Bachmann, P., & Ernst, R. R. (1979). Investigation of exchange processes by two-dimensional NMR spectroscopy. *The Journal of Chemical Physics, 71*, 4546–4553.
Jensen, F. (2015). Segmented contracted basis sets optimized for nuclear magnetic shielding. *Journal of Chemical Theory and Computation, 11*, 132–138.
Joo, S., Cho, I. J., Seo, H., Son, H. F., Sagong, H.-Y., Shin, T. J., et al. (2018). Structural insight into molecular mechanism of poly(ethylene terephthalate) degradation. *Nature Communications, 9*, 382.
Kenwright, A. M., Peace, S. K., Richards, R. W., Bunn, A., & MacDonald, W. A. (1999). End group modification in poly(ethylene terephthalate). *Polymer, 40*, 2035–2040.
Lee, C., Yang, W., & Parr, R. G. (1988). Development of the Colle-Salvetti correlation-energy formula into a functional of the electron density. *Physical Review B, 37*, 785–789.
Liu, S.-F., Mao, J.-D., & Schmidt-Rohr, K. (2002). A robust technique for two-dimensional separation of undistorted chemical-shift anisotropy powder patterns in magic-angle-spinning NMR. *Journal of Magnetic Resonance, 155*, 15–28.

Ma, Y., Agarwal, U. S., Vekemans, J. A. J. M., & Sikkema, D. J. (2003). NMR based determination of minute acid functionality: End-groups in PET. *Polymer*, *44*, 4429–4434.
Metz, G., Wu, X., & Smith, S. O. (1994). Ramped-amplitude cross polarization in magic-angle-spinning NMR. *Journal of Magnetic Resonance, Series A*, *110*, 219–227.
Paul, S., Thakur, R. S., Goswami, M., Sauerwein, A. C., Mamone, S., Concistrè, M., et al. (2009). Supercycled homonuclear dipolar decoupling sequences in solid-state NMR. *Journal of Magnetic Resonance*, *197*, 14–19.
Peez, N., Janiska, M.-C., & Imhof, W. (2019). The first application of quantitative ^{1}H NMR spectroscopy as a simple and fast method of identification and quantification of microplastic particles (PE, PET, and PS). *Analytical and Bioanalytical Chemistry*, *411*, 823–833.
Perrin, C. L., & Dwyer, T. J. (1990). Application of two-dimensional NMR to kinetics of chemical exchange. *Chemical Reviews*, *90*, 935–967.
Pritchard, B. P., Altarawy, D., Didier, B., Gibson, T. D., & Windus, T. L. (2019). New basis set exchange: An open, up-to-date resource for the molecular sciences community. *Journal of Chemical Information and Modeling*, *59*, 4814–4820.
Qian, R., Shen, D., Sun, F., & Wu, L. (1996). The effects of physical ageing on conformational changes of poly(ethylene terephthalate) in the glass transition region. *Macromolecular Chemistry and Physics*, *197*, 1485–1493.
Reichert, D., Bonagamba, T. J., & Schmidt-Rohr, K. (2001). Slow-down of ^{13}C spin diffusion in organic solids by fast MAS: A CODEX NMR study. *Journal of Magnetic Resonance*, *151*, 129–135.
Roberts, J. E., Vega, S., & Griffin, R. G. (1984). Two-dimensional heteronuclear chemical shift correlation spectroscopy in rotating solids. *Journal of the American Chemical Society*, *106*, 2506–2512.
Ronkvist, Å. M., Xie, W., Lu, W., & Gross, R. A. (2009). Cutinase-catalyzed hydrolysis of poly(ethylene terephthalate). *Macromolecules*, *42*, 5128–5138.
Roth, C., Wei, R., Oeser, T., Then, J., Föllner, C., Zimmermann, W., et al. (2014). Structural and functional studies on a thermostable polyethylene terephthalate degrading hydrolase from *Thermobifida fusca*. *Applied Microbiology and Biotechnology*, *98*, 7815–7823.
Schaefer, J., & Stejskal, E. O. (1976). Carbon-13 nuclear magnetic resonance of polymers spinning at the magic angle. *Journal of the American Chemical Society*, *98*, 1031–1032.
Schaefer, J., Stejskal, E. O., Steger, T. R., Sefcik, M. D., & McKay, R. A. (1980). Carbon-13 $T_{1\rho}$ experiments on solid glassy polymers. *Macromolecules*, *13*, 1121–1126.
Schmidt-Rohr, K., Hu, W., & Zumbulyadis, N. (1998). Elucidation of the chain conformation in a glassy polyester, PET, by two-dimensional NMR. *Science*, *280*, 714–717.
Schmidt-Rohr, K., & Spiess, H. W. (1994). *Multidimensional solid-state NMR and polymers*. Academic Press.
Sefcik, M. D., Schaefer, J., Stejskal, E. O., & McKay, R. A. (1980). Analysis of the room-temperature molecular motions of poly(ethylene terephthalate). *Macromolecules*, *13*, 1132–1137.
Steffen, C., Thomas, K., Huniar, U., Hellweg, A., Rubner, O., & Schroer, A. (2010). TmoleX—A graphical user interface for TURBOMOLE. *Journal of Computational Chemistry*, *31*, 2967–2970.
Sulaiman, S., Yamato, S., Kanaya, E., Kim, J.-J., Koga, Y., Takano, K., et al. (2012). Isolation of a novel cutinase homolog with polyethylene terephthalate-degrading activity from leaf-branch compost by using a metagenomic approach. *Applied and Environmental Microbiology*, *78*, 1556–1562.
Thakur, R. S., Kurur, N. D., & Madhu, P. K. (2006). Swept-frequency two-pulse phase modulation for heteronuclear dipolar decoupling in solid-state NMR. *Chemical Physics Letters*, *426*, 459–463.
Tournier, V., Topham, C. M., Gilles, A., David, B., Folgoas, C., Moya-Leclair, E., et al. (2020). An engineered PET depolymerase to break down and recycle plastic bottles. *Nature*, *580*, 216–219.

VanderHart, D. L., & Garroway, A. N. (1979). ^{13}C NMR rotating frame relaxation in a solid with strongly coupled protons: Polyethylene. *The Journal of Chemical Physics*, *71*, 2773–2787.

Vertommen, M. A. M. E., Nierstrasz, V. A., van der Veer, M., & Warmoeskerken, M. M. C. G. (2005). Enzymatic surface modification of poly(ethylene terephthalate). *Journal of Biotechnology*, *120*, 376–386.

Vinod Chandran, C., Madhu, P. K., Kurur, N. D., & Bräuniger, T. (2008). Swept-frequency two-pulse phase modulation (SW_f-TPPM) sequences with linear sweep profile for heteronuclear decoupling in solid-state NMR. *Magnetic Resonance in Chemistry*, *46*, 943–947.

Vinogradov, E., Madhu, P. K., & Vega, S. (2001). Phase modulated Lee-Goldburg magic angle spinning proton nuclear magnetic resonance experiments in the solid state: A bimodal Floquet theoretical treatment. *Journal of Chemical Physics*, *115*, 8983–9000.

Wang, K., Zhang, Z., Ding, X., Tian, F., Huang, Y., Chen, Z., et al. (2018). Spin-echo based diagonal peak suppression in solid-state MAS NMR homonuclear chemical shift correlation spectra. *Journal of Magnetic Resonance*, *287*, 91–98.

Wei, R., Breite, D., Song, C., Gräsing, D., Ploss, T., Hille, P., et al. (2019). Biocatalytic degradation efficiency of postconsumer polyethylene terephthalate packaging determined by their polymer microstructures. *Advanced Science*, *6*, 1900491.

Wei, R., Song, C., Gräsing, D., Schneider, T., Bielytskyi, P., Böttcher, D., et al. (2019). Conformational fitting of a flexible oligomeric substrate does not explain the enzymatic PET degradation. *Nature Communications*, *10*, 5581.

Wei, R., & Zimmermann, W. (2017). Biocatalysis as a green route for recycling the recalcitrant plastic polyethylene terephthalate. *Microbial Biotechnology*, *10*, 1302–1307.

Wilhelm, M., & Spiess, H. W. (1996). Detection of slow 180° phenylene flips in PET fibers via ^{13}C two-dimensional solid-state exchange NMR. *Macromolecules*, *29*, 1088–1090.

Zemke, K., Chmelka, B. F., Schmidt-Rohr, K., & Spiess, H. W. (1991). Conformational exchange near the glass transition: Two-dimensional ^{13}C NMR study of atactic polypropylene. *Macromolecules*, *24*, 6874–6876.

CHAPTER TWELVE

Fluorimetric high-throughput screening method for polyester hydrolase activity using polyethylene terephthalate nanoparticles

Lara Pfaff[a], Daniel Breite[b], Christoffel P.S. Badenhorst[a], Uwe T. Bornscheuer[a], and Ren Wei[a,*]

[a]Department of Biotechnology & Enzyme Catalysis, Institute of Biochemistry, University of Greifswald, Greifswald, Germany

[b]Leibniz Institute of Surface Engineering (IOM), Leipzig, Germany

*Corresponding author: e-mail address: ren.wei@uni-greifswald.de

Contents

Abstract

Biocatalysis has recently emerged as a powerful and eco-friendly technology in waste plastic recycling, especially for the widely used polyethylene terephthalate (PET). So far, however, a high-throughput screening assay specifically toward PET-hydrolyzing activity has rarely been applied. This hinders the identification of new polyester hydrolases

Methods in Enzymology, Volume 648
ISSN 0076-6879
https://doi.org/10.1016/bs.mie.2020.11.003

and their variants with adequate activities fulfilling the requirements for industrial applications. This chapter describes the detailed procedure for assaying terephthalate as a major product of enzymatic PET hydrolysis in a 96-well microtiter plate format. Using PET nanoparticles derived readily from waste food packaging as a substrate, an active thermophilic PET hydrolase was clearly distinguished from an inactive variant by a Fenton chemistry-mediated fluorimetric detection. The assay uses enzymes in crude cell lysates, obtained by a simple freeze–thaw protocol. The experimental work validates the applicability of this method for screening mutant libraries of novel PET hydrolases and will thus facilitate the identification of promising variants useful for effective plastic waste recycling.

1. Introduction

Plastic pollution poses a serious risk to terrestrial and aquatic ecosystems as well as human health (Eriksen et al., 2014; Rochman, Hoh, Kurobe, & Teh, 2013; Wright & Kelly, 2017). The global production of plastics, excluding synthetic fibers, reached 359 million metric tons p.a. in 2018 (PlasticsEurope, 2019) and is rapidly increasing. Therefore, efficient disposal strategies are urgently needed for sustainable treatment of plastic waste. Recycling of the existing plastic waste presents the most sustainable and viable solution in context of decreased CO_2 emission as well as of the fuel and energy savings (Rahimi & García, 2017; Vollmer et al., 2020). In 2015, only 9% of plastic waste was recycled worldwide (Geyer, Jambeck, & Law, 2017). Research achievements in the last two decades empowered the applicability of selected enzymes and microbes as an alternative eco-friendly plastic recycling technology (Wei et al., 2020). Especially for polyethylene terephthalate (PET), which is widely used to produce food packaging, beverage bottles, and synthetic fibers, the biotechnological recycling of plastics that allows the recovery of their monomeric building blocks has been demonstrated both at laboratory and industrially relevant scales (Kawai, Kawabata, & Oda, 2019, 2020; Tournier et al., 2020; Wei & Zimmermann, 2017). Since the identification of an efficient PET hydrolase from *Thermobifida fusca* in 2005 (Müller, Schrader, Profe, Dresler, & Deckwer, 2005), the discovery of novel PET hydrolases has become an important field of research (Kawai et al., 2020). So far, the enzymatic hydrolysis of PET has been performed using a variety of PET hydrolases obtained from both eukaryotic and prokaryotic microbes. Besides different hydrolases reported from actinomycetes, including different strains of *Thermobifida* species (Herrero Acero et al., 2011; Wei, Oeser, & Zimmermann, 2014), a commercial cutinase from the fungus

Thermomyces insolens (formerly *Humicola insolens*) has been reported to efficiently degrade low-crystalline PET (Ronkvist, Xie, Lu, & Gross, 2009). Another major discovery was the identification of a bacterium, *Ideonella sakaiensis,* which is able to grow on and assimilate PET at 30°C under laboratory conditions, albeit very slowly (Yoshida et al., 2016). This discovery is also interesting with respect to the speed of natural evolution of enzymes/microbes to be able to access a man-made carbon source less than 80 years after PET was first synthesized by chemists (Bornscheuer, 2016). Recently, the efficient biocatalytic depolymerization of pretreated postconsumer PET bottles within 10 h has been reported and well accepted as a breakthrough enabling the future application of biotechnological plastic recycling (Tournier et al., 2020). A variant of the plant metagenome-derived cutinase LCC which showed inherently high thermal stability and activity against amorphous PET at up to 70°C (Sulaiman et al., 2012; Sulaiman, You, Kanaya, Koga, & Kanaya, 2014; Wei, Song, et al., 2019) was generated and revealed markedly increased PET-hydrolyzing activity, compared to the wild-type enzyme. The discovery of LCC in a metagenomic library as well as further PET hydrolases found from published metagenome databases (Danso et al., 2018) emphasized the value of environmental metagenomes for the identification of more plastic-degrading hydrolases. To identify such novel polyester hydrolases, a powerful high-throughput screening (HTS) assay specifically targeting a real-world plastic substrate is required.

In this chapter, we describe the detailed procedure for assaying terephthalate, the main product from the enzymatic hydrolysis of PET nanoparticles, in a 96-well microtiter plate format (Fig. 1). This assay enables the rapid identification of outstanding PET hydrolase variants from large mutant or metagenome libraries.

2. PET nanoparticles: Generation, application, and characterization

Nanoplastics (size <100 nm) are omnipresent pollutants in natural environments as a result of the fragmentation of larger plastic debris. Nanoplastics may pose a serious threat to human and animal health due to their small size and larger relative surface area, which corresponds to an enhanced ability to penetrate tissues and adsorb pathogens and toxic chemicals (Lehner, Weder, Petri-Fink, & Rothen-Rutishauser, 2019; Wagner & Reemtsma, 2019). For investigations of their ecotoxicity and fate in the biological degradation

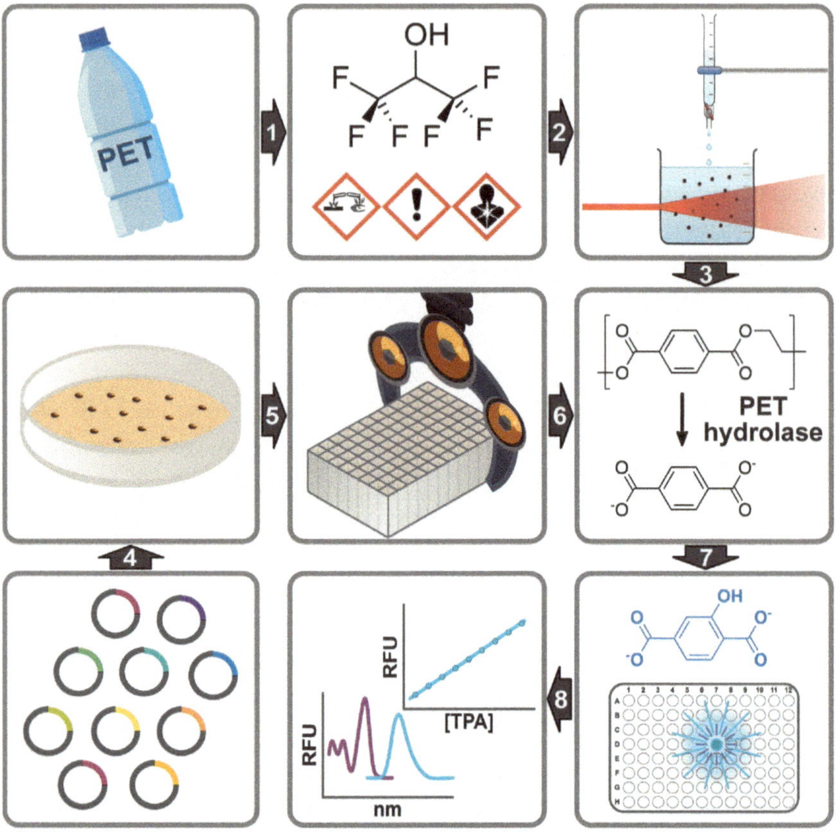

Fig. 1 Workflow of the fluorimetric high-throughput screening method to detect PET hydrolase activity. (1) Postconsumer PET is cut and the pieces are dissolved in hexafluoro-2-propanol. (2) The dissolved PET is dropped into rapidly stirring ultrapure water, resulting in the formation of nanoparticles that are characterized by scanning electron microscopy and dynamic light scattering. (3) After removal of the solvent, the nanoparticles are ready to be used in the enzymatic assays. (4) A plasmid library is transformed into an expression strain like *Escherichia coli* BL21(DE3). (5) The resulting colonies are used to inoculate 96-deep-well plates for protein expression, which is followed by preparation of crude cell lysates. (6) The crude cell lysates are used with PET nanoparticles to assay PET-hydrolyzing activity. (7–8) Formation of the product terephthalate is detected by reaction with hydroxyl radicals, generated by a Fenton-like reaction. Terephthalate is converted to 2-hydroxyterephthalate and detected by fluorescence measurements (excitation at 315 nm and emission at 421 nm).

process, artificial polymer nanoparticles (NPs) are often utilized (Liu et al., 2018; Vogel et al., 2020). The impact of artificially generated PET NPs on human macrophages (Rodríguez-Hernández, Muñoz-Tabares, Aguilar-Guzmán, & Vazquez-Duhalt, 2019), and their degradation by various polyester hydrolases, have been studied (Barth et al., 2015; Danso et al., 2018; Weber et al., 2019; Wei, Oeser, Barth, et al., 2014; Wei, Oeser, Billig, & Zimmermann, 2012; Wei, Oeser, Then, et al., 2014; Vogel et al., 2020).

Nanometer and submicrometer PET particles have been produced using different sources of polymers dissolved in various solvents dripped to water as an anti-solvent (Rodríguez-Hernández et al., 2019; Welzel, Müller, & Deckwer, 2002). Hexafluoro-2-propanol (HFIP) has been found to be the best solvent for PET (Berkowitz, 1984). It can rapidly dissolve PET materials with different thermal histories and crystallinities by simple incubation at room temperature (Wei, Breite, et al., 2019; Wei, Oeser, Barth, et al., 2014). Dissolved PET polymers revealed molecular weight distributions comparable to the literature value, indicating that the dissolution of PET in HFIP does not lead to significant chain scission (Wei, Breite, et al., 2019). Postconsumer PET packaging (trays, clamshells, and boxes) for fresh vegetables and fruits, which is a mass-produced low-crystalline form of PET available in the market, could be readily used to prepare PET NPs without any pretreatment. With help of an ULTRA TURRAX mixer (Protocol 5.1), HFIP droplets containing dissolved PET were rapidly distributed in ultrapure water (Fig. 1, steps 2–3), resulting in a homogenous suspension of NPs with an average size well below the 100 nm threshold defined for nanoplastics (Fig. 2A). An average size of 76.6 nm was determined using dynamic light scattering (DLS) and verified by scanning electron microscopy (SEM; Fig. 2B), which also demonstrated the spherical shape of the PET NPs. The diameter determined by DLS is in good agreement with that previously measured by nanoparticle tracking analysis (Vogel et al., 2020).

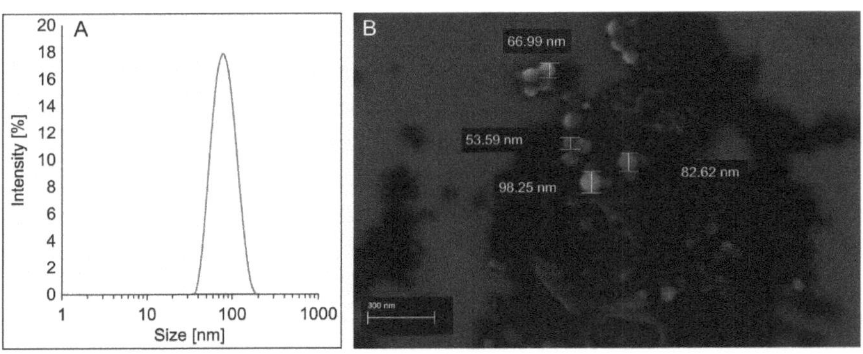

Fig. 2 Evaluation of the size distribution of PET NPs prepared using HFIP-dissolved postconsumer PET packaging by (A) dynamic light scattering and (B) scanning electron microscopy.

3. Fluorimetric high-throughput screening assay

Terephthalic acid (TPA) is one of the main products released by the enzymatic degradation of PET. Based on their characteristic absorbance at 241 nm, TPA and related esters such as mono(2-hydroxyethyl) terephthalate (MHET) and bis(2-hydroxyethyl) terephthalate (BHET) have been monitored by reversed phase high-performance liquid chromatography (RP-HPLC) with UV detection (Eberl et al., 2009; Herrero Acero et al., 2011; Palm et al., 2019). Although most modern HPLCs are equipped with autosamplers, sequential measurements are time consuming and will complicate the analysis of large mutant and metagenome libraries.

In alkaline buffer (pH above 8), TPA is present in its deprotonated species as terephthalate. Terephthalate possesses a symmetrical structure, and can thus be monohydroxylated in the presence of hydroxyl radicals to a single product, 2-hydroxyterephthalate (2-HOTP; Fig. 3), exhibiting stable fluorescence at 421 nm (excitation at 315 nm) for up to 36 h at room temperature (Saran & Summer, 1999). Dihydroxyterepthalic acid (DHT) is another fluorescent hydroxylated derivate of TPA with absorption (excitation) maximum at 375 nm (Cho et al., 2019). However, no literature has described the possibility of DHT synthesis via direct radical hydroxylation of TPA, and it will thus not interfere with this assay.

For terephthalate obtained from depolymerized PET, a high concentration (up to 30% v/v) of hydrogen peroxide and an incubation at 90°C for 30 min has been utilized in an earlier study to generate hydroxyl radicals for the conversion to fluorescent 2-HOTP (O'Neill & Cavaco-Paulo, 2004; Silva et al., 2011). It is hard to apply this temperature to microtiter plates in a high-throughput manner. Furthermore, undesirable loss of liquid by evaporation needs to be avoided, complicating sample handling. To avoid this susceptibility to interferences, the iron autoxidation-mediated hydroxylation of terephthalate has been successfully conducted at room temperature to quantify the enzymatic hydrolysis products of PET (Wei et al., 2012). The chelator ethylenediaminetetraacetic acid (EDTA) was used instead of hydrogen peroxide required for the canonical Fenton reaction. EDTA can facilitate the formation of hydroxyl radicals through stabilizing iron(III) and thus promotes the hydroxylation of terephthalate (Welch, Davis, & Aust, 2002). This Fe(II)-EDTA complex-based method has been previously established for the fluorimetric quantification of the hydroxyl radical scavenging ability of various compounds in the presence of a sodium terephthalate solution (Yang & Guo, 2001).

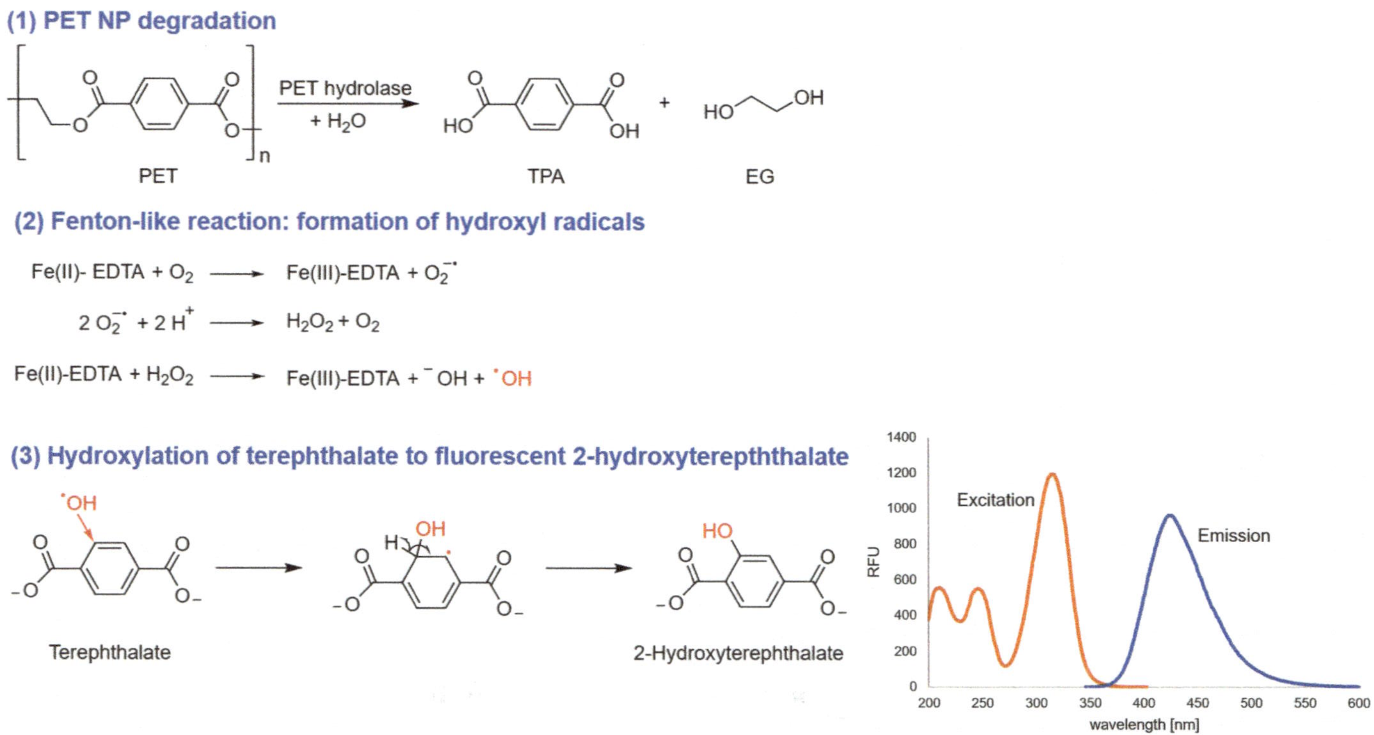

Fig. 3 Fluorimetric detection of terephthalate as the PET hydrolysis product. (1) Enzymatic hydrolysis of PET yields TPA and EG as monomeric products. (2) The formation of hydroxyl radicals mediated by the Fe(II)-EDTA complex. (3) The hydroxylation of terephthalate to fluorescent 2-HOTP in the presence of hydroxyl radicals. The excitation and emission maxima of 2-HOTP are 315 nm and 421 nm, respectively. *RFU*: relative fluorescence units.

This fluorimetric assay has proven its high sensitivity in the monitoring of TPA resulting from PET hydrolysis using purified TfCut2, a thermophilic polyester hydrolase from *Thermobifida fusca*, in a 96-well microtiter plate format (Wei et al., 2012). The assay is robust at different phosphate buffer concentrations from 100 to 500 mM and is not inhibited by residual purified enzyme in the reaction supernatant, but is not compatible with Tris buffer, which scavenges the hydroxyl radicals. The assay was further validated by comparing the amounts of TPA determined by the fluorimetric assay and RP-HPLC, which showed an excellent agreement between the two quantification methods (Wei et al., 2012).

As high-throughput purification of target proteins (for example, in a 96-well microtiter plate format) is currently very cost-intensive, we adapted the fluorimetric assay for use with unpurified enzymes in crude cell lysates. We expressed recombinant TfCut2 as well as its inactive variant TfCut2_S130A (Vogel et al., 2020) in ZYM-5052 autoinduction medium in 96-deep-well plates (Protocol 5.3) based on a recently published protocol (Tournier et al., 2020). Crude cell lysates obtained by repeated freeze–thaw cycles (Protocol 5.3) were directly used for the hydrolysis of PET NPs at 60°C (the optimal temperature for TfCut2). Furthermore, crude cell lysates without PET NPs were also incubated in the same manner and the resulting soluble supernatants were subjected to fluorimetric detection after adding terephthalate standards (final concentrations of 0–200 μM). As shown in Fig. 4A, linear calibration curves ($R^2 > 0.992$) were obtained independent of the incubation time from 10 to 60 min for the hydroxylation of the terephthalate. Following the enzymatic hydrolysis of PET NPs with both active and inactive enzymes at 60°C for 30–60 min, 150 μL reaction supernatants were removed after centrifugation and added to a 96-well microtiter plate for the fluorimetric estimation of the terephthalate released. By subtracting the background autofluorescence resulting from the inactive enzyme, which has proven to cause no release of TPA from PET NP by RP-HPLC, the relative fluorescence units (RFU) determined with the active TfCut2 revealed a good agreement with the corresponding RP-HPLC analysis (Fig. 4B). This validates the applicability of the complete procedure schematically shown in Fig. 1 which allows sequentially the expression of enzyme libraries, the cell lysis, the hydrolysis reaction using PET NPs, as well as the final estimation of the terephthalate release by fluorimetry to be successfully performed in a 96-well microtiter plate format.

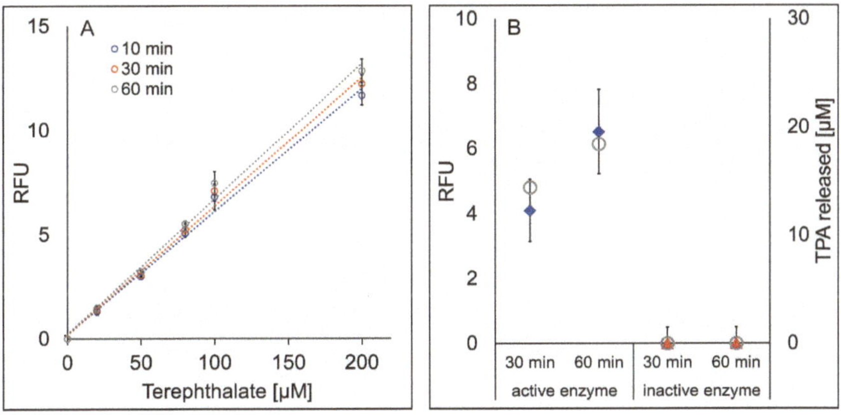

Fig. 4 Fluorimetric estimation of the terephthalate concentrations. (A) Calibration curves determined with the cell lysates following incubation at 60°C for 60 min and the subsequent hydroxylation of various terephthalate standards in the presence of Fe(II)-EDTA (0.625 mM) for 10–60 min at room temperature. Incubation time with Fe(II)-EDTA did not have a significant effect, and 10 min was used for further assay development. (B) Comparison of the relative fluorescence units (RFU) determined in the supernatants of the enzyme-catalyzed hydrolysis of PET NPs (after 30 and 60 min shaking at 60°C, respectively) as an example for identifying hits. With the active TfCut2, significantly higher RFU values (*blue diamonds*) were monitored in a good agreement with the terephthalic acid (TPA) amounts determined by RP-HPLC (*gray circles*, secondary y-axis) than those obtained with the inactive enzyme (*red triangles*). Error bars indicate the standard deviations obtained with at least three independent measurements.

4. Materials, equipment, and reagents

4.1 PET NP production

- Postconsumer PET packaging materials (for example, clamshell boxes used as containers for fresh vegetables and fruits)
- 1,1,1,3,3,3-hexafluoro-2-propanol (HFIP)
- Ultrapure water (for example, obtained with the Milli-Q® Direct Water Purification System)
- Folded filter paper (Rotilabo®- folded filters, type 600P, Carl Roth, Karlsruhe, Germany)
- Thick-walled glass test tube with screw cap
- 1 L (narrow) glass beaker
- Glass burette
- 500 mL round-bottomed glass flask

- Disperser (IKA® ULTRA TURRAX® T-25 basic, Staufen, Germany)
- Rotary evaporator (Laborota 4000 WB, Heidolph, Schwabach, Germany)
- Centrifuge (Micro Star 17, VWR, Darmstadt, Germany)

4.2 PET NP characterization

- Scanning electron microscope (Ultra 55 SEM, Carl Zeiss Ltd., Göttingen, Germany)
- Chromium sputter system (Leybold LH Z400, Leybold, Hanau, Germany)
- Ethanol, technical grade
- Zetasizer Nano ZS with multipurpose titrator MPT-2 (Malvern Instruments, Worcestershire, UK)
- Disposable cuvettes (PMMA, 5 mL)

4.3 Protein production

- Gene of interest (for example, GenBank: FR727681.1 for the wild-type TfCut2) cloned into an appropriate expression vector (for example, pET-26b(+))
- *Escherichia coli* BL21(DE3) (New England Biolabs, Ipswich, MA, US)
- ZYM-5052 medium for autoinduction (Studier, 2005)
- 50 mM sodium phosphate (pH 8.0)
- Lysogeny Broth (LB)-agar plates
- Masterblock®, 96-well, 2 mL (manufactured from polypropylene), V-bottom, sterile, single-packed (Greiner Bio-One, Frickenhausen, Germany)
- Sterile toothpicks
  - Alternatively, an automated colony-picking robot can be employed
- Multichannel pipette, Transferpette®-8, 20–200 μL (VWR, Darmstadt, Germany)
- Incubation shaker with 96-well plate tray (Infors HT Minitron, Bottmingen, Switzerland)
- Centrifuge (Heraeus® Multifuge® 3SR, Kendro Laboratory Products GmbH, Osterode, Germany)
- Low-temperature (−80°C) freezer
- Ice

4.4 Enzymatic degradation of PET NPs

- 500 mM sodium phosphate (pH 8.0)
- Ultrapure water

- PET NP suspension (1 mg/mL)
- 96-well microtiter plate
- Microtiter plate shaker (iEMS® Incubator Shaker HT, ThermoFisher Scientific, Darmstadt, Germany)

4.5 Determination of polyester hydrolysis activity

4.5.1 Fluorimetric HTS assay

- 500 mM sodium phosphate (pH 8.0)
- Terephthalic acid (TPA), reagent grade, 98% (Sigma-Aldrich, Steinheim, Germany)
- 100 mM sodium hydroxide
- 5 mM sodium EDTA solution (pH 8.0)
- 5 mM freshly prepared $Fe(II)SO_4$ solution
- FLUOTRAC™ 96-well microtiter plate (manufactured from polystyrene), F-bottom (chimney well), black, medium binding (Greiner Bio-One, Frickenhausen, Germany)
- Microtiter plate reader/spectrophotometer (for example, Varioskan™ LUX, ThermoFisher Scientific, Darmstadt, Germany)

4.5.2 RP-HPLC

- Acetonitrile, analytical grade
- Ultrapure water (degassed)
- Formic acid (FA), reagent grade, ≥95% (Sigma- Aldrich, Steinheim, Germany)
- Terephthalic acid (TPA), 98%, (Sigma-Aldrich, Steinheim, Germany)
- PET NP degradation reaction solution
- HPLC glass vials
- HPLC glass inlets
- HPLC vial lids
- Hitachi LaChrom Elite HPLC (VWR, Darmstadt, Germany)
  - Column oven: L-2300
  - Autosampler: L-2200
  - UV detector: L-2400
  - Pump: L-2130
  - HPLC Degasser 2005
- Column: Kinetex®, 5 μM EVO C18 100 Å, 150 × 4.6 mm (Phenomenex®, Aschaffenburg, Germany)

5. Protocols

5.1 PET NP production

1. Dissolve 300 mg of postconsumer PET packaging materials in 20 mL HFIP in a closed thick-walled glass test tube with screw cap and incubate at room temperature for approximately 12 h until completely dissolved.
   a. HFIP is considered a harmful solvent, with GHS codes GHS05, GHS07, and GHS08 (GHS hazard pictograms are shown in Fig. 1).
   b. Appropriate personal protective equipment (face shields, gloves, goggles, and a multipurpose combination respirator cartridge) should be used.
   c. The solvent should be carefully handled in a fume hood and the waste obtained by rotary evaporation (halogenated organic waste) should be properly disposed of according to local guidelines.
   d. It is important to use a strong and well-sealable glass vessel for incubation of HFIP.
2. Fill the solution into a burette and drip (drop speed of approximately 0.5 mL/min) into 300 mL ultrapure water under strong stirring at 13,000 rpm using the IKA® ULTRA TURRAX® (T-25 basic).
3. Remove large aggregates of the precipitated polymer by passing the liquid through a Rotilabo®-folded filter (type 600P).
4. Remove HFIP from the NP suspension using a rotary evaporator at 190 mbar and 50°C for approximately 1 h. Continue until the volume is reduced to a third of the initial volume.
5. Determine the PET NP concentration gravimetrically. Centrifuge 6 mL of suspension at room temperature and 17,000 × *g* for >5 min and dry the resulting pellet using a SpeedVac or by incubation at 50°C for 24 h.
6. Determine the mass of the PET nanoparticle pellet and calculate the concentration of the original suspension in mg/mL.

5.2 PET NP characterization

5.2.1 Analysis of particle size and morphology by SEM

1. Prepare PET NP samples for SEM by diluting the PET NP suspension with ethanol (1:1).
2. Spread the suspension on a silicon waver and leave to dry at room temperature for 24 h.
3. Mount the SEM samples on a sample holder using metal clamps.
4. Transfer the sample holder to the sputter system.

5. Apply a sputtered layer of chromium (30 nm) to ensure conductivity of the sample.
6. Transfer the sample holder to the SEM system.
7. SEM is then carried out under vacuum at different magnifications.

5.2.2 Analysis of particle size distribution by DLS

1. Fill a disposable cuvette (5 mL) with 2.5 mL of the PET NP suspension and mount the cuvette in the measurement system.
2. Particle size is determined following the software instructions using standard settings for DLS. Particle size can be measured based on the hydrodynamic radius, which can be calculated from the light scattering of the used laser light.
3. Use the dip cell unit (cuvette cap with two electrodes) on the cuvette in the measurement system, so that zeta potential measurements can be carried out.
4. Zeta potential is determined following the software instructions using standard settings. The zeta potential can be measured using the same solution applying the principle of ELS via the electrophoretic mobility in an applied electrical field, using the Laser-Doppler effect.

5.3 Protein production

The ZYM-5052 autoinduction medium (Studier, 2005) for protein production by autoinduction consists of the following components: 1% tryptone, 0.5% yeast extract, 25 mM Na_2HPO_4, 25 mM KH_2PO_4, 50 mM NH_4Cl, 5 mM Na_2SO_4, 0.5% glycerol, 0.05% glucose, 0.2% α-lactose and 2 mM $MgSO_4$.

The recombinant expression of desired enzymes in *E. coli* BL21(DE3) is performed in ZYM-5052 autoinduction medium at 21°C for 23 h.

1. Transform the gene of interest into competent *E. coli* BL21(DE3) using either chemically competent or electrocompetent cells.
2. Add the appropriate amount of antibiotics to the ZYM-5052 medium and transfer 1 mL of medium to every well of the 96-deep-well plate.
3. Using a toothpick or colony-picking robot, transfer freshly transformed colonies from the LB-agar plate to the medium in the wells.
4. Incubate the 96-deep-well plate in an incubation shaker at 250 rpm and 21°C for 23 h.
5. Harvest *E. coli* cells by centrifugation at 3600 × *g* for 30 min.

The *E. coli* BL21(DE3) cells are disrupted by repeated freeze–thaw cycles (Johnson & Hecht, 1994).

1. Freeze the 96-deep-well plate containing the cell pellets at −80°C for 30 min.
2. Incubate the plate in ice water for 10 min to thaw.
3. Repeat steps 1 and 2 three times.
4. Resuspend lysed cell pellets in 250 μL of 50 mM sodium phosphate (pH 8.0).
5. Incubate cell suspensions at 100 rpm for 30 min at room temperature to allow the release of recombinant enzymes.
6. Centrifuge the 96-deep-well plate at 3600 × *g* and 4°C for 30 min to separate the insoluble debris from the soluble fraction containing recombinant enzymes.

5.4 Enzymatic degradation of PET NP

The enzymatic degradation of PET NP is performed in 96-well microtiter plates shaken at 900 rpm and 60°C. Recombinant TfCut2 and an inactive S130A variant are used as positive and negative controls, respectively. The reaction temperature and the duration of incubation should be adapted with respect to the optimal temperature of the applied enzyme (in the case of metagenome screening, use a temperature matching the desired properties of the polyester hydrolase to be found).

1. Mix 10 μL of a 1 mg/mL PET NP solution (10 μg) with 40 μL of 500 mM sodium phosphate (final concentration 100 mM, pH 8.0) and 50 μL ultrapure water.
2. Incubate the assay mixture with 100 μL crude cell lysate (final volume 200 μL) at 60°C for 30 and 60 min in a microtiter plate shaker (900 rpm).
3. Stop the degradation reaction by centrifugation at 2844 × *g* and 4°C for 15 min to remove the (precipitated) protein and unhydrolyzed NP.
4. Remove the soluble supernatant to determine the amount of degradation products using the fluorimetric HTS assay and RP-HPLC.

5.5 Quantification of the PET degradation products

5.5.1 Fluorimetric HTS assay

The emission and excitation maxima of the fluorescent product 2-HOTP are determined using a commercial standard. All measurements should be taken at least in triplicate.

1. Dissolve 1.82 mg 2-HOTP in 1 mL 100 mM sodium hydroxide to produce a 10 mM stock solution.

2. Prepare a 2-HOTP standard by diluting the 10 mM stock solution in ultrapure water to a final concentration of 100 μM.
3. Mix 10 μL of the 2-HOTP (100 μM) with 40 μL of 500 mM sodium phosphate (pH 8.0, final buffer concentration of 100 mM) and 150 μL ultrapure water in a black FLUOTRAC™ 96-well microtiter plate (final volume of 200 μL in each well).
4. Determine the excitation maximum by recording fluorescence emission at 421 nm while scanning excitation wavelengths between 200 and 403 nm using a Varioskan LUX microtiter plate reader.
5. Determine the emission maximum by recording fluorescence emission between 346 and 600 nm while exciting the sample at 315 nm using a Varioskan LUX microtiter plate reader.

The fluorimetric HTS assay is performed in black FLUOTRAC™ 96-well microtiter plates at 25 °C in a Varioskan LUX microtiter plate spectrophotometer.

6. Dissolve 16.6 mg TPA in 10 mL 100 mM sodium hydroxide to produce a 10 mM terephthalate stock solution
7. Prepare a terephthalate standard series by diluting the stock solution in ultrapure water to a final concentration in the range of 0–200 μM
8. Incubate 120 μL crude cell lysates at the desired reaction temperature for the same duration to resemble the degradation reaction without the addition of PET NPs in a microtiter plate shaker
9. Mix 10 μL of the terephthalate standard solutions with 40 μL of 500 mM sodium phosphate (pH 8.0) and 100 μL soluble supernatant (step 8) in a black FLUOTRAC™ 96-well microtiter plate (final volume of 150 μL in each well)
10. Initiate hydroxylation of terephthalate by adding 25 μL of 5 mM EDTA and 25 μL of 5 mM Fe(II)SO_4 (final concentration of 625 μM Fe(II)-EDTA) and incubating for at least 10 min at room temperature
11. Measure the fluorescence emission at 421 nm upon excitation at 315 nm using the microtiter plate reader
12. Plot the fluorescence obtained in step 11 against the terephthalate concentrations applied to obtain a calibration curve
13. Similarly as described above in step 9, mix 150 μL of the reaction supernatant following the enzymatic hydrolysis of PET NP instead of the terephthalate standard solution to perform the fluorimetric determination from steps 9 to 11 and estimate the amount of terephthalate based on the calibration curves obtained in step 12.

5.5.2 RP-HPLC

1. Separate the degradation products on a Kinetex® (5 μM EVO C18 100 Å, 150 × 4.6 mm) column by injecting 10 μL sample.
2. Use 0.1% (v/v) formic acid in water (A) and acetonitrile (B) as mobile phases. Run a gradient from 5% to 44% (B) over 12 min, then increase to 70% (B) over 3 min. Use a flow rate of 0.8 mL/min.
3. Detect the amounts of UV-absorbing compounds by measuring absorbance at 240 nm.

Acknowledgments

The authors acknowledge the financial support received from the European Union's Horizon 2020 research and innovation program under the grant agreement 870294 (MIX-UP project). The authors would like to thank the analytical team of the IOM (Andrea Prager, Nadja Schönherr, and Ingrid Reinhard) for their contributions to the analysis of PET NPs.

References

Barth, M., Oeser, T., Wei, R., Then, J., Schmidt, J., & Zimmermann, W. (2015). Effect of hydrolysis products on the enzymatic degradation of polyethylene terephthalate nanoparticles by a polyester hydrolase from *Thermobifida fusca*. *Biochemical Engineering Journal*, *93*, 222–228.

Berkowitz, S. (1984). Viscosity–molecular weight relationships for poly(ethylene terephthalate) in hexafluoroisopropanol–pentafluorophenol using SEC–LALLS. *Journal of Applied Polymer Science*, *29*, 4353–4361.

Bornscheuer, U. T. (2016). Feeding on plastic. *Science*, *351*, 1154–1155.

Cho, E., Choi, J., Jo, S., Park, D.-H., Hong, Y. K., Kim, D., et al. (2019). A single-benzene-based fluorophore: Optical waveguiding in the crystal form. *ChemPlusChem*, *84*, 1130.

Danso, D., Schmeisser, C., Chow, J., Zimmermann, W., Wei, R., Leggewie, C., et al. (2018). New insights into the function and global distribution of polyethylene terephthalate (PET)-degrading bacteria and enzymes in marine and terrestrial metagenomes. *Applied and Environmental Microbiology*, *84*, e02773-17.

Eberl, A., Heumann, S., Brückner, T., Araujo, R., Cavaco-Paulo, A., Kaufmann, F., et al. (2009). Enzymatic surface hydrolysis of poly(ethylene terephthalate) and bis (benzoyloxyethyl) terephthalate by lipase and cutinase in the presence of surface active molecules. *Journal of Biotechnology*, *143*, 207–212.

Eriksen, M., Lebreton, L. C. M., Carson, H. S., Thiel, M., Moore, C. J., Borerro, J. C., et al. (2014). Plastic pollution in the world's oceans: More than 5 trillion plastic pieces weighing over 250,000 tons afloat at sea. *PLoS One*, *9*, e111913.

Geyer, R., Jambeck, J. R., & Law, K. L. (2017). Production, use, and fate of all plastics ever made. *Science Advances*, *3*, e1700782.

Herrero Acero, E., Ribitsch, D., Steinkellner, G., Gruber, K., Greimel, K., Eiteljoerg, I., et al. (2011). Enzymatic surface hydrolysis of PET: Effect of structural diversity on kinetic properties of cutinases from *Thermobifida*. *Macromolecules*, *44*, 4632–4640.

Johnson, B. H., & Hecht, M. H. (1994). Recombinant proteins can be isolated from *E. coli* cells by repeated cycles of freezing and thawing. *Nature Biotechnology*, *12*, 1357–1360.

Kawai, F., Kawabata, T., & Oda, M. (2019). Current knowledge on enzymatic PET degradation and its possible application to waste stream management and other fields. *Applied Microbiology and Biotechnology*, *103*, 4253–4268.

Kawai, F., Kawabata, T., & Oda, M. (2020). Current state and perspectives related to the polyethylene terephthalate hydrolases available for biorecycling. *ACS Sustainable Chemistry & Engineering*, *8*, 8894–8908.

Lehner, R., Weder, C., Petri-Fink, A., & Rothen-Rutishauser, B. (2019). Emergence of nanoplastic in the environment and possible impact on human health. *Environmental Science & Technology*, *53*, 1748–1765.

Liu, J., Ma, Y., Zhu, D., Xia, T., Qi, Y., Yao, Y., et al. (2018). Polystyrene nanoplastics-enhanced contaminant transport: Role of irreversible adsorption in glassy polymeric domain. *Environmental Science & Technology*, *52*, 2677–2685.

Müller, R.-J., Schrader, H., Profe, J., Dresler, K., & Deckwer, W.-D. (2005). Enzymatic degradation of poly(ethylene terephthalate): Rapid hydrolyse using a hydrolase from *T. fusca*. *Macromolecular Rapid Communications*, *26*, 1400–1405.

O'Neill, A., & Cavaco-Paulo, A. (2004). Monitoring biotransformations in polyesters. *Biocatalysis and Biotransformation*, *22*, 353–356.

Palm, G. J., Reisky, L., Böttcher, D., Müller, H., Michels, E. A. P., Walczak, M. C., et al. (2019). Structure of the plastic-degrading *Ideonella sakaiensis* MHETase bound to a substrate. *Nature Communications*, *10*, 1717.

PlasticsEurope. Plastics—the Facts. (2019). An analysis of European plastics production. In *Demand and waste data*. https://www.plasticseurope.org/en/resources/publications/1804-plastics-facts-2019.

Rahimi, A., & García, J. M. (2017). Chemical recycling of waste plastics for new materials production. *Nature Reviews Chemistry*, *1*, 0046.

Rochman, C. M., Hoh, E., Kurobe, T., & Teh, S. J. (2013). Ingested plastic transfers hazardous chemicals to fish and induces hepatic stress. *Scientific Reports*, *3*, 3263.

Rodríguez-Hernández, A. G., Muñoz-Tabares, J. A., Aguilar-Guzmán, J. C., & Vazquez-Duhalt, R. (2019). A novel and simple method for polyethylene terephthalate (PET) nanoparticle production. *Environmental Science: Nano*, *6*, 2031–2036.

Ronkvist, Å. M., Xie, W., Lu, W., & Gross, R. A. (2009). Cutinase-catalyzed hydrolysis of poly(ethylene terephthalate). *Macromolecules*, *42*, 5128–5138.

Saran, M., & Summer, K. H. (1999). Assaying for hydroxyl radicals: Hydroxylated terephthalate is a superior fluorescence marker than hydroxylated benzoate. *Free Radical Research*, *31*, 429–436.

Silva, C., Da, S., Silva, N., Matamá, T., Araújo, R., Martins, M., et al. (2011). Engineered *Thermobifida fusca* cutinase with increased activity on polyester substrates. *Biotechnology Journal*, *6*, 1230–1239.

Studier, F. W. (2005). Protein production by auto-induction in high density shaking cultures. *Protein Expression and Purification*, *41*, 207–234.

Sulaiman, S., Yamato, S., Kanaya, E., Kim, J.-J., Koga, Y., Takano, K., et al. (2012). Isolation of a novel cutinase homolog with polyethylene terephthalate-degrading activity from leaf-branch compost by using a metagenomic approach. *Applied and Environmental Microbiology*, *78*, 1556–1562.

Sulaiman, S., You, D.-J., Kanaya, E., Koga, Y., & Kanaya, S. (2014). Crystal structure and thermodynamic and kinetic stability of metagenome-derived LC-cutinase. *Biochemistry*, *53*, 1858–1869.

Tournier, V., Topham, C. M., Gilles, A., David, B., Folgoas, C., Moya-Leclair, E., et al. (2020). An engineered PET depolymerase to break down and recycle plastic bottles. *Nature*, *580*, 216–219.

Vogel, K., Wei, R., Al-Fatih, H., Pfaff, L., Ortmann, C., Estrela-Lopis, I., et al. (2020). Enzymatic degradation of polyethylene terephthalate nanoplastics analyzed in real time by isothermal titration calorimetry. submitted.

Vollmer, I., Jenks, M. J. F., Roelands, M. C. P., White, R. J., van Harmelen, T., de Wild, P., et al. (2020). Beyond mechanical recycling: Giving new life to plastic waste. *Angewandte Chemie, International Edition, 59*, 15402–15423.

Wagner, S., & Reemtsma, T. (2019). Things we know and don't know about nanoplastic in the environment. *Nature Nanotechnology, 14*, 300–301.

Weber, J., Petrović, D., Strodel, B., Smits, S. H. J., Kolkenbrock, S., Leggewie, C., et al. (2019). Interaction of carbohydrate-binding modules with poly(ethylene terephthalate). *Applied Microbiology and Biotechnology, 103*, 4801–4812.

Wei, R., Breite, D., Song, C., Gräsing, D., Ploss, T., Hille, P., et al. (2019). Biocatalytic degradation efficiency of postconsumer polyethylene terephthalate packaging determined by their polymer microstructures. *Advanced Science, 6*, 1900491.

Wei, R., Oeser, T., Barth, M., Weigl, N., Lübs, A., Schulz-Siegmund, M., et al. (2014). Turbidimetric analysis of the enzymatic hydrolysis of polyethylene terephthalate nanoparticles. *Journal of Molecular Catalysis B: Enzymatic, 103*, 72–78.

Wei, R., Oeser, T., Billig, S., & Zimmermann, W. (2012). A high-throughput assay for enzymatic polyester hydrolysis activity by fluorimetric detection. *Biotechnology Journal, 7*, 1517–1521.

Wei, R., Oeser, T., Then, J., Kühn, N., Barth, M., Schmidt, J., et al. (2014). Functional characterization and structural modeling of synthetic polyester-degrading hydrolases from *Thermomonospora curvata*. *AMB Express, 4*, 44.

Wei, R., Oeser, T., & Zimmermann, W. (2014). Synthetic polyester-hydrolyzing enzymes from thermophilic actinomycetes. *Advances in Applied Microbiology, 89*, 267–305.

Wei, R., Song, C., Gräsing, D., Schneider, T., Bielytskyi, P., Böttcher, D., et al. (2019). Conformational fitting of a flexible oligomeric substrate does not explain the enzymatic PET degradation. *Nature Communications, 10*, 5581.

Wei, R., Tiso, T., Bertling, J., O'Connor, K., Blank, L. M., & Bornscheuer, U. T. (2020). Possibilities and limitations of biotechnological plastic degradation and recycling. *Nature Catalysis, 3*, 867–871.

Wei, R., & Zimmermann, W. (2017). Biocatalysis as a green route for recycling the recalcitrant plastic polyethylene terephthalate. *Microbial Biotechnology, 10*, 1302–1307.

Welch, K. D., Davis, T. Z., & Aust, S. D. (2002). Iron autoxidation and free radical generation: Effects of buffers, ligands, and chelators. *Archives of Biochemistry and Biophysics, 397*, 360–369.

Welzel, K., Müller, R.-J., & Deckwer, W.-D. (2002). Enzymatischer Abbau von Polyester-Nanopartikeln. *Chemie Ingenieur Technik, 74*, 1496–1500.

Wright, S. L., & Kelly, F. J. (2017). Plastic and human health: A micro issue? *Environmental Science & Technology, 51*, 6634–6647.

Yang, X. F., & Guo, X. Q. (2001). Fe(II)-EDTA chelate-induced aromatic hydroxylation of terephthalate as a new method for the evaluation of hydroxyl radical-scavenging ability. *The Analyst, 126*, 928–932.

Yoshida, S., Hiraga, K., Takehana, T., Taniguchi, I., Yamaji, H., Maeda, Y., et al. (2016). A bacterium that degrades and assimilates poly(ethylene terephthalate). *Science, 351*, 1196–1199.

CHAPTER THIRTEEN

Anchor peptides promote degradation of mixed plastics for recycling

Yu Ji[a], Yi Lu[a], Hendrik Puetz[a], and Ulrich Schwaneberg[a,b,*]
[a]Institute of Biotechnology, RWTH Aachen University, Aachen, Germany
[b]DWI-Leibniz Institute for Interactive Materials, Aachen, Germany
*Corresponding author: e-mail address: u.schwaneberg@biotec.rwth-aachen.de

Contents

Abstract

Resource stewardship and sustainable use of natural resources is mandatory for a circular plastic economy. The discovery of microbes and enzymes that can selectively degrade mixed-plastic waste enables to recycle plastics. Knowledge on how to achieve efficient and selective enzymatic plastic degradation is a key prerequisite for biocatalytic recycling of plastics. Wild-type natural polymer degrading enzymes such as cellulases pose often selective non-catalytic binding domains that facilitate a targeting and efficient degradation of polymeric substrates. Recently identified polyester hydrolases with synthetic polymer degrading activities, however, lack in general such selective domains. Inspired by nature, we herein report a protocol for the identification and engineering of anchor peptides which serve as non-catalytic binding domains specifically toward

Methods in Enzymology, Volume 648
ISSN 0076-6879
https://doi.org/10.1016/bs.mie.2020.12.027

synthetic plastics. The identified anchor peptides hold the promise to be fused to known plastic degrading enzymes and thereby enhance the efficiency of biocatalytic plastic recycling processes.

1. Introduction

Polymer manufacturing has grown into a thriving industry primed by the synthesis of the first fully synthetic plastic in the early 20th century (Baekeland, 1909). Polymers with the highest economical values comprise polyethylene (PE), polypropylene (PP), polyvinyl chloride (PVC), polystyrene (PS), polyethylene terephthalate (PET), polyurethane (PU), and polyamides (PA). Today, at least 368 million (Plastics - the Facts 2020, 2020) metric tons of plastics are produced annually (Danso, Chow, & Streit, 2019). In fact, plastic usage and disposal increased with a constant growth rate of 3%–4%/year, well in accordance with the growing world population and economic expansion in developing countries (Kawai, Kawabata, & Oda, 2020). With a rising awareness of the ecological impact of human actions, concerns toward detrimental effects of plastic accumulation especially in the marine environment become apparent (Anjana, Hinduja, Sujitha, & Dharani, 2020). Particularly micro- (MP) and nanoplastics (NP) harbor risks due to their high surface/mass ratio promoting the binding of hydrophobic pollutants such as polychlorinated biphenyls (PCBs) and polycyclic aromatic hydrocarbons (PHAs) (Liu, Fokkink, & Koelmans, 2016). Second, the small size enables the penetration of tissue and transport through biological membranes such as the blood–brain barrier (Kashiwada, 2006; Von Moos, Burkhardt-Holm, & Köhler, A., 2012). As a result, synthetic polymers accumulate through the food chain and can eventually be taken up by humans.

Yet, efficient collection and recycling solutions are scarce. Globally, 72% of plastic packaging materials are not recovered as 40% end up in landfills and another 32% leak into the environment and especially the oceans (MacArthur, Waughray, & Stuchtey, 2016). Another 14% are sent toward incineration and energy recovery, while the residual 14% of packaging waste deducting 4% of process loss are recycled. Finally, 8% of recovered plastics go downwards the cascading cycle and are repurposed for lower value applications with only 2% being reused for similar-quality applications. Hence, innovative technologies in the plastic recycling sector are in demand

and besides pyrolysis, solvolysis, dissolution/precipitation and mechanical degradation, biocatalytic approaches are gaining attention (Deguchi, Kitaoka, Kakezawa, & Nishida, 1998; Kakudo, Negoro, Urabe, & Okada, 1993; Negoro et al., 2007; Takehara et al., 2018; Wei et al., 2020).

From a chemical perspective, plastics may be divided into two groups. The first category's (comparably hydrophilic) backbone consists of hydrolysable monomers linked by ester, carbamate or amide bonds (PET, PU, PA). In contrast, the second group (hydrophobic) represents non-hydrolysable polymers harboring a rigid carbon-carbon backbone (PE, PP, PVC, PS). To this point, a considerable amount of articles covering the enzymatic degradation of hydrolysable polymers has been published, whereas for non-hydrolysable polymers solely microbial degradation by certain bacterial strains, bacterial consortia and fungi was reported (Inderthal, Tai, & Harrison, 2020; Krueger, Harms, & Schlosser, 2015; Restrepo-Flórez, Bassi, & Thompson, 2014). For hydrolysable polymers, the enzymatic modes of action are better understood. As an example, the degradation of PET is mainly assigned to α/β-fold serine hydrolases (Ollis et al., 1992; Wei et al., 2014). Especially, most enzymes characterized for PET degradation were derived from gram-positive *Actinobacteria* with the most promising examples originating from the thermophile genera *Thermobifida* and *Thermomonospora* (Hu, Thumarat, Zhang, Tang, & Kawai, 2010; Kawai et al., 2020; Wei, Oeser, & Zimmermann, 2014; Wei & Zimmermann, 2017). Recently, an promising PET hydrolase from *Ideonella sakaiensis*was (*Is*PETase) was reported to efficiently degrade amorphous PET at ambient temperature (Yoshida et al., 2016). Moreover, *Is*PETase showed activity toward other polymers such as polyethylene-2,5-furandicarboxylate (PEF) (Austin et al., 2018). Thus, besides the generally promiscuous *Is*PETase also other cutinases were reported to degrade PEF (Pellis et al., 2016; Weinberger et al., 2017). Interestingly, some thermophile cutinases such as TfCut2, Tcur0390 and Tcur1278 show comparably high promiscuity, accepting not only PET but also other polyesters like ester-based PU (Schmidt, Wei, Oeser, et al., 2017; Wei, Oeser, Then, et al., 2014). On a molecular level, enzymatic degradation of ester bonds in ester-based PU was assigned to lipases, proteases and metallo-hydrolases (Howard & Blake, 1998; Russell et al., 2011; Stern & Howard, 2000).

Generally, the elucidation of enzyme sequences and structures opens doors for protein engineers to modify enzyme properties matching process needs. As a result, general parameters such as pH, temperature and organic solvent tolerance but also more specific properties such as substrate specificity may be adjusted by either rational design or random mutagenesis approaches (Knott et al., 2020). However, enzyme engineering is not limited to sole alteration of the primary sequence but offers the possibility to design complex proteins by fusion of, e.g., fluorescent or substrate binding promoting domains. Especially for complex substrates such as polymers, substrate accessibility is essential. Facing this issue, nature developed binding-promoting proteins for naturally occurring polymers such as cellulose, chitin and polyhydroxybutyrate (Beintema, 1994; Bolam et al., 1998; Kasuya, Ohura, Masuda, & Doi, 1999; Linder & Teeri, 1997). Those binding domains are genetically fused to catalytic domains, ensuring that the depolymerase is more efficiently bound to the substrate surface and thereby enabling an efficient degradation. Consequently, first examples of fusion enzymes with non-natural binding domains were designed to enhance enzyme activity on complex substrates. More precisely, cutinases fused to carbohydrate-binding modules (originated from certain cellulases) or polymer-binding modules (originated from certain polyhydroxyalkanoate depolymerase) showed enhanced enzymatic activity on cotton fiber as well as PET (Ribitsch et al., 2013; Zhang et al., 2010; Zhang, Wang, Chen, & Wu, 2013). Accordingly, an emerging trend in protein engineering comprises the fusion of adhesion-promoting anchor peptides to enhance enzyme efficiency or specifically binding enzymes to polymer surfaces (Care, Bergquist, & Sunna, 2015; Zernia et al., 2016). This method is eminently important for evolutionary young enzyme classes acting on manmade substances, which did not yet optimize throughout millennia. For instance, cutinase-anchor peptide fusion enzymes acting on polyester-PU proved to increase the degradation rate of PU nanoparticles by 6.6 times compared to the anchor-free wild type (Shohana, Apitius, Jakob, & Schwaneberg, 2019). Notably, applied concentrations were as low as 0.04% *w/v*. Considering that only low micro- and nanoplastics concentrations can be found apart laboratory conditions, e.g., in wastewater treatment plants, anchor peptides may help to compensate for low substrate concentrations and insufficient enzyme affinity.

2. Engineering the adhesion peptide binding modules for enhanced polymer absorption and enzymatic degradation

Anchor peptides (APs) bind to various synthetic polymers including PS, PP, PVC, PET, PU, Nylon, poly(methyl methacrylate) (PMMA), chlorine-doped polypyrrole (PPyCl), and poly(L-lactide) (PLLA)(Table 1)

Table 1 Anchor peptides for selective polymer binding.

Polymer	Anchor peptide	Protein sequence
PP	LCI	*AIKLVQSPNGNFAASFVLDGTKWIFKSKYYDSSKGYWVGIYEVWDRK* (Rübsam, Stomps, Böker, Jakob, & Schwaneberg, 2017)
	LCI-M1-PP(I24T, Y29H, E42K)	*AIKLVQSPNGNFAASFVLDGTKWTFKSKHYDSSKGYWVGIYKVWDRK* (Rübsam, Weber, Jakob, & Schwaneberg, 2018)
	LCI-M2-PP(D31V, E42G)	*AIKLVQSPNGNFAASFVLDGTKWIFKSKYYVSSKGYWVGIYGVWDRK* (Rübsam, Weber, et al., 2018)
	LCI-KR-2(Y29R and G35R)	*AIKLVQSPNGNFAASFVLDGTKWIFKSKRYDSSKRYWVGIYEVWDRK* (Rübsam, Davari, Jakob, & Schwaneberg, 2018)
	LCI-CSD-3(E42V/D45H)	*AIKLVQSPNGNFAASFVLDGTKWIFKSKYYDSSKGYWVGIYVVWHRK* (Apitius, Rübsam, Jakesch, Jakob, & Schwaneberg, 2019)
PU (Impranil® DLN-SD)	TA2	*YSRCQLQGFNCVVRSYGLPTIPCCRGLTCRSYFPGSTYGRCQRY* (Shohana et al., 2019)
PS	TA2-M1-PS(R3S, L6P, V12 K, S15P, C29R, R30L, F33S, Y44H)	*YSSCQPQGFNCKVRPYGLPTIPCCRGLTRLSYSPGSTYGRCQRH* (Rübsam, Weber, et al., 2018)
	TA2-M2-PS (F9C, C24S, G26D, S31G, C41S, Y44Q)	*YSRCQLQGCNCVVRSYGLPTIPCSRDLTCRGYFPGSTYGRSQRQ* (Rübsam, Weber, et al., 2018)
PCL	LCI	*AIKLVQSPNGNFAASFVLDGTKWIFKSKYYDSSKGYWVGIYEVWDRK* (Apitius et al., 2019)
PC	PC-OMP6	*NSNFFGLVDGLNFAVQYLGK* (Kumada, Murata, Ishikawa, Nakatsuka, & Kishimoto, 2012)
PMMA	PM-OMP25	*DVEGIGDVDLVNYFEVGATYTFNK* (Kumada et al., 2012)
PET	TA2	*YSRCQLQGFNCVVRSYGLPTIPCCRGLTCRSYFPGSTYGRCQRY* (Dedisch et al., 2020)
PLLA	c22 peptide	*QLMHDYR* (Matsuno, Sekine, Yajima, & Serizawa, 2008)
PMMA	c02 peptide	*ELWRPTR* (Serizawa, Sawada, Matsuno, Matsubara, & Sato, 2005)

(Adey, Mataragnon, Rider, Carter, & Kay, 1995; Büscher et al., 2019; Cunningham, Lowe, O'brien, Wang, & Wilkins, 2010; Matsuno et al., 2008; Rübsam et al., 2017; Sanghvi, Miller, Belcher, & Schmidt, 2005; Serizawa, Sawada, & Kitayama, 2007; Serizawa, Techawanitchai, & Matsuno, 2007; Shohana et al., 2019). Mechanisms how peptides recognize and bind to polymers surfaces rely on interaction of functional groups which are diverse in nature and are often not understood on a molecular level. Hydrogen bonding, hydrophobic and electrostatic interactions as well as π-π stacking have been identified as dominating interactions in peptide-polymer binding (Ejima, Matsuno, & Serizawa, 2010; Serizawa et al., 2005; Serizawa, Sawada, & Kitayama, 2007; Swaminathan & Cui, 2012). For example, peptides harboring aromatic amino acids (tryptophan, tyrosine, phenylalanine) especially exposing WXXW-, FHXXW- and WXXWXXXW-motifs, bind to aromatic polymers due to aromatic ring interaction (Anni, Nikolaeva, & Israel, 2001; Qiang et al., 2017; Serizawa, Matsuno, & Sawada, 2011). Main contributors in interactions of peptides binding to PMMA are hydrogen bonds with PMMA ester groups (Serizawa et al., 2011). In addition to the interaction between peptide and polymer, it seems that solvent-peptide interactions, temperature and pH modulate the binding strength between anchor peptide and especially hydrophobic polymers (Tsai, Ruaan, & Liu, 2012).

Phage display libraries gave rise to adhesion peptides for many of the above-mentioned materials, which however, often consist of only a few amino acids limiting the peptide sequence (Serizawa et al., 2011; Serizawa, Sawada, & Kitayama, 2007; Vodnik, Zager, Strukelj, & Lunder, 2011). Naturally occurring anchor peptides (e.g., Liquid chromatography peak I (LCI, 47 aa), Tachystatin A2 (TA2, 44 aa)) are "longer" than peptides identified from phage display screening experiments. They have often a defined 3D-structure and are therefore excellent starting molecules to tune their binding strength/preference for application demands by protein engineering (Gong, Wang, Chen, Xia, & Lu, 2011; Osaki et al., 1999). Accordingly, we developed a peptide-polymer evolution protocol (PeP*evo* protocol) enabling a quick and reliable way to improve AP binding to polymer surfaces (Rübsam, Weber, et al., 2018). Directed evolution approaches generally comprise three steps (Güven, Prodanovic, & Schwaneberg, 2010). First, a set of genetically diverse variants of the target gene is produced via error-prone PCR (epPCR). In the second step, variants are screened for the desired property under application of a selection pressure. Finally, promising variants harboring beneficial mutations are isolated. Generally, epPCRs generate only little variance with mutation rates of 1–3 mutations/kb (Kang Lan & Ulrich, 2007). Considering that the average peptide gene length is around a hundred base pairs, most "variants" will carry the wild type. Thus, a modified epPCR protocol with a very high mutation rate of up to 60 mutations/kb was necessary to efficiently evolve adhesion promoting peptides (Rübsam, Weber, et al., 2018). In order to minimize experimental efforts and maximize property improvements with a molecular understanding of each amino acid substitution, the protein engineering strategy KnowVolution was developed several years ago and employed in >15 successful directed evolution campaigns (Contreras et al., 2020; Islam et al., 2018; Mandawe et al., 2018; Novoa et al., 2019; Rübsam, Davari, Jakob, & Schwaneberg, 2018; Wallraf et al., 2018). In brief, key positions for property improvement (e.g., peptide-polymer binding strength) are identified, beneficial substitutions evaluated via computational analysis and finally synergistically combined.

In this chapter, we report the protocol used to identify anchor peptides with enhanced binding strength to target polymers and therefore accelerate the degradation of polymers. The selection of anchor peptides as well as how to engineer the anchor peptide with strong binding to a target polymer will help scientist to obtain a selective binder. Moreover, generation of the identified anchor-respective enzyme construct and how to monitor the degradation are also described in this protocol.

3. Equipment

1. PCR thermocycler (Eppendorf Mastercycler proS, Germany) for DNA polymerization
2. Centrifuge (Centrifuge 5810R, Eppendorf, Germany)
3. Agarose gel electrophoresis system (Bio-Rad, Germany)
4. Shaker (Microtron/Minitron, Infors HT, Switzerland)
5. Ultrasonic processor for interrupt cells (VCX 130, Vibra-Cell, USA)
6. Ultrasound cleaner for degas of protein purification reagents (Bandelin Sonorex, Germany)
7. ÄKTA prime plus for purifying protein (GE healthcare, USA)
8. Scanning system for DNA/protein visualization (UGenius, Syngene, UK)

9. Fluorescence microscope (EX51, Olympus, Japan)
10. Multi-mode Microplate Reader (FLUOstar Omega, BMG Labtech, Germany)
11. Eporator (Eporator®, Eppendorf, Germany)

4. Materials

1. Program for visualization of DNA sequences, primer design, and evaluation of sequencing data (e.g., Clone Manager 9 Professional Edition, Scientific & Educational Software, USA)
2. DNA polymerases (New England Biolabs, USA)
3. Oligonucleotides (Eurofins MWG Operon, Germany)
4. *E. coli* strains DH5α and BL21 Gold (DE3) (Agilent Technologies, USA)
5. Plasmid extraction kit (Macherey-Nagel GmbH & Co.KG, Germany)
6. PCR purification kit (Qiagen GmbH, Germany)
7. BCA Protein Assay kit (Thermo Fisher Scientific, USA)
8. Protino Ni-IDA 2000 kit for His-tag protein purification (Macherey-Nagel GmbH & Co.KG, Germany)
9. Lysogeny broth medium (LB) for the expression of EGFP-anchor peptide fusion protein (Table 2)
10. Terrific broth medium (TB) for the expression of EGFP-anchor peptide fusion protein (Table 3)
11. Buffers for EGFP-anchor peptide fusion protein purification by His-tag affinity chromatography (Table 4)
12. Yeast extract–Peptone–Dextrose (YPD) medium for the expression of cutinase Tcur1278 (Table 5)
13. Super Optimal broth (SOC) medium used for recovery step of *E. coli* competent cell transformations (Table 6)

Table 2 Preparation of LB medium.

Composition	Amount (g)
Tryptone	10
Yeast extract	5
NaCl	10

Dissolve the components in 1000 mL double-distilled water (ddH_2O) and autoclave at 121 °C for 21 min.

Table 3 Preparation of TB medium.

Medium	Composition	Amount
TB-Solution A	Peptone	12 g
	Yeast extract	24 g
	Glycerol	4 g
	ddH_2O	800 mL
TB-Solution B	K_2HPO_4	12.5 g
	KH_2PO_4	2.3 g
	ddH_2O	200 mL

Autoclave TB-solutions A and B separately at 121 °C for 21 min and mix them together under sterile conditions.

Table 4 Preparation of buffers for protein purification.

Buffer	Composition	Concentration (g/L)
Buffer A (50 mM Tris buffer, 5 mM imidazole, pH 8.0)	Tris-HCl	4.44
	Tris base	2.65
	Imidazole	0.34
Buffer B (50 mM Tris buffer, 500 mM imidazole, pH 8.0)	Tris-HCl	4.44
	Tris base	2.65
	Imidazole	34.04
Buffer C (50 mM Tris buffer, pH 8.0)	Tris-HCl	4.44
	Tris base	2.65

Table 5 Preparation of YPD medium.

Medium	Composition	Amount
YPD-Solution A	Yeast extract	10 g
	Peptone	20 g
	(Agar)	20 g
	ddH_2O	900 mL
YPD-Solution B	D-glucose	20 g
	ddH_2O	100 mL

Autoclave solution A and B separately at 121 °C for 21 min and mix them together afterwards under sterile conditions.

Table 6 Preparation of SOC medium.

Composition	Concentration
Tryptone	20 g/L
NaCl	2 g/L
Yeast extract	5 g/L
KCl	2.5 g/L
$MgCl_2$	10 mM
$MgSO_4$	10 mM
glucose	20 mM

Autoclave SOC medium at 121 °C for 21 min.

5. Protocol

5.1 Generation of EGFP-anchor peptide fusion constructs

EGFP (Enhanced green fluorescent protein) is used as a reporter protein to monitor the anchor peptide binding on polymer surfaces.

1. The construction of an EGFP-anchor peptide gene is designed as shown in Fig. 1. His-EGFP is fused to the N-terminus of the anchor peptide using a stiff helical spacer (17 amino acids: AEAAAKEAAAKEAAAKA) (Arai, Ueda, Kitayama, Kamiya, & Nagamune, 2001) and a tobacco etch virus (TEV) protease cleavage site (7 amino acids: ENLYFQG) (Kapust et al., 2001) as functional separators of both units and then cloned into the pET28a plasmid (Novagen, La, USA)
2. Generate the constructions by phosphorothioate-based ligase-independent gene cloning (PLICing) method (Blanusa, Schenk, Sadeghi, Marienhagen, & Schwaneberg, 2010) and transform the PLICing products into electro-competent *E. coli* DH5α (Rübsam et al., 2017)
3. Incubate the transformed cells on fresh prepared LB Kanamycin agar plates overnight

5.2 Expression of EGFP-anchor peptide fusion proteins

The generated fusion proteins (EGFP-APs and the negative control EGFP) can be expressed in *E. coli* BL21 (DE3) golden cells.

1. Pick single colonies from a fresh LB Kanamycin agar plate from 5.1
2. Grow the cells overnight (16 h) by incubation at 37 °C, 250 rpm

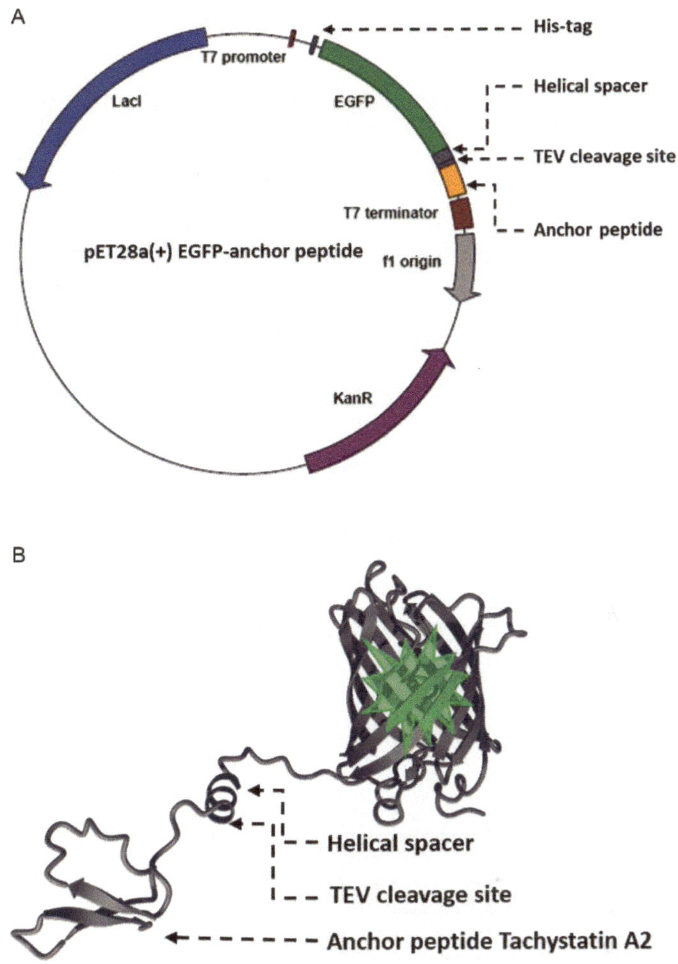

Fig. 1 (A) Vector map of pET28a(+)-EGFP-anchor peptide. (B) Schematic representation of the fusion protein EGFP-Tachystatin A2 (TA2) consisting of the fluorescent reporter protein EGFP, a spacer sequence including 17 × amino acids and a TEV-protease cleavage site, which separates EGFP from the anchor peptide TA2 (determined by YASARA (Yet Another Scientific Artificial Reality Application; v19.12.14)).

3. Inoculate 100 mL TB medium, add 100 μL of 100 mg/mL kanamycin in a 500 mL Erlenmeyer flask with the overnight preculture to an initial OD_{600} of 0.025
4. Incubate main culture at 37 °C, 250 rpm until OD_{600} reaches 0.8 (around 4 h)

5. Induce expression by adding 100 µL 0.1 M IPTG (final concentration: 0.1 mM)
6. Incubate main culture at 16 °C, 250 rpm for 18 h
7. Harvest cells in 50 mL tubes by centrifugation at 4 °C, 4000 rpm (30 min)
8. Discard the supernatant and store cell pellets at −20 °C for further use

5.3 Purification of EGFP-anchor peptide fusion proteins

1. Dissolve cell pellet with buffer A (6 mL per gram) as well as 0.1 mM PMSF, 0.1 mM DTT (add just before use)
2. Disrupt cells by ultrasonic processor (amplititude: 40%; time: 20 min; 30 s on, 30 s off)
3. Centrifuge for 1 h at 4 °C, 10,000 rpm to obtain lysate
4. Filtrate supernatant with 0.45 µm non-sterile cellulose acetate membrane filters to remove insoluble particles
5. Install a 5 mL Ni-IDA affinity column on ÄKTA prime
6. Wash the Ni-IDA column with 5 column volumes of filtered ddH_2O and 5 column volumes of buffer A
7. Load the filtered supernatant with the flow rate of 0.5 mL/min (maximal pressure: 0.6 MPa)
8. Gradient elute with 100 mL 0%–50% buffer B (100%–50% buffer A)
9. Collect the eluate in 2–5 mL fractions
10. Check fractions by SDS-PAGE (12% acrylamide)
11. Dialyze protein against buffer C at 4 °C, 24 h
12. Determine purity by SDS-PAGE and measure concentration by BCA-assay
13. Store purified protein at 4 °C for further use

5.4 Anchor peptide binding assay

This assay is used to detect anchor peptide binding to the surface material (e.g., PET) employing purified samples. The anchor peptide can attach to the tested surface and the fluorescence of eGFP can be used as detection signal on the surface.

1. Prepare polymer granules with smooth surface (wash with ddH_2O and ethanol before use) or coat the glass slides with polymer nanoparticle and dried by nitrogen (Fig. 2)
2. Prepare EGFP and EGFP-anchor peptide fusion peptides stock with buffer C (10.0 µM)

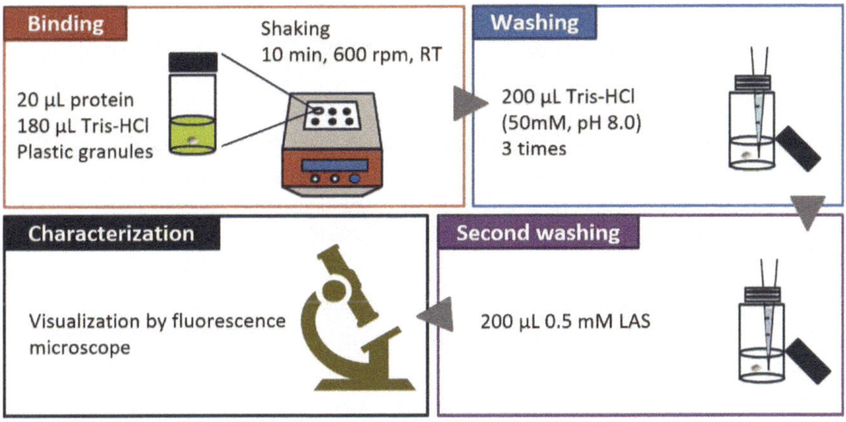

Fig. 2 Schematic graph of anchor peptide binding procedures. The procedures consist of four steps: binding, washing, second washing, and characterization.

3. Add 20 µL EGFP fusion peptides and 180 µL buffer C in a clean glass tube and mix with pipette (3 replicates)
4. Place the polymer granules into the glass tube by using a tweezer
5. Incubate at room temperature for 10 min with shaking (600 rpm)
6. Remove liquid phase with pipette
7. Wash polymer granules with 200 µL buffer C for 3 times
8. Wash polymer granules with 200 µL 0.5 mM LAS (sodium dodecylbenzene sulfonate) to further remove the non-specific binding
9. Visualize the remaining fluorescence on polymer granules through fluorescence microscopy (objective 10 ×, observation method GFP)
10. Repeat the binding assay three times for each anchor peptide and select the best binder

5.5 Directed evolution of the anchor peptide for improved binding affinity

To further improve the binding affinity to target polymers, peptide-polymer evolution protocol (PeP*evo* protocol) and *E. coli* ultrahigh-throughput cell surface screening system can be applied (Apitius, Rübsam, et al., 2019; Rübsam, Davari, et al., 2018).

1. The constructs of anchor peptide-cell surface display (CSD) and CSD control is designed as shown in Fig. 3
2. Order synthetic gene *ompA-anchor peptide-17xhelix-e-eptitope-estA** and *ompA-17xhelix-e-eptitope-estA** and clone these two genes into the pET 22b(+) vector

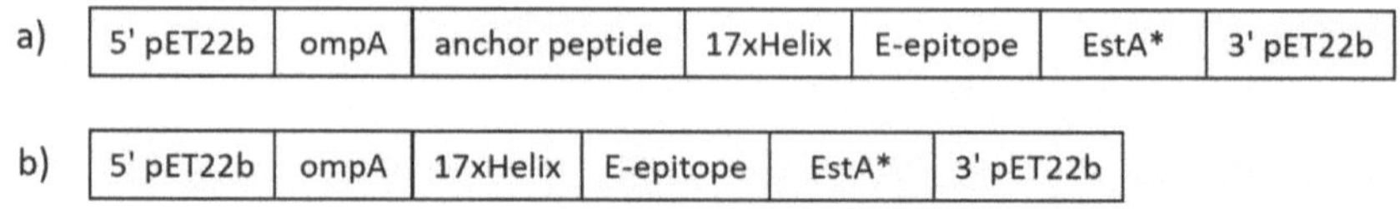

Fig. 3 Schematic illustration of anchor peptide-CSD and CSD control constructs. (A) The anchor peptide-CSD construct consisted of (5′-3′) *ompA, anchor peptide, 17xhelix, e-eptitope,* and *estA**; (B) The CSD control consisted of (5′-3′) *ompA, 17xhelix, e-eptitope,* and *estA** (Apitius, Buschmann, et al., 2019).

3. Generate anchor peptide random mutagenesis library by the Casting epPCR (cepPCR) method (Yang, Ruff, Arlt, & Schwaneberg, 2017)
4. Transform plasmids of CSD control, AP-CSD-WT, and AP-CSD library into electrocompetent *E. coli* BL21 Gold (DE3) cells by electroporation (1700 V)
5. Recover cells in SOC medium (45 min, 500 μL) and transfer into 200 mL LB medium supplemented with ampicillin (100 μg/mL)
6. Grow cells for 20 h (30 °C, 200 rpm, 70% humidity) until an OD_{600} of 0.8 is reached
7. Induce with 0.1 mM IPTG and cultivate for 2 h (37 °C, 200 rpm, 70% humidity)
8. Incubate cells with polymer beads for 10 min, then transfer polymer beads with attached cells into disposable column (Fig. 4)
9. Wash column with 50 mL phosphate buffered saline (PBS) buffer (pH 7.4) to remove weakly attached cells
10. Wash column with 10 mL 0.05 mM LAS buffer (LAS in PBS buffer, pH 7.4) and 5 mL 0.25 mM LAS buffer (LAS in PBS buffer, pH 7.4) to remove weakly immobilized cells
11. Wash column with 25 mL PBS buffer (pH 7.4) to remove detached cells and remaining LAS
12. Transfer polymer beads with immobilized cells to LB agar plates and recover cells overnight at 37 °C
13. Pick single colonies to 5 mL LB medium and incubate overnight at 37 °C
14. Extract plasmids and send for sequencing

5.6 Production of enzyme-anchor peptide fusion protein

The depolymerase for plastic degradation will be illustrated using the cutinase Tcur1278 gene from *Thermomonospora curvata* DSM43183 (Wei, Oeser, Then, et al., 2014) cloned into the pBSYA1S1Z vector which has a zeocin resistance. For expression, *Pichia pastoris* BSYBG11 will be used.

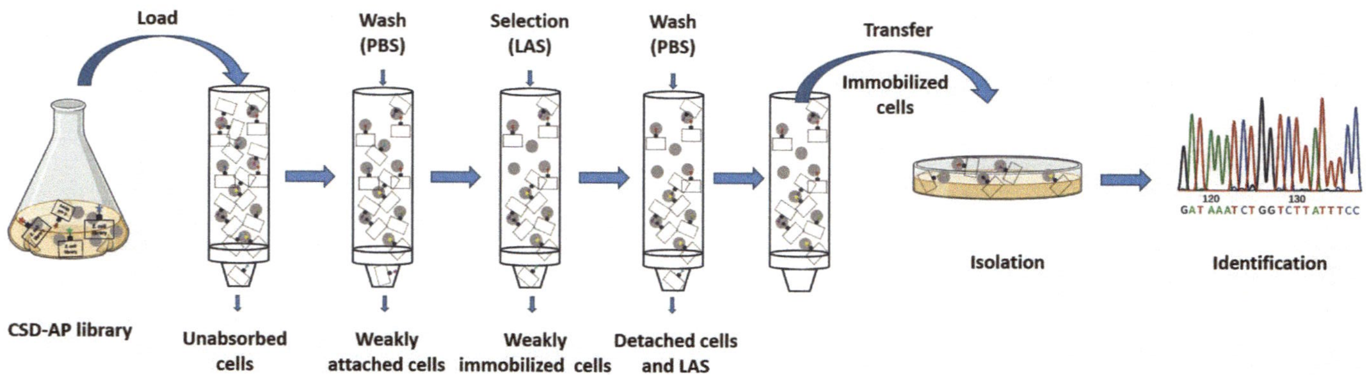

Fig. 4 Schematic overview of CSD-based screening system for improved polymer binding anchor peptides. First, polymer beads were incubated with cells of a CSD-AP library and transferred into columns, then unabsorbed and unspecific binding anchor peptides were washed out by applying PBS and LAS buffer. Afterward, polymer beads were transferred to LB agar plates and immobilized cells were recovered. Finally, the single colonies were picked for sequencing.

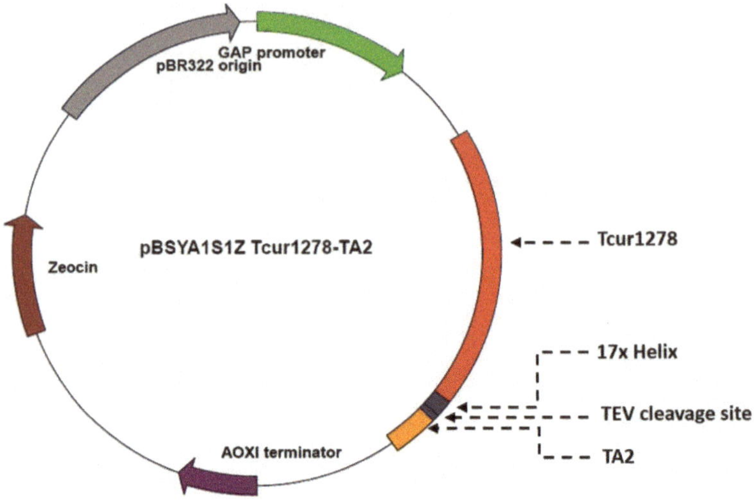

Fig. 5 Vector map of pBSYA1S1Z-Tcur1278-TA2.

1. Design the cutinase Tcur1278-anchor peptide Tachystatin A2 (TA2) gene construct as shown in Fig. 5 and clone into pBSYA1S1Z vector
2. Transform into electro-competent *E. coli* DH5α
3. Plate transformants on an YPD_{Zeocin} agar plate and incubate the plate at 30 °C for 48 h
4. Inoculate 10 mL YPD_{Zeocin} (supplemented with 10 μL of 100 mg/mL Zeocin) medium with a colony from the agar plate
5. Grow preculture for 24 h at 30 °C, 200 rpm
6. Inoculate the main culture containing 25 mL YPD_{Zeocin} (supplemented with 10 μL of 100 mg/mL Zeocin) in a 250 mL Erlenmeyer flask to an OD_{600} of 0.5
7. Express protein at 30 °C, 200 rpm for 72 h
8. Centrifuge the cultivated culture in falcon tubes at 4 °C, 4000 rpm for 30 min
9. Store the supernatant at 4 °C

5.7 Plastic degradation promoted by anchor peptides

Polyester-PUR (polyester-polyurethane) degradation assay is a continuous turbidimetric measurement. Cutinase degrades the ester bond in polyester-PUR, resulting in time dependent decrease of the turbidity of reaction mixture containing polyester-PUR. The degradation is traced at 600 nm.

1. Pipette 10 μL culture supernatant and 85 μL 50 mM Tris buffer (pH 8.0) by a multichannel pipette using the order in a fresh F-bottom microtiter plate (MTP) (6–8 replicates)
2. Incubate the MTP at 800 rpm for 5 min using an MTP shaker
3. Initiate the reaction by adding 5 μL of 10 mM *p*NPB (4-nitorphenyl butyrate dissolved in DMSO)
4. Measure activity directly for 5 min using the TecanSunrise (405 nm, 30 cycles, minimum interval)
5. Calculate activity in U/mL using the following formula:

$$Total\ activity\left(\frac{U}{mL}\right) = \frac{\Delta E \cdot V_T}{\epsilon_{405nm} \cdot d_{MTP} \cdot V_{lysate}}$$

ΔE: calculated slope of initial linear part of the kinetic curve in absorbance per min.

V_{Lysate}: lysate volume (10 μL).

V_T: total reaction volume (100 μL).

ε_{405nm}: molar extinction coefficient of *p*-NP (16.6 mM^{-1} cm^{-1}).

d_{MTP} represents the diameter of a well of MTP (96-well, flat bottom, 0.2855 cm)

6. In order to normalize activity, dilute culture supernatant of Tcur1278-WT and Tcur1278-Anchor to the same *p*NPB activity (U/mL) with 50 mM Tris buffer (pH 8.0).
7. Pipette 20 μL activity normalized culture supernatant, 130 μL polyester-PUR dispersion (1:200 dilution) with a multichannel pipette in a fresh F-bottom MTP (6–8 replicates):
8. Measure activity directly for 30–60 min using a Tecan Sunrise according to the following set-up (Table 7):

Table 7 The settings of Tecan Sunrise.

Measurement mode	**Absorbance**
Measurement wavelength	600 nm
Read mode	Accuracy
Number of kinetic cycles	60
Kinetic interval	60 s
Shake duration (Outside Normal)	2 s
Shake between cycles	45 s

9. Visualize the degradation of polyester-PUR by field emission scanning electron microscopy (FE-SEM)

6. Summary and outlook

Significant progress has been achieved in building a toolbox of enzymes and anchor peptides to depolymerize plastics. In addition, methodologies based on protein engineering through the KnowVolution strategy and an ultra-high-throughput screening system have been developed. The latter enables to tune binding strength, selectivity and generate a molecular understanding of interactions that control the adhesion of anchor-peptide fusion proteins to polymer surfaces. In summary, we are convinced that in the future selective degradation of mixed plastics through anchor peptides technology will become economically feasible and enable a circular plastics economy, which significantly reduces CO_2 and contributes to low carbon industries.

Acknowledgments

This work was supported by the European Union's Horizon 2020 research and innovation program under grant agreement no. 870294 for the project MIX-UP.

References

Adey, N. B., Mataragnon, A. H., Rider, J. E., Carter, J. M., & Kay, B. K. (1995). Characterization of phage that bind plastic from phage-displayed random peptide libraries. *Gene*, *156*(1), 27–31.

Anjana, K., Hinduja, M., Sujitha, K., & Dharani, G. (2020). Review on plastic wastes in marine environment–Biodegradation and biotechnological solutions. *Marine Pollution Bulletin*, *150*, 110733.

Anni, H., Nikolaeva, O., & Israel, Y. (2001). Selection of phage-display library peptides recognizing ethanol targets on proteins. *Alcohol*, *25*(3), 201–209.

Apitius, L., Buschmann, S., Bergs, C., Schönauer, D., Jakob, F., Pich, A., et al. (2019). Biadhesive peptides for assembling stainless steel and compound loaded micro-containers. *Macromolecular Bioscience*, *19*(9), 1900125.

Apitius, L., Rübsam, K., Jakesch, C., Jakob, F., & Schwaneberg, U. (2019). Ultrahigh-throughput screening system for directed polymer binding peptide evolution. *Biotechnology and Bioengineering*, *116*(8), 1856–1867.

Arai, R., Ueda, H., Kitayama, A., Kamiya, N., & Nagamune, T. (2001). Design of the linkers which effectively separate domains of a bifunctional fusion protein. *Protein Engineering*, *14*(8), 529–532.

Austin, H. P., Allen, M. D., Donohoe, B. S., Rorrer, N. A., Kearns, F. L., Silveira, R. L., et al. (2018). Characterization and engineering of a plastic-degrading aromatic polyesterase. *Proceedings of the National Academy of Sciences*, *115*(19), E4350–E4357.

Baekeland, L. H. (1909). United States Patent Office.

Beintema, J. J. (1994). Structural features of plant chitinases and chitin-binding proteins. *FEBS Letters*, *350*(2–3), 159–163.

Blanusa, M., Schenk, A., Sadeghi, H., Marienhagen, J., & Schwaneberg, U. (2010). Phosphorothioate-based ligase-independent gene cloning (PLICing): An enzyme-free and sequence-independent cloning method. *Analytical Biochemistry, 406*(2), 141–146.

Bolam, D. N., Ciruela, A., Mcqueen-Mason, S., Simpson, P., Williamson, M. P., Rixon, J. E., et al. (1998). Pseudomonas cellulose-binding domains mediate their effects by increasing enzyme substrate proximity. *Biochemical Journal, 331*(3), 775–781.

Büscher, N., Sayoga, G. V., Rübsam, K., Jakob, F., Schwaneberg, U., Kara, S., et al. (2019). Biocatalyst immobilization by anchor peptides on an additively manufacturable material. *Organic Process Research & Development, 23*(9), 1852–1859.

Care, A., Bergquist, P. L., & Sunna, A. (2015). Solid-binding peptides: smart tools for nanobiotechnology. *Trends in Biotechnology, 33*(5), 259–268.

Contreras, F., Thiele, M. J., Pramanik, S., Rozhkova, A. M., Dotsenko, A. S., Zorov, I. N., et al. (2020). KnowVolution of a GH5 cellulase from penicillium verruculosum to improve thermal stability for biomass degradation. *Acs Sustainable Chemistry & Engineering, 8*(33), 12388–12399.

Cunningham, S. D., Lowe, D. J., O'brien, J. P., Wang, H., & Wilkins, A. E. (2010). Nylon binding peptides and methods of use: Google Patents.

Danso, D., Chow, J., & Streit, W. R. (2019). Plastics: Microbial degradation, environmental and biotechnological perspectives. *Applied and Environmental Microbiology, 85*(19), e01095–19.

Dedisch, S., Wiens, A., Davari, M. D., Söder, D., Rodriguez-Emmenegger, C., Jakob, F., et al. (2020). Matter-tag: A universal immobilization platform for enzymes on polymers, metals, and silicon-based materials. *Biotechnology and Bioengineering, 117*(1), 49–61.

Deguchi, T., Kitaoka, Y., Kakezawa, M., & Nishida, T. (1998). Purification and characterization of a nylon-degrading enzyme. *Applied and Environmental Microbiology, 64*(4), 1366–1371.

Ejima, H., Matsuno, H., & Serizawa, T. (2010). Biological identification of peptides that specifically bind to poly (phenylene vinylene) surfaces: recognition of the branched or linear structure of the conjugated polymer. *Langmuir, 26*(22), 17278–17285.

Gong, W., Wang, J., Chen, Z., Xia, B., & Lu, G. (2011). Solution structure of LCI, a novel antimicrobial peptide from bacillus subtilis. *Biochemistry, 50*(18), 3621–3627.

Güven, G., Prodanovic, R., & Schwaneberg, U. (2010). Protein engineering - An option for enzymatic biofuel cell design. *Electroanalysis, 22*(7–8), 765–775.

Howard, G. T., & Blake, R. C. (1998). Growth of Pseudomonas fluorescens on a polyester–polyurethane and the purification and characterization of a polyurethanase–protease enzyme. *International Biodeterioration & Biodegradation, 42*(4), 213–220.

Hu, X., Thumarat, U., Zhang, X., Tang, M., & Kawai, F. (2010). Diversity of polyester-degrading bacteria in compost and molecular analysis of a thermoactive esterase from Thermobifida alba AHK119. *Applied Microbiology and Biotechnology, 87*(2), 771–779.

Inderthal, H., Tai, S. L., & Harrison, S. T. (2020). Non-hydrolyzable plastics—An interdisciplinary look at plastic bio-oxidation. *Trends in Biotechnology, 39*(1), 12–23.

Islam, S., Laaf, D., Infanzon, B., Pelantova, H., Davari, M. D., Jakob, F., et al. (2018). KnowVolution Campaign of an Aryl Sulfotransferase Increases Activity toward Cellobiose. *Chemistry-a European Journal, 24*(64), 17117–17124.

Kakudo, S., Negoro, S., Urabe, I., & Okada, H. (1993). Nylon oligomer degradation gene, nylC, on plasmid pOAD2 from a Flavobacterium strain encodes endo-type 6-aminohexanoate oligomer hydrolase: purification and characterization of the nylC gene product. *Applied and Environmental Microbiology, 59*(11), 3978–3980.

Kang Lan, T., & Ulrich, S. (2007). Directed evolution of oxygenases: Screening systems, success stories and challenges. *Combinatorial Chemistry & High Throughput Screening, 10*(3), 197–217.

Kapust, R. B., Tözsér, J., Fox, J. D., Anderson, D. E., Cherry, S., Copeland, T. D., et al. (2001). Tobacco etch virus protease: Mechanism of autolysis and rational design of stable mutants with wild-type catalytic proficiency. *Protein Engineering*, *14*(12), 993–1000.

Kashiwada, S. (2006). Distribution of nanoparticles in the see-through medaka (Oryzias latipes). *Environmental Health Perspectives*, *114*(11), 1697–1702.

Kasuya, K.-i., Ohura, T., Masuda, K., & Doi, Y. (1999). Substrate and binding specificities of bacterial polyhydroxybutyrate depolymerases. *International Journal of Biological Macromolecules*, *24*(4), 329–336.

Kawai, F., Kawabata, T., & Oda, M. (2020). Current state and perspectives related to the PET hydrolases available for biorecycling. *ACS Sustainable Chemistry & Engineering*, *8*(24), 8894–8908.

Knott, B. C., Erickson, E., Allen, M. D., Gado, J. E., Graham, R., Kearns, F. L., et al. (2020). Characterization and engineering of a two-enzyme system for plastics depolymerization. *Proceedings of the National Academy of Sciences*, *117*(41), 25476–25485.

Krueger, M. C., Harms, H., & Schlosser, D. (2015). Prospects for microbiological solutions to environmental pollution with plastics. *Applied Microbiology and Biotechnology*, *99*(21), 8857–8874.

Kumada, Y., Murata, S., Ishikawa, Y., Nakatsuka, K., & Kishimoto, M. (2012). Screening of PC and PMMA-binding peptides for site-specific immobilization of proteins. *Journal of Biotechnology*, *160*(3–4), 222–228.

Linder, M., & Teeri, T. T. (1997). The roles and function of cellulose-binding domains. *Journal of Biotechnology*, *57*(1–3), 15–28.

Liu, L., Fokkink, R., & Koelmans, A. A. (2016). Sorption of polycyclic aromatic hydrocarbons to polystyrene nanoplastic. *Environmental Toxicology and Chemistry*, *35*(7), 1650–1655.

MacArthur, D. E., Waughray, D., & Stuchtey, M. (2016). The new plastics economy, rethinking the future of plastics. In *Paper presented at the World Economic Forum, Ellen MacArthur Foundation*. https://www.ellenmacarthurfoundation.org/publications/the-new-plastics-economy-rethinking-the-future-of-plastics.

Mandawe, J., Infanzon, B., Eisele, A., Zaun, H., Kuballa, J., Davari, M. D., et al. (2018). Directed evolution of hyaluronic acid synthase from pasteurella multocida towards high-molecular-weight hyaluronic acid. *Chembiochem: A European Journal of Chemical Biology*, *19*(13), 1414–1423.

Matsuno, H., Sekine, J., Yajima, H., & Serizawa, T. (2008). Biological selection of peptides for poly (L-lactide) substrates. *Langmuir*, *24*(13), 6399–6403.

Negoro, S., Ohki, T., Shibata, N., Sasa, K., Hayashi, H., Nakano, H., et al. (2007). Nylon-oligomer degrading enzyme/substrate complex: Catalytic mechanism of 6-Aminohexanoate-dimer hydrolase. *Journal of Molecular Biology*, *370*(1), 142–156.

Novoa, C., Dhoke, G. V., Mate, D. M., Martinez, R., Haarmann, T., Schreiter, M., et al. (2019). KnowVolution of a Fungal Laccase toward Alkaline pH. *Chembiochem*, *20*(11), 1458–1466.

Ollis, D. L., Cheah, E., Cygler, M., Dijkstra, B., Frolow, F., Franken, S. M., et al. (1992). The α/β hydrolase fold. *Protein Engineering, Design and Selection*, *5*(3), 197–211.

Osaki, T., Omotezako, M., Nagayama, R., Hirata, M., Iwanaga, S., Kasahara, J., et al. (1999). Horseshoe crab hemocyte-derived antimicrobial polypeptides, Tachystatins, with Sequence Similarity to Spider Neurotoxins. *The Journal of Biological Chemistry*, *274*(37), 26172–26178.

Pellis, A., Haernvall, K., Pichler, C. M., Ghazaryan, G., Breinbauer, R., & Guebitz, G. M. (2016). Enzymatic hydrolysis of poly (ethylene furanoate). *Journal of Biotechnology*, *235*, 47–53.

Plastics - the Facts 2020. (2020). from https://nam03.safelinks.protection.outlook.com/?url=https%3A%2F%2Fwww.plasticseurope.org%2Fen%2Fresources%2Fpublications%

2F4312-plastics-facts-2020&data=04%7C01%7Cj.selvam.1%40elsevier.com%7C7cfded cab7f545b8392908d8aca67a2a%7C9274ee3f94254109a27f9fb15c10675d%7C0%7C0%7C 637449178306577192%7CUnknown%7CTWFpbGZsb3d8eyJWIjoiMC4wLjAwMDAi LCJQIjoiV2luMzIiLCJBTiI6Ik1haWwiLCJXVCI6Mn0%3D%7C1000&sdata=ED8dfZ %2FsDduR8oxIalQg4UzgSwITlX7Tj62xUb6IgzU%3D&reserved=0.

Qiang, X., Sun, K., Xing, L., Xu, Y., Wang, H., Zhou, Z., et al. (2017). Discovery of a polystyrene binding peptide isolated from phage display library and its application in peptide immobilization. *Scientific Reports*, 7(1), 2673.

Restrepo-Flórez, J.-M., Bassi, A., & Thompson, M. R. (2014). Microbial degradation and deterioration of polyethylene–A review. *International Biodeterioration & Biodegradation*, *88*, 83–90.

Ribitsch, D., Yebra, A. O., Zitzenbacher, S., Wu, J., Nowitsch, S., Steinkellner, G., et al. (2013). Fusion of binding domains to thermobifida cellulosilytica cutinase to tune sorption characteristics and enhancing PET hydrolysis. *Biomacromolecules*, *14*(6), 1769–1776.

Rübsam, K., Davari, M., Jakob, F., & Schwaneberg, U. (2018). KnowVolution of the polymer-binding peptide LCI for improved polypropylene binding. *Polymers*, *10*(4), 423.

Rübsam, K., Stomps, B., Böker, A., Jakob, F., & Schwaneberg, U. (2017). Anchor peptides: A green and versatile method for polypropylene functionalization. *Polymer*, *116*, 124–132.

Rübsam, K., Weber, L., Jakob, F., & Schwaneberg, U. (2018). Directed evolution of polypropylene and polystyrene binding peptides. *Biotechnology and Bioengineering*, *115*(2), 321–330.

Russell, J. R., Huang, J., Anand, P., Kucera, K., Sandoval, A. G., Dantzler, K. W., et al. (2011). Biodegradation of polyester polyurethane by endophytic fungi. *Applied and Environmental Microbiology*, 77(17), 6076–6084.

Sanghvi, A. B., Miller, K. P.-H., Belcher, A. M., & Schmidt, C. E. (2005). Biomaterials functionalization using a novel peptide that selectively binds to a conducting polymer. *Nature Materials*, *4*(6), 496–502.

Schmidt, J., Wei, R., Oeser, T., Dedavid e Silva, L. A., Breite, D., Schulze, A., et al. (2017). Degradation of polyester polyurethane by bacterial polyester hydrolases. *Polymers*, *9*(2), 65.

Serizawa, T., Matsuno, H., & Sawada, T. (2011). Specific interfaces between synthetic polymers and biologically identified peptides. *Journal of Materials Chemistry*, *21*(28), 10252.

Serizawa, T., Sawada, T., & Kitayama, T. (2007). Peptide motifs that recognize differences in polymer-film surfaces. *Angewandte Chemie International Edition*, *46*(5), 723–726.

Serizawa, T., Sawada, T., Matsuno, H., Matsubara, T., & Sato, T. (2005). A peptide motif recognizing a polymer stereoregularity. *Journal of the American Chemical Society*, *127*(40), 13780–13781.

Serizawa, T., Techawanitchai, P., & Matsuno, H. (2007). Isolation of peptides that can recognize syndiotactic polystyrene. *ChemBioChem*, *8*(9), 989–993.

Shohana, I., Apitius, L., Jakob, F., & Schwaneberg, U. (2019). Targeting microplastic particles in the void of diluted suspensions. *Environment International*, *123*, 428–435.

Stern, R. V., & Howard, G. T. (2000). The polyester polyurethanase gene (pueA) from Pseudomonas chlororaphis encodes a lipase. *FEMS Microbiology Letters*, *185*(2), 163–168.

Swaminathan, S., & Cui, Y. (2012). Recognition of poly (dimethylsiloxane) with phage displayed peptides. *RSC Advances*, *2*(33), 12724–12727.

Takehara, I., Fujii, T., Tanimoto, Y., Kato, D.-I., Takeo, M., & Negoro, S. (2018). Metabolic pathway of 6-aminohexanoate in the nylon oligomer-degrading bacterium Arthrobacter sp. KI72: Identification of the enzymes responsible for the conversion of 6-aminohexanoate to adipate. *Applied Microbiology and Biotechnology*, *102*(2), 801–814.

Tsai, C.-W., Ruaan, R.-C., & Liu, C.-I. (2012). Adsorption of antimicrobial indolicidin-derived peptides on hydrophobic surfaces. *Langmuir*, *28*(28), 10446–10452.
Vodnik, M., Zager, U., Strukelj, B., & Lunder, M. (2011). Phage display: Selecting straws instead of a needle from a haystack. *Molecules*, *16*(1), 790–817.
Von Moos, N., Burkhardt-Holm, P., & Köhler, A. (2012). Uptake and effects of microplastics on cells and tissue of the blue mussel Mytilus edulis L. after an experimental exposure. *Environmental Science & Technology*, *46*(20), 11327–11335.
Wallraf, A.-M., Liu, H., Zhu, L., Khalfallah, G., Simons, C., Alibiglou, H., et al. (2018). A loop engineering strategy improves laccase lcc2 activity in ionic liquid and aqueous solution. *Green Chemistry*, *20*(12), 2801–2812.
Wei, R., Oeser, T., Then, J., Kühn, N., Barth, M., Schmidt, J., et al. (2014). Functional characterization and structural modeling of synthetic polyester-degrading hydrolases from Thermomonospora curvata. *AMB Express*, *4*(1), 44.
Wei, R., Oeser, T., & Zimmermann, W. (2014). Synthetic polyester-hydrolyzing enzymes from thermophilic actinomycetes. *Advances in Applied Microbiology*, *89*, 267–305. Elsevier.
Wei, R., Tiso, T., Bertling, J., O'Connor, K., Blank, L. M., & Bornscheuer, U. T. (2020). Possibilities and limitations of biotechnological plastic degradation and recycling. *Nature Catalysis*, *3*(11), 867–871.
Wei, R., & Zimmermann, W. (2017). Microbial enzymes for the recycling of recalcitrant petroleum-based plastics: How far are we? *Microbial Biotechnology*, *10*(6), 1308–1322.
Weinberger, S., Canadell, J., Quartinello, F., Yeniad, B., Arias, A., Pellis, A., et al. (2017). Enzymatic degradation of poly (ethylene 2, 5-furanoate) powders and amorphous films. *Catalysts*, 7(11), 318.
Yang, J., Ruff, A. J., Arlt, M., & Schwaneberg, U. (2017). Casting epPCR (cepPCR): A simple random mutagenesis method to generate high quality mutant libraries. *Biotechnology and Bioengineering*, *114*(9), 1921–1927.
Yoshida, S., Hiraga, K., Takehana, T., Taniguchi, I., Yamaji, H., Maeda, Y., et al. (2016). A bacterium that degrades and assimilates poly (ethylene terephthalate). *Science*, *351*(6278), 1196–1199.
Zernia, S., Ott, F., Bellmann-Sickert, K., Frank, R., Klenner, M., Jahnke, H.-G., et al. (2016). Peptide-mediated specific immobilization of catalytically active cytochrome P450 BM3 variant. *Bioconjugate Chemistry*, *27*(4), 1090–1097.
Zhang, Y., Chen, S., Xu, M., Cavoco-Paulo, A., Wu, J., & Chen, J. (2010). Characterization of Thermobifida fusca cutinase-carbohydrate-binding module fusion proteins and their potential application in bioscouring. *Applied and Environmental Microbiology*, *76*(20), 6870–6876.
Zhang, Y., Wang, L., Chen, J., & Wu, J. (2013). Enhanced activity toward PET by site-directed mutagenesis of Thermobifida fusca cutinase–CBM fusion protein. *Carbohydrate Polymers*, *97*(1), 124–129.

CHAPTER FOURTEEN

Tuning of adsorption of enzymes to polymer

Doris Ribitsch[a,b,*] and Georg M. Guebitz[a]

[a]Institute for Environmental Biotechnology, IFA Tulln, University of Natural Resources and Life Sciences, Vienna, Austria

[b]ACIB—Austrian Centre of Industrial Biotechnology, Graz, Austria

*Corresponding author: e-mail address: doris.ribitsch@boku.ac.at

Contents

Methods in Enzymology, Volume 648
ISSN 0076-6879
https://doi.org/10.1016/bs.mie.2020.12.010

Abstract

In the past years, several serine hydrolases such as cutinases, esterases and lipases have shown the ability to degrade not only natural polymers but also synthetic polyesters, even aromatic representatives like polyethylene terephthalate (PET). Hence, cutinases and related ester hydrolases have become very important to be applied in the biocatalytic plastic recycling as green alternative to chemical recycling as well as to the functionalization of polyester surfaces in order to change superficial properties like hydrophobicity or hydrophilicity. Sorption characteristics of the enzymes to the polymers have turned out to be a crucial process for efficient polymer hydrolysis. Hence, special attention was paid on tuning the sorption of the enzymes to the hydrophobic polymers. Engineering of the enzyme surface, fusion of hydrophobic substrate-binding domains or truncation of domains hindering the access of the polymer to the enzyme has led to significant improvement of sorption processes and consequently increased activity on the bulky substrate. Finally, the combination of engineering approaches has proved that they can bring additional advantages in improving the enzyme activity when used in a synergistic manner.

1. Introduction

Intensive research on polymer biodegradation has led to identification of several hydrolases which have proven ability to degrade synthetic polyesters like PET or PBAT (Danso, Chow, & Streita, 2019; Papadopoulou, Hecht, & Buller, 2019; Salvador et al., 2019). However, these enzymes were originally designed from nature for completely different substrates and hence, specific activities toward the bulky, insoluble and non-natural polymers were quite low. Soon after the first hydrolase with PET activity was identified (Müller, Schrader, Profe, Dresler, & Deckwer, 2005), it became clear that there was a strong need to improve these enzymes by protein engineering tools. Different approaches have been applied to adapt the active site architecture to the bulky polymers, reduce inhibition effects caused by release products and improve stability (Wei & Zimmermann, 2017). In this chapter, focus is placed on the improvement of the adsorption/desorption process of polymer hydrolases on polyesters which turned out to be a crucial and rate-limiting step in enzymatic polymer hydrolysis. For improvement of sorption characteristics, polymer hydrolyzing enzymes were subjected to different approaches of rational design and effects were demonstrated by various analytical methods (Quartinello, Guebitz, & Ribitsch, 2019).

The term enzyme engineering refers to the process in which the natural sequence of an enzyme is altered to tailor the activity and properties of an enzyme for a specific reaction (Quin & Schmidt-Dannert, 2011). In most cases, protein engineering aims to clarify and improve structure-function relationships by drawing attention to the geometry of the active site and its immediate vicinity, whereas the remaining part of the protein such as the protein surface is often neglected (Petersen, Harald Jonson, Fojan, Petersen, & Neves Petersen, 1998). However, it has been shown that even the surface far from the active site plays a crucial role in substrate selectivity and catalytic efficiency by e.g. directing the substrate to the active site (Wilding, Scott, & Warden, 2018). The surface represents the interface between an enzyme and the mainly aqueous environment. It contains polar, nonpolar and charged residues and is essential for the stability of the protein. Around 95% of all titratable residues are located at the surface of soluble proteins and changes in the physical chemical environment are sensed through the surface. Consequently, interaction of a substrate with the enzyme can be strongly influenced by the surface properties and hence, the surface is a highly important target for enzyme engineering.

In general, there are two main strategies for protein engineering, namely directed evolution and rational design, but the two strategies can also be combined to semi-rational design in order to improve an enzyme in a less time and cost intensive manner (Steiner & Schwab, 2012).

Directed evolution, as it is used today, goes back to the early 1990s and resembles natural evolution, the gradual adaptation of organisms to a selection pressure (Arnold, 1998, 2018). The evolutionary principle is based on repetitive cycles in which mutations are introduced into the genome, beneficial phenotypes are selected, and selected mutations are transferred to the offspring. Directed evolution, for which the Nobel Prize in Chemistry 2018 was awarded to Prof. Frances Arnold, follows the same principle, but in a much faster time than evolution occurs in nature. Evolution in the lab is based on random mutagenesis of the enzyme sequence which generates large mutant libraries typically consisting of 10^3–10^6 mutants. By high-throughput screening of the mutant libraries, the mutant coding for the desired phenotype is selected and subjected to another round of directed evolution till the enzyme with the desired properties is obtained. Error-prone PCR is probably the most common technique to randomize a gene and allows insertion of mutations by the usage of a Taq polymerase which lacks the 3′–5′ exonuclease proofreading activity. However, the success of this method is inevitably associated with the time-consuming screening of a large number of mutants which requires suitable high-throughput assays.

A more direct and resource-efficient approach of protein engineering is *rational design* (Bornscheuer & Pohl, 2001). The increasing number of three-dimensional structures or structure models and knowledge about the relations between sequence, structure and mechanism or function of enzymes has made it possible to predict beneficial mutations which are introduced into the gene by site-directed mutagenesis. The targeted generation of mutants significantly reduces the number of phenotypes which has to be expressed and characterized and is especially useful if no high-throughput assays are available. However, the rational approach primarily targets mutations that appeared logical based on the 3D structure and "irrational" mutations e.g. far from the active site are not considered.

Semi-rational design is another knowledge-based engineering method that combines the advantages of random mutagenesis and rational design which tremendously reduces the size of the mutant library to be screened (Quin & Schmidt-Dannert, 2011). This approach uses available information about the primary sequence, structure and function of an enzyme and computational algorithms to predict sites and domains in the protein which are randomly mutated. In this approach, high throughput screenings of small mutant libraries with typically 10^2–10^3 mutants are advantageous but not essential and random mutagenesis make it possible to generate also "irrational mutations" which offer the probability to identify synergistic mutations.

Sorption properties of enzymes on polyesters can be characterized and quantified by using quartz crystal microbalance analysis with dissipation monitoring (QCM-D) which allows drawing conclusions on the binding strength of enzymes to the polymer and a real-time monitoring of the hydrolytic mass loss from polyester thin films (Rieckmann & Völker, 2004). This method is based on the measurement of frequency changes of mass deposited on the crystal. In comparison to other methods, the QCM-D approach is unique since it provides insights into the dynamics of the enzymatic polyester hydrolysis at the nanometer scale. Adsorbed protein can also be detected by chemiluminescence using a nickel (Ni^{2+})-activated derivative of horseradish peroxidase (HRP) that enables direct, IMAC (immobilized metal affinity chromatography)-based detection of His-tagged proteins (Ribitsch et al., 2013). There are a number of other methods which can be used to investigate and visualize enzyme sorption on polymeric substrates, including scanning electron microscopy (SEM) (Weinberger et al., 2017) and confocal laser scanning microscopy (after labelling of enzymes) or atomic force microscopy (AFM) (Greimel et al., 2013). SAXS can additionally provide information on enzyme agglomeration which is obviously likewise influenced by enzyme surface properties (Biundo, Ribitsch, Steinkellner, Gruber, & Guebitz, 2017).

Apart from direct methods to investigate enzyme sorption, any other method for quantification of the reaction products will provide information about the impact of enzyme sorption engineering on enzyme efficiency. Different protocols for qualitative and quantitative analysis of the hydrolysis products generated by polyester hydrolyzing enzymes are described in order to give a real proof of enzymatic degradation instead of just referring to the weight loss of polymers. Quantification of water-soluble release products (monomers and oligomers) can be performed by high-performance liquid chromatography (HPLC) while liquid chromatography coupled to time of flight mass spectrometry (LC–MS/TOF) can be used to identify reaction products. Changes of polymer molecular weight are analyzed e.g. by gel permeation chromatography (GPC) while there are various methods to quantify changes in surface properties. Changes in hydrophobicity/hydrophilicity of polyester surfaces due to introduction of hydroxyl and carboxyl groups via enzymatic hydrolysis can be quantified via water contact angle (WCA) analysis (Giovambattista, Debenedetti, & Rossky, 2007). These novel functional groups on the surface can also be quantified using derivatization (Vecchiato et al., 2017) and confocal laser scanning microscopy (Sagong et al., 2020). This chapter will present methods for engineering the sorption features of polyester hydrolyzing enzymes while the above methods to quantify resulting changes in the enzymatic activities against polyester substrates have been described recently elsewhere (Quartinello et al., 2019).

2. Expression and purification of polymer degrading hydrolases

Plant pathogenic bacteria and fungi are able to cleave the ester bonds of the natural polyester cutin, the major component of most plant cuticles which cover leaves, fruits, flowers, and other non-woody parts of higher plants (Chen, Su, Chen, & Wu, 2013). Cutin is an insoluble, aliphatic polyester consisting of predominantly primarily long-chain hydroxy fatty acids and known to be chemically stable and resistant to degradation. Plant pathogens must therefore first cleave the polyester cutin to penetrate the plant. This cleavage is performed by enzymes called cutinases which belong to the group of carboxylic ester hydrolases (Nikolaivits, Kanelli, Dimarogona, & Topakas, 2018). In the past, cutinases have shown the ability to degrade not only natural polymers but also several synthetic polyesters, even aromatic ones like polyethylene terephthalate (Wei & Zimmermann, 2017). Hence, cutinases and related α/β hydrolases like esterases and lipases have become very important as green alternative to chemical recycling as well as functionalization of polyester surfaces in order to change superficial properties like hydrophobicity or hydrophilicity. The enzymes can be produced in soluble form in *E. coli* with high yield and activity.

2.1 Materials and reagents

- Restriction endonucleases *Hind*III and *Nde*I
- Phusion DNA polymerase (Finnzymes)
- T4 DNA-ligase
- Plasmid Mini Kit
- DNA purification Kit
- Vector pET26b(+) (Novagen)
- *E. coli* BL21-Gold(DE3) (Stratagene, La Jolla, CA)
- LB-medium
- Kanamycin (40 mg/mL)
- Isopropyl β-D-1-thiogalactopyranoside (IPTG) (0.1 M)

2.2 Buffers

- Ni-NTA Lysis Buffer: 50 mM NaH_2PO_4, 300 mM NaCl, 10 mM imidazole, pH 8
- Tris–HCl, 100 mM, pH 7.0

2.3 Strains

- *Thermobifida cellulosilytica* DSM44535 (DSMZ, Germany)

2.4 Equipment

- Gene Amp PCR 2200 thermocycler (Applied Biosystems, USA)
- Electroporator MicroPulser™ (Bio-Rad Laboratories, Inc., Hercules, USA)
- HT Multitron Pro (Infors, Switzerland)
- Vibra-Cell™ Ultrasonic Liquid Processor (Sonics Materials, Switzerland)
- Eppendorf centrifuge 5920 R(Germany)
- Äkta pure (GE Healthcare, Chalfont St. Giles, UK)
- PD-10 columns (GE Healthcare, Vienna, Austria)

2.5 Protocols

1. Amplify genes coding for cutinases Thc_Cut1 and Thc_Cut2 from genomic DNA of *T. cellulosilytica* DSM44535 by standard polymerase chain reaction (PCR) (Herrero Acero et al., 2011) using primers Thc_cut.fw and Thc_cut.rev (Table 1) and the following PCR conditions (35 cycles): 30s at 98 °C (denaturation), 30s at 63 °C (annealing), and 2 min at 72 °C (extension).
2. Purify PCR-products by the DNA purification Kit and digest with *Nde*I and *Hind*III.
3. Ligate the restricted and purified PCR-products into pET26b(+) using T4 DNA ligase.
4. Transform plasmids in *E. coli* BL21-Gold(DE3).
5. Sequence genes by custom service using primers pET.fw and pET.rv (Table 1).
6. Condon optimize the gene (GenBank accession number CAL82416.1) coding for lipase from *Clostridium botulinum* (strain ATCC 3502) for expression in *E. coli* (Perz et al., 2016). Synthesize genes (without signal peptide) by a commercial service.
7. Digest plasmids containing the synthetic gene using *Nde*I and *Hind*III.
8. Purify and ligate into pET26b(+).
9. Transform plasmids in *E. coli* BL21-Gold(DE3).
10. Cultivate freshly transformed *E. coli* BL21-Gold(DE3) cells in 20 mL of LB-medium supplemented with 40 μg/mL kanamycin. Let the cells grow at 37 °C and 160 rpm overnight.
11. Inoculate 200 mL LB-medium containing 40 μg/mL kanamycin to $OD_{600}=0.1$ and let the cells grow to OD_{600} of 0.6–0.8 at 37 °C and 160 rpm.

Table 1 Primer sequences for cloning of *T. cellulosilytica* cutinases.

Primer name	Nucleotide sequence
Thc_cut.fw	5‘-CCCCCGCTCATATGGCCAACCCCTACGAGCG-3‘
Thc_cut.rev	5‘-GTGTTCTAAGCTTCAGTGGTGGTGGTGGTGGTG CTCGAGTGCCAGGCA CTGAGAGTAGT-3‘
pET.fw	5‘-GAGCGGATAACAATT CCCCTCTAGAA-3‘
pET.rv	5‘-CAGCTTCCTTTCGGGCTTTGT-3‘

12. Cool the cell culture down to 20 °C and induce with 0.05 mM IPTG.
13. Express hydrolases for 20 h at 20 °C and 160 rpm.
14. Harvest cells by centrifugation for 25 min at 10 °C and 4000 rpm.
15. Resuspend pellets from 100 mL of culture in 10 mL Ni-NTA Lysis Buffer and sonicate on ice with three 30-s pulses.
16. Centrifugate cell lysates (30 min, 10 °C, 4000 rpm) and purify as described in the manufacturer's instructions (IBA GmbH, Goettingen, Germany).
17. Change buffer to 100 mM Tris–HCl pH 7.0 by the use of PD-10 columns.

3. Surface engineering by site-directed mutagenesis

Thermobifida cellulosilytica DSM44535 expresses two extracellular cutinases, namely Thc_Cut1 and Thc_Cut2, which differ in only 18 amino acids corresponding to 93% identity. X-ray crystallography at resolutions of 1.5 Å for Thc_Cut1 (PDB ID:5LUI) and 2.20 Å for Thc_Cut2 (PDB ID:5LUJ) revealed that out of these 18 amino acids, eight are located at the enzyme surface. Comparison of the structures showed no significant differences close or at the proposed active site but interestingly, differences of the electrostatic potential and hydrophobicity on the surface in two distinct regions close to the active site were observed. Despite the high similarity, the enzymes showed distinct hydrolytic activities on PET films (Herrero Acero et al., 2011). Terephthalic acid (TA) was the major hydrolysis product for Thc_Cut1 whereas for Thc_Cut2, mono-(2-hydroxyethyl) terephthalic acid (MHET) was the most abundant product. In order to find out, whether the two identified surface regions are possibly interacting with the polymer chain, docking studies were performed to mimic the covalently bound tetrahedral intermediate of Thc_Cut2 and the polymer substrate (Fig. 1). According to the docking studies it seemed possible that the polymeric substrate interacts with the two identified regions. Indeed, introduction of mutations Arg29Asn and Ala30Val (Fig. 1) resulted in improved hydrolysis of bis(benzoyloxyethyl) terephthalate (3PET) and PET, from the latter polymer TA was released as the main product, whereas native Thc_Cut2 released comparable amounts of TA and MHET.

Combination of the beneficial mutations succeeded in a double (Arg29Asn, Ala30Val) and a triple (Arg19Ser, Arg29Asn, Ala30Val) mutant which showed higher activities than both the native Thc_Cut2 and Thc_Cut1 thereby suggesting synergistic effects of the introduced mutations.

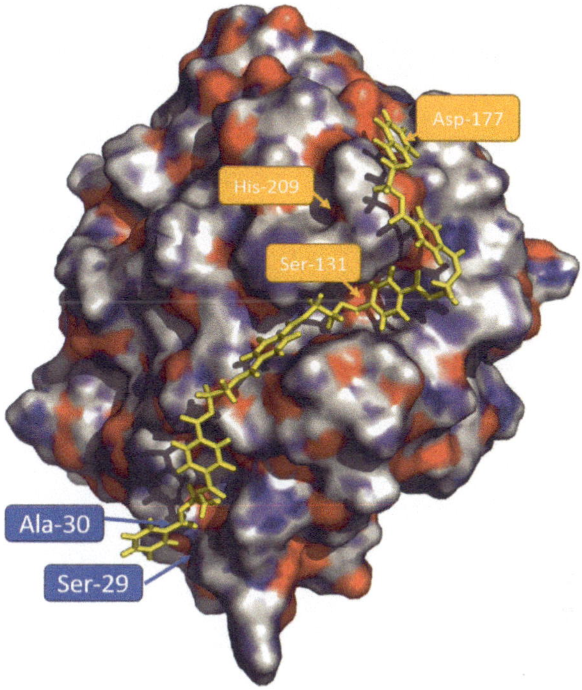

Fig. 1 Identification of key amino acids involved in the binding of 3PET onto cutinase Thc_Cut2. Yellow boxes: amino acids of the active site. Blue boxes: Amino acids on the surface confirmed to be involved in the binding of 3PET.

3.1 Materials and reagents

- QuikChange multisite-directed mutagenesis kit (Stratagene)
- Restriction enzyme *DpnI*
- GeneRuler™ 1 kb DNA Ladder (Fermentas)
- 1.5 mm glass capillaries

3.2 Buffers

- TAE buffer (Merck)

3.3 Equipment

- 2720 Thermal Cycler (Applied Biosystems, USA)

3.4 Protocols

1. Perform mutagenesis of Thc_Cut2 by using the QuikChange multisite-directed mutagenesis kit with pET26b(+)_Thc_cut2 as template (Herrero Acero et al., 2013) and megaprimers carrying the appropriate mutation (Table 2).

Table 2 Primer used for site-directed mutagenesis.

Primer name	Nucleotide sequences	T_M (°C)	Length (bp)
Thc_Cut2_ Arg29Asn.FW	5'-CCC TTC TCC GTG AGT GAA GAA AAC GCC TCC CGC TTC GGT GCT GAC GGT TTC GGC GGC GGC ACC-3'	76	63
Thc_Cut2_ Arg29Asn.REV	5′-GGT GCC GCC GCC GAA ACC GTC AGC ACC GAA GCG GGA GGC GTT TTC TTC ACT CAC GGA GAA GGG-3'	76	63
Thc_Cut2_ Ala30Val.FW	5'-CCC TTC TCC GTG AGT GAA GAA CGG GTC TCC CGC TTC GGT GCT GAC GGT TTC GGC GGC GGC-3′	76	60
Thc_Cut2_ Ala30Val.REV	5'-GCC GCC GCC GAA ACC GTC AGC ACC GAA GCG GGA GAC CCG TTC TTC ACT CAC GGA GAA GGG-3'	76	60
Thc_Cut2_Asn29_ Val30.FW	5'-CCC TTC TCC GTG AGT GAA GAA AAC GTC TCC CGC TTC GGT GCT GAC GGT TTC GGC GGC GGC ACC-3′	75.2	63
Thc_Cut2_Asn29_ Val30.Rev	5′-GGT GCC GCC GCC GAA ACC GTC AGC ACC GAA GCG GGA GAC GTT TTC TTC ACT CAC GGA GAA GGG-3'	75.2	63
Thc_Cut2_Asn29_ Val30_Ser19.FW	5'-ACC CGA CCG ACG CCC TGC TCG AAG CCA GCA GCG GCC CCT TCT CCG TGA GTG AAG AAA ACG TCT-3'	76.2	63
Thc_Cut2_Asn29_ Val30_Ser19.Rev	5′-AGA CGT TTT CTT CAC TCA CGG AGA AGG GGC CGC TGC TGG CTT CGA GCA GGG CGT CGG TCG GGT-3'	76.2	63

2. Prepare two PCR mixtures (Table 3) per each mutated enzyme for a two stage PCR. One mixture includes the forward primer, the other one the reverse primer in a total volume of 50 μL.
3. Amplify these mixtures in the first stage in four cycles.

Table 3 Composition of PCR mixtures.

Reagent	**Volume (μL)**
Pfu buffer (10 ×)	5
dNTP (each 10 mM)	1
Template DNA	2
Primer [10 pmol/μL] rev or. fw	2
Pfu DNA polymerase [2.5 U/μL] (Promega)	1
H_2O (nuclease free)	39
Total volume	50

Table 4 Thermal cycling parameters for PCR (2720 Thermal Cycler, Applied Biosystems).

Initial denaturation	95 °C	1 min	
Denaturation	95 °C	50 s	4 × // 30 ×
Annealing	60 °C	50 s	
Extension	72 °C	2 min/kb = 12 min	
Final extension	72 °C	7 min	
	4 °C	∞	

4. Pool 25 μL of each reaction mixture for the second stage and run for another 30 cycles. Amplification conditions are listed in Table 4
5. Incubate 25 μL of the PCR products with 1 μL *DpnI* for 1 h at 37 °C.
6. Inactivate *Dpn*I at 70 °C for 10 min.
7. Load 1 μL of *Dpn*I digested PCR products on an agarose gel (1%) in 1 × TAE buffer and a size marker. Use PCR products appearing at the expected size of approximately 5 kb for transformation in *E. coli* XL-1.
8. Determine sequence as custom service.

4. Fusion of hydrophobic binding domains

In nature, the activity of several polymer modifying enzymes such as cellulases or polyhydroxyalkanoate (PHA) depolymerases depends on the interaction with polymeric substrates via C-terminal binding modules with discrete folding, which are located outside of the catalytic domain.

In cellulases, the binding modules (CBMs) are typically discrete protein domains which are normally joined by relatively unstructured linker sequences to the catalytic protein (Shoseyov, Shani, & Levy, 2006). Structural investigation has shown that CBMs have a hydrophobic surface, which is composed of several aromatic amino acids, particularly tryptophan and tyrosine. At the beginning of research on CBMs it was assumed that they solely increase the effective concentration of the catalytic domain on the carbohydrates. However, it turned out that CMBs can have also other functions like the disruption of the polymeric substrate. For instance, the CBM from *Cellulomonas fimi* (Armenta, Moreno-Mendieta, Sánchez-Cuapio, Sánchez, & Rodríguez-Sanoja, 2017) was able to disrupt the structure of cellulose fibers releasing small particles without any detectable hydrolytic activity and to prevent the flocculation of microcrystalline bacterial cellulose. Similar to cellulases, PHA degrading bacteria are producing extracellular PHA depolymerases which can bind over substrate-binding domains, consisting of 40–60 amino acids, to the highly hydrophobic polymeric substrate. Mimicking the natural approach for hydrolysis of water insoluble polymers, Thc_Cut1 was C-terminally fused to the CBM from *Hypocrea jecorina* cellobiohydrolase I (Thc_Cut1+CBM) which belongs to family 1 of CBMs and to the polymer-binding module from *Alcaligenes faecalis* polyhydroxyalkanoate depolymerase which is a fibronectin type 3 domain (Thc_Cut1+PBM) (Schöber, Thiel, & Jendrossek, 2000). In both variants, the binding modules were fused by the linker of the cellobiohydrolase I to the catalytic domain. The fusion enzymes were active both on the insoluble PET model substrate 3PET and on PET films. In general, both fusion variants showed lower activity compared to the native cutinase when incubated on 3PET (Fig. 2).

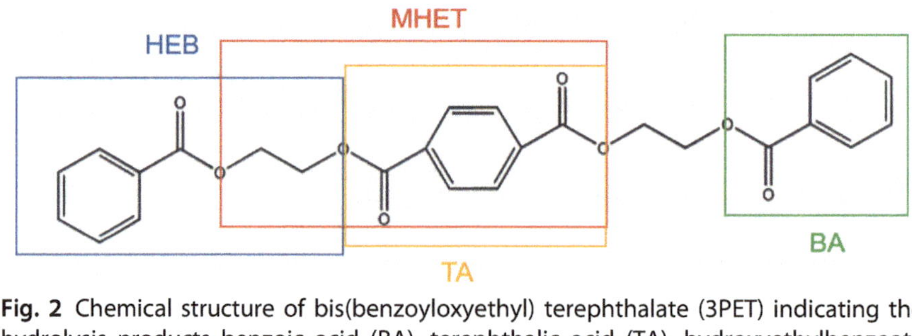

Fig. 2 Chemical structure of bis(benzoyloxyethyl) terephthalate (3PET) indicating the hydrolysis products benzoic acid (BA), terephthalic acid (TA), hydroxyethylbenzoate (HEB) and mono(2-hydroxyethyl) terephthalate (MHET).

In the earlier reaction stage, the native cutinase released comparable amounts of all possible hydrolysis products whereas in the later reaction stage of more than 24h the amounts of benzoic acid (BA) and terephthalic acid (TA) increased most likely because exotype hydrolysis of hydroxyethylbenzoate (HEB) and mono(2-hydroxyethyl) terephthalate (MHET). For both variants, however, the ratios of HEB/BA and MHET/TA were significantly higher at longer incubation times compared to those obtained with wildtype Thc_Cut1. That indicated that 3PET was too small to be adsorbed via the external binding domains. On the contrary, hydrolysis of PET films clearly showed the expected positive effect of the binding domains. Compared to the native cutinase, Thc_Cut1 + PBM released 3-times more TA and 11-times more MHET from PET after 72h of incubation whereas Thc_Cut1+CMB resulted in similar amounts of TA but only 4-times more MHET.

The effect of PBM was investigated in more detail using quartz crystal microbalance with dissipation monitoring (QCM-D) (Perz et al., 2015). For the measurements, catalytically inactive mutations were constructed by site-directed mutagenesis of the catalytic Ser-131 to alanine in the native cutinase and Thc_Cut1 _PBM. In this study, poly (1,4-butylene adipate) (PBA) was used as model polymer for aliphatic polyesters. Like before, fusion of PBM led to around 2-times higher hydrolysis rates than the native cutinase as determined by the released adipic acid. The real-time monitoring of the hydrolytic mass loss by QCM-D clearly demonstrated that the initial mass during polymer hydrolysis was larger for the inactive enzyme fused to the PBM than for the inactive native cutinase and hence, the PBM increased the affinity of the cutinase to the PBA films.

4.1 Software

- Molecular modeling suite Rosetta (RosettaCommons)
- GROMACS 4.5.5

4.2 Protocols

1. For gen design of Thc_Cut1+CBM, attach the cellulose binding module of 1,4-β-cellobiohydrolase I from *Trichoderma reesei* (CBH I; Thr288-Leu323 of CBH I) (GenBank P62694.135) to the cutinase via a linker region (Pro263-Pro287 of CBH I).
2. For gen design of Thc_Cut1+PBM, attach the polymer binding module (PMB; Ala428-Pro488) of polyhydroxybutyrate (PHB) depolymerase

from *Alcaligenes faecalis* (PhaZAfa, GenBank AAA21974.120) to the cutinase via the linker region of CBH I from *T. reesei*.

3. Fuse a C-terminal 6xHisTag for simply purification by affinity chromatography to both fusion enzymes for simply purification by affinity chromatography.
4. Synthesize the genes of Thc_Cut1+CBM and Thc_Cut1+PBM and clone them over the *Nde*I and *Hind*III restriction sites into pET26b(+) by commercial service.
5. Express and purify the fusion enzymes according to Section 2.

4.3 Application of hydrophobins

Hydrophobins are small, cysteine-rich proteins produced by filamentous fungi and able to spontaneously self-assemble into amphipathic monolayers at hydrophobic/hydrophilic interfaces resulting in a reversal of the surface wettability (Ren, Kwan, & Sunde, 2013). The structure of hydrophobins is characterized by a small β-structured core formed by eight conserved cysteine residues that establish four disulfide bridges. On the surface, the proteins have large hydrophobic regions which are responsible for the high surface activity. Hydrophobins are grouped into class I and class II proteins. Class I hydrophobins assemble into monolayers with fibrillar substructures known as rodlets which are extremely stable and structurally similar to amyloid fibrils. Therefore, they are considered to be functional amyloids and important for the formation of biofilms and in attachment of microorganisms to surfaces. Class II hydrophobins, on the contrary, form highly ordered 2D crystalline amphipathic monolayers that can be easily depolymerized by treatment with e.g. detergents. In the past, it has been shown that hydrophobins can influence the biodegradation of polymers. For instance, growth of the mold fungus *Aspergillus oryzae* on the biodegradable polyester polybutylene succinate-coadipate (PBSA) induced not only the expression of a cutinase but also biosynthesis of a hydrophobin, which adsorbed to the hydrophobic surface of PBSA, recruited the cutinase, and stimulated its hydrolysis of PBSA. Another study has proven, that a polyester hydrolase from *Coprinopsis cinerea* cleaved cutin and suberin more efficiently in the presence of hydrophobin HFB2 from *Trichoderma reesei* (Kontkanen et al., 2009). This finding caused us to presume that hydrophobins, like class II hydrophobins from *Trichoderma reesei*, can also stimulate the enzymatic modification of synthetic polymers such as PET. For our studies, two novel class II HFBs from *Trichoderma* (HFB4 and HFB7) were identified and subjected to further investigation (Espino-Rammer et al., 2013). Indeed, incubation of *Humicola insolens* cutinase (HiC) on PET resulted in more than 2.5-times higher amounts of release products in the presence of HFB4 compared to the reaction without hydrophobin. HFB7, on the other hand, stimulated cutinase activity on PET at low concentrations but inhibited hydrolysis at higher concentrations.

The fact, that hydrophobins can accelerate polyester hydrolysis has encouraged us to fuse the small hydrophobic proteins to the cutinase from *T. cellulosilytica* (Ribitsch et al., 2015). Three hydrophobins from *Trichoderma*, the class II hydrophobins HFB4 and HFB7 and the pseudo-class I hydrophobin HFB9b, were C-terminally fused over a linker region to Thc_Cut1. Amazingly, fusion of HFB4 or HFB7 resulted in more than 16-fold increase PET hydrolysis compared to the free cutinase whereas a mixture of Thc_Cut1 and the free hydrophobins led only to a 4-fold improvement. In contrary, fusion with the pseudo-class II hydrophobin HFB9b did not accelerate the rate of hydrolysis over that of the enzyme-hydrophobin mixture. However, HFB9b performed best when PET was preincubated with the hydrophobins before enzyme treatment resulting six times more release products compared to the non-preincubated cutinase.

4.4 Materials and reagents

- Vector pGEX4T-2v (Merck)
- Vector pET26b(+)
- IPTG
- Restriction enzymes *Nde*I and *Hind*III

4.5 Strains

- *E. coli* BL21

4.6 Protocols

1. Express the *hfb4* and *hfb7* genes as fusion proteins at the C-terminus to glutathione *S*-transferase (GST).
2. Use *E. coli* BL21 as host employing the *lacZ* promoter as previously described (Pail et al., 2004).
3. Express GST alone in the same vector.
4. Amplify the corresponding cDNA fragments and ligate them.
5. Amplify the corresponding cDNA fragments using the primers listed in (Espino-Rammer et al., 2013) and ligate them into vector pGEX4T-2v according to the manufacturer's instructions.

6. Induce the cells by IPTG addition.
7. Harvest cells by centrifugation at 4000 × *g* for 10 min at 4 °C.
8. Resuspended the cells in 5 mL of phosphate-buffered saline containing 1% Triton X-100 and disrupt them by sonication.
9. Centrifuge (10,000 rpm, 4 °C, 10 min) the supernatant and purify it by affinity chromatography using glutathione-sepharose 4B beads (GE Healthcare).
10. For expression in *E. coli*, codon optimize the genes coding for the individual fusion cutinases (Thc_Cut) with hydrophobins and synthesize them by commercial service.
11. Connect hydrophobins hfb4, hfb7 and hfb9b to Thc_Cut1 from *Thermobifida cellulosilytica* (GenBank accession no. ADV92526.1) over the *T. reesei* cellobiohydrolase I linker region (P263 to P287; GenBank accession no. P62694.1).
12. Clone the resulting genes into pET26b(+) over the *Nde*I and *Hind*III restriction sites enabling expression of the fusion cutinases without the pelB signal peptide.
13. Attach a C-terminal 6xHis tag to all fusion-proteins for purification by affinity chromatography.

5. Enzyme truncation and mutagenesis of the metal-binding site

Clostridium botulinum esterase Cbotu_EstA was the first enzyme from an anaerobic microorganism which showed ability to hydrolyse the polyester PBAT (Perz et al., 2016). The 3D structure of the enzyme determined at a resolution of 1.2 Å (PDB ID: 5AH1) by X-ray crystallography revealed that the enzyme belongs to the family of α/β hydrolases. The catalytic triad is formed by Ser-182, His-426 and Asp-384 with the serine embedded in the highly conserved Gly-X-Ser-X-Gly motif and located at the top of the nucleophile elbow. The region surrounding Ser-182 consists of mainly non-polar side chains which indicated the hydrophobicity of the substrate binding site. In contrast to known polymer cleaving cutinases, the structure of Cbotu_EstA exhibited a cap domain that covers the active site. The N-terminal residues 28–63 form an "arm" that is wrapped around this cap, similar to the "lid" of lipases. The cap blocks the entrance to the active site in the closed conformation of the enzyme and requires conformational changes to enable polymer hydrolysis. Furthermore, the structure disclosed a metal-binding site close to the enzyme surface comprising a zinc ion which is tetragonally coordinated by Asp-130, His-150, His-156, and Asp-302. A very similar site was described for a *Geobacillus* lipase which was assumed to be involved in the thermostability of the protein but also important in the regulation of the lid opening (Jeong et al., 2002). The latter hypothesis could also apply to Cbotu_EstA since one of the zinc coordinating residues (Asp-302) is part of the cap domain. More precise investigation of the cap has pointed out that it covers two hydrophobic surface patches which may become accessible upon removal of the domain thereby improving the sorption of the enzyme on hydrophobic surfaces. Consequently, we have truncated Cbotu_EstA for the 71 N-terminal amino acids yielding del71Cbotu_EstA and characterized the truncated enzyme regarding activity towards PET which was only poorly hydrolyzed by the native esterase (Biundo, Ribitsch, et al., 2017). Indeed, an impressive improvement of more than eight times higher yield in total release-products was measured with del71Cbotu_EstA on PET films, compared to that obtained by the wild-type enzyme. Measurement of chemiluminescence of the esterases on PET confirmed our hypothesis that truncation of the cap domain improved adsorption of del71Cbotu_EstA on PET.

Following the investigation of the cap domain, we focused on the elucidation of the function of the zinc-binding site with respect to polyester hydrolysis (Biundo et al., 2017). First, the four zinc-coordinating amino acids were exchanged by side chains without the native functional group, obtaining mutants D130L, H150F, H156F, and D302L. All mutations resulted in a significant loss of stability above 50 °C which clearly proved the significance of the domain for the thermostability of the entire protein. Additionally, residues with a distance cut-off of 4.5 Å from the zinc ion were subjected to mutagenesis in order to evaluate their influence on polymer hydrolysis. Out of a set of mutations, W129A and S199A showed eightfold increased activity on PBAT compared to the native enzyme. Both residues located at the entrance of the zinc-cavity were changed to Alanine for more hydrophobic residues.

5.1 Materials and reagents

- QuickChange kit (Stratagene)
- Vector pET26b(+)
- Restriction enzymes *Nde*I and *Hind*III
- Ligation Kit Mighty mix (Takara Bio Inc., Japan)
- PureYield™ Plasmid Miniprep System (Promega GmbH, Germany)

5.2 Strains

- *E. coli* XL-10
- *E. coli* BL21-Gold(DE3)

5.3 Equipment

- Gene Amp PCR 2200 thermocycler (Applied Biosystems, USA).
- Electroporator MicroPulser™ (Bio-Rad Laboratories, Inc., Hercules, USA).

5.4 Protocols

1. Carry out the site-directed mutagenesis of the gene of Cbotu_EstA using the QuickChange kit from Stratagene with pET26b(+)_Cbotu_EstA as the template (Perz et al., 2016) and megaprimers carrying the specific mutation (Table 5). Conditions for the PCR are given in Table 6.
2. Transform amplified PCR-Product into chemically competent *E. coli* XL-10 cells.
3. Verify sequences by commercial sequencing (LGC Genomics, Germany).
4. Clone PCR products into the pET26b(+) vector for further expression.
5. Transform plasmids into *E. coli* BL21-Gold(DE3).
6. For N-terminal truncation of Cbotu_EstA, synthesize the gene coding for truncated esterase del71Cbotu_EstA by commercial service.
7. Digest separately the vector pET26b(+) and the gene del71Cbotu_EstA using the restriction enzymes *NdeI* and *HindIII* following the manufacturer's instructions (New England Biolabs, USA).
8. Perform agarose gel electrophoresis with 0.8% agarose to recover digested DNA.
9. Purify DNA using a QIAQuick Gel Extraction Kit.
10. Clone the digested gene into the linearized vector using the DNA Ligation kit according to the manufacturer's protocol.
11. Transform the product into competent *E. coli* XL1 cells.
12. Screen for positive clones on LB-agar plates containing 40 μg mL-1 kanamycin.
13. Extract plasmid DNA from the positive clones by the PureYield™ Plasmid Miniprep System.
14. Sequence genes by commercial service.

Table 5 Primer for site-directed mutagenesis of Cbotu_EstA.

Primer name	Nucleotide sequences
Cbotu_EstA_D130L_F	5′-ACC GTT GGT CCG GTT AGC AGC AAT TGG CTG CGT GCA TGT GAA CTG ATG CAT ATA TTG T-3′
Cbotu_EstA_D130L_R	5′-ACA ATA TAT GCA TAC AGT TCA CAT GCA CGC AGC CAA TTG CTG CTA ACC GGA CCA ACG GT-3′
Cbotu_EstA_H150F_F	5′-GAT TAT GGT GAA GCA TTT GCC AAA AAA TTC-3′
Cbotu_EstA_H150F_R	5′-GAA TTT TTT GGC AAA TGC TTC ACC ATA ATC-3′
Cbotu_EstA_H156F_F	5′-CCA AAA AAT TCA AAT TTA ATC GTT ACG GTC G-3′
Cbotu_EstA_H156F_R	5′-CGA CCG TAA CGA TTA AAT TTG AAT TTT TTG G-3′
Cbotu_EstA_D302L_F	5′-ACA GCA CCA AAC TGA TTG CAA CCT ATG-3′
Cbotu_EstA_D302L_R	5′-CAT AGG TTG CAA TCA GTT TGG TGC TGT-3′
Cbotu_EstA_S127A_F	5′-CCT ATA CCG CAA CCG TTG GTC CGG TTA GCG CCA ATT GGG ATC GTG CAT GTG AAC TGT ATG CA-3′
Cbotu_EstA_S127A_R	5′-TGC ATA CAG TTC ACA TGC ACG ATC CCA ATT GGC GCT AAC CGG ACC AAC GGT TGC GGT ATA GG-3′
Cbotu_EstA_W129A_F	5′-CGC AAC CGT TGG TCC GGT TAG CAG CAA TGC GGA TCG TGC ATG TGA ACT GTA TGC ATA TAT TGT TGG T-3′
Cbotu_EstA_W129A_R	5′-ACC AAC AAT ATA TGC ATA CAG TTC ACA TGC ACG ATC CGC ATT GCT GCT AAC CGG ACC AAC GGT TGC G-3′
Cbotu_EstA_F154Y_F	5′-CAC CGT TGA TTA TGG TGA AGC ACA TGC CAA AAA ATA CAA ACA TAA TCG TTA CGG TCG TAC CTA TCC-3′
Cbotu_EstA_F154Y_R	5′-GGA TAG GTA CGA CCG TAA CGA TTA TGT TTG TAT TTT TTG GCA TGT GCT TCA CCA TAA TCA ACG GTG-3′

Continued

Table 5 Primer for site-directed mutagenesis of Cbotu_EstA.—cont'd

Primer name	Nucleotide sequences
Cbotu_EstA_W274H_F	5′-AGC AGC ATC TAT GAT CTG AAA CTG GAT CAG CAC GGT CTG AAA AAA CAG AAT GGT GAA AGC-3′
Cbotu_EstA_W274H_R	5′-GCT TTC ACC ATT CTG TTT TTT CAG ACC GTG CTG ATC CAG TTT CAG ATC ATA GAT GCT GCT-3′
Cbotu_EstA_S199A_F	5′-ATT CGT ACC CTG ACC CAG CTG CTG AGC GAA GGT GCC GAA GAA GAA ATT AAC TGC GGT CAA GAA AAC AT-3′
Cbotu_EstA_S199A_R	5′-ATG TTT TCT TGA CCG CAG TTA ATT TCT TCT TCG GCA CCT TCG CTC AGC AGC TGG GTC AGG GTA CGA AT-3′

Table 6 Thermal cycling parameters for site-directed mutagenesis.

Initial denaturation	95 °C	1 min	
Denaturation	95 °C	50 s	
Annealing	58 °C	30 s	4 × // 30 ×
Extension	72 °C	9 min	
Final extension	72 °C	5 min	
	4 °C	∞	

6. Summary

In recent years, several microorganisms and their enzymes have been identified with ability to hydrolyze polymers. Beside cutinases, which are primarily involved in nature in the composting of organic matter, hydrolases from typical microorganisms living in soil, aquatic systems or under anaerobic conditions were investigated in respect to their ability to degrade polyesters with different chemical structures. However, since enzymes are built from nature for totally different molecules, their activity on the synthetic substrates were very low. Hence, there is a strong need to improve the polymer degrading enzymes by protein engineering. Initial studies have shown that the interaction between the enzyme and the substrate plays a crucial role and hence, improvement of adsorption/desorption of the enzymes on synthetic polyesters by genetic engineering was performed. Application of different approaches of rational design based on their resolved 3D structures and structure models succeeded in multiple increase of the hydrolytic activity on the recalcitrant PET and novel biodegradable polyesters. Both the mutagenesis of the enzyme surface and the fusion of the catalytic domain with "anchor domains" facilitated better adsorption of the enzyme on the polymer. However, it always has to be taken into account that the enzyme is able to bind in the correct orientation onto the bulky substrate which requires that a certain mobility of the enzyme must be retained. It can be expected that new developments in the field of protein engineering will also improve the interaction between enzyme and substrate in the future.

References

Armenta, S., Moreno-Mendieta, S., Sánchez-Cuapio, Z., Sánchez, S., & Rodríguez-Sanoja, R. (2017). Advances in molecular engineering of carbohydrate-binding modules. *Proteins: Structure, Function, and Bioinformatics*, *85*(9), 1602–1617.

Arnold, F. H. (1998). Design by directed evolution. *Accounts of Chemical Research*, *31*(3), 125–131.

Arnold, F. H. (2018). Directed evolution: Bringing new chemistry to life. *Angewandte Chemie - International Edition*, *57*(16), 4143–4148.

Biundo, A., Ribitsch, D., Steinkellner, G., Gruber, K., & Guebitz, G. M. (2017). Polyester hydrolysis is enhanced by a truncated esterase: Less is more. *Biotechnology Journal*, *12*(8).

Biundo, A., Steinkellner, G., Gruber, K., Spreitzhofer, T., Ribitsch, D., & Guebitz, G. M. (2017). Engineering of the zinc-binding domain of an esterase from Clostridium botulinum towards increased activity on polyesters. *Catalysis Science and Technology*, 7(6), 1440–1447.

Bornscheuer, U. T., & Pohl, M. (2001). Improved biocatalysts by directed evolution and rational protein design. *Current Opinion in Chemical Biology*, *5*(2), 137–143.

Chen, S., Su, L., Chen, J., & Wu, J. (2013). Cutinase: Characteristics, preparation, and application. *Biotechnology Advances*, *31*(8), 1754–1767.

Danso, D., Chow, J., & Streita, W. R. (2019). Plastics: Environmental and biotechnological perspectives on microbial degradation. *Applied and Environmental Microbiology*, *85*(19), 1–14.

Espino-Rammer, L., Ribitsch, D., Przylucka, A., Marold, A., Greimel, K. J., Acero, E. H., et al. (2013). Two novel class ii hydrophobins from Trichoderma spp. stimulate enzymatic hydrolysis of poly(ethylene terephthalate) when expressed as fusion proteins. *Applied and Environmental Microbiology*, *79*(14), 4230–4238.

Giovambattista, N., Debenedetti, P. G., & Rossky, P. J. (2007). Effect of surface polarity on water contact angle and interfacial hydration structure. *Journal of Physical Chemistry B*, *111*(32), 9581–9587.

Greimel, K. J., Perz, V., Koren, K., Feola, R., Temel, A., Sohar, C., et al. (2013). Banning toxic heavy-metal catalysts from paints: Enzymatic cross-linking of alkyd resins. *Green Chemistry*, *15*(2), 381–388.

Herrero Acero, E., Ribitsch, D., Dellacher, A., Zitzenbacher, S., Marold, A., Steinkellner, G., et al. (2013). Surface engineering of a cutinase from Thermobifida cellulosilytica for improved polyester hydrolysis. *Biotechnology and Bioengineering*, *110*(10), 2581–2590.

Herrero Acero, E., Ribitsch, D., Steinkellner, G., Gruber, K., Greimel, K., Eiteljoerg, I., et al. (2011). Enzymatic surface hydrolysis of PET: Effect of structural diversity on kinetic properties of cutinases from Thermobifida. *Macromolecules*, *44*(12), 4632–4640.

Jeong, S. T., Kim, H. K., Kim, S. J., Chi, S. W., Pan, J. G., Oh, T. K., et al. (2002). Novel zinc-binding center and a temperature switch in the Bacillus stearothermophilus L1 lipase. *Journal of Biological Chemistry*, *277*(19), 17041–17047.

Kontkanen, H., Westerholm-Parvinen, A., Saloheimo, M., Bailey, M., Rättö, M., Mattila, I., et al. (2009). Novel Coprinopsis cinerea polyesterase that hydrolyzes cutin and suberin. *Applied and Environmental Microbiology*, *75*(7), 2148–2157.

Müller, R.-J., Schrader, H., Profe, J., Dresler, K., & Deckwer, W.-D. (2005). Enzymatic degradation of poly(ethylene terephthalate): Rapid hydrolyse using a hydrolase from *T. fusca*. *Macromolecular Rapid Communications*, *26*(17), 1400–1405.

Nikolaivits, E., Kanelli, M., Dimarogona, M., & Topakas, E. (2018). A middle-aged enzyme still in its prime: Recent advances in the field of cutinases. *Catalysts*, *8*(12), 612.

Pail, M., Peterbauer, T., Seiboth, B., Hametner, C., Druzhinina, I., & Kubicek, C. P. (2004). The metabolic role and evolution of L-arabinitol 4-dehydrogenase of *Hypocrea jecorina*. *European Journal of Biochemistry*, *271*(10), 1864–1872.

Papadopoulou, A., Hecht, K., & Buller, R. (2019). Enzymatic PET degradation. *Chimia*, *73*(9), 743–749.

Perz, V., Baumschlager, A., Bleymaier, K., Zitzenbacher, S., Hromic, A., et al. (2016). Hydrolysis of synthetic polyesters by *Clostridium botulinum* esterases. *Biotechnology and Bioengineering*, *113*(5), 1024–1034.

Perz, V., Zumstein, M. T., Sander, M., Zitzenbacher, S., Ribitsch, D., et al. (2015). Biomimetic approach to enhance enzymatic hydrolysis of the synthetic polyester poly (1,4-butylene adipate): Fusing binding modules to esterases. *Biomacromolecules*, *16*(12), 3889–3896.

Petersen, S. B., Harald Jonson, P., Fojan, P., Petersen, E. I., Neves Petersen, M. T., Hansen, S., et al. (1998). Protein engineering the surface of enzymes. *Journal of Biotechnology*, *66*(1), 11–26.

Quartinello, F., Guebitz, G. M., & Ribitsch, D. (2019). Surface functionalization of polyester. *Methods in Enzymology*, *627*, 339–360.

Quin, M. B., & Schmidt-Dannert, C. (2011). Engineering of biocatalysts: From evolution to creation. *ACS Catalysis*, *1*(9), 1017–1021.

Ren, Q., Kwan, A. H., & Sunde, M. (2013). Two forms and two faces, multiple states and multiple uses: Properties and applications of the self-assembling fungal hydrophobins. *Biopolymers*, *100*(6), 601–612.

Ribitsch, D., Acero, E. H., Przylucka, A., Zitzenbacher, S., Marold, A., Gamerith, C., et al. (2015). Enhanced cutinase-catalyzed hydrolysis of polyethylene terephthalate by covalent fusion to hydrophobins. *Applied and Environmental Microbiology*, *81*(11), 3586–3592.

Ribitsch, D., Yebra, A. O., Zitzenbacher, S., Wu, J., Nowitsch, S., Steinkellner, G., et al. (2013). Fusion of binding domains to *Thermobifida cellulosilytica* cutinase to tune sorption characteristics and enhancing PET hydrolysis. *Biomacromolecules*, *14*(6), 1769–1776.

Rieckmann, T., & Völker, S. (2004). Poly(ethylene terephthalate) polymerization–Mechanism, catalysis, kinetics, mass transfer and reactor design. In *Modern polyesters: Chemistry and technology of polyesters and copolyesters* (pp. 29–115). John Wiley & Sons.

Sagong, H. Y., Seo, H., Kim, T., Son, H. F., Joo, S., Lee, S. H., et al. (2020). Decomposition of the PET film by MHETase using Exo-PETase function. *ACS Catalysis*, *10*(8), 4805–4812.

Salvador, M., Abdulmutalib, U., Gonzalez, J., Kim, J., Smith, A. A., Faulon, J. L., et al. (2019). Microbial genes for a circular and sustainable bio-PET economy. *Genes*, *10*(5).

Schöber, U., Thiel, C., & Jendrossek, D. (2000). Poly(3-hydroxyvalerate) depolymerase of *Pseudomonas lemoignei*. *Applied and Environmental Microbiology*, *66*(4), 1385–1392.

Shoseyov, O., Shani, Z., & Levy, I. (2006). Carbohydrate binding modules: Biochemical properties and novel applications. *Microbiology and Molecular Biology Reviews*, *70*(2), 283–295.
Steiner, K., & Schwab, H. (2012). Recent advances in rational approaches for enzyme engineering. *Computational and Structural Biotechnology Journal*, *2*(3).
Vecchiato, S., Ahrens, J., Pellis, A., Scaini, D., Mueller, B., Herrero Acero, E., et al. (2017). Enzymatic functionalization of HMLS-polyethylene terephthalate fabrics improves the adhesion to rubber. *ACS Sustainable Chemistry and Engineering*, *5*(8), 6456–6465.
Wei, R., & Zimmermann, W. (2017). Microbial enzymes for the recycling of recalcitrant petroleum-based plastics: How far are we? *Microbial Biotechnology*, *10*(6), 1308–1322. John Wiley and Sons Ltd. https://doi.org/10.1111/1751-7915.12710.
Weinberger, S., Canadell, J., Quartinello, F., Yeniad, B., Arias, A., Pellis, A., et al. (2017). Enzymatic degradation of poly(ethylene 2,5-furanoate) powders and amorphous films. *Catalysts*, 7(11), 318.
Wilding, M., Scott, C., & Warden, A. C. (2018). Computer-guided surface engineering for enzyme improvement. *Scientific Reports*, *8*(1), 1–7.

CHAPTER FIFTEEN

Characterization of the enzymatic degradation of polyurethanes

Audrey Magnin, Eric Pollet, and Luc Avérous*
BioTeam/ICPEES-ECPM, UMR CNRS 7515, Université de Strasbourg, Strasbourg Cedex 2, France
*Corresponding author: e-mail address: luc.averous@unistra.fr

Contents

Abstract

For decades, polyurethanes (PUR) have mainly been synthesized for long-term applications and are therefore highly persistent in the environment. Proper waste disposal approaches, including recycling techniques, must be developed to limit the accumulation of PUR in the environment. Evaluation of enzymatic polyurethane degradation is needed for the development of enzymatic recycling. A series of techniques has been carefully implemented to monitor the biotic and abiotic degradation of PUR. Both the degraded polymer and the degradation products are analyzed to obtain a complete overview of the degradation.

Methods in Enzymology, Volume 648
ISSN 0076-6879
https://doi.org/10.1016/bs.mie.2020.12.011

1. Introduction

Polyurethanes (PUR) are a family of versatile polymers mainly tailored for long-term applications. Consequently, PUR are designed to be highly resistant to degradation in different environments. Then, if PUR waste is not properly disposed, persistent and potentially toxic compounds can be released in the environment, and then contribute to global plastic pollution (Min, Cuiffi, & Mathers, 2020; Turner & Lau, 2016). Consequently, large amounts of microplastics are formed by fragmentation after initial abiotic degradation steps (UV, temperature, oxidation, etc.). In the last decades, great efforts have been made to decrease the environmental impact of PUR. The development of efficient recycling techniques is particularly challenging, considering the versatility of PU material. Recycling paths involving efficient enzymes are currently being studied for polyethylene terephthalate (PET) (Blank, Narancic, Mampel, Tiso, & O'Connor, 2019; Tournier et al., 2020; Yoshida et al., 2016). The applicability of this sustainable recycling process is currently evaluated for PUR. Characterization of PUR enzymatic degradation is thus a critical step for this development.

1.1 PU structures and applications

PUR represent the 6th most produced synthetic polymer (Furtwengler, Perrin, Redl, & Avérous, 2017) with a production of 17 million tons in 2017, which is expected to reach 21 million tons in 2022 (Garside, 2019). The range of applications of PU is broad. The main PU markets are furniture/mattresses (30%), automotive (15%), construction (13%), technical insulation (10%) and shoes (3%) (Ionescu, 2005).

The urethane bond is the common resistant chemical group of all polyurethanes. Although different chemical pathways for synthesis can be used, this linkage is mainly obtained by polyaddition between isocyanate and hydroxyl groups. Aromatic isocyanates such as 4,4′-methylenediphenyl diisocyanate (MDI) or toluene diisocyanate (TDI) are most commonly used. Aliphatic isocyanates such as hexane diisocyanate (HDI), 4,4′-methylene-bis(cyclohexylisocyanate) (HMDI) and isophorone diisocyanate can also be used for the synthesis of PUR. The main long polyols are polyesters or polyethers. More rarely, polycarbonate-based polyols can be also used (Fig. 1).

Thermoplastics can be dissolved in organic solvents and melt at high temperature. For the synthesis of thermoplastic PUR, building blocks with a functionality of 2 ($f=2$) are involved (diisocyanates and diols) to obtain a

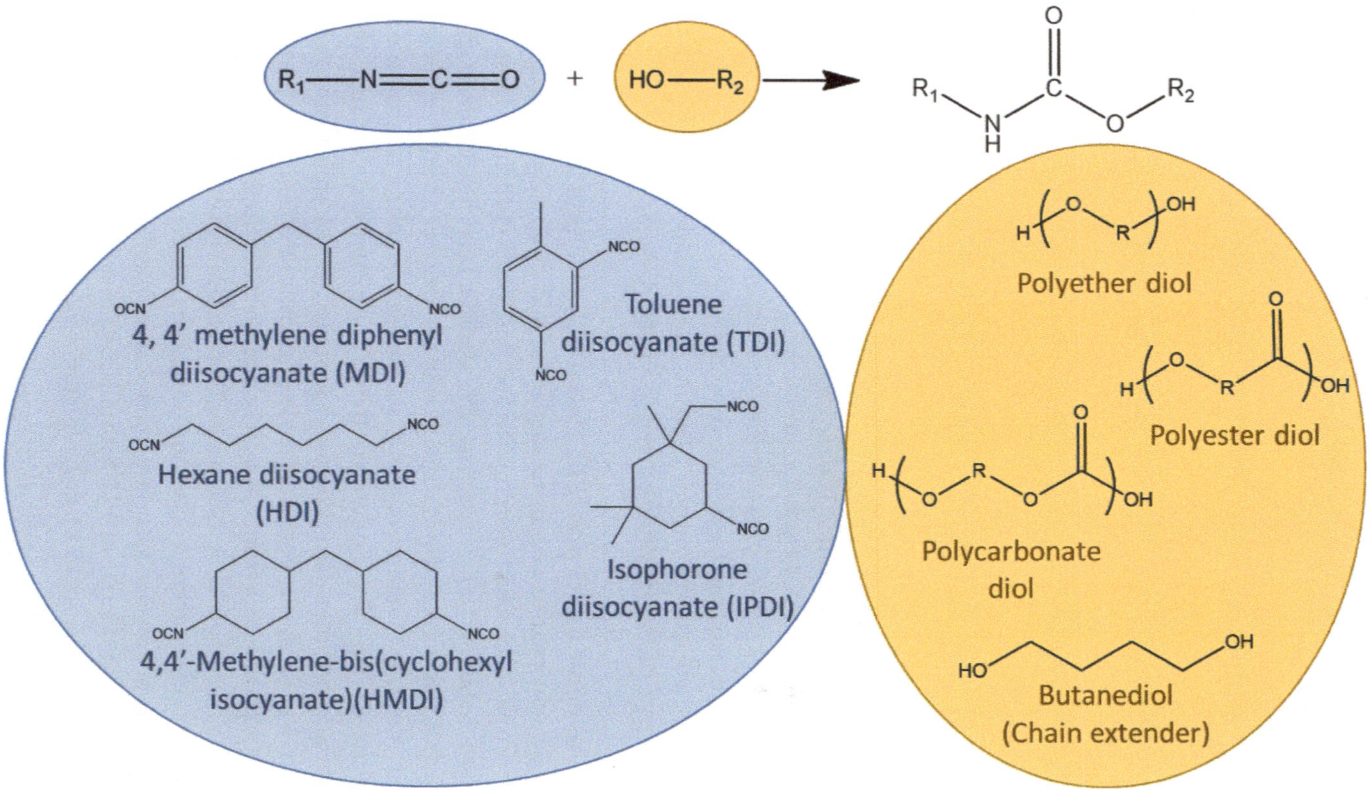

Fig. 1 Synthesis and main building blocks of polyurethanes.

Fig. 2 Different schematic polymeric structures of polyurethanes.

linear or slightly branched structure (Fig. 2). In a first step, the diisocyanate and the long diol are reacted to obtain a prepolymer with NCO ending chain. Then, in a second step, a short-chain diol, such as 1–4-butanediol is added as chain extender to obtain PU of high molar mass. In a one-step process, polyol and short-chain diol can be added simultaneously under equimolar conditions. Rarely, the prepolymer isocyanates can react with a diamine as chain-extender, forming urea moieties (Tang, Santerre, Labow, & Taylor, 1997). It is commonly accepted that thermoplastic PUR shows highly ordered rigid segments, consisting of the isocyanate and the chain extender part and amorphous soft segments, corresponding to the polyester or polyether part of the polymer backbone (Tang, Labow, & Santerre, 2001). Thermoplastic PUR (TPU) are, for example, elastomers convenient for sportswear, shoes or medical devices. PUR intended for biomedical applications are frequently subjected to enzymatic degradation in contact with living tissue, in order to evaluate their biocompatibility and bioresorbability (Christenson, Anderson, & Hiltner, 2007; Santerre, Woodhouse, Laroche, & Labow, 2005).

Thermosets cannot dissolve in solvents and degrade at high temperature without melting. Thermoset materials have crosslinked macromolecular architectures (Fig. 2). To obtain these 3D networks, building blocks containing more than two functional moieties ($f > 2$) are involved in the synthesis. PU foam is a typical example of thermoset PU. Foams have alveolar structures resulting from physical or chemical blowing generated by carbon dioxide or isopentane. Isocyanurate or urea moieties are present in PU foam due to side-reactions occurring during the synthesis. Additives such as antioxidants, processing aids, anti-microbial compounds or fire retardants are generally considered in PU formulation. Foams are divided into two categories, rigid foams and flexible foams, which are based on closed and open cells, respectively. Rigid foams represent 36% of the global PU production. They are mainly produced for building insulation, for energy saving in constructions. Representing 32% of the global PU production, flexible foams are used for cushioning, furniture, or automotive seats (Magnin, Pollet, Phalip, & Avérous, 2019).

Since they are commonly used as model substrates, when focusing on PU biological degradation, it is necessary to also introduce polyurethane dispersions, particularly waterborne polyurethane dispersions (WPUD). In WPUD, polymer particles of nanometric size are surrounded by hydrophilic moieties to obtain homogeneous dispersions in aqueous media (Fig. 2). PU dispersions are mainly used to form coatings that provide resistance to abrasion, weathering, or corrosion.

1.2 Mechanism of enzymatic PUR degradation

The chemical compositions and macromolecular architectures previously described directly impact the mechanism and kinetics of enzymatic degradation of PUR materials.

Polymer biodegradation is a complex process involving different steps. Abiotic (UV attack, abrasion, etc.) and biotic factors (biofilm formation) will together contribute to the fragmentation of the material into small pieces to form micro- or nano-plastics (Shim, Song, Hong, & Jang, 2016). Then, microorganisms will produce an enzymatic arsenal to depolymerize the polymer. Monomers (or small building blocks) resulting from this step will be assimilated and mineralized by microorganisms, thus contributing to microbial growth (Lucas et al., 2008). In the literature, the biological degradation of PUR is described with different biological entities. Microbial communities such as those from soil or compost consist of several microorganisms, which act symbiotically. Single strain microorganisms such as bacteria or fungi are also able to degrade PUR. For all these types of degradation, the functional entity is the enzymes. Enzymatic degradation is the most desired path for polymer recycling (Wei & Zimmermann, 2017; Wierckx et al., 2015). Indeed, microbial degradation involves assimilation and mineralization of the degradation products. By contrast, these products remain accessible after enzymatic treatment and can be valorized into second generation products.

Two classes of enzymes are largely described in the literature for PUR degradation: oxidoreductases and hydrolases. However, enzymatic oxidation of PU has only been described once. In this study, a laccase and a horseradish peroxidase displayed some degradation activity on a thermoplastic poly(ester-urethane) (Ignat, Ignat, Ciobanu, Doroftei, & Popa, 2011).

Different mechanisms of enzymatic hydrolysis are involved in PU degradation. A majority of publications describes the hydrolysis of ester bonds (of the polyester polyol) in poly(ester-urethane). This reaction leads to the release of compounds with hydroxyl and carboxylic acid groups (Fig. 3A). Within the esterases class, lipases are well described as well as cutinases (Biffinger et al., 2015; Schmidt et al., 2017).

Urea bonds are present in the PUR synthesized using diamines as chain extenders and in PU foams. Ureases are enzymes that specifically recognize urea moieties. Phua et al., described for instance the degradation of a poly(ether-urethane) with a urease (Phua, Castillo, Anderson, & Hiltner, 1987) (Fig. 3B).

The hydrolysis of the urethane bond is mainly described with amidases and proteases. Indeed, amide or peptide linkages are chemically similar to urethane bonds. Molecules resulting from urethane bond cleavage are alcohols, amines and carbon dioxide (Fig. 3C). Rarely, esterases were described for the hydrolysis of the urethane linkage (Santerre, Labow, Duguay, Erfle, & Adams, 1994; Smith, Williams, & Oliver, 1987) (Fig. 3D). These studies dealt with poly(ether-urethane). Consequently, the hydrolysis can only occur on the urethane moieties. In this report, enzymes able to hydrolyze the urethane linkage will be called urethanases, independent of the enzyme classification which they belong to (Mohapatra & Bapuji, 1997).

2. Materials, equipment and reagents

2.1 Equipment

- Plate press (Labtech Hot Press)
- Standard analogue shaker (VWR)
- Incubator BD56 (Binder GmbH)
- Scale (Sartorius analytical)
- Sputter coater Quorum Q150R S (Quorum Technologies)
- Scanning electron microscope VEGA3 (Tescan)
- Size exclusion chromatography Acquity-APC (Waters)
- Fourier transform infrared spectrometer Nicolet 380 (Thermo Scientific)
- Thermogravimetric analysis Hi-Res TGA Q5000 (TA instruments)
- Differential scanning calorimetry Q200 calorimeter (TA instruments)
- Nuclear magnetic resonance UltraShield 400 MHz (Bruker)
- Liquid chromatography/mass spectrometer Acquity UPLC Synapt G2-Si HDMS (Waters)

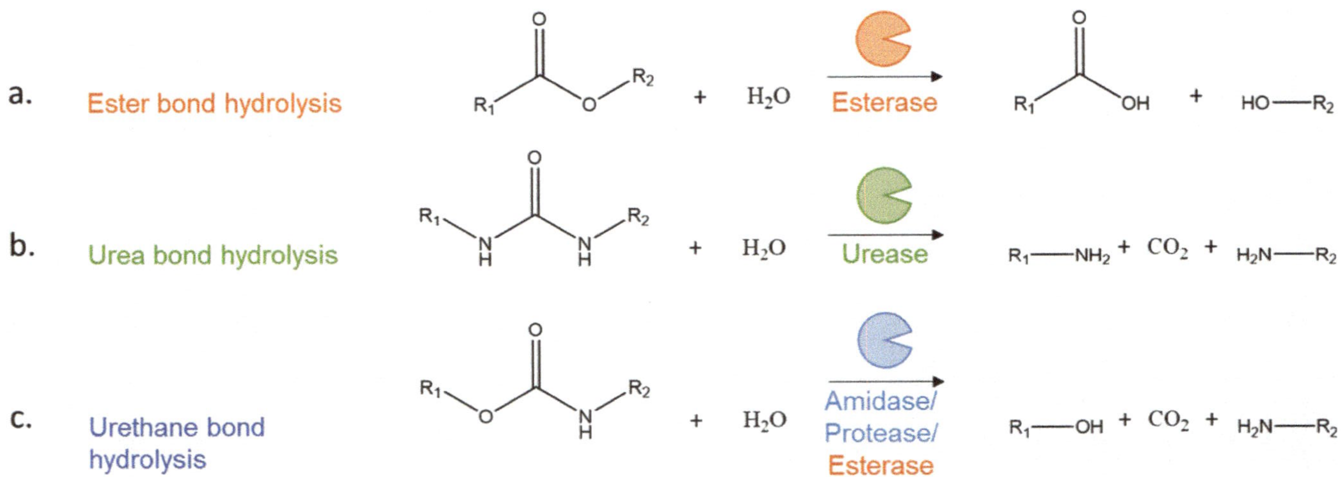

Fig. 3 Different mechanisms of the enzymatic PUR hydrolysis. *Adapted from Magnin, A., Pollet, E., Phalip, V., & Avérous, L. (2019). Evaluation of biological degradation of polyurethanes.* Biotechnology Advances, *107457.*

2.2 Materials

- Square teflon mold of 12 × 12 × 1 cm
- 500 mL three-necked round-bottom flask
- 4 mL-glass vials
- 1.5 mL-glass vials for chromatography
- PTFE filter of 0.2 μm
- C18 Kinetex EVO core-shell column (ID 3.0 mm × 150 mm, 4.6 mm, 100 Å, Phenomenex)
- Acquity APC XT 4.6 × 150 mm columns (450 Å/2.5 μm, 200 Å/2.5 μm and 45 Å/1.7 μm, Waters)

2.3 Reagents

When no further information is provided, reagents and solvents were purchased from Sigma and used as received.

- 4,4′-methylene diphenyl diisocyanate (4,4'-MDI) (Suprasec 1306—M_w = 250.25 g/mol, Huntsman)
- polycaprolactone (PCL, M_n = 2000)
- 1,4-butanediol (1,4-BDO, Invista)
- Enzymatic solution
- Phosphate buffer (pH 7.0, 100 mM)
- Distilled water
- Ethanol
- Methanol
- Tetrahydrofuran
- Deuterated chloroform ($CDCl_3$)
- Acetonitrile
- Formic acid

3. Protocol

The protocol below is adapted and modified from the publication (Magnin et al., 2019) in which the enzymatic degradation of a thermoplastic poly(ester-urethane) synthesized from PCL, 1,4-BDO and 4,4'-MDI was described. However, it is possible to change these building blocks to test enzymatic activity on other types of material like poly(ether-urethane). Control of the polymer synthesis instead of working directly on PUR products is beneficial to simplify the analysis. Because of the low volumes required (1 mL reactions) and the small amount of polymer needed, this protocol presents a relatively high-throughput scale, ideal for performing several repetitions, determining optimal conditions or screening enzymes. This protocol is a four-step procedure presenting polymer synthesis, degradation, analysis and finally degradation products analysis.

3.1 Polymer synthesis

The TPU synthesis followed a one-step bulk process for a batch of 50 g, without catalyst, under stoichiometric conditions (NCO/OH ratio of 1). Because of the one-step bulk process, the polyol and the chain extender are added in the same time for the reaction.

- PCL diol and 1,4-BDO are pre-heated at 80 °C in an oven prior to use.
- In a previously flame-dried, 250 mL three-necked round-bottom flask, the 4,4'-MDI (9.5 g, 1 M equivalent) is first added in an oil bath at 80 °C under mechanical stirring and constant nitrogen flow.
- The proper amounts of 1,4-BDO (1.7 g, 0.5 M equivalent), directly followed by PCL diol (38.8 g, 0.5 M equivalent) are added.
- After 2 h of reaction, the reaction mixture is poured in a Teflon mold and let to cure overnight in an oven at 75 °C.
- Finally, the resulting TPU is cut into pieces and compression-molded at 145 °C (Labtech Hot Press) to obtain 0.3 mm thin films.
- Films obtained were cut into small pieces of 50–80 mg (about $0.5 \times 0.5\,cm^2$).

3.2 Polymer degradation

Degradation studies are performed with, at least, three repetitions. Negative controls without enzymes must be performed.

- Polymer pieces are weighed and placed in 4 mL glass vials.
- 1 mL of liquid medium is added to the vials. This volume contains the enzymatic solution and the phosphate buffer (pH 7.0, 100 mM).
- 37 °C and rotary shaking at 150 rpm for 48 h to 72 h.
- The liquid fraction is removed and stored at −20 °C for analysis.
- Polymer pieces are then washed twice with water and once with ethanol. The ethanol washing fraction can be stored at −20 °C as well, as it might contain degradation products that have not been solubilized in water.
- Polymer pieces are dried over night at room temperature and weighed.
- If no degradation is detected, the procedure can be repeated several times on the same polymer to increase degradation and be able to detect it by the techniques described below.

After degradation, two fractions can be analyzed: (i) the solid fraction corresponding to the residual material of polymer and (ii) the liquid fraction corresponding to degradation products that were solubilized in water or ethanol. A set of analytical procedures for both is presented in the following sections.

3.3 Polymer analysis

3.3.1 Weight loss measurement

The weight loss measurement indicates the proportion of material that was degraded and solubilized in water or ethanol compared to the residual material of polymer. It is calculated according to the following formula:

$$\text{Weight loss }(\%) = \frac{\text{Initial mass} - \text{final mass}}{\text{Initial mass}}$$

3.3.2 Scanning electron microscopy (SEM)

Changes in polymer surfaces can often be observed visually. Observation can be deepened using scanning electron microscopy (SEM). This technique allows evaluation of the evolution of the surface morphology of the degraded PU material.

- Before examination, samples are coated with a thin layer of gold using a sputter coater (Quorum Q150R S, Quorum Technologies).
- A VEGA3 (Tescan) scanning electron microscope in high vacuum mode, with an acceleration voltage of 5 kV and working distances in the range of 6–8 mm, is used.

3.3.3 Size exclusion chromatography (SEC)

Size Exclusion Chromatography (SEC) is a powerful tool to evaluate the average polymer chain length of thermoplastics.

- TPU are dissolved in tetrahydrofuran (THF) at a concentration of 2 mg/mL and filtered with 0.2 μm PTFE filters directly into vials.
- 10 μL of dissolved polymer solution are injected in the Acquity-APC (Waters) in THF at a flow rate of 0.6 mL/min. SEC is equipped with three columns maintained at 40 °C (Acquity APC XT 4.6 × 150 mm: 450 Å/2.5 μm, 200 Å/2.5 μm and 45 Å/1.7 mm, Waters).
- Detection is performed with PDA-UV detector at 254 nm. If the polymer does not contain aromatic moieties, detection can be performed with a RI detector at 35 °C.

- A calibration curve using linear polystyrene (PS) standards was built and is used for molar mass determination.

3.3.4 Fourier-transform infrared spectroscopy (FTIR)

This surface analysis can highlight changes in chemical moieties and allows mechanisms of degradation to be suggested.

- Fourier-transform infrared spectrometer Nicolet 380 from Thermo Scientific (US) is used in reflection mode and is equipped with an ATR diamond module (FTIR-ATR).
- An atmospheric background is collected before each sample analysis (32 scans, resolution $4\,cm^{-1}$).

3.3.5 Thermal analysis

Thermogravimetric analysis (TGA) aims at analyzing the thermal stability and degradation of a polymer.

- Between 1 and 3 mg of sample is weighed.
- TGA was performed by using a TA Instrument Hi-Res TGA Q5000 device at a heating rate of 10 °C/min from RT to 700 °C under a nitrogen atmosphere (flow rate 25 mL/min).

Differential scanning calorimetry provides, among others, information about the polymer crystallinity. This parameter is important since it might evolve during enzymatic degradation.

- Samples of 1–3 mg are sealed in standard aluminum pans.
- DSC is performed by using a TA Instrument Q200 calorimeter under a nitrogen flow (50 mL/min) on a temperature range from −50 to 450 °C at a heating/cooling rate of 10 °C/min.

3.4 Degradation product analysis

Analysis of degradation products is the key for understanding the degradation mechanism. Likewise, identification of these products is the first step toward feedstock recovery. Two techniques are described below as examples. The ^{1}H NMR is suitable for analysis of degradation products at concentrations in the range of, at least, grams per liter. Liquid chromatography—mass spectrometry (LC-MS) is particularly convenient to detect low concentrations of degradation products.

3.4.1 ^{1}H nuclear magnetic resonance (^{1}H NMR)

Identification of the 6-hydroxycaproic acid monomer resulting from PCL segment hydrolysis has been performed using this technique.

- The liquid fraction collected after polymer degradation is dried overnight by lyophylization or under vacuum at 40 °C. The resulting products are solubilized in $CDCl_3$.
- Spectra are recorded using a Bruker UltraShield 400 MHz at room temperature. The relaxation delay is set to 10 s with 32 scans.
- Spectra are calibrated using the $CHCl_3$ peak at 7.26 ppm.

3.4.2 Liquid chromatography—Mass spectrometry (LC-MS)

The separation and identification of degradation products can be performed by LC-MS. For the analysis of diamines and derivatives resulting from the hydrolysis of the hard segment, a technique, involving liquid chromatography electrospray ionization quadrupole time-of-flight mass spectrometry (LC/ESI/qTOF-MS) is implemented. The analytical protocol is inspired by the publication of Gamerith et al. (2016).

- Samples of liquid fractions collected along the degradation experiments are pooled.
- An equivalent volume of methanol is added in each sample for protein precipitation. Filtration at 0.2 μm allows precipitated proteins to be removed.
- Reversed-phase chromatography using an ACQUITY UPLC system (Waters) is equipped with a C18 Kinetex EVO core-shell column (Phenomenex, ID 3.0 mm × 150 mm, 4.6 mm, 100 Å) for separation. A gradient elution system using a mobile phase (A) of 65 mM ammonium formate and mobile phase (B) of acetonitrile (containing 0.1% formic acid) with a flow rate of 0.75 mL/min. Run started with 5% acetonitrile, gradually increased to 10% in 1 min, to 50% in 8 min and to 95% in 10 min. Column was equilibrate 3 more minutes at 95% acetonitrile then returned to 5% acetonitrile in 5 min. The HPLC eluent is directly electrosprayed at an applied voltage of 3 kV, using a desolvation gas (N_2) flow of 600 L/h, a nebulizer gas pressure of 6.5 bar and source desolvation temperatures of 80 and 150 °C, respectively.
- The chromatographic system is coupled to a SYNAPT G2-Si mass spectrometer (Waters). The full MS survey scans are acquired at a 2000 (FWHM) resolving power, over the mass range of 20–2000 *m/z*. The method is set to analyze the most intense ions from the survey scan. Peaks are analyzed using Mass Lynx software (Waters).
- Using 4,4'-MDA as standard, concentrations as low as 0.1 mg/L can be quantified.

4. Safety considerations and standards

The most hazardous step of this procedure is the synthesis of polymer as it involves isocyanates. In the present example, 4,4'-MDI must be used carefully under a fume hood. Other isocyanates such as 2,4-toluene diisocyanate (2,4 TDI) are much more dangerous. 4,4' MDA standard solution must be prepared cautiously as this molecule is carcinogenic.

5. Analysis and statistics

The first analysis proposed in this procedure is the weight loss measurement. This indicator is very easy to implement and it provides a yield of degradation by considering both remaining polymer material and soluble building blocks released. However, low degradation performance might not lead to detectable weight loss.

It is recommended to perform complementary analyses such as scanning electron microscopy (SEM) or size exclusion chromatography (SEC) with the residual polymer materials, even when the polymers have not shown weight loss. Indeed, polymer chains may have been cut but remain in the solid fraction. After incubation, both the solid fraction that might have been altered by the enzymatic degradation and the liquid fraction can be analyzed.

For example, a yellowing of a poly(ester-urethane) was observed after 50 days of incubation with an esterase (Magnin, Pollet, Perrin, et al., 2019). As SEM is a qualitative analysis, only one sample of each condition is generally presented. We observed two poly(ester-urethane) pieces after degradation with a mixture of an esterase E3576 (provided by Protéus-Seqens) and an amidase E4143 (provided by Protéus-SeqensProtéus): a PCL-based PUR and a fatty acid dimer polyester-based PUR (Fig. 4). Two magnifications were displayed: the lowest (x50, top of the figure) in order to evaluate the homogeneity of the degradation on the surface and the highest (x150, bottom of the figure) to get more detailed information on the type of defect. This figure highlights the need of the global overview of the polymer surface. Indeed, for the fatty acid dimer polyester-based PU, the "defect" is located on a single place while cracks cover all the piece of PCL-based PU.

The evolution of the molar mass can be measured with SEC analysis. For this purpose, the average molar mass in number (M_n), average molar mass in mass (M_w) and the dispersity (Đ), corresponding to the M_w/M_n ratio, are

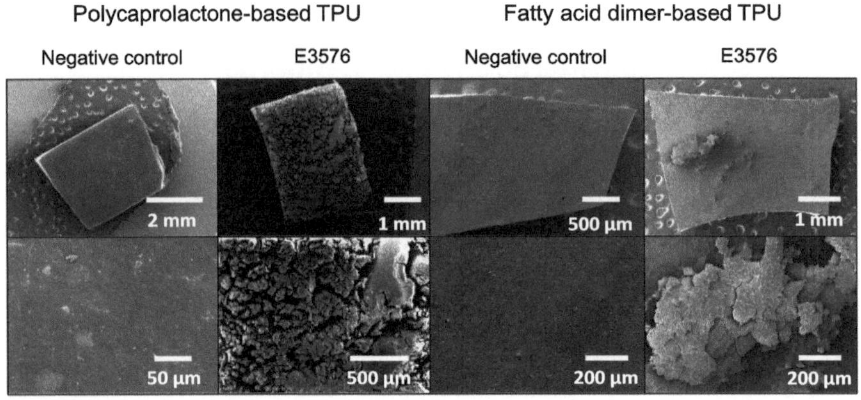

Fig. 4 SEM image capture of a PCL-based PUR and a fatty acid dimer polyester-based PUR incubated with a buffer (negative control) and a mixture of an esterase E3576 (Protéus-Seqens) and an amidase E4143 (Protéus-Seqens). Two magnifications are presented: x50 on the top and x150 on the bottom. *Adapted from Magnin, A., Pollet, E., Perrin, R., Ullmann, C., Persillon, C., Phalip, V., et al. (2019). Enzymatic recycling of thermoplastic polyurethanes: Synergistic effect of an esterase and an amidase and recovery of building blocks.* Waste Management, 85, *141–150.*

determined. Chromatograms also provide information on the degradation. For instance, peaks corresponding to low molar mass polymer chains may appear after degradation.

In the Fig. 5, elution profiles of a polyester TPU incubated with three enzymatic solutions and with a buffer (negative control) are displayed. TPU incubated with enzymatic solution containing the esterase E3576 (Protéus-Seqens) shows a bimodal profile with the main polymer chains and oligomers at low molar masses. Moreover, a shift of the molar mass distribution peak toward lower molar masses can be observed in comparison to the negative control. Finally, calculation of molar masses revealed a decrease of the M_n from 120,000 to 105,000 g/mol. All the parameters considered by SEC highlighted the effective degradation of the polymer by the enzymatic solutions containing the esterase E3576.

Even if it was following a fungal degradation, we performed FTIR analysis on a PCL-based PU. Polymer pieces of polyester urethane were incubated 2 months at 30 °C with strains of *Aspergillus* section *Flavi*, *Alternaria* sp. and *Penicillium* section *Lanata-divaricata*. Several differences were observed between the negative control and the polymer incubated with the fungi. The large signal around 3300 cm^{-1} corresponds to the –OH and –NH stretching. At 1700 cm^{-1} are the C=O of ester and at 1720 cm^{-1} the

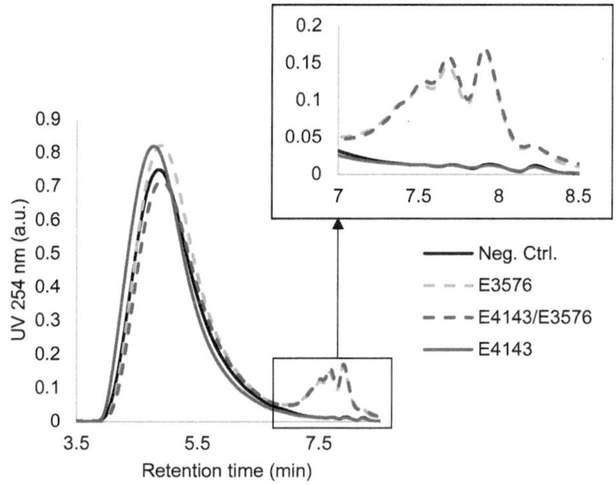

Fig. 5 Size exclusion chromatography elution profiles of a poly(ester-urethane) degraded by enzymes E3576 and E4143 (Magnin, Pollet, Perrin, et al., 2019).

C=O of carboxylic acids. Finally, the signal at $1530\,cm^{-1}$ corresponds to the urethane/amide NH bending vibration in the secondary amide. All peak heights are normalized according to the peak at $2860\,cm^{-1}$, corresponding to $-CH_2$, moieties that are not affected by the degradation. Hydrolysis of the ester bonds leads to the release of an alcohol and a carboxylic acid. This results in the decrease of the signal of ester bonds at $1700\,cm^{-1}$ and an increase of the signals at $3300\,cm^{-1}$ and the shoulder of the signal at $1730\,cm^{-1}$, confirming the appearance of alcohol and carboxylic acid moieties (Fig. 6).

Procedure described for TGA analysis can also be useful to detect and characterize biological degradation. For instance, Mathur et al. described the degradation of a poly(ester-urethane) with a strain of *Aspergillus flavus* (incubation 30 days in M3 media at 30 °C on a rotary shaker at 120 rpm) (Mathur & Prasad, 2012). The degraded polymer shows that degradation occurs much more between 250 °C and 400 °C and less between 400 °C and 500 °C in comparison to the non-degraded polymer. The first zone corresponds to the thermal degradation of urethane bonds while the second is the thermal degradation of esters. This analysis then reveals that the polymer incubated with the fungi contains more urethane and less ester than the control polymer. Thus, degradation occurs on the ester part.

Thermal analysis performed by differential scanning calorimetry (DSC) (Umare & Chandure, 2008) revealed a higher melting temperature after enzymatic degradation of a poly(ester-urethane) with *Rhizopus delemar* lipase

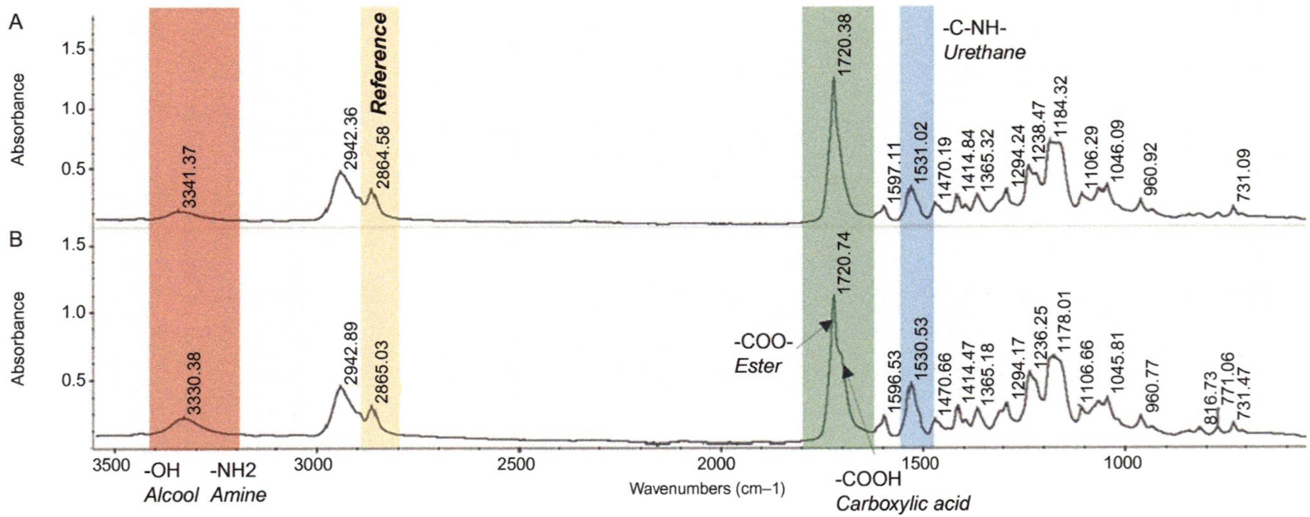

Fig. 6 FTIR analysis of PCL-based PU incubated 2 months at 30 °C (a) in minimal medium and (b) with a strain of *Alternaria* sp. *Adapted from Magnin, A., Hoornaert, L., Pollet, E., Laurichesse, S., Phalip, V., & Avérous, L. (2018). Isolation and characterization of different promising fungi for biological waste management of polyurethanes.* Microbial Biotechnology.

(1 mg of lipase in 10 mL of phosphate buffer pH 7.2, 144 h of incubation at 37 °C) in comparison to incubation with a buffer. This can be explained by an increased ratio of hard segments containing urethane linkages resulting from the preferential degradation of the soft segments containing the more sensitive ester bonds. Similarly, Kuang et al., showed an increased crystallinity after enzymatic degradation of a poly(ester-urethane) (Kuang & Mather, 2018).

After 50 days of degradation of PU based on PCL and 4,4'-MDI with a mixture of enzyme (E3576 and E4143, already described above), three degradation products were detected: 6-hydroxycaproic acid, 4,4'-MDA and 4,4'-MDA linked to a unit of 6-hydroxycaproic acid by a urethane bond (Magnin, Pollet, Perrin, et al., 2019). 6-Hydroxycaproic acid was detected by ^{1}H NMR at concentration level of gram per liter. 4,4'-MDA was detected as trace in the liquid fraction (below 0.5 mg/L). 4,4'-MDA linked to a unit of 6-hydroxycaproic acid by a urethane linkage was quantified as 4,4'-MDA equivalents, at a concentration of 3 mg/L. Quantification of this product suggests that, after the hydrolysis of ester linkages, low molar mass urethane molecules are accessible to the amidase and can be further degraded.

6. Pros and cons

The strength of the present procedure is to rely on the degradation of self-made model polyurethanes. Composition is controlled and can be varied to evaluate the impact of the chemical structure on the degradation. Procedures described for commercial PUR products are limited to information provided by the supplier then limiting interpretation of the results, especially when looking for degradation products.

The other strength of the procedure is the richness and diversity of analysis proposed. Tools described provide information on:

- The amount of degradation product released (Weight loss)
- The morphological aspect of the polymer piece after degradation (SEM)
- The evolution of the polymer molar mass (SEC)
- The moieties affected by the degradation (FTIR)
- The thermal degradation (TGA)
- The crystallinity (DSC)
- The nature and amount of degradation products released (^{1}H NMR and LC-MS)

One of the equipment suggested is particularly powerful. Indeed, the SEC is a UPLC (Ultra Performance Liquid Chromatography) allowing high flow rate and thus, fast analysis. Each run lasts 11 min in this configuration. High throughput of analysis can be implemented under this condition.

The procedure described is mainly adapted to TPU. Other techniques such as evaluation of the mechanical properties should be developed for foams.

7. Alternative methods/procedures

The present procedure is adapted to thermoplastic. Some analyses are suitable the same way for PU foams that are thermoset (SEM, FT-IR, thermal analysis or degradation products analysis). Others, such as SEC, are not suitable as thermoset polymers cannot dissolve in solvent. Weight loss can be applied on foams but with caution as foam pieces can get away from the foam because of physical constraints, shaking for example, rather than because of biological action.

8. Troubleshooting and optimization

Parameters of incubation with enzymes are critical to obtain an optimal degradation. First, the time of incubation is a key parameter. Degradation of high molar mass and hydrophobic substrates is a long process that can take several days. Enzymes are proteins that naturally lose their 3D structure and consequently their activity over time, especially at high temperature (Pastorino, Pioli, Zilli, Converti, & Nicolini, 2004; Tournier et al., 2020). To cope with this limitation and allow experiments of several days or weeks, it was suggested to remove the enzymatic solution and add fresh enzymes at regular time interval (Phua et al., 1987). pH and molarity of the buffer as well as temperature and shaking of the reactions should be adapted to the enzymes' optimal conditions.

SEC analysis and ^{1}H NMR analysis are conditioned by the solubility of the polymer and the degradation products respectively. Chloroform is also frequently used as mobile phase for SEC analysis of TPU. ^{1}H NMR can be performed in any deuterated solvent (water, THF, etc.).

9. Summary

For now, enzymatic degradation of PU is still at the stage of active enzymes discovery, mechanistic studies on model polymers and tentative optimization of the degradation. For instance, protein engineering can help improving enzyme anchoring at the polymer surface for a better depolymerization (Gamerith et al., 2016). Interestingly, all analytical tools described in this chapter can easily be adapted to a wide range of PUR materials.

The subsequent challenge will be the scale-up and the valorization of the obtained building blocks that are likely to be a mixture of highly diverse molecules. For that, efficient separation, purification and recovery of degradation products should also be developed.

Acknowledgments

The work for this review was fully supported by the Fondation pour la Recherche en Chimie (the Frontier Research in Chemistry Foundation) with the project "Biocycling" (2019-2020), in the continuation of a previous H2020 project (P4SB).

References

Biffinger, J. C., Barlow, D. E., Cockrell, A. L., Cusick, K. D., Hervey, W. J., Fitzgerald, et al. (2015). The applicability of Impranil®DLN for gauging the biodegradation of polyurethanes. *Polymer Degradation and Stability, 120*, 178–185. https://doi.org/10.1016/j.polymdegradstab.2015.06.020.

Blank, L. M., Narancic, T., Mampel, J., Tiso, T., & O'Connor, K. (2019). Biotechnological upcycling of plastic waste and other non-conventional feedstocks in a circular economy. *Current Opinion in Biotechnology, 62*, 212–219. https://doi.org/10.1016/j.copbio.2019.11.011.

Christenson, E., Anderson, J., & Hiltner, A. (2007). Biodegradation mechanisms of polyurethane elastomers. *Corrosion Engineering, Science and Technology, 42*(4), 312–323.

Furtwengler, P., Perrin, R., Redl, A., & Avérous, L. (2017). Synthesis and characterization of polyurethane foams derived of fully renewable polyester polyols from sorbitol. *European Polymer Journal, 97*, 319–327. https://doi.org/10.1016/j.eurpolymj.2017.10.020.

Gamerith, C., Herrero Acero, E., Pellis, A., Ortner, A., Vielnascher, R., Luschnig, D., et al. (2016). Improving enzymatic polyurethane hydrolysis by tuning enzyme sorption. *Polymer Degradation and Stability, 132*, 69–77. https://doi.org/10.1016/j.polymdegradstab.2016.02.025.

Garside, M. (2019). *Polyurethane demand worldwide from 2012 to 2022*. Retrieved from https://www.statista.com/statistics/747004/polyurethane-demand-worldwide/.

Ignat, L., Ignat, M., Ciobanu, C., Doroftei, F., & Popa, V. I. (2011). Effects of flax lignin addition on enzymatic oxidation of poly(ethylene adipate) urethanes. *Industrial Crops and Products, 34*(1), 1017–1028. https://doi.org/10.1016/j.indcrop.2011.03.010.

Ionescu, M. (2005). *Chemistry and technology of polyols for polyurethanes*. iSmithers Rapra Publishing.

Kuang, W., & Mather, P. T. (2018). A latent crosslinkable PCL-based polyurethane: Synthesis, shape memory, and enzymatic degradation. *Journal of Materials Research, 33*(17), 2463–2476.

Lucas, N., Bienaime, C., Belloy, C., Queneudec, M., Silvestre, F., & Nava-Saucedo, J. E. (2008). Polymer biodegradation: Mechanisms and estimation techniques. *Chemosphere, 73*(4), 429–442. https://doi.org/10.1016/j.chemosphere.2008.06.064.

Magnin, A., Pollet, E., Perrin, R., Ullmann, C., Persillon, C., Phalip, V., et al. (2019). Enzymatic recycling of thermoplastic polyurethanes: Synergistic effect of an esterase and an amidase and recovery of building blocks. *Waste Management, 85*, 141–150.

Magnin, A., Pollet, E., Phalip, V., & Avérous, L. (2019). Evaluation of biological degradation of polyurethanes. *Biotechnology Advances, 107457*, 1–24.

Mathur, G., & Prasad, R. (2012). Degradation of polyurethane by *Aspergillus flavus* (ITCC 6051) isolated from soil. *Applied Biochemistry and Biotechnology, 167*(6), 1595–1602. https://doi.org/10.1007/s12010-012-9572-4.

Min, K., Cuiffi, J. D., & Mathers, R. T. (2020). Ranking environmental degradation trends of plastic marine debris based on physical properties and molecular structure. *Nature Communications, 11*(1), 727. https://doi.org/10.1038/s41467-020-14538-z.

Mohapatra, B., & Bapuji, M. (1997). Characterization of urethanase from Micrococcus species associated with the marine sponge (Spirasfrella species). *Letters in Applied Microbiology, 25*(6), 393–396.

Pastorino, L., Pioli, F., Zilli, M., Converti, A., & Nicolini, C. (2004). Lipase-catalyzed degradation of poly(ε-caprolactone). *Enzyme and Microbial Technology, 35*(4), 321–326. https://doi.org/10.1016/j.enzmictec.2004.05.005.

Phua, S., Castillo, E., Anderson, J., & Hiltner, A. (1987). Biodegradation of a polyurethane in vitro. *Journal of Biomedical Materials Research Part A, 21*(2), 231–246.

Santerre, J., Labow, R., Duguay, D., Erfle, D., & Adams, G. (1994). Biodegradation evaluation of polyether and polyester-urethanes with oxidative and hydrolytic enzymes. *Journal of Biomedical Materials Research Part A, 28*(10), 1187–1199.

Santerre, J. P., Woodhouse, K., Laroche, G., & Labow, R. S. (2005). Understanding the biodegradation of polyurethanes: From classical implants to tissue engineering materials. *Biomaterials, 26*(35), 7457–7470. https://doi.org/10.1016/j.biomaterials.2005.05.079.

Schmidt, J., Wei, R., Oeser, T., Dedavid e Silva, L., Breite, D., et al. (2017). Degradation of polyester polyurethane by bacterial polyester hydrolases. *Polymers, 9*(12), 65. https://doi.org/10.3390/polym9020065.

Shim, W. J., Song, Y. K., Hong, S. H., & Jang, M. (2016). Identification and quantification of microplastics using Nile red staining. *Marine Pollution Bulletin, 113*(1–2), 469–476. https://doi.org/10.1016/j.marpolbul.2016.10.049.

Smith, R., Williams, D., & Oliver, C. (1987). The biodegradation of poly (ether urethanes). *Journal of Biomedical Materials Research Part A, 21*(9), 1149–1165.

Tang, Y., Labow, R., & Santerre, J. (2001). Enzyme-induced biodegradation of polycarbonate polyurethanes: Dependence on hard-segment concentration. *Journal of Biomedical Materials Research Part A, 56*(4), 516–528.

Tang, Y., Santerre, J., Labow, R., & Taylor, D. (1997). Application of macromolecular additives to reduce the hydrolytic degradation of polyurethanes by lysosomal enzymes. *Biomaterials, 18*(1), 37–45.

Tournier, V., Topham, C. M., Gilles, A., David, B., Folgoas, C., Moya-Leclair, E., et al. (2020). An engineered PET depolymerase to break down and recycle plastic bottles. *Nature, 580*(7802), 216–219. https://doi.org/10.1038/s41586-020-2149-4.

Turner, A., & Lau, K. S. (2016). Elemental concentrations and bioaccessibilities in beached plastic foam litter, with particular reference to lead in polyurethane. *Marine Pollution Bulletin, 112*(1–2), 265–270. https://doi.org/10.1016/j.marpolbul.2016.08.005.

Umare, S. S., & Chandure, A. S. (2008). Synthesis, characterization and biodegradation studies of poly(ester urethane)s. *Chemical Engineering Journal, 142*(1), 65–77. https://doi.org/10.1016/j.cej.2007.11.017.

Wei, R., & Zimmermann, W. (2017). Microbial enzymes for the recycling of recalcitrant petroleum-based plastics: How far are we? *Microbial Biotechnology, 10*(6), 1308–1322. https://doi.org/10.1111/1751-7915.12710.

Wierckx, N., Prieto, M. A., Pomposiello, P., de Lorenzo, V., O'Connor, K., & Blank, L. M. (2015). Plastic waste as a novel substrate for industrial biotechnology. *Microbial Biotechnology, 8*(6), 900–903. https://doi.org/10.1111/1751-7915.12312.

Yoshida, S., Hiraga, K., Takehana, T., Taniguchi, I., Yamaji, H., Maeda, Y., et al. (2016). A bacterium that degrades and assimilates poly (ethylene terephthalate). *Science, 351*(6278), 1196–1199.

CHAPTER SIXTEEN

Structural analysis of PET-degrading enzymes PETase and MHETase from *Ideonella sakaiensis*

Leonie G. Graf[a], Emil A.P. Michels[a], Yelwin Yew[b], Weidong Liu[c,d], Gottfried J. Palm[a,*], and Gert Weber[a,*]

[a]Macromolecular Crystallography, Institute of Biochemistry, University of Greifswald, Greifswald, Germany
[b]Biotechnology, University of Applied Sciences Bremerhaven, Bremerhaven, Germany
[c]Department of Biotechnology & Enzyme Catalysis, Institute of Biochemistry, University of Greifswald, Greifswald, Germany
[d]Industrial Enzymes National Engineering Laboratory, Tianjin, Institute of Industrial Biotechnology, Chinese Academy of Sciences, Tianjin, China
[*]Corresponding authors: e-mail address: palm@uni-greifswald.de; gert.weber@helmholtz-berlin.de

Contents

Methods in Enzymology, Volume 648
ISSN 0076-6879
https://doi.org/10.1016/bs.mie.2020.12.015

Abstract

The concept of biocatalytic PET degradation for industrial recycling processes had made a big step when the bacterium *Ideonella sakaiensis* was discovered to break PET down to its building blocks at ambient temperature. This process involves two enzymes: cleavage of ester bonds in PET by PETase and in MHET, the resulting intermediate, by MHETase. To understand and further improve this unique capability, structural analysis of the involved enzymes was aimed at from early on. We describe a repertoire of methods to this end, including protein expression and purification, crystallization of apo and substrate-bound enzymes, and modeling of PETase complexed with a ligand.

1. Introduction

With the discovery of *Ideonella sakaiensis* in 2016, the first organism was described that could grow on PET as its major carbon source. With the activities of two enzymes, extracellular PETase and presumably membrane anchored MHETase, the polymer is hydrolyzed to its building blocks ethylene glycol (EG) and terephthalate (TPA) (Bornscheuer, 2016; Yoshida et al., 2016) via the intermediate mono(2-hydroxyethyl) terephthalate (MHET). Substrate-bound PETase and MHETase protein structures were soon after described (Han et al., 2017; Palm et al., 2019). Both enzymes possess the α/β-hydrolase fold, which is widely spread among esterases (Nardini & Dijkstra, 1999). The fold's core elements are conserved. Especially the catalytic triad, responsible for the hydrolytic reaction, Ser160-His237-Asp206 for PETase (including the signal peptide) and Ser225-His528-Asp492 for MHETase, are well superimposable (Fig. 1). The serine attacks the ester bond on the acyl side and forms a covalent intermediate while releasing the alcohol. In a second step, the intermediate is hydrolyzed, and the acid is released as well. This esterase activity has been described for many homologous enzymes. Whereas structural alignment of PETase and MHETase is possible (DALI Server, (Holm & Sander, 1995)) with a root mean square deviation of 3.4 Å for 262 residues (11% identity), pairwise sequence alignment tools fail.

The difference between the two enzymes is in the substrate-binding sites. Whereas MHETase has been crystallized with the amide analog of its substrate MHET, the ligands in PETase structures are all low molecular weight compounds (Table 2) in contrast to their natural polymeric substrate. PETase has a shallow binding cleft fully accessible to the solvent (Fig. 1A). Additionally, several loops, including residues 204–213 with Asp206 of the

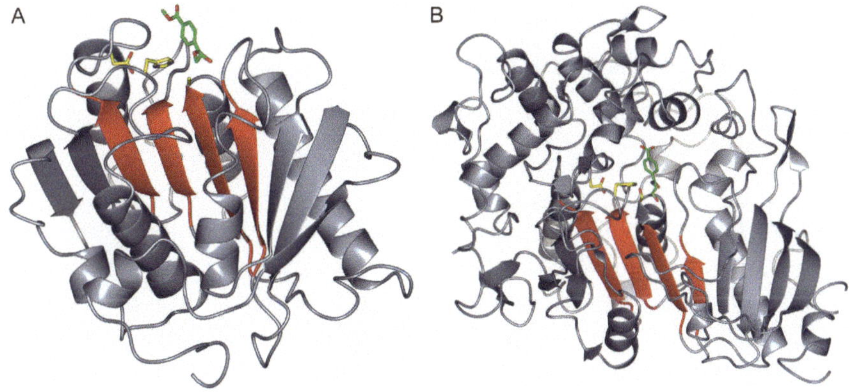

Fig. 1 Ribbon diagram of crystal structures of (A) PETase (5XH3, Han et al., 2017) and (B) MHETase (6QGA, Palm et al., 2019), each with a representative ligand. The same orientation of the homologous residues was chosen for both hydrolases. Central β-strands of the α/β-hydrolase fold are emphasized in red. Cylinder models are shown for the catalytic residues in yellow and the ligands 1-(2-hydroxyethyl) 4-methyl-terephthalate and *N*-(2-hydroxyethyl) terephthalic acid monoamide, the amide analog of MHET, in green carbons.

catalytic triad, are calculated to be more flexible at 298 K than in related thermophilic enzymes (Fecker et al., 2018). This allows the protein to adjust to the conformation required to bind to a given PET conformation on the plastic surface. On the other hand, ligand binding to exposed and flexible sites is often weaker. This may well have prevented detailed structural characterization of larger substrates mimicking substantial stretches of PET.

MHETase has a large lid domain of 23 kDa (Leu254–Gly468) that binds MHET snuggly and specifically (Fig. 1B). Accordingly, K_M values for MHETase are much smaller, in the μM range for MHET (Yoshida et al., 2016), whereas PETase K_M values for PET are difficult to define due to the solid substrate, but the kinetics (e.g., Cui et al., 2019) can only be simulated with much larger K_M values.

Recently, progress in devising more processive PETase mutants has been made. For example, compared to the wild-type enzyme, a thermostable mutant for the treatment of semicrystalline (23%) PET film showed 300-fold PET degradation in a 10-day experiment (Cui et al., 2019), another mutant 14-fold degradation activity at 40 °C (Son et al., 2020). Different enzymes with high PET-degrading activity, e.g., leaf-branch compost cutinases, LCC, are formidable alternatives (Tournier et al., 2020; Wei et al., 2019).

2. Methods background

2.1 Expression and purification

Both enzymes were expressed recombinantly in *Escherichia coli*. To aid purification, constructs encoding N-terminally fused His tags were used for both enzymes. Common protocols were used for expression and purification (Section 3). It is worth to point out that using a strain that supports the correct formation of disulfide bonds (see Section 3.2) was critical for sufficient yields. Metal affinity chromatography enriched the enzyme to a large extent, so that cleavage of the His-tag with TEV protease (only for PETase) and gel filtration chromatography rendered the purification successful.

2.2 Crystallization

Structural information for proteins is currently predominantly obtained by X-ray diffraction of crystals. Although success of crystallization experiments is not predictable, screening of a few thousand conditions or less is much faster compared to nuclear magnetic resonance (NMR). Furthermore, it has a high success rate for well-folded enzymes compared to Cryo-EM which is particularly difficult for proteins smaller than 100–150 kDa. Among all available structural information 90% of the deposited structures were thus determined by X-ray diffraction, for enzymes, the fraction is even larger. Crystallization conditions can vary depending on even small sequence changes, protein preparation and purity, crystallization plates, temperature or volume. Specific crystallization conditions for PETase / MHETase from the experience of the authors are given in Section 4.3 and a compilation of successful conditions from other groups in Section 4.4. Structures of protein-ligand complexes can be obtained either by co-crystallization or by soaking the protein crystals with ligands. For PETase and MHETase, soaking was advantageous. For PETase, we have tried tens of ligands in hundreds of co-crystallization drops. In over 50 crystals measured, no bound ligand was found (observations made in preparation of Palm et al., 2019). Only a few of the soaked ligands bound successfully in a different study (Han et al., 2017). For MHETase, focusing on ligands known to bind by other methods (differential scanning fluorimetry, Palm et al., 2019) gave a much higher success rate. A note of caution should be mentioned here, though. Screening with different methods may miss promising ligands and can result in a small fraction of overlapping hits as experienced in fragment screening:

The hits determined by the combination of six biophysical methods (fluorescence-based high concentration biochemical screening, saturation-transfer difference NMR, a reporter-displacement assay, native MS, micro-scale thermophoresis and thermal shift assay) comprised only 56% of the hits that were found by crystallography (Schiebel et al., 2016).

2.3 Modeling

Besides the analysis of crystal structures with a bound ligand, modeling of a desired complex can be attempted. In silico docking predicts binding modes and is used to help understanding binding-site occupation by a certain ligand. The docking approach is based on computer programs, like SwissDock (Bitencourt-Ferreira & de Azevedo, 2019), AutoDock Vina (Trott & Olson, 2009) or Autodock 4.2 (Morris et al., 2009) to name three, and protein and ligand as input files. In Section 5, the focus will be on the use of SwissDock, Chimera ViewDock, and PDBePISA, as these programs were used to obtain a docking-mode of BHET (PET fragment EG-TPA-EG) (Fig. 2) and PET fragment TPA-EG-TPA-EG (TE2) (Fig. 3) to PETase (Yew, 2020). For comparison of the mentioned docking programs, critical elements for ease of usage and docking targets are listed (Table 1). Autodock 4.2 and Autodock Vina dock the ligand at a location predefined by the user and can be run locally without an internet connection. In contrast, SwissDock docks the ligand around the whole molecule and is run on a web service. Therefore, the calculation takes longer, and more possible

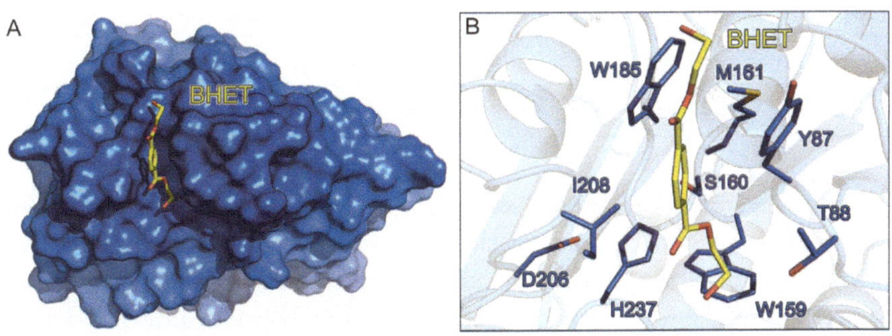

Fig. 2 BHET-molecule docked into the active site of PETase. Docking was done with SwissDock. Details are presented as stick model with colored polar atoms (nitrogens in blue, oxygens in red). (A) Whole molecule of PETase, presented as surface model, with BHET at the active site. (B) Close-up view of the active site of PETase as cartoon model with interfacing residues (marine) and BHET (yellow) presented as stick models. Interfacing residues are Tyr87, Thr88, Trp159, Ser160, Met161, Trp185, Ile208 and His237, calculated with PDBePISA.

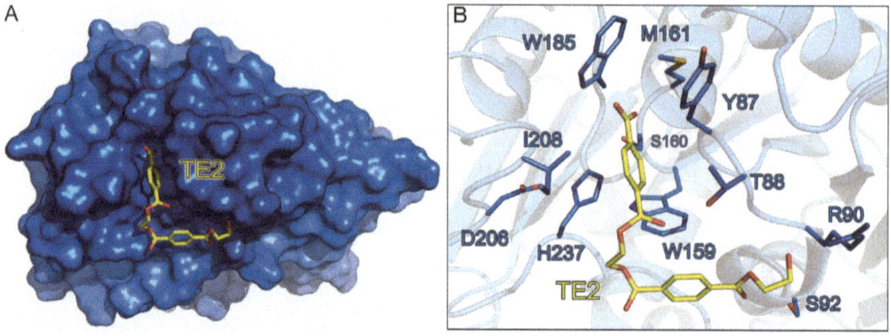

Fig. 3 PET fragment docked into the active site of PETase. Docking was done with Autodock Vina. Details are presented as stick model with colored polar atoms (nitrogens in blue, oxygens in red) . (A) Whole molecule of PETase, presented as surface model, with TPA-EG-TPA-EG (TE2) in the active site. (B) Close-up view of the active site of PETase as cartoon model with interfacing residues (marine) and TE2 (yellow) presented as stick models. Interfacing residues are Tyr87, Thr88, Arg90, Ser92, Trp159, Ser160, Met161, Trp185, Ile208 and His237, calculated with PDBePISA.

Table 1 Docking programs: Comparison of three different docking programs (SwissDock, Autodock 4.2, Autodock Vina).

	SwissDock	**Autodock 4.2**	**Autodock Vina**
Graphical user interface	Web service	MGLTools	Chimera and MGLTools
Offline modus	No	Yes	Yes
Docking location	Whole molecule	Predefined	Predefined
Preparation of ligand	No	Yes	Yes
Preparation of receptor	Yes	Yes	Yes
Ease of use	Beginner	Advanced	Advanced

dockings are generated. As SwissDock is run online, a local program (like Chimera) is needed to analyze the output data. For Autodock 4.2 and Autodock Vina, the output data can be evaluated using the graphical user interfaces (see Table 1). For best results, a comparison of different docking programs should be considered.

3. Expression and purification

I. Sakaiensis PETase and MHETase were recombinantly expressed in *Escherichia coli*. They contain disulfide bonds (2 in PETase, 5 in MHETase). An *E. coli* strain expressing disulfide bond isomerase DsbC supports correct disulfide bond formation and increases the yield of soluble protein (Lobstein et al., 2012). For easier purification, the enzymes were N-terminally His_6-tagged. In the MHETase construct, the signal peptide is replaced by a His_6-tag (expressed sequence MNHKVHHHHHHMGG...PP, GG...PP are residues 20–603 of UniProtKB entry A0A0K8P8E7). All chromatographic purifications were monitored at 280 nm. Once SDS PAGE identified the fractions with sufficiently pure protein, they were pooled in later runs based on the chromatogram.

3.1 Equipment

1. Sonicator (e.g., Sonopuls HD 4100 with UW 100, TS113V, SH100G and tip TT213 (⌀ 12.7 mm), BANDELIN electronic GmbH & Co. KG, Berlin, Germany)
2. French Press (Thermo Spectronic, USA)
3. Centrifugal ultrafiltration units (10.000 Da MW cutoff, Sartorius Vivaspin®, Sartorius Lab Instruments, Göttingen, Germany)
4. Medium pressure (low MPa range) chromatography system with UV detection (280 nm) (e.g., Äkta Purifier, GE Healthcare, Solingen, Germany) and fraction collector
5. Preparative gel filtration column (e.g., HiLoad® 16/600 Superdex® 75 pg column (GE Healthcare, Solingen, Germany)
6. UV/Vis spectrophotometer (e.g., NanoDrop ND 1000, VWR International GmbH, Hannover, Germany)
7. Liquid nitrogen. Handle with appropriate care, in well-ventilated areas and with suitable protective equipment for ultralow temperatures

3.2 Chemicals

1. Lysogeny broth (LB) media (sterilize by autoclaving)
2. *E. coli* SHuffle T7 express cells (New England Biolabs, Frankfurt, Germany)
3. Isopropyl β-D-1-thiogalactopyranoside (IPTG) (1 M stock in H_2O)
4. Buffer 1: 25 mM TRIS-HCl, pH 8.0, 500 mM NaCl and 20 mM imidazole (filtered)
5. Buffer 2: 25 mM TRIS-HCl, pH 8.0, 500 mM NaCl and 250 mM imidazole (filtered)
6. Buffer 3: 25 mM TRIS-HCl, pH 8.0, 500 mM NaCl (filtered)
7. Buffer R: 50 mM TRIS-HCl, pH 7.5, 150 mM NaCl, 10 mM imidazole and 1 mM dithiothreitol

8. Buffer E: 50 mM TRIS-HCl, pH 7.5, 150 mM NaCl, 200 mM imidazole and 1 mM dithiothreitol
9. Buffer S: 20 mM TRIS pH 7.5, 150 mM NaCl, 1 mM dithiothreitol (filtered and degassed)

3.3 Protocol

The following protocols describe expression and purification based on Han et al. (2017) and Palm et al. (2019) for PETase and MHETase, respectively.

3.3.1 Expression of PETase

1. Transform *E. coli* BL21trxB(*DE3*) cells with *pET32a-PETase* plasmid.
2. Select the cells after transformation on LB agar plates with 100 μg mL^{-1} ampicillin and incubate over night at 37 °C.
3. Take one colony of the agar plate to inoculate a LB-overnight culture supplemented with appropriate antibiotics (typically 10 mL for 1 L culture). Grow the cells overnight at 37 °C and 200 rpm in a shaking incubator.
4. Express the cells in 2 L baffled Erlenmeyer flasks. Every Erlenmeyer flask contained 400 mL LB medium supplemented with 100 μg mL^{-1} ampicillin. The inoculation of 400 mL medium was done with 5 mL overnight culture. Grow the cells in a shaking incubator at 37 °C and 200 rpm to an optical density at 600 nm (OD_{600}) of 0.8 and induce the target protein expression with 0.6 mM IPTG and continue the incubation at 16 °C for 24 h.
5. Harvest the cells by centrifugation at 5000 × *g*, 4 °C for 15 min, discard the supernatant and immediately use cell pellets, or store at −20 °C for further use. Typically, 5 g wet cells were obtained from 1 L broth.

3.3.2 Purification of PETase

PETase was purified in two steps with Ni-NTA affinity chromatography and size exclusion chromatography.

1. Resuspend the cell pellets on ice in buffer 1 (5 g wet cell weight in 30 mL buffer 1) and disrupt the cells by French Press (15,000 psi) twice at 4 °C.
2. Centrifuge the cell lysate at 17000 × *g*, 4 °C for 60 min to remove the cell debris.
3. Load the supernatant on a 10 mL NTA sepharose affinity column pre-charged with nickel using an FPLC system (GE Healthcare) at 4 °C. Wash the column to remove unbound proteins with 15 column volumes (cv) buffer 1 supplemented with 20 mM imidazole. PETase was eluted at ~70 mM imidazole when using a 20–250 mM imidazole gradient from buffer 1 to buffer 2.

4. Pool target protein from those fractions that were identified by SDS PAGE or a peak in the chromatogram and dialyze against buffer 3. Exchange the buffer after 1 h against fresh buffer 3.
5. Add TEV protease to the target protein fraction (1 mg protease: 50 mg PETase) for the removal of the His_6-tag and incubate over night at 4 °C.
6. Pass the TEV-digested mixture through a second Ni-NTA column pre-equilibrated with buffer 3 to remove the TEV protease, un-digested target protein and other contaminants.
7. Concentrate the protein fraction from the second Ni-NTA affinity chromatography to 1 mL by ultrafiltration with a 10 kDa molecular weight cut off centrifugal ultrafiltration unit.
8. Size exclusion chromatography was used for the final polishing step of purification. Equilibrate a HiLoad® 16/600 Superdex® 75 pg column (GE Healthcare) with buffer 3. Load concentrated target protein onto the column using a sample loop and elute from the column with 200 mL buffer 3. The flow rate was 0.5 mL min^{-1} and the eluate was collected in 1 mL fractions.
9. Pool PETase fractions identified by SDS PAGE, concentrate to circa 20 mg/mL, split the protein solution to aliquots of maximal 50 μL, freeze in liquid nitrogen and store at −80 °C. The concentration of protein was determined spectrophotometrically with a NanoDrop ND 1000 (VWR International GmbH, Hannover, Germany). An extinction coefficient ε_{280} of 39,670 $M^{-1} cm^{-1}$ was used.

3.3.3 Expression of MHETase

1. Transform *E. coli* SHuffle T7 express cells with the *pColdII* expression vector containing the MHETase sequence.
2. Select the cells after transformation on LB agar plates with 100 μg mL^{-1} ampicillin and incubate over night at 30 °C.
3. Take one colony of the agar plate to inoculate the overnight culture (typically 15 mL for 1 L culture). Grow the cells overnight at 37 °C and 130 rpm in a shaking incubator.
4. Use 1 L baffled Erlenmeyer flasks for the expression. Every Erlenmeyer flasks contained 200 mL LB medium supplemented with 100 μg mL^{-1} ampicillin. The inoculation of 200 mL medium was done with 2 mL overnight culture. Grow the cells in a shaking incubator at 33 °C and 130 rpm to an optical density at 600 nm (OD_{600}) of 1. At an OD_{600} of 1 induce the protein expression with 1 mM IPTG and at an OD_{600} of 2.5 lower the temperature to 16 °C. The total time for the expression was typically 18 h.

5. Harvest the cells by centrifugation at 10000 × *g*, 4 °C for 20 min. Discard the supernatant and store the cell pellets at −20 °C pending on further use. Typically, 6 g wet cells were obtained from 1 L broth.

3.3.4 Purification of MHETase

The affinity tag allowed standard chromatographic procedures. The protein was purified in a two-step procedure comprising nickel-affinity- and size exclusion chromatography.

1. Resuspend a cell pellet on ice in buffer R (6 g wet cell weight in 20 mL buffer R) and disrupt by sonification at 4 °C (Sonopuls HD 4100 with UW 100, TS113V, SH100G and tip TT213 (⌀ 12.7 mm), BANDELIN electronic GmbH & Co. KG, Berlin, Germany), typically six cycles with 2 min, intensity 70%, pulse 0.5 s/1.5 s.
2. Centrifuge the cell lysate at 39.000 × *g*, 4 °C for 60 min to remove the cell debris.
3. Load the supernatant on a gravity flow affinity column with 2 mL Ni-NTA sepharose at 4 °C. Wash with 15 column volumes (cv) buffer R supplemented with 20 mM imidazole to remove unbound proteins. MHETase was eluted with 2.5 cv buffer E.
4. For the last step of purification, size exclusion chromatography is used. To this end, equilibrate a HiLoad® 16/600 Superdex® 75 pg column (GE Healthcare, Solingen, Germany) with 160 mL buffer S. Concentrate the protein fractions from the nickel-affinity chromatography to 2 mL and apply them to the equilibrated HiLoad® 16/600 Superdex® 75 pg column. Isocratic elution from the column was performed with 140 mL buffer S at 4 °C. The flow rate was 1 mL min^{-1} and the eluate was collected in 1.5 mL fractions.
5. Pool the pure protein fractions as identified by SDS PAGE, concentrate to circa 10 mg/mL, freeze in liquid nitrogen and store at −80 °C. The concentration of protein was determined spectrophotometrically with a NanoDrop ND 1000 (VWR International GmbH, Hannover, Germany). An extinction coefficient ε_{280} of 102,955 M^{-1} cm^{-1} was used.

4. Crystallization

4.1 Equipment

1. Centrifugal filtration units (10 kDa MW cutoff)
2. Sitting drop crystallization plates with cover slides and sealant (e.g., high vacuum silicon grease) or with sealing film

3. Liquid nitrogen. Handle with appropriate care, in well-ventilated areas and with suitable protective equipment for ultralow temperatures

4.2 Chemicals

Suggested stock solutions are given in parentheses.

1. 2-[4-(2-hydroxyethyl) piperazin-1-yl] ethanesulfonic acid (HEPES) (1 M)
2. 2-Methylpentane-2,4-diol (2,4-MPD) (100%)
3. Ammonium sulfate (4 M)
4. 2-Morpholin-4-ylethanesulfonic acid (MES) (1 M)
5. 2-Amino-2-(hydroxymethyl)propane-1,3-diol (TRIS) (1 M)
6. PEG 8000 (50%)
7. Zinc acetate (1 M)
8. PEG 400 (50%)
9. Lithium sulfate (2 M)

4.3 Protocol

The following protocols describe the crystallization in Han et al. (2017) and Palm et al. (2019) for PETase and MHETase, respectively. Further conditions are described in the literature cited in Tables 2 and 3.

4.3.1 Crystallization of PETase

1. Prepare the following reservoir solution: 0.1 M MES pH 6.0, 30% (v/v) PEG 6000.
2. Adjust the PETase to a final concentration of 10 mg/mL in buffer 3 (Section 3.2).
3. Arrange a 24-well sitting drop crystallization plate using 200 μL solution in the reservoir well, mix 1 μL of the respective reservoir solution and 1 μL of the protein solution.
4. Seal the crystallization plate airtight with the corresponding sealant.
5. Incubate the crystallization plate for 2–3 days at 25 °C.
6. For the soaking of the crystals with substrates, saturate the respective cryo-protectant solution (reservoir solution) with ligands like *p*-nitrophenol or 1-(2-hydroxyethyl) 4-methyl-terephthalate (HEMT). Transfer the crystals to their respective saturated solution for 2–5 days at room temperature and reseal the well.
7. Harvest each crystal with a loop and flash cool in liquid nitrogen.

Table 2 Published PETase structures from www.rcsb.org (09.2020) sorted by resolution.

PDB code	Resolution (Å)	R_{work} (%)	R_{free} (%)	R_{free}- R_{work} (%)	RMSD (angles)[a]	RMSD (bonds)[a]	Citation
6EQE	0.92	10.0	11.0	1.0	1.43	0.008	Austin et al. (2018)
to be dep.	1.14	13.6	15.5	1.9	1.46	0.018	Yew (2020)
5XH2	1.20	9.8	12.5	2.7	1.55	0.013	Han et al. (2017)
5XH3	1.30	11.8	14.5	2.7	1.54	0.012	Han et al. (2017)
5YNS	1.36	16.1	19.0	2.9	2.03	0.022	Joo et al. (2018)
5YFE	1.39	23.0	23.5	0.5	1.91	0.025	Liu et al. (2018)
5XFY	1.40	13.4	15.9	2.5	1.71	0.029	Han et al. (2017)
6IJ3	1.40	19.4	23.3	3.9	2.00	0.016	Son et al. (2019)
6ILX	1.45	16.9	19.5	2.6	1.36	0.008	Liu, Shi, Zhu, Wei, and Yin (2019)
5XJH	1.54	16.0	19.2	3.2	2.12	0.026	Joo et al. (2018)
5XFZ	1.55	13.9	17.0	3.1	1.88	0.030	Han et al. (2017)
6ILW	1.57	16.0	17.6	1.6	1.82	0.030	Liu et al. (2019)
5XG0	1.58	13.7	16.2	2.5	1.52	0.012	Han et al. (2017)
6EQH	1.58	16.9	18.4	1.5	1.20	0.007	Austin et al. (2018)
6KY5	1.63	16.9	18.2	1.3	1.29	0.009	Cui et al. (2019)
6EQD	1.70	18.8	20.5	1.7	1.01	0.004	Austin et al. (2018))
6EQF	1.70	14.9	18.5	3.6	1.21	0.008	Austin et al. (2018)
6IJ5	1.72	20.6	26.2	5.6	1.98	0.018	Son et al. (2019)
6EQG	1.80	15.1	18.4	3.3	1.01	0.004	Austin et al. (2018)
6KUS	1.80	16.7	19.8	3.1	1.84	0.015	(to be pub.)
6IJ4	1.86	20.5	24.7	4.2	1.90	0.016	Son et al. (2019)
6KUO	1.90	16.3	18.8	2.5	1.86	0.015	Son et al. (2020)
6KUQ	1.91	15.6	19.1	3.5	1.88	0.016	(to be pub.)
6IJ6	1.95	15.1	19.3	4.2	1.88	0.016	Son et al. (2019)
6QGC	2.00	21.3	26.1	4.8	1.88	0.017	Palm et al. (2019)
6ANE	2.02	17.7	21.7	4.0	1.50	0.013	Fecker et al. (2018)

Table 2 Published PETase structures from www.rcsb.org (09.2020) sorted by resolution.—cont'd

PDB code	Meth.	T/K	Space group	Mol/au	Ligand	Crystallization condition
6EQE	SD	290	$P2_12_12_1$	1	–	0.2 M $MgCl_2$, 0.1 M MES, pH 6.0, 20% PEG 6000
to be dep.	SD	293	$P2_12_12_1$	3	–	1–2 M NaOAc, 10 mM sodium cacodylate, pH 6.5
5XH2	SD	298	$P2_12_12_1$	1	pNP[b]	$(NH_4)_2SO_4$, NaCl, HEPES, pH 7.5
5XH3	SD	298	$P2_12_12_1$	1	HEMT[c]	$(NH_4)_2SO_4$, NaCl, HEPES, pH 7.5
5YNS	HD	293	$P2_12_12_1$	1	–	Bis-Tris, NH_4OAc, PEG 10000, pH 5.5
5YFE	HD	293	$P2_12_12_1$	1	–	25% PEG 4000, NaOAc, pH 4.6, 0.2 M $(NH_4)_2SO_4$
5XFY	SD	298	C2	1	–	$(NH_4)_2SO_4$, NaCl, Bis-Tris, pH 6.5
6IJ3	HD	293	C2	1	–	Tris, PEG 3000, NaCl, pH 6.5
6ILX	HD	289	$P2_12_12_1$	1	–	0.1 M Bis-Tris, pH 5.5, 3 M NaCl
5XJH	HD	293	$P2_12_12_1$	1	–	Bis-Tris, NH_4OAc, PEG 10000, pH 5.0
5XFZ	SD	298	$P2_12_12_1$	1	–	$(NH_4)_2SO_4$, NaCl, HEPES, pH 7.5
6ILW	HD	298	$P2_12_12_1$	1	–	0.1 M Bis-Tris, pH 5.5, 3 M NaCl
5XG0	SD	298	$P2_12_12_1$	3	–	PEG 6000, glycerol, MES, pH 6.0
6EQH	SD	290	$C222_1$	3	–	1.6 M $MgSO_4$, 0.1 M MES, pH 6.5
6KY5	SD	295	$P2_12_12$	1	–	$(NH_4)_2SO_4$, pH 8.0
6EQD	SD	290	$C222_1$	3	–	0.1 M malonate/imidazole/borate, 25% PEG1500, pH 5.0
6EQF	SD	290	$P2_12_12_1$	1	–	0.1 M NH_4Cl, 0.1 M MES, pH 6.0, 20% PEG 6000
6IJ5	HD	293	C2	1	–	Tris, PEG 3000, NaCl, pH 6.5
6EQG	SD	290	$P2_1$	3	–	0.2 M Li_2SO_4, 0.1 M Bis-Tris, pH 5.5, 25% PEG 335

Continued

Table 2 Published PETase structures from www.rcsb.org (09.2020) sorted by resolution.—cont'd

PDB code	Meth.	T/K	Space group	Mol/au	Ligand	Crystallization condition
6KUS	HD	293	C2	1	–	17% PEG 10000, 0.1 M Bis-Tris, pH 5.0, 0.15 M NH_4OAc
6IJ4	HD	293	C2	1	–	Tris, PEG 3000, NaCl, pH 6.5
6KUO	HD	293	C2	1	–	17% PEG 10000, 0.1 M Bis-Tris, pH 5.0, 0.15 M NH_4OAc
6KUQ	HD	293	$P3_12$	1	–	17% PEG 10000, 0.1 M Bis-Tris, pH 5.0, 0.15 M NH_4OAc
6IJ6	HD	293	C2	1	–	Bis-Tris, NH_4OAc, PEG 10000, pH 5.0
6QGC	SD	293	$P2_1$	4	–	0.1 M NaCitrate or NaOAc, pH 5.0, 15% (*w*/*v*) PEG 8000, 0.5 M Li_2SO_4
6ANE	SD	277	$C222_1$	3	–	0.1 M MES, pH 6.5, 1.6 M $MgSO_4$, 20% glycerol

[a]RMSD(angles) and RMSD(bonds) are calculated separately with Phenix and not taken from publication.
[b]*p*-Nitrophenolate.
[c]1-(2-Hydroxyethyl) 4-methyl-terephthalate.
SD: sitting drop, HD: hanging drop.

Table 3 Published MHETase structures from www.rcsb.org (09.2020) sorted by resolution.

PDB code	Resolution (Å)	R_{work} (%)	R_{free} (%)	R_{free}-R_{work} (%)	RMSD (angles)[a]	RMSD (bonds)[a]	Citation
6QZ4	1.74	15.9	20.2	4.3	1.05	0.005	Knott et al. (2020)
6QG9	2.05	21.0	26.2	5.2	1.41	0.012	Palm et al. (2019)
6QGA	2.10	17.3	19.8	2.5	1.36	0.010	Palm et al. (2019)
6JTU	2.10	17.0	19.8	2.8	1.99	0.016	Sagong et al. (2020)
6QZ3	2.10	18.0	21.9	3.9	1.06	0.004	Knott et al. (2020)
6QGB	2.20	18.5	21.4	2.9	1.12	0.007	Palm et al. (2019)
6JTT	2.51	19.0	22.6	3.6	1.99	0.015	Sagong et al. (2020)

PDB code	Method	T/K	Space group	Mol/au	Ligand	Crystallization condition
6QZ4	SD	277	$P2_1$	2	–	20.4% PEG 8000, 0.1 M HEPES pH 8.0
6QG9	SD	293	P1	10	–	0.1 M MES, pH 6.5, 10% (*w/v*) PEG 8000, 0.1 M ZnOAc
6QGA	SD	293	$P2_12_12_1$	6	MHETA[b]	0.1 M HEPES, pH 7.5, 30% (w/v) 2,4-MPD, 0.12 M $(NH_4)_2SO_4$
6JTU	HD	293	C2	3	–	PEG 3350, HEPES, Tacsimate, pH 7.0
6QZ3	SD	289	I422	3	Benz[c]	0.1 M NaCl, 0.1 M Bis-Tris, 1.5 M $(NH_4)_2SO_4$, pH 6.5
6QGB	SD	293	$P2_12_12_1$	6	Benz[c]	0.1 M HEPES, pH 7.5, 30% (w/v) 2,4-MPD, 0.12 M $(NH_4)_2SO_4$
6JTT	HD	293	C2	3	MHET[d], BHET[e]	PEG 3350, HEPES, Tacsimate, pH 7.0

[a]RMSD(angles) and RMSD(bonds) are calculated separately with Phenix and not taken from publication.
[b]*N*-(2-hydroxyethyl) terephthalic acid monoamide.
[c]Benzoic acid.
[d]mono(2-hydroxyethyl) terephthalic acid.
[e]bis(hydroxyethyl) terephthalic acid.
SD: sitting drop, HD: hanging drop.

4.3.2 Crystallization of MHETase

1. Prepare the following stock solutions: 0.1 M HEPES pH 7.5, 30% (v/v) 2-Methylpentane-2,4-diol (2,4-MPD), 0.12 M ammonium sulfate (reservoir solution A, for crystals in space group $P2_12_12_1$)
   Cryo protectant solution A = reservoir solution A.
   0.1 M 2-morpholino-4-ylethanesulfonic acid (MES) pH 6.5, 10% (v/v) PEG 8000, 0.1 M zinc acetate (reservoir solution B, for crystals in space group P1).
   0.1 M TRIS pH 8.5, 5% (v/v) PEG 8000, 20% (v/v) PEG 400, 0.5 M lithium sulfate (cryo protectant solution B, for crystals from reservoir solution B)
2. Adjust the MHETase to a final concentration of 10 mg/mL in buffer R.

3. Set up a 96-well sitting drop crystallization plate using 40 μL reservoir solution A or B in the reservoir well and a mixture of 1 μL of the respective reservoir solution and 1 μL of the protein solution (10 mg/mL protein in buffer R) in the upper platform well.
4. Cover the crystallization plate airtight with a sealing film.
5. Incubate the crystallization plate for 5–10 days at 20 °C.
6. Soaking was only performed for crystals grown from reservoir solution A in space group $P2_12_12_1$. Saturate the reservoir solution A with a small bit of solid N-(2-hydroxyethyl) terephthalic acid monoamide (MHETA). Transfer the crystals to the saturated solution for 24 h at room temperature.
7. Transfer crystals to their corresponding cryoprotecting solution (see 1), if not done yet for soaking. The reservoir solution containing 30% (v/v) 2-Methylpentane-2,4-diol can be directly used as cryo-protectant. Harvest each crystal with a loop and flash cool in liquid nitrogen.

4.4 Successful crystallization conditions

Crystallization conditions can vary depending on even small changes, but similar systems can also be a good start. For comparison, we have therefore compiled successful crystallization conditions reported in the literature for PETase (Table 2) and MHETase (Table 3).

5. Modeling of a PETase-BHET complex

The ligand molecule is placed around the receptor, and for each placement the values for free enthalpy (ΔG) and fitness are calculated by the SwissDock server. Based on ΔG and fitness, the different docking modes are validated, and the most likely docking mode can be achieved. ΔG and fitness are based on the properties of atoms, covalent bonds, and functional groups. They are values for the estimated energy, with which a ligand (here BHET) binds to the active site of a receptor molecule (here PETase). Therefore, in silico docking will generate docking modes close to the natural state. In the end, it is still required to create co-crystals, but in silico docking can help to understand enzymatic mechanisms until co-crystals are achieved.

5.1 Protocol

5.1.1 Protocol for in silico docking with SwissDock

1. Prepare the PETase PDB-file by deleting solvent and ligands, previously added during refinement. Preparation can be done via Chimera Dock Prep.

2. Download a ".mol2"-file for the ligand of interest from the PRODRG server (Van Aalten A et al., 1996), in this case BHET.
3. Upload the PETase PDB-file and BHET mol2-file to the web service of SwissDock (http://www.swissdock.ch/docking#).
4. Press "Start docking" on SwissDock web service.
5. After docking is performed (takes a few minutes), download prediction files zip folder and unpack it.
6. Evaluate all docking modes found, by loading your Protein PDB-file in Chimera and start Chimera ViewDock with the SwissDock "clusters.dock4.pdb" output file.
7. Delete/Purge all docking modes not at the binding-site.
8. Review all docking modes at the binding-site and evaluate the most fitting docking mode for PETase and BHET, considering the values for ΔG and FullFitness (clusters.dock4.csv output of SwissDock) and the orientation of BHET at binding-site.
9. Save the best fitting docking mode in one file (PETase + BHET-docking).
10. For determination of all interfacing residues, upload the docking PDB-file to the web service of PDBePISA and press the "interfaces"-button.
11. After PDBePISA successfully calculated all interfering residues, review the results and prepare a docking-model for publication, for example using Pymol or Chimera.

5.1.2 Protocol for in silico docking with Autodock Vina

Another protein docking program is Autodock Vina (Trott & Olson, 2009) and the following instruction is based on the implemented version in Chimera. As described in Section 5.1.1 (steps 1 and 2), the PETase receptor protein (pdb-file) and the ligand (pdb-file) are needed.

1. Open the receptor file in Chimera.
2. Prepare the PETase PDB-file by deleting solvent and ligands, previously added during refinement. Preparation can be done via Chimera Dock Prep.
3. Download a "pdb"-file for the ligand of interest from the PRODRG server (Van Aalten A' et al., 1996), in this case the deprotonated ethylenglycol-terephthalate-ethylenglycol-terephthalate (TE2), and open it in Chimera.
4. Under "Tools → Surface/Binding Analysis" the implemented "Autodock Vina" version can be found.
5. After choosing Autodock Vina, a new window will open. Specify the output directory, receptor molecule and ligand molecule.

6. For Autodock Vina a 3D-box of the docking site has to be assigned. This can be done by selecting "Resize search volume using:" and the given mouse button. For ease of use, change from "button 2" to "ctrl + button 1."
7. Go to the main window again and assign a 3D-box around the binding site of your receptor protein by pressing "ctrl + button 1". The 3D-box should be big enough to rotate your ligand freely, so the best docking mode can be found.
8. Go back to the "Autodock Vina"-window and adjust the parameters for receptor options, ligand options and advanced options. Receptor and ligand options are based on the given pdb-files and the advanced options can be adjusted freely.
9. Advanced options: Higher values will generate better results, but will take longer for calculation. Used parameters were:
   Number of binding modes: 9.
   Exhaustiveness of search: 8.
   Maximum energy difference (kcal/mol): 3.
10. Press "Ok" to start the docking. After a few minutes the docking results are saved in your output directory and "ViewDock" will start automatically by opening a new window with the docking results.
11. Review all docking modes at the binding-site and evaluate the most fitting docking mode for PETase and TE2, considering the values for "Score" and "RMSD" and the orientation of TE2 at binding-site.
12. Save the most fitting docking mode in one file (PETase + TE2-docking).
13. For determination of all interfacing residues, upload the docking PDB-file to the web service of PDBePISA and press the "interfaces"-button.
14. After PDBePISA successfully calculated all interfering residues, review the results and prepare a docking-model for publication, for example using Pymol or Chimera.

6. Summary

Structural information is the most used single source of information to understand the mechanism of action of the PET-degrading enzymes PETase and MHETase from *I. sakaiensis* and thereby to improve their properties for biotechnological applications. This chapter describes the preparation of both proteins and their crystals for X-ray diffraction structure determination. PETase was crystallized much faster at many conditions with various known crystal forms to date. Nonetheless, complexes with the native substrate or larger oligomeric fragments resembling PET segments are hard to crystallize.

The substrate may have no preferred conformation, and the protein surface is wide open, possibly without specific binding pockets for remote monomers. An example of creating a theoretical model by molecular modeling is, therefore, also given. MHETase crystallizes more reluctantly but binds suitable substrates tightly and well defined. No flexible regions or disordered domains could be identified that could have caused poor crystallizability.

Acknowledgments

This work was supported by start-up funding from the University of Greifswald and support by the Helmholtz-Zentrum Berlin to G.W. The China Scholarship Council (CSC) supported W.L. The authors thank Imke Lang, Insa Mannott and Carsten Harms for the support of Y.Y.

References

Austin, H. P., Allen, M. D., Donohoe, B. S., Rorrer, N. A., Kearns, F. L., Silveira, R. L., et al. (2018). Characterization and engineering of a plastic-degrading aromatic polyesterase. *Proceedings of the National Academy of Sciences*, *115*(19), E4350–E4357. National Academy of Sciences.

Bitencourt-Ferreira, G., & de Azevedo, W. F. (2019). Docking with SwissDock. *Methods in Molecular Biology*, *2053*, 189–202. Humana Press Inc.

Bornscheuer, U. T. (2016). Feeding on plastic. *Science*, *351*(6278), 1154–1155. American Association for the Advancement of Science.

Cui, Y., Chen, Y., Liu, X., Dong, S., Tian, Y., Qiao, Y., et al. (2019). Computational redesign of a PETase for plastic biodegradation by the GRAPE strategy. *bioRxiv*.

Fecker, T., Galaz-Davison, P., Engelberger, F., Narui, Y., Sotomayor, M., Parra, L. P., et al. (2018). Active site flexibility as a hallmark for efficient PET degradation by *I. sakaiensis* PETase. *Biophysical Journal*, *114*(6), 1302–1312.

Han, X., Liu, W., Huang, J.-W., Ma, J., Zheng, Y., Ko, T.-P., et al. (2017). Structural insight into catalytic mechanism of PET hydrolase. *Nature Communications*, *8*(1), 2106. Nature Publishing Group.

Holm, L., & Sander, C. (1995). Dali: A network tool for protein structure comparison. *Trends in Biochemical Sciences*, *20*(11), 478–480. Elsevier Current Trends.

Joo, S., Cho, I. J., Seo, H., Son, H. F., Sagong, H. Y., Shin, T. J., et al. (2018). Structural insight into molecular mechanism of poly(ethylene terephthalate) degradation. *Nature Communications*, *9*(1), 382.

Knott, B. C., Erickson, E., Allen, M. D., Gado, J. E., Graham, R., Kearns, F. L., et al. (2020). Characterization and engineering of a two-enzyme system for plastics depolymerization. *Proceedings of the National Academy of Sciences of the United States of America*, *117*(41), 25476–25485. https://doi.org/10.1073/PNAS.2006753117.

Liu, B., He, L., Wang, L., Li, T., Li, C., Liu, H., et al. (2018). Protein crystallography and site-direct mutagenesis analysis of the poly(ethylene terephthalate) hydrolase petase from *Ideonella sakaiensis*. *Chembiochem*, *19*(14), 1471–1475.

Liu, C., Shi, C., Zhu, S., Wei, R., & Yin, C. C. (2019). Structural and functional characterization of polyethylene terephthalate hydrolase from *Ideonella sakaiensis*. *Biochemical and Biophysical Research Communications*, *508*(1), 289–294. Elsevier B.V.

Lobstein, J., Emrich, C. A., Jeans, C., Faulkner, M., Riggs, P., & Berkmen, M. (2012). SHuffle, a novel *Escherichia coli* protein expression strain capable of correctly folding disulfide bonded proteins in its cytoplasm. *Microbial Cell Factories*, *11*(1), 753. BioMed Central.

Morris, G. M., Ruth, H., Lindstrom, W., Sanner, M. F., Belew, R. K., Goodsell, D. S., et al. (2009). Software news and updates AutoDock4 and AutoDockTools4: Automated docking with selective receptor flexibility. *Journal of Computational Chemistry*, *30*(16), 2785–2791. NIH Public Access.

Nardini, M., & Dijkstra, B. W. (1999, December 1). α/β hydrolase fold enzymes: The family keeps growing. *Current Opinion in Structural Biology*. Current Biology Ltd.

Palm, G. J., Reisky, L., Böttcher, D., Müller, H., Michels, E. A. P., Walczak, M. C., et al. (2019). Structure of the plastic-degrading *Ideonella sakaiensis* MHETase bound to a substrate. *Nature Communications*, *10*(1), 1717. Nature Publishing Group.

Sagong, H.-Y., Seo, H., Kim, T., Son, H. F., Joo, S., Lee, S. H., et al. (2020). Decomposition of the PET Film by MHETase using Exo-PETase. *ACS Catalysis*, *10*, 4805–4812. https://doi.org/10.1021/ACSCATAL.9B05604.

Schiebel, J., Radeva, N., Krimmer, S. G., Wang, X., Stieler, M., Ehrmann, F. R., et al. (2016). Six biophysical screening methods miss a large proportion of crystallographically discovered fragment hits: A case study. *ACS Chemical Biology*, *11*(6), 1693–1701.

Son, H. F., Cho, I. J., Joo, S., Seo, H., Sagong, H. Y., Choi, S. Y., et al. (2019). Rational protein engineering of thermo-stable PETase from *Ideonella sakaiensis* for highly efficient PET degradation. *ACS Catalysis*, *9*(4), 3519–3526. American Chemical Society.

Son, H. F., Joo, S., Seo, H., Sagong, H. Y., Lee, S. H., Hong, H., et al. (2020). Structural bioinformatics-based protein engineering of thermo-stable PETase from *Ideonella sakaiensis*. *Enzyme and Microbial Technology*, *141*, 109656. Elsevier Inc.

Tournier, V., Topham, C. M., Gilles, A., David, B., Folgoas, C., Moya-Leclair, E., et al. (2020). An engineered PET depolymerase to break down and recycle plastic bottles. *Nature*, *580*(7802), 216–219. Nature Research.

Trott, O., & Olson, A. J. (2009). AutoDock Vina: Improving the speed and accuracy of docking with a new scoring function, efficient optimization, and multithreading. *Journal of Computational Chemistry*, *31*(2). NA-NA. Wiley.

Van Aalten A', D. M. F., Bywater, R., Findlay, J. B. C., Hendlich, M., Hooft, R. W. W., & Vriend, G. (1996). PRODRG, a program for generating molecular topologies and unique molecular descriptors from coordinates of small molecules*. Journal of Computer-Aided Molecular Design 10 255-62.

Wei, R., Song, C., Gräsing, D., Schneider, T., Bielytskyi, P., Böttcher, D., et al. (2019, December 1). Conformational fitting of a flexible oligomeric substrate does not explain the enzymatic PET degradation. *Nature Communications. Nature Research.*

Yew, Y. (2020). *Crystal structure of I. sakaiensis PETase, a plastic digesting protein.* Thesis Germany: Applied University of Bremerhaven.

Yoshida, S., Hiraga, K., Takanaha, T., Taniguchi, I., Yamaji, H., Maeda, Y., et al. (2016). A bacterium that degrades and assimilates poly(ethyleneterephthalate). *Science*, *351*(6278), 1196–1199.

CHAPTER SEVENTEEN

Structural and functional characterization of nylon hydrolases

Seiji Negoro[a,*], Dai-ichiro Kato[b], Taku Ohki[a], Kengo Yasuhira[a], Yasuyuki Kawashima[a], Keisuke Nagai[a], Masahiro Takeo[a], Naoki Shibata[c], Katsumasa Kamiya[d], and Yasuteru Shigeta[e]

[a]Department of Applied Chemistry, Graduate School of Engineering, University of Hyogo, Himeji, Japan
[b]Department of Science, Graduate School of Science and Engineering, Kagoshima University, Kagoshima, Japan
[c]Department of Picobiology, Graduate School of Life Science, University of Hyogo, Ako-gun, Hyogo, Japan
[d]Education Development Center, Kanagawa Institute of Technology, Atsugi, Japan
[e]Center for Computational Sciences, University of Tsukuba, Tsukuba, Japan
*Corresponding author: e-mail address: negoro@eng.u-hyogo.ac.jp

Contents

Methods in Enzymology, Volume 648
ISSN 0076-6879
https://doi.org/10.1016/bs.mie.2020.11.004

Abstract

Biodegradation of synthetic polymers is recognized as a useful way to reduce their environmental load and pollution, loss of natural resources, extensive energy consumption, and generation of greenhouse gases. The potential use of enzymes responsible for the degradation of the targeted polymers is an effective approach which enables the conversion of the used polymers to original monomers and/or other useful compounds. In addition, the enzymes are expected to be applicable in industrial processes such as improving the surface structures of the polymers. Especially, conversion of the solid polymers to soluble oligomers/monomers is a key step for the biodegradation of the polymers. Regarding the hydrolysis of polyamides, three enzymes, 6-aminohexanoate-cyclic-dimer hydrolase (NylA), 6-aminohexanoate-dimer hydrolase (NylB), and 6-aminohexanoate-oligomer *endo*-hydrolase (nylon hydrolase, NylC), are found in several bacterial strains. In this chapter, we describe our approach for the screening of microorganisms which degrade nylons and related compounds; preparation of substrates; assay of hydrolytic activity for soluble and insoluble substrates; and X-ray crystallographic and computational approaches for analysis of structure and catalytic mechanisms of the nylon-degrading enzymes.

1. Introduction

A polyamide is a polymer which contains recurring amide groups (R-CO-NH-R′) as integral parts of the main polymer chain. Nylons are the major polyamides and are made from (a) a dicarboxylic acid and a diamine, (b) an amino acid that is able to undergo self-condensation, or (c) the corresponding lactam, such as ε-caprolactam. Because of the high strength, rigidity, elasticity, and chemical and thermal resistance, nylons have been used for the production of various fibers and other plastic products. Particularly, nylon-66 (containing adipic acid and 1,6-hexamethylene diamine as the monomeric units) and nylon-6 (containing 6-aminohexanoate, Ahx, as the monomeric unit) represent approximately 90% of the worldwide production of nylons.

As for the degradation of nylons to smaller metabolites, two modes of degradation, i.e., hydrolytic metabolisms by specific hydrolases and oxidative metabolisms by peroxidases, are known. Among the hydrolytic degradations, three enzymes, i.e., 6-aminohexanoate-cyclic-dimer hydrolase (NylA, EC 3.5.2.12), 6-aminohexanoate-dimer hydrolase (NylB, EC 3.5.1.46), and 6-aminohexanoate-oligomer *endo*-hydrolase (nylon hydrolase) (NylC, EC3.5.1.117) have been found in bacterial strains (Negoro, 2000, 2002) (Fig. 1A). NylA specifically hydrolyzes one of the two equivalent amide bonds in the Ahx-cyclic dimer, generating an Ahx linear dimer (Yasuhira et al., 2010). NylB hydrolyzes Ahx oligomers by an exo-type mode (Negoro et al., 2005, 2007). NylC degrades Ahx-cyclic and -linear oligomers with a degree of polymerization greater than three by an endo-type mode. A NylC mutant has activity for the various aliphatic nylons (nylon-6, nylon-66) (Nagai et al., 2014; Negoro et al., 2012, 2018). The monomeric substrate (Ahx) obtained from byproducts of a nylon-6 factory (nylon oligomer mixture) is further metabolized to adipate via adipate semialdehyde by successive reactions using 6-aminohexanoate aminotransferase (NylD, EC 1.2.1.–) and adipate semialdehyde dehydrogenase (NylE, EC 2.6.1.–) (Takehara et al., 2018) (Fig. 1A). In contrast, in the oxidative metabolism of nylons by fungal strains, the methylene group adjacent to the nitrogen atom in the polymer chains is attacked by peroxidases as characterized by the NMR analysis of the degradation products (Deguchi, Kitaoka, Kakezawa, & Nishida, 1998).

The efficient depolymerization of synthetic polymers depends on the isolation of enzymes with high activity against the targeted substrates.

Therefore, in this chapter, we describe our approach for screening of microorganisms which degrade nylons and the related compounds by hydrolytic enzymes, along with enzyme assays using soluble and insoluble substrates. Moreover, since knowledge of the three-dimensional structure and catalytic mechanisms of the enzymes are important to improve their activities by protein engineering, we refer to the methodology for the X-ray crystallographic and computational analysis, focusing on a nylon degrading enzyme, NylB.

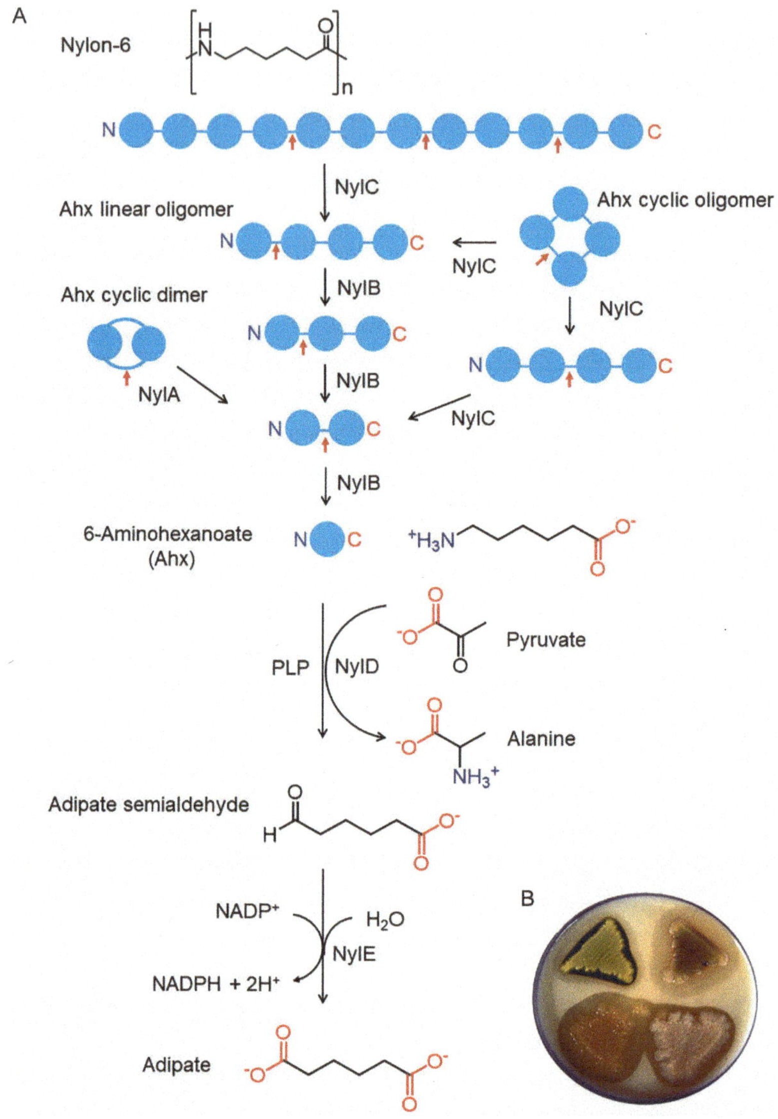

Fig. 1 Enzymatic metabolism of nylon-6 related compounds. (A) Three enzymes responsible for the degradation of nylon-6, 6-aminohexanoate-cyclic and linear oligomers to 6-aminohexanoate (Ahx) are schematically illustrated. NylA, 6-aminohexanoate-cyclic dimer hydrolase; NylB, 6-aminohexanoate dimer hydrolase; NylC, 6-aminohexanoate-oligomer *endo*-hydrolase (nylon hydrolase). The monomeric unit of Ahx is illustrated as a *"light blue"* circle. Terminal amino N and carboxyl C are shown as *"light blue"* and *"red" letters*, respectively. Red arrows indicate the hydrolytic sites catalyzed by each enzyme. Ahx is further metabolized to adipate by NylD and NylE. NylD, 6-aminohexanoate aminotransferase, catalyzes the conversion of Ahx to adipate semialdehyde using α-keto acid (pyruvate, etc.) as an amino acceptor and pyridoxal phosphate (PLP) as a cofactor. NylE, adipate semialdehyde dehydrogenase, catalyzes the conversion of adipate semialdehyde to adipate using $NAD(P)^+$ as a cofactor. (B) Growth of nylon oligomer-degrading microorganisms on an LB plate containing insoluble 6-aminohexanoate cyclic oligomers.

2. Screening of microorganisms and enzymes degrading nylon-related compounds

The degradability of polymeric nylons by microorganisms is generally very low. Therefore, prior to the direct selection for the targeted polymers, prescreening using model compounds having similar structures is effective. Nylon-6 consists of more than 100 units of Ahx (Fig. 1A). During the polymerization reaction, some molecules fail to polymerize and remain linear oligomers, whereas others undergo head-to-tail condensation to form cyclic oligomers. Ahx-cyclic oligomers are only minimally solubilized in water and cause turbidity in an agar plate. Therefore, microorganisms with hydrolytic activities form a clear zone around the colony on the turbid agar plate, in which the insoluble substrates are homogeneously dispersed in the medium (Fig. 1B, Protocol 6.1) (Yasuhira et al., 2007). By now, we have isolated four bacterial strains, *Arthrobacter* sp. KI72, *Pseudomonas* sp. NK87, *Agromyces* sp. KY5R, and *Kocuria* sp. KY2 strains, which grow on Ahx oligomers as a sole source of carbon and nitrogen (Negoro, 2002; Yasuhira et al., 2007). Agar medium containing *N*-carbobenzoxy Ahx-trimer is also used for the similar selection of *E. coli* cells including the cloned *nylC* gene (Negoro, Kakudo, Urabe, & Okada, 1992).

To prepare various substrates used for enzyme assays and crystallographic analysis of enzyme-substrate complex, the following four basic routes are attempted: i.e., (i) fractionation of cyclic and linear oligomers from by-products of nylon-6 factories, (ii) stepwise chemical synthesis from monomers, (iii) limited hydrolysis of polymeric nylons by formic acid treatment, and (iv) interfacial polymerization using diamine and monochlorinated dicarboxylic acids (e.g., hexamethylenediamine and adipoyl chloride for the preparation of nylon-66). In Section 3, we describe the outlines of the procedures and typical protocols for the preparation of these substrates.

3. Preparation of oligomeric and polymeric substrates for enzyme assays

3.1 Preparation of 6-aminohexanoate oligomers

Nylon oligomer mixture (NOM) obtained as a by-product of nylon-6 manufacture is a mixture of monomeric Ahx, Ahx-cyclic and linear oligomers with various degrees of polymerization (Fig. 2).

Ahx-cyclic dimer (substrate for NylA assay) and Ahx-cyclic oligomer (degree of polymerization >3; designated as Aco3; substrate for NylC) are obtained on the basis of the specific enzymatic hydrolysis of NOM and differences in the solubility of the reaction components in water. The protocol of the fractionation is summarized in Fig. 2. Briefly, monomeric Ahx and its linear oligomers are soluble in cold water. In contrast, cyclic oligomers are solubilized in hot water, because solubilities of cyclic oligomers are quite low in cold water. Therefore, Ahx cyclic oligomers (degree of

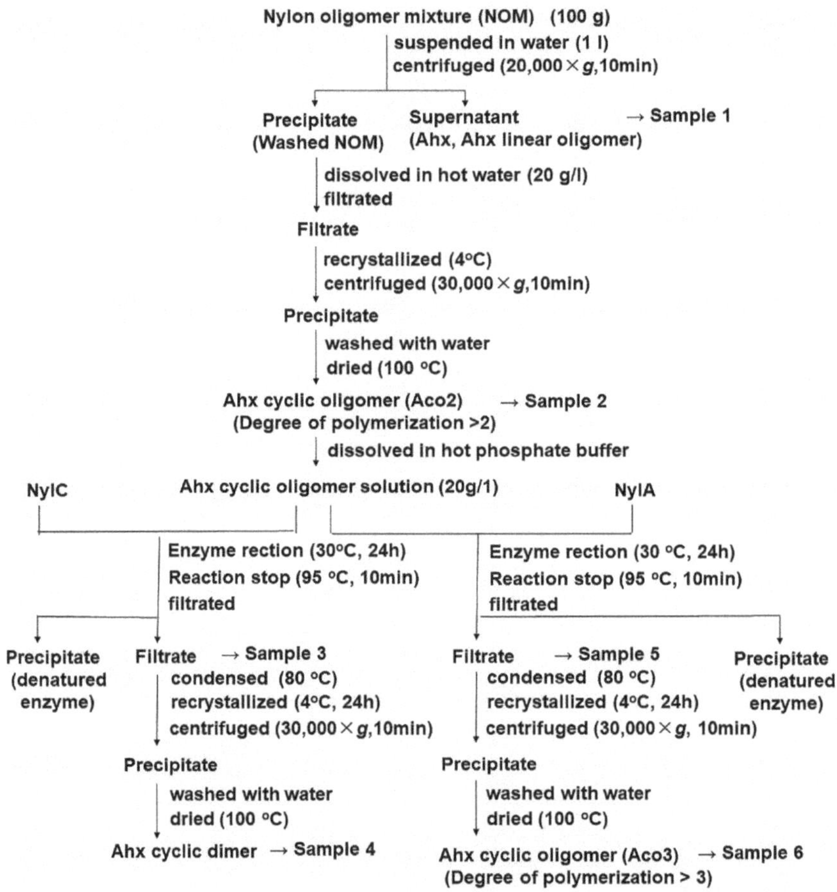

Fig. 2 Protocol of the enzymatic fractionation of 6-aminohexanoate oligomers. 6-Aminohexanoate cyclic dimer (substrate for NylA) and cyclic oligomer (degree of polymerization >3, substrate for NylC) are fractionated from a nylon oligomer mixture (NOM) (waste material from a nylon-6 factory) according to the differences in the solubility in water and specific hydrolysis by NylA and NylC enzymes. Samples at different fractionation steps (samples 1–6) were analyzed by TLC (Fig. 3A).

polymerization >2; sample 2) are roughly fractionated by filtration of the NOM suspended in cold water (Yasuhira et al., 2007).

To fractionate Ahx-cyclic dimer, sample 2 suspended in phosphate buffer is treated with NylC to hydrolyze the cyclic oligomers to water-soluble linear oligomers. Homogeneous Ahx-cyclic dimer (sample 4) is obtained by crystallization in water. Similarly, sample 2 suspended in phosphate buffer is treated with NylA, and soluble linear oligomers are removed. Ahx cyclic oligomers (degree of polymerization >3; sample 6) are purified by crystallization in water. Thin-layer chromatography (TLC) analyses (described in Protocol 6.6) revealed that sample 4 (having no free amino groups) and sample 4 treated with NylC are undetectable by the ninhydrin reaction, since the cyclic dimer is not hydrolyzed by NylC (Fig. 3A, slots 4 and 6). However, sample 4 treated with NylA gives a spot corresponding to authentic Ahx linear dimer (Fig. 3A, slot 5). Similarly, sample 6 and the sample treated with NylA give no spots by the ninhydrin reaction, since the cyclic oligomers (degree of polymerization >3) is not hydrolyzed by NylA (Fig. 3A, slots 8 and 9). In contrast, sample 6 treated with NylC gives spots corresponding to Ahx and linear dimer (Fig. 3A, slot 10).

In addition, mass spectrometric analyses revealed that the molecular weights of the fractionated samples are almost identical to the theoretical values (Kakudo, Negoro, Urabe, & Okada, 1995; Negoro et al., 2012).

To prepare the homogeneous linear oligomers (Ahx dimer, trimer, tetramer) used for enzyme assay and X-ray crystallographic analysis of enzyme-substrate complex, stepwise chemical synthesis is required. Typical liquid phase synthesis of the Ahx linear dimer (Ald) is described in Protocol 6.2 (Fig. 4).

3.2 Limited hydrolysis of nylons by formic acid

Nylon samples with lower molecular weights are obtained from the original samples by partial hydrolysis with formic acid treatment. For example, powder (3 g) of commercial nylon-6 is incubated with 10 ml formic acid (26.07 M) in a glass tube at 100 °C for 10–50 h (see Protocol 6.3). After the reaction, formic acid is removed by lyophilization. From GPC analysis, we have confirmed that weight-average molecular weights (M_w) and number-average molecular weight (M_n) of the commercial nylon samples ($M_w = 56{,}300$; $M_n = 11{,}000$) are reduced with increasing reaction period of formic acid treatment: 10 h ($M_w = 9500$; $M_n = 4400$), 30 h ($M_w = 7200$; $M_n = 3400$), and 50 h ($M_w = 3800$, $M_n = 1500$). The sample was then

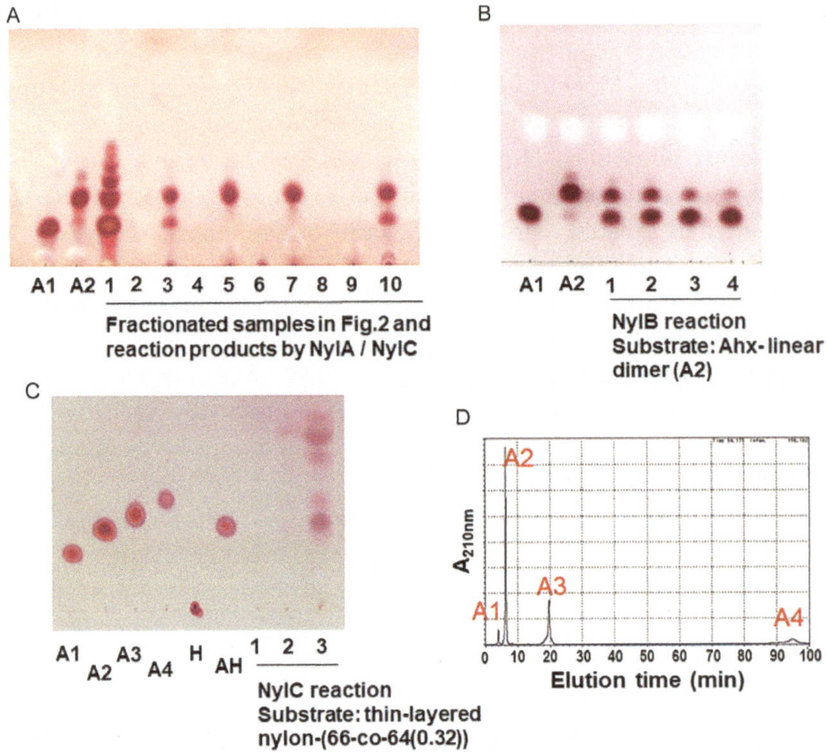

Fig. 3 TLC and HPLC analysis of 6-aminohexanoate oligomers. (A) Samples at fractionation steps (samples 1–6) in Fig. 2 were analyzed by TLC. Spots were detected by a ninhydrin reaction. *Slot 1*, sample 1 (Ahx, Ahx linear oligomer); *slot 2*, sample 2 (Ahx cyclic oligomer with degree of polymerization >2, Aco2); *slot 3*, sample 3 at the fractionation steps in Fig. 2; *slot 4*, sample 4 (Ahx-cyclic dimer); *slot 5*, sample 4 treated with NylA; *slot 6*, sample 4 treated with NylC; *slot 7*, sample 5 at the fractionation steps in Fig. 2; *slot 8*, sample 6 (Ahx-cyclic oligomer with degree of polymerization >3, Aco3); *slot 9*, sample 6 treated with NylA; *slot 10, sample 6* treated with NylC. (B) Reaction products obtained by NylB reaction of chemically synthesized Ahx-linear dimer (Ald) were analyzed. *Slots 1–4*, reaction products after incubating 10 mM Ald with wild-type NylB (0.05 mg/mL) at 30 °C for 15 min (*slot 1*), 30 min (*slot 2*), 1 h (*slot 3*), and 2 h (*slot 4*). (C) Reaction products obtained from thin-layered nylon-(66-co-64(0.32)) copolymer were analyzed. *Slots 1–3*, reaction products after incubating with nylon hydrolase (1 mg/mL) at 30 °C for 0 h (*slot 1*), 40 h (*slot 2*), and 100 h (*slot 3*). *Slot A1*, authentic 6-aminohexanoate (Ahx); *slot A2*, chemically synthesized Ahx-dimer; *slot A3*, Ahx-trimer; *slot A4*, Ahx-tetramer; *slot H*, hexamethylene diamine; *slot AH*, adipoyl-hexamethylene diamine (66-nylon dimer). (D). HPLC analysis of 6-aminohexanoate (Ahx) linear oligomers. Ahx oligomers synthesized from Ahx by reverse reaction of NylB in 90% *t*-butyl alcohol/10% phosphate buffer were analyzed using C_{18} reversed-phase HPLC. Peaks for Ahx (A1), Ahx-dimer (A2), Ahx-trimer (A3), and Ahx-tetramer (A4) are shown.

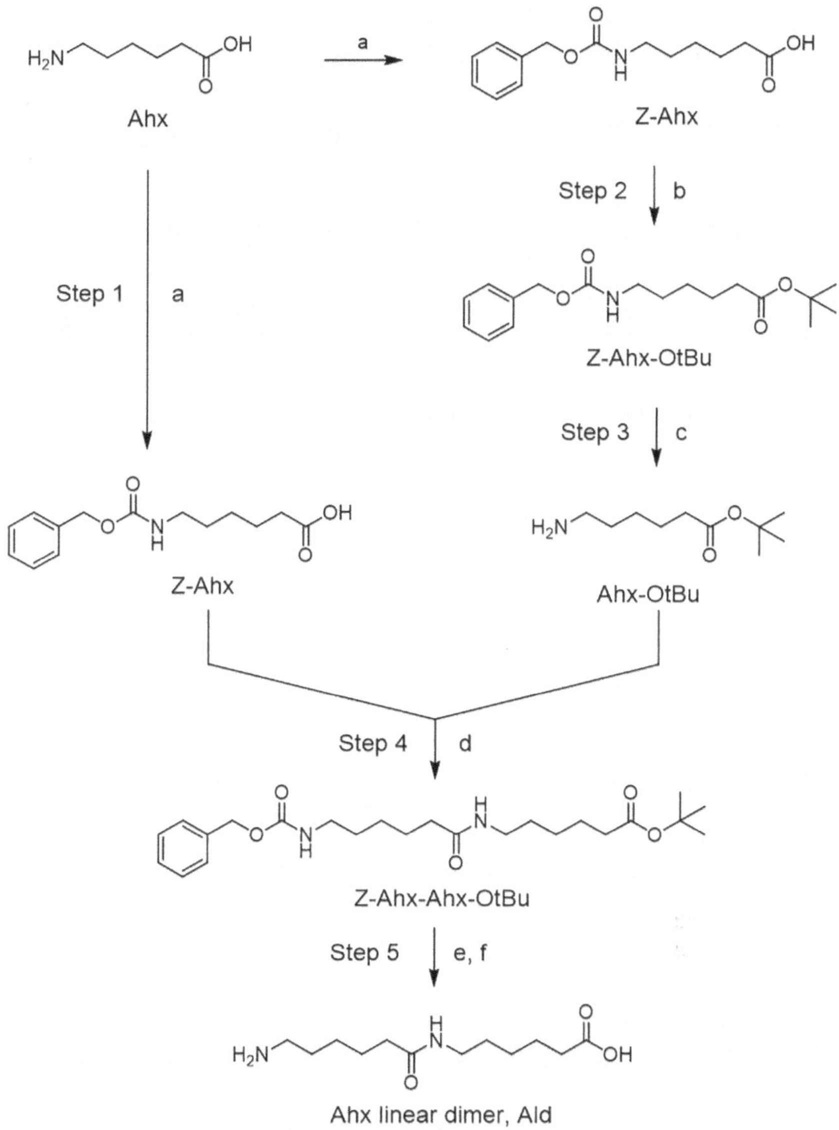

Fig. 4 Chemical synthesis of 6-aminohexanoate linear dimer. 6-aminohexanoate linear dimer (Ald) is chemically synthesized by a liquid phase synthetic method according to the following steps: Step 1, preparation of Z-Ahx; Step 2, preparation of Z-Ahx-OtBu; Step 3, preparation of Ahx-OtBu; Step 4, synthesis of Z-Ahx-Ahx-OtBu: Step 5, synthesis of Ahx linear dimer. Reagents and conditions: (A) Z-Cl, 0 °C; (B) *tert*-BuOH, DCC, DMAP, CH_2Cl_2, 0 °C; (C) H_2, Pd/C, MeOH, room temperature (rt); (D) HOBt, EDC, *N*-ethyldiisopropylamine, DMF, rt.; (E) H_2, Pd/C, MeOH, room temperature; (F) TFA, rt. Details are described in "Protocol 6.2."

disintegrated to micropowder by freezing disruption (Protocol 6.3). This sample is used for the enzymatic reaction.

3.3 Interfacial polymerization of nylon-66, nylon-66/nylon-64 copolymer

On the laboratory scale, nylon-66 and its copolymer with another monomer are chemically synthesized by interfacial polymerization. The fibrous polymer produced at the interface between the hexane and water phases is obtained by rolling the fibers around a glass rod. To produce nylon-66/nylon-64 copolymer, in which a portion of the adipoyl units are replaced with succinyl units, we performed interfacial polymerization using hexamethylenediamine and a mixture of adipoyl chloride (C6) and succinyl chloride (C4) (Protocol 6.4). This copolymer is suitable for enzyme assays using thin-layered nylon films, since the extent of the reaction is monitored by imaging analysis of the decrease in the turbidity of the films (Protocol 6.12) (Fig. 5C) (Nagai et al., 2014).

4. Enzymatic hydrolysis of nylons and related substrates

4.1 Enzyme assay for nylon oligomers

For qualitative detection of the NylA activity, an enzymatic reaction is performed using Ahx-cyclic dimer as substrate in phosphate buffer, and then the reaction product (Ahx linear dimer) is detected on a TLC plate by ninhydrin reagents (Fig. 3A, slot 5) (Protocol 6.6). Similarly, for the detection of NylC activity, Ahx-cyclic oligomer (Aco3) is used as substrate, and the reaction products (Ahx and dimer) are detected (Fig. 3A, slot 10). For the detection of NylB activity, chemically synthesized Ahx-linear dimer is used as the substrate, and the reaction product Ahx is detected (Fig. 3B).

For quantitative analysis of NylB activity, reaction components are fractionated using a C_{18} reversed-phase HPLC column. Enzyme activity is estimated from the decrease in Ahx linear dimer or the increase in Ahx by monitoring the absorbance at 210 nm (Fig. 3D) (Protocol 6.7). Alternatively, it is possible to quantify the activity from the assay of the released amino groups by the colorimetric method using trinitrobenzene sulfonic acid (TNBS) (Protocol 6.8).

For the solid samples, the activity is estimated either from quantification of the soluble components released to the water layer, alteration of mass size, or thickness of the thin-layered nylons, as described in the following sections.

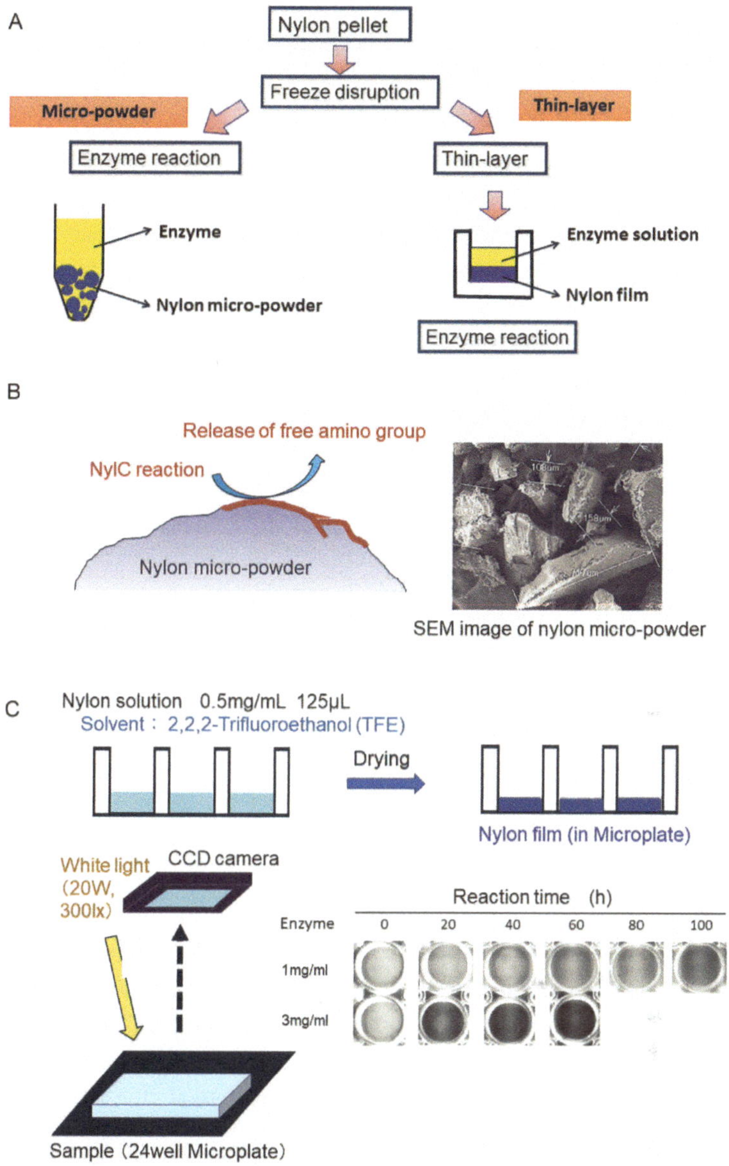

Fig. 5 Enzymatic hydrolysis of polymeric nylons. (A) Scheme of an enzymatic reaction of nylon powder (*left*) and thin-layered nylon (*right*). (B) Illustrated model of the enzymatic hydrolysis of nylon powder. Hydrolytic reactions are performed using nylon hydrolase (0.1–3 mg/mL) in a 20 mM phosphate buffer (pH 7.3) containing 5–50 mg/mL of the nylon powder at 30 °C or 60 °C. Free amino groups released by the hydrolysis of the amide linkage are assayed using TNBS. Calibration curves using the different concentrations of Ahx are constructed, and the amount of free amino groups produced during the hydrolysis of nylon is quantified. (C) Preparation of thin-layered nylon and time course of the enzymatic degradation. *Upper*, Preparation of thin-layered nylon in the wells of a microassay plate. *Lower*, Photograph of nylon layer after incubating with nylon hydrolase (1 and 3 mg/mL). Enzyme reactions are initiated by adding nylon hydrolase (250 μL) to wells of a microassay plate, where the nylon-(66-co-64(0.32)) copolymer is attached as thin-layers. After incubating the plates at 30 °C, all of the soluble fractions in each well are recovered at different incubation periods (0 – 100 h). Enzyme reactions are monitored by the imaging analysis of the thin layer (insoluble fraction) and release of free amino groups (soluble fraction).

4.2 Enzyme assay for powdered nylons

Disintegration of the polymer samples (pellets, fibers) to micrometer sized powder increases the polymer surface, and therefore the degradability of the polymer. Actually, degradation testing using powdered samples is the most practical procedure to evaluate the degradability of the targeted polymers (Fig. 5A). However, it should be noted that particles of nylon powders obtained by the conventional freeze disruption method have large distributions in sizes. SEM analysis of the nylon powders revealed that shapes of the powders are irregular and that the maximum length of 157 particles observed in electron microphotographs was found to range from 68 to 720 μm (average ± standard deviation, 270 ± 140 μm; Fig. 5B) (Negoro et al., 2012). A typical protocol of the hydrolytic reactions using nylon hydrolase is described in Protocol 6.9, in which the reaction is performed in phosphate buffer (pH 7.3) containing 5–50 mg/mL of the nylon powders at 30 °C or 60 °C. Free amino groups released by the hydrolysis are assayed (Protocol 6.8). Since the degradation rate should be dependent on the particles sizes, the degradation is actually decreased with the preferential usage of smaller micropowders over the course of the reaction (Nagai et al., 2014).

4.3 Enzyme assay for thin-layered nylons

To measure the enzyme activity for a solid sample more quantitatively, assay methods using thin-layered nylons are preferable. We made a thin-layer of nylon at the bottom surface of each well in a polystyrene-based microassay plate (Fig. 5C; Protocol 6.10) (Nagai et al., 2014). A layer is made by spreading the nylon solutions dissolved in 2,2,2-trifluoroethanol (TFE) in each well. The nylon layers are dried by incubation in a draft chamber. We described the protocols to prepare thin-layers with average thickness of 260 nm (approximately 600 layers of nylon strands). The enzymatic hydrolysis of thin-layered nylons is described in Protocol 6.11, where the reaction is performed using nylon hydrolase (0.1–3.0 mg/mL in phosphate buffer). To quantify the activity, released amino groups are monitored by the TNBS method (Protocol 6.8). For monitoring the reaction of the insoluble biopolymer at the solid surface, several in situ methods using high-speed atomic force microscopy (AFM) (Igarashi et al., 2012) and quartz crystal microbalance (QCM) (Turon, Rojas, & Deinhammer, 2008) have been reported.

5. Structural analysis of 6-aminohexanoate dimer hydrolase NylB

X-ray crystallographic and computational approaches are effective tools to analyze the structure and catalytic mechanisms of enzymes. By now, we have identified the three-dimensional structures of wild-type NylA, NylB and NylC and various mutants (PDB ID codes: 3A2P, 3A2Q (NylA); 1WYB, 1WYC, 2DCF, 2E8I, 2ZM0, 2ZM7, 2ZMA, 2ZLY, 2ZM2, 2ZM8, 2ZM9 (NylB); 3AXG, 5XYG, 5XYO, 5XYP, 5XYQ, 5XYS, 5XYT, 5Y0L, 5Y0M (NylC)). Basic approaches for gene cloning, high expression and purification of the enzymes, and detailed conventional techniques of the crystallographic and computational analyses have been described in previous references (Bujnicki, 2008; Drenth, 2007; Fuxreiter, 2014). Therefore, in the following sections, we describe the specific techniques used in the X-ray crystallographic and computational analysis, and provide the results of the structure and catalytic mechanisms, focusing on NylB and its variants.

5.1 Crystallization

Crystallization of proteins requires appropriate alignment of the target molecules with the adjacent molecules in solution fostered by suitable precipitants (polyethylene glycol, ammonium sulfate, 2-methyl-2,4-pentanediol, etc.). Since formation of crystals is dependent on the surface structures of protein molecules in the selected solvent environments, suitable crystallization conditions differ among protein molecules and are generally very stringent. For the NylB enzyme, in initial crystallization trials for the wild-type enzyme from *Arthrobacter* sp. KI72, no crystals were obtained under any crystallization conditions tested. Through sparse matrix screening for NylB mutants and various crystallization conditions using screening kits Wizard I and Wizard II (Emerald Biostructures) and implementing the sitting-drop vapor-diffusion technique in 96-well plates (Protocol 6.13), we found a suitable crystallization condition using 2.0–2.2 M ammonium sulfate to induce crystal formation of a NylB mutant (designated as Hyb-24: a hybrid protein composed of wild-type NylB and its homologous carboxylesterase NylB′ protein; Fig. 6A) (Ohki et al., 2005).

5.2 Crystallographic analysis of substrate-unbound enzyme

We performed X-ray crystallographic analysis of the NylB mutants (Hyb-24 and its mutant) according to the following steps. First, X-ray diffraction data

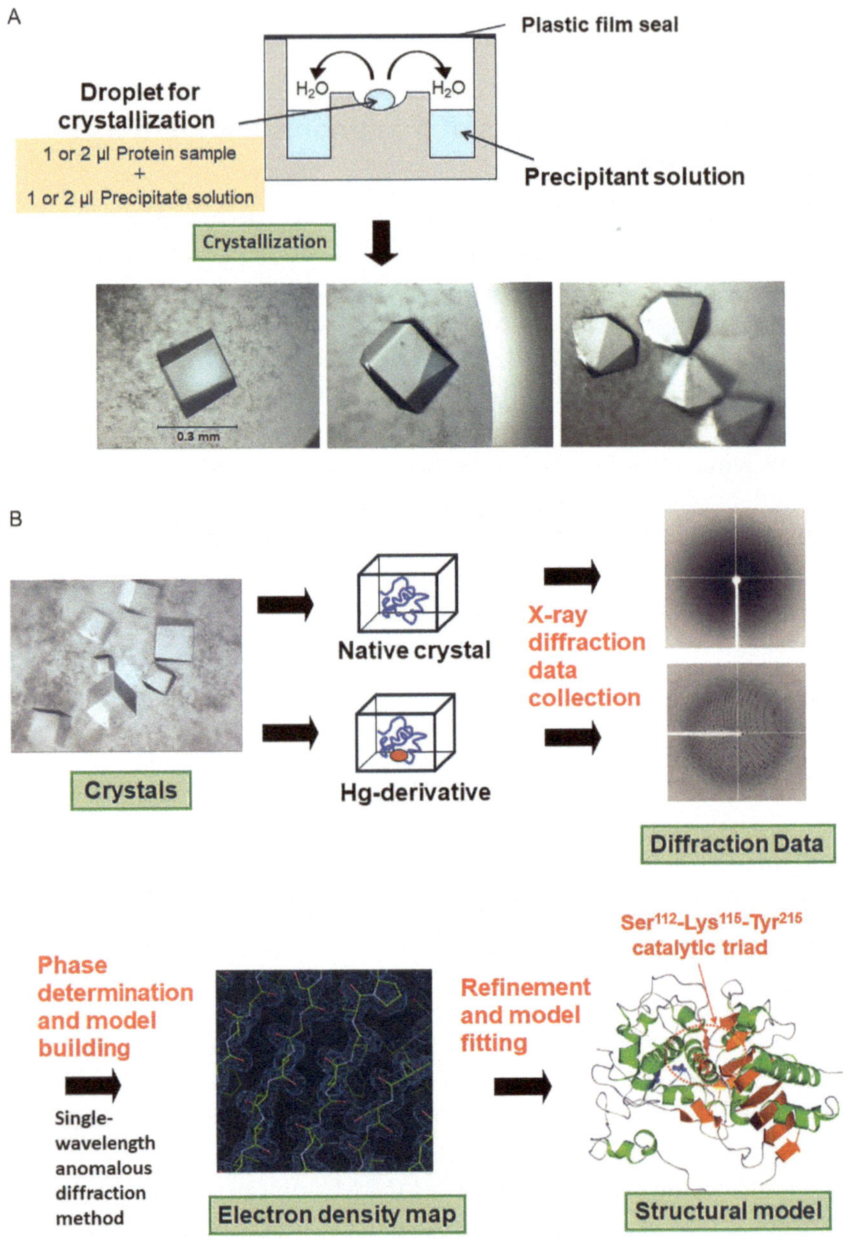

Fig. 6 Crystallization and X-ray crystallographic analysis of the NylB enzyme. (A) Crystallization of purified Hyb-24 by the sitting-drop vapor-diffusion method. Crystals are obtained after 1 week. (B) X-ray crystallographic analysis of NylB. X-ray diffraction data sets from the native and Hg-derivative crystals using synchrotron beamlines. Phase determination was performed by the SAD method using the Hg-derivative data set, followed by the initial model building. The final structure was obtained by iterations of refinement and fitting. Scheme of the preparation of Hg-derivative, X-ray radiation, collection of diffraction data, phase determination, and model building of the three-dimensional structure are shown. Details are described in Protocols 6.13–6.16.

were collected for the native enzyme and the methylmercuric chloride derivative crystals (Fig. 6B) (Protocol 6.14). Second, phase determination was performed by the single wavelength anomalous diffraction (SAD) method using the methylmercuric chloride derivative. Initial phase parameters and electron density map of sufficient quality were obtained to trace the entire protein molecule. Finally, the initial model was then used for positional and B-factor refinement with several cycles of manual model rebuilding (Negoro et al., 2005) (Protocol 6.15).

From these analyses, the mutant NylB enzymes were found to be composed of a two-domain structure that is similar to the folds of the penicillin-recognizing family of serine-reactive hydrolases. Structural analysis of the enzyme suggested that a Ser^{112}-Lys^{115}-Tyr^{215} triad constitutes the catalytic center of the enzyme. This suggestion is further confirmed by site-directed mutagenesis for these amino acid residues. Namely, single S112A, K115A or Y215A substitution in the enzyme by site-directed mutagenesis decreased the enzyme activity to undetectable levels, indicating that the triad is essential for catalysis (Fig. 6B) (Negoro et al., 2005).

5.3 Crystallographic analysis of enzyme-substrate complex

To analyze the catalytic mechanisms and interaction of enzyme with substrate in detail, X-ray crystallographic analysis for the enzyme complexed with substrate is essential. However, it is frequently the case that the enzyme reaction proceeds even in the crystal form for the active enzyme. To prevent the enzymatic hydrolysis of the substrate during soaking of crystals in substrate solution (6-aminohexanoate linear dimer, Ald), the enzyme-substrate complex for the inactive S112A mutant was prepared by soaking the crystals in the cryoprotectant (2.0 M ammonium sulfate, 30% (v/v) glycerol, and 0.1 M MES buffer, pH 6.5) containing 100 mM substrate (Ald) for 3 h (Protocol 6.16). The initial model of the mutant enzyme/Ald complex was obtained from the diffraction data with a similar method using the refined Hyb-24 structure (Negoro et al., 2007) (Fig. 6B).

5.4 Computational analysis

Computational analysis of an enzyme's structure provides us with important information regarding (i) dynamic motion of the protein molecules (fluctuation of loop regions and side chains), (ii) interaction between the enzyme, substrate, coenzyme, metal ion, and water molecules, (iii) transient structure of the protein molecules (from enzyme-substrate complex to enzyme-product complex via various intermediates in the entire catalytic cycle), and (iv) the energetic alterations among the intermediate structures (activation energy of the rate-limiting step, etc.).

In the field of protein engineering, molecular dynamics (MD) simulations are widely used to estimate the possible stable structures of mutated proteins on the basis of the crystallographic structure of the parental protein. Actually, from the initial model of the crystallographic structure of the inactive NylB-Ala^{112} mutant/Ald complex (PDB ID code: 2DCF), we have estimated the structure of active Ser^{112} enzyme complexed with substrate Ald by MD analysis using the computer software package AMBER (http://amber.scripps.edu/) (Fig. 7A) (Baba et al., 2011). A similar approach is applicable for various mutant enzymes, once the equilibrated structure of the enzyme molecule is obtained. From the MD analysis of the behavior of the loop region and water molecules around the catalytic center, we found that movement of the loop segment fills key roles not only in allowing the substrate to be bound by the induced-fit mechanism but also in forming a water-exclusive environment (Baba et al., 2011).

By applying a reactive quantum mechanics/molecular mechanics (QM/MM) approach to analyze the catalytic mechanism of the NylB enzyme (Fig. 7B), we showed that the involvement of Tyr^{170} in the enzyme-substrate interactions is responsible for structural variation in the substrate-binding state (Baba et al., 2015; Kamiya et al., 2014). In addition, we have illustrated the free energy diagram of the catalytic reaction, and revealed that the catalysis is driven by the catalytic triad Ser^{112}, Lys^{115}, and Tyr^{215}, acting as a nucleophile, general base, and general acid, respectively. The acylation step via a tetrahedral intermediate is the rate-limiting step in the catalytic cycle of the NylB reaction, with the free-energy barrier of approximately 88 kJ/mol. The functional interaction of Tyr^{170} with this triad leads to an efficient disruption of the tetrahedral intermediate, promoting a conformational change of the substrate which is favorable for proton donation from the general acid (Kamiya et al., 2014).

Typical protocols for MD and QM/MM analyses of NylB and mutant enzymes are described in Protocols 6.17 and 6.18.

6. Protocol

6.1 Screening of microorganisms degrading nylon-related compounds

1.1 Collect source samples (e.g., waste water from a nylon factory, activated sludge from a sewage disposal plant, soils) to isolate the desired microorganisms.

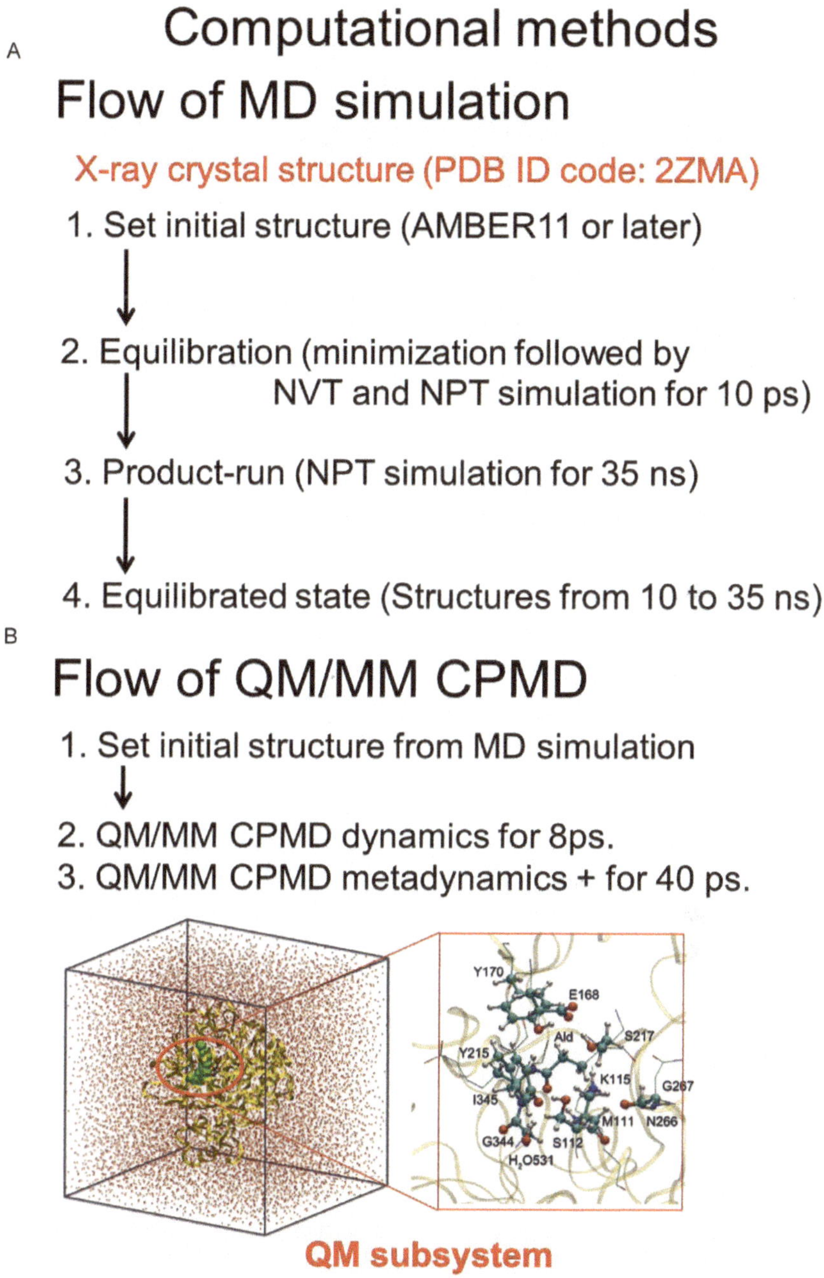

Fig. 7 Computational method for the structural analysis of the NylB enzyme. (A) Molecular dynamic (MD) analysis of NylB and its mutant enzyme. MD analysis is performed by the computer software, "AMBER" (http://amber.scripps.edu/). The atomic coordinates of the NylB/Ald complex are constructed from that of the Hyb-24DN-A^{112}/Ald complex (PDB ID code: 2DCF). Flow of procedures (complementation of deleted region *(Continued)*

1.2 Perform enrichment culture using a minimal medium which includes Ahx cyclic oligomer (washed NOM in Fig. 2) as the sole source of carbon and nitrogen to increase the targeted microorganisms in the collected samples. Compositions of liquid media used for neutrophilic and alkaliphilic microorganisms are shown (Negoro et al., 1992; Yasuhira et al., 2007):

Example 1 (for *Arthrobacter* sp. KI72 and *Pseudomonas* sp. NK87 strains).

Composition: 2 g of Ahx cyclic oligomer (washed NOM), 3 g of KH_2PO_4, 1 g of K_2HPO_4, 2 g of NaCl, 0.25 g of $MgSO_4 \cdot 7H_2O$, 0.8 mg of $FeCl_3$ (pH 6.3) per liter.

Example 2 (for alkalophilic *Agromyces* sp. KY5R and *Kocuria* sp. KY2 strains).

Composition: 2 g of washed NOM, 2 g of $Na_3PO_4 \cdot 12H_2O$, 1 g of K_2HPO_4, 2.6 g of Na_2CO_3, 2 g of $NaHCO_3$, 0.2 g of $MgSO_4 \cdot 7H_2O$, 5 g of NaCl (pH 10) per liter.

0.1 g of yeast extract is added as supplementary nutrients (vitamins etc.), if cell growth is at low level.

1.3 Select nylon oligomer-degrading microorganisms growing on Ahx cyclic oligomer minimal plate (Ahx cyclic oligomer medium containing 1.5% agar). If colonies appear after several days of incubation at 30 °C, each colony is independently picked up, and then purified on the same agar medium.

1.4 Otherwise, select the microorganisms on LB-NOM agar medium (10 g of Bacto Tryptone, 5 g of yeast extract, 10 g of NaCl, 20 g of washed NOM, 15 g of agar per liter). Strains exhibiting degradation activity are selected as colonies which produce a clear zone around the colonies on the turbid agar plate through the degradation of the insoluble nylon oligomers (Fig. 1B).

Fig. 7—Cont'd in the crystallographic structure, setting of protonation state, water molecules, force field, equilibration, energy minimization, introduction of amino acid substitution) are described in Protocols 6.17–6.18. (B) Quantum mechanics/molecular mechanics (QM/MM) analysis. At first, an all-atom model is constructed from the PDB file by adding missing residues and mutation back to the wild-type sequence. Then, classical molecular dynamics (MD) simulation is performed to relax the modeled structure. Unbiased QM/MM Car-Parrinello MD (CPMD) is conducted to obtain the initial structure for QM/MM enzymatic reaction analysis, and then QM/MM CPMD with metadynamics is performed to obtain the free energy map for the enzymatic reaction.

6.2 Chemical synthesis of 6-aminohexanoate linear dimer

2.1 Preparation of *N*-benzyloxycarbonyl-Ahx (*Z*-Ahx).

1. *N*-Benzyloxycarbonyl chloride (Z-Cl, 5 g, 29 mmol) in toluene (30 mL) was added dropwise to the solution of Ahx (3 g, 23 mmol) in aqueous sodium hydrogen carbonate solution (3.6 g, 43 mmol in 60 mL water) at 0 °C (Fig. 4).
2. The reaction mixture was vigorously stirred overnight under 0 °C. To remove unreacted *Z*-Cl, the reaction mixture was washed by $CHCl_3$ three times.
3. After washing, the aqueous phase was acidified by the addition of 1 M aqueous HCl, and extracted by ethyl acetate (EtOAc), washed with brine, dried over Na_2SO_4 and concentrated.
4. Solid *Z*-Ahx obtained was subjected to the next reaction without further purification (4.5 g, 75%).

2.2 Preparation of Z-Ahx-OtBu.

1. *N,N'*-Dicyclohexylcarbodiimide (DCC, 2.43 g, 12 mmol) in CH_2Cl_2 (6 mL) was added dropwise to the solution of Z-Ahx (3 g, 11 mmol), 4-dimethylaminopyridine (DMAP, 0.225 g, 2 mmol) and *tert*-butyl alcohol (tBuOH) (1.68 g, 23 mmol) in CH_2Cl_2 (6 mL) at 0 °C under a nitrogen atmosphere.
2. The reaction was stirred overnight at 0 °C. The precipitate was filtered through a pad of celite, 0.5 M HCl was added to the filtrate, then the filtrate was extracted by EtOAc, washed with saturated aqueous $NaHCO_3$, washed with brine, dried over Na_2SO_4 and concentrated.
3. Obtained crude solid was purified by silica gel column chromatography (hexane/EtOAc = 2:1) to provide white solid Z-Ahx-OtBu.

2.3 Preparation of Ahx-OtBu.

1. The mixture of Z-Ahx-OtBu (1.6 g, 5 mmol) in methanol (6 ml) containing 10% palladium on carbon (100 mg) was stirred vigorously in a H_2 atmosphere under balloon pressure at room temperature.
2. After 3 h, the mixture was diluted with methanol, filtered and concentrated in vacuo to obtain Ahx-OtBu (0.49 g, 53% 2 steps) as a white solid.

2.4 Synthesis of Z-Ahx-Ahx-OtBu.

1. 1-Hydroxybenzotriazole monohydrate (HOBt, 0.38 g, 2.5 mmol) and *N*-ethyldiisopropylamine (0.52 g, 4 mmol) in DMF (3 mL) 1-ethyl-3-(3-dimethylaminopropyl)carbodiimide hydrochloride (EDC, 0.38 g, 2 mmol) in *N,N*-dimethylformamide (DMF) (3 mL) were

added to the solution of Z-Ahx (0.27 g, 1 mmol), Ahx-OtBu (0.19 g, 1 mmol) at 0 °C under a nitrogen atmosphere.

2. The reaction was stirred overnight at room temperature. After removal ofDMF in vacuo, the residual solution was diluted by H_2O, extracted by EtOAc, washed with saturated aqueous $NaHCO_3$, washed with brine, dried over Na_2SO_4 and concentrated.
3. Obtained crude solid was purified by silica gel column chromatography ($CHCl_3$/MeOH = 9:1) to obtain Z-Ahx-Ahx-OtBu as a white solid (0.33 g, 75%).

2.5 Synthesis of Ahx linear dimer.

1. The mixture of Z-Ahx-Ahx-OtBu (0.2 g, 0.5 mmol) in methanol (6 mL) containing 10% palladium on carbon (10 mg) was stirred vigorously in a H_2 atmosphere under balloon pressure at room temperature. After 3 h, the mixture was diluted with methanol, filtered and concentrated in vacuo.
2. Trifluoroacetic acid (TFA) (1.5 mL) was subsequently added and stirred at room temperature. After 6 h, the mixture was concentrated in vacuo to yield Ahx linear dimer (0.12 g, 98%) as a white solid.

6.3 Limited hydrolysis of polymeric nylons by formic acid

3.1 Put nylon pellets (3 g) in grinding jar (35 mL) with a stainless ball (diameter = 20 mm) (Retsch GmbH, Haan, Germany), and cool the jar in liquid nitrogen for 3 min.

3.2 Disintegrate nylon pellets to powder by a Mixer Mill at 20 Hz (3 min) (freezing disruption method).

3.3 Incubate nylon powder obtained (3 g) with 10 mL formic acid (26.07 M) in glass tubes at 100 °C for 10 to 50 h.

3.4 Freeze the samples and store at −80 °C.

3.5 Remove formic acid by lyophilization.

3.6 Add a small amount of water followed by lyophilization to remove residual formic acid.

3.7 Disintegrate the dried sample to micropowder by the freezing disruption method using a Mixer Mill.

Instrument: Mixer Mill (model MM400, Retsch GmbH, Haan, Germany).

6.4 Interfacial polymerization of nylon-66/nylon-64 copolymer

4.1 Prepare a mixture of adipoyl chloride (C6) and succinyl chloride (C4) at the C6/C4 molar ratio of 7/3 to produce nylon-66 in which

30% of the adipoyl units are replaced with succinyl units (510 mg adipoyl chloride and 190 mg succinyl chloride in 20 mL hexane).

4.2 Gradually add the mixture (20 mL) into 20 mL of a hexamethylenediamine solution (460 mg per 20 mL water, containing 0.64 g of sodium hydroxide).

4.3 Recover the fibrous polymer produced at the interface between the hexane and water phases by rolling the fibers around the glass rod.

4.4 Remove the unreacted monomers and soluble oligomer fractions, in which the polymerization reaction was terminated at the oligomer stages, by the following washing processes.

1. Wash with 100 mL of water (containing 2% acetic acid).
2. Wash by 50 mL each of methanol, acetone, and ethyl acetate.
3. Dry the washed sample (100–200 mg) by flowing air.

Note: 1. Confirmation of the structure of the copolymer by ^{1}H NMR spectrometry is recommended, since the actual content of C6/C4 units in the copolymer is not equivalent to the ratio in the monomer compounds. A new peak specific to the succinyl unit was found at 2.7 ppm in the copolymer, and the exact molar ratio of C6/C4 units in the polymer is estimated to be 1/0.32 from the peak height specific for each monomer unit. The copolymer is designated as nylon-(66-co-64(0.32)) and used for the enzyme assay (see Protocols 6.10–6.12).

Instrument: 500 MHz NMR spectrometer (Bruker Biospin K.K., model DRX500, Kanagawa, Japan) at 300 K.

2. GPC analysis of the synthesized polymer is recommended to estimate the weight average molecular weight (M_w) and number average molecular weight (M_n) (see Protocol 6.5) (Nagai et al., 2014).

6.5 Gel permeation chromatography (GPC) analyses of nylons

5.1 Dissolve the nylon powder sample (2 mg) in 4 mL hexafluoroisopropanol/10 mM sodium trifluoroacetate.

5.2 Perform gel permeation chromatography (GPC) analysis.

Instrument: GPC EcoSEC HLC-8220 (Tosoh Co., Tokyo, Japan).

Column: TSK gel SuperHM-H and SuperH2000 columns (Tosoh Co.).

Flow rate: 0.25 mL/min Injection volume: 20 μL.

Temperature: 40 °C Detection: refractive index.

Molecular weight standard: poly(methyl methacrylate) (PMMA).

6.6 Qualitative detection of the enzyme activity by thin-layer chromatography (TLC)

6.1 Perform the enzyme reaction using Ahx-cyclic dimer (assay for NylA activity), linear dimer (NylB activity), or cyclic oligomer (NylC activity) as substrate (in 20 mM phosphate buffer, pH 7.3) in a plastic tube. Enzyme concentration and reaction time should be optimized depending on the enzyme activity

Reaction volume: 50 μL in 1.5 mL Eppendorf tube.

6.2 Spot the reaction mixture (2 μL) onto a silica gel plate, and dry by blowing with a dryer.

6.3 Develop the sample by a solvent mixture (1-propanol:water:ethyl acetate:ammonia = 24:12:4:1.3).

6.4 Detect the degradation products by spraying with 0.2% ninhydrin solution (in 1-butanol saturated with water).

Note: A time course of enzyme reaction using 10 mM Ahx-linear dimer and wild-type NylB (specific activity: 4 U/mg protein) at 30 °C is shown in Fig. 3B. Approximately 50% of the substrate is hydrolyzed in the reaction time of 15–30 min (Negoro et al., 2005).

6.7 Quantitative analysis of reaction products by high-performance liquid chromatography (HPLC)

7.1 Perform the enzyme reaction at 30 °C using Ahx-linear dimer as substrate and suitable concentration of enzyme (depending on the enzyme activity) (see Protocol 6.6)

Reaction volume: 200 μL in 1.5 mLEppendorf tube.

7.2 Sample 20 μL aliquots sequentially from the Eppendorf tube, and stop the enzyme reaction by keeping the tube in a 100 °C bath (boiling water) for 3 min.

7.3 Centrifuge at 15,000 r.p.m. for 5 min to remove the denatured insoluble enzyme, and recover the supernatant (reaction products).

7.4 Analyze the reaction products by C_{18} reversed-phase HPLC.
Injection volume: 10 μL.
Mobile phase: 20 mM phosphate buffer (pH 7.3):CH_3CN = 95:5.
Column temperature: 40 °C.
Flow rate: 0.6 mL/min.
Detection: Absorbance at 210 nm (A_{210nm}).
Note: 1. Detection of Ahx-linear dimer is recommended, since the molar absorption coefficient of the dimer is higher than Ahx, and

the dimer is clearly separated from contaminated hydrophilic cellular components.

2. Analysis of reaction mixture using Ahx-cyclic dimer and oligomer is inappropriate, since these substrates are easily crystallized.
3. It is possible to analyze the Ahx-linear oligomers (up to tetramers) synthesized from Ahx by reverse reaction of NylB in organic solvent (Fig. 3D) (Negoro et al., 2016).

6.8 Quantification of amino groups by a colorimetric method

8.1 Perform the enzyme reaction at 30 °C using suitable concentration of enzyme (depending on the enzyme activity) (see Protocol 6.6)
Reaction volume: 1 mL in 1.5 mL Eppendorf tube.

8.2 Sample 60 μL aliquots sequentially (at 1 min to 16 h) (depending on the enzyme activity) into new Eppendorf tubes.

8.3 Stop the enzyme reaction by keeping the sample tube at 100 °C (in boiling water) for 10 min.

8.4 Add 0.1 M sodium borate solution (pH 8.0, 0.75 mL), 0.3 ml TNBS solution (0.05% trinitrobenzene sulfonic acid sodium salt, 0.065% sodium sulfite) to the samples.

8.5 Keep the tube at 40 °C for 60 min.

8.6 Measure the absorbance at 420 nm (A_{420nm}) using a spectrophotometer.

8.7 Estimate the released amino groups from the calibration curve (A_{420nm} versus Ahx concentration) obtained by various concentrations of Ahx (60 μL) (at steps 8.4–8.6).

6.9 Enzymatic degradation assay with nylon powder

9.1 Disintegrate nylon-6 pellets to powder by a freezing disruption method (Fig. 5) (see Protocol 6.3).

9.2 Add the nylon powder (5–50 mg) into plastic tubes (Eppendorf Co.).

9.3 Add the 20 mM phosphate buffer (pH 7.3) (0.9 mL), and then add NylC enzyme (1–10 mg/mL, 0.1 mL) to each tube (reaction volume: 1 mL).

9.4 Incubate the sample at 60 °C, and remove 60 μL aliquots sequentially (at 5 min to 16 h) (depending on the enzyme activity) into new Eppendorf tubes.

9.5 Detect the reaction products by a colorimetric method using TNBS (see Protocol 6.8).

Note: 1. Reaction volume is reduced to 50 μL for the TLC analysis (see Protocol 6.6).

2. Enzyme concentration, substrate concentration and reaction time are optimized depending on the enzyme activity

6.10 Preparation of thin-layered nylons

10.1 Prepare nylon solution (0.5 mg/mL) dissolved in 2,2,2-trifluoroethanol (TFE).

10.2 Spread 125 μL of the nylon solutions in each well in the polystyrene-based microassay plates (for 24-well plate: $\phi = 16$ mm, Asahi Glass Co., Ltd. Type 1820–024, Tokyo, Japan).

10.3 Dry the nylon layers by incubating at 25 °C for 60 min in a draft chamber.

Notes: 1. Under this condition, thin layers with average thickness of 260 nm are expected to be formed in the wells.

2. To produce the desired thin-layer thickness, alter the amount of nylon (0.015–0.25 mg)

Safety considerations: Handling of TFE should be performed in draft chamber due to the environmental toxicity and inflammable characteristics of TFE.

6.11 Enzymatic hydrolysis of thin-layered nylons

11.1 Create a thin layer of nylons at the bottom surface of each well in the polystyrene-based microassay plates (Fig. 5C) (see Protocol 6.10).

11.2 Start the enzyme reactions by adding nylon hydrolase (0.1–3.0 mg/mL in 250 μL phosphate buffer).

11.3 Incubate the plates at 30 °C for different incubation periods (4–150 h).

11.4 Recover all of the soluble fractions in each well.

11.5 Quantify the concentrations of soluble amino compounds released to aqueous fractions by TNBS methods (Protocol 6.8).

11.6 Quantify the thickness of nylon films if possible. Since thin-layered nylon-(66-co-64(0.32)) copolymer is turbid, it is possible to estimate the thickness by imaging analysis (Fig. 5C, Nagai et al., 2014) (see Protocol 6.12).

Note: The reaction rate of hydrolysis should be essentially constant during the entire reaction, if the surface area of the thin layer is essentially maintained at the same level. In addition, the thickness of the nylon thin films (260 nm) is approximately 1/1000 of the average size of nylon powders.

6.12 Estimation nylon film thickness by imaging analysis

12.1 Set the microplate on dark background in an imaging analyzer equipped with a CCD camera (Fig. 5C).

12.2 Instrument: imaging analyzer model Lumi Vision PRO 400EX (Aishin Seiki Co., Ltd., Kariya, Japan).

12.3 Illuminate the plate by a white light source (20 W, approximately 300lx) from the upward direction.

12.4 Obtain photographic images of the thin-layer.

12.5 Quantify the intensities by image-analyzing programs (Lumi Viewer 2.0 and Lumi Vision Analyzer).

12.6 To analyze the background levels of the intensity, add 125 μL of TFE to an empty well of the micro-assay plate followed by drying as described above.

12.7 Analyze the intensities of the empty well (no treatment) and TFE-treated well.

12.8 Plot the intensity (I_A–I_B) for the thickness of the nylon film, where I_A is the intensity of each nylon film; I_B is the background level of intensity for TFE-treated well.

12.9 Calculate the relative turbidity normalized for nylon-6 film from the following equation: Relative turbidity $= \frac{I_A - I_B}{I_{PA6} - I_B}$, where I_{PA6} is the intensity of nylon-6 film with a thickness of 260 nm.

12.10 Estimate the thickness of the nylon films on the basis of the calculated relative turbidity (Nagai et al., 2014).

Note: The wet nylon-films were more transparent than the dried films, the surfaces of the thin-layers were dried by keeping the plate at 60 °C for 20 min, and their densities were quantified (step 12.4).

Safety considerations: see Protocol 6.10.

6.13 Crystallization of NylB

13.1 Test the crystallization conditions of the targeted proteins using commercially available sparse matrix screening kits. The crystallization condition of NylB mutant (designated as Hyb-24: a hybrid protein composed of wild-type NylB and its homologous carboxylesterase NylB′ protein) was tested using screening kits (Wizard I and Wizard II) (Emerald Biostructures) by implementing the sitting-drop vapor-diffusion technique in 96-well plates.

1. Place 100 μL of the precipitate solution in the reservoir.
2. Mix 1 μL protein sample (17 mg/mL) and 1 μL precipitate solution in the well.

3. Keep the crystallization plates at 10 °C for several days (in some cases, more than 1 month), and check the formation of crystals. Crystals of Hyb-24 were obtained after 1 week.

13.2 After crystallization hits are obtained, optimize the screening conditions by implementing the sitting-drop vapor-diffusion technique in 24-well plates (Fig. 6A).

1. Place 800 μL of the precipitate solution in the reservoir.
2. Mix 2 μL protein sample (34 mg/mL) and 2 μL precipitate solution in the central well.
3. Keep the crystallization plates at 10 °C for several days.

13.3 We finally determined the following crystallization conditions which produce hexagonal-shaped crystals (typical dimensions: 0.3 × 0.3 × 0.3 mm) suitable for X-ray data collection (Fig. 6A).

Precipitant: Ammonium sulfate (2.0–2.2 M).

Additive: Lithium sulfate (0.1–0.2 M) and Glycerol (15–25% (v/v)).

Buffer: 0.1 M MES buffer (pH 6.5).

6.14 Collection of diffraction data in X-ray crystallographic analysis

14.1 For the native data set, soak the crystals in a cryoprotectant solution containing 30% (v/v) glycerol and the other components of the above crystallization condition (Fig. 6B).

14.2 Prepare heavy atom derivatives by soaking the crystals for 24 h in cryoprotectant solution containing 0.1 mM methylmercuric chloride.

Safety considerations: Methylmercuric chloride is very toxic. Wear protective gloves, eye or face protection, and protective clothing to avoid contact with skin and eyes.

14.3 Harvest crystals using a cryoloop (Hampton Research, Inc.) to mount and freeze them on the goniometer head of the X-ray diffractometer in a synchrotron beamline for protein crystallography.

14.4 Perform cryocooling by blowing cold nitrogen gas steam (100 K) using a cryostream cooler (Oxford Cryosystems) onto the mounted crystals.

14.5 Collect diffraction datasets of the frozen crystals at 100 K at 1.8 Å resolution for both the native and the heavy atom derivative crystals. The following parameters were chosen for data collection:

Wavelength: 1.0000 Å.

Crystal to detector distance: 150 mm.

Oscillation range per image: 1°.

Integration of reflections was performed using the HKL2000 program package (Otwinowski & Minor, 1997).

For NylB mutants, resolutions range between 1.05 and 2.0 Å.

6.15 Phase determination, model building and crystallographic refinement

15.1 Perform the phase determination for the collected diffraction data sets. For Hyb-24, the structure was determined by the SAD method using a methylmercuric chloride derivative (Fig. 6B).

15.2 Determine the initial phase parameters by maximum-likelihood heavy-atom parameter refinement using the program SHARP (de La Fortelle & Bricogne, 1997). The electron density map was of sufficient quality to trace almost the entire molecule with ARP/wARP (Langer, Cohen, Lamzin, & Perrakis, 2008) automatically.

15.3 For the region of which the amino acids cannot be traced by ARP/wARP, the remaining structure was manually built by fitting the amino acid residues to the electron density map.

15.4 Perform structure refinement. For Hyb-24, the initial model was used for positional and B-factor refinement with the program CNS (Brünger et al., 1998) with several cycles of manual model rebuilding by XFIT (McRee, 1993) or COOT (Emsley & Cowtan, 2004) to yield an *R*-factor and *R*-free value of 18.5% and 20.3%, respectively.

15.5 Create the Ramachandran plot to verify the modeled structure. For Hyb-24, the Ramachandran plot showed no residues in a disallowed region.

6.16 X-ray crystallographic analysis of enzyme-substrate complex

16.1 Prepare enzyme-substrate complex by soaking the crystals in the cryoprotectant containing substrate. For NylB, an inactive S112A mutant was constructed by site-directed mutagenesis to prevent the enzymatic degradation of the substrate (6-aminohexanoate linear dimer: Ald) in the crystal. Enzyme-Ald complex was prepared by soaking the crystals in the cryoprotectant (2.0 M ammonium sulfate, 30% (v/v) glycerol, and 0.1 M MES buffer at pH 6.5) containing 100 mM Ald for 3 h.

16.2 Collect the diffraction data sets at 100 K using a synchrotron beamline.

16.3 A rigid-body refinement was performed using the coordinates of Hyb-24 to fit the unit cell of the Hyb-24DN crystal, followed by positional and B-factor refinement with CNS.

6.17 Molecular dynamic (MD) analysis

17.1 At first, the all-atom models were constructed in the following manner. The initial structure was taken from the X-ray structure of the S112A mutant of the nylon-6 byproduct-degrading enzyme (resolution 1.51 Å, PDB ID: 2ZMA) (Fig. 7A). To model full length sequence protein structure, eight missing amino acid residues of Met^1, Asn^2, Ala^3, Arg^4, Val^{53}, Asn^{54}, Ala^{55}, and Pro^{56} were added to the X-ray structure and Ala^{112} was mutated back to Ser^{112}. For the protonation state of K^{115}, which was regarded as a general base, we assumed a deprotonated side chain (NH_2). Note here that other Lys residues were protonated. Histidines 31, 108, 180 and 309 were set to be protonated, and the remaining ones were set to neutral after examination of local hydrogen-bonding networks. All Asp and Glu are deprotonated, and all Arg are protonated. By using the model wild-type (WT) protein structure, the E168Q and the Y170F mutants were modeled by replacing E168 with Q168 and Y170 with F170, respectively. The atomic coordinates of Hyb-24 (closed form)/Ald complex and of Hyb-24DN/Ald complex were constructed from those of the Hyb-24DN-K112A/Ald complex (PDB ID code: 2DCF) in the same manner. All protein models are solvated with the TIP-3P water model and neutralized by Na^+ ions by replacing them with water molecules to make the net charge of the total system being 0. These protocols were executed by LEaP modules in AMBER tools.

17.2 The charges of atoms in Ald were obtained using the geometry extracted from the enzyme-substrate complex (PDB ID: 2ZMA) using the electrostatic potential fit calculated with Gaussian 03 (HF/6-31G* level of theory) (Frisch et al., 2004). The general AMBER force field (GAFF) (Wang, Wolf, Caldwell, Kollman, & Case, 2004) and AMBER ff99 (Cornell et al., 1995; Wang, Cieplak, & Kollman, 2000) were used to construct the enzyme and Ald force field using Antechamber modules in AMBER tools (Wang, Wang, Kollman, & Case, 2006), since Ald exhibits a quite similar structure to the amino acids, where the amid bond and C- and N-terminus exist.

17.3 Molecular dynamic (MD) analysis of proteins was performed by the computer software package AMBER ver. 8 (http://amber.scripps.edu/) using the Amber ff99 force field (Pearlman et al., 1995). The periodic boundary condition, where the system is placed in an orthorhombic cell of size $a=83.66$ Å, $b=82.64$ Å, and $c=74.75$ Å, was used. The time step was 2 fs using the SHAKE algorithm (Ryckaert, Ciccotti, & Berendsen, 1977). For the long-range interaction, cut-off length was set to 12 Å. Temperature and pressure were set to 300 K and 1 atm by using the weak-coupling schemes (Berendsen, Postma, Gunsteren, Dinola, & Haak, 1984). (i) First, geometric optimization was performed until the absolute gradient of the system was 0.0001 kcal/mol/Å or less. (ii) Next, 10 ps MD simulation was performed for solvent equilibration. (iii) Then, we performed 14 ps MD simulations for all of the system by heating gradually from 0 to 300 K.

17.4 MD analysis was performed at 300 K using data from 10 to 35 ns with 1 ps intervals (25,000 configurations) to estimate the radial distribution functions between substrate and specific amino acid residues and counting water molecules around the active site.

6.18 Quantum mechanics/molecular mechanics (QM/MM) analysis

18.1 Before starting QM/MM analysis (Warshel & Levitt, 1976), model structures were constructed, and QM/MM settings were required (Fig. 7B). Structural data for QM/MM analysis were taken from the results obtained by MD simulations for the WT and Y170F mutant (see Section 6.16 in detail). The QM region consists of Met^{111}, Ser^{112}, Lys^{115}, Glu^{168}, Tyr^{170}, Tyr^{215}, Ser^{217}, Asn^{266}, Gly^{267}, Gly^{344}, Ile^{345}, two crystallographic water molecules (H_2O^{515} and H_2O^{531} in the PDB ID: 2ZMA), and the Ald substrate. To compensate for the electronic charge of the cut bond of each CH_2 group connecting the QM subsystem to the MM region, an H-like monovalent link atom was used. The QM/MM coupling adopted here is the Hamiltonian scheme that makes use of the restrained electrostatic potential.

18.2 The Hamprecht, Cohen, Tozer, and Handy (HCTH) generalized-gradient corrected exchange-correlation functional (Hamprecht, Cohen, Tozer, & Handy, 1998) was adopted in density functional theory (Kohn & Sham, 1965) to describe the electronic structure of the QM part. Valence-core interactions were described by Troullier-Martins norm-conserving pseudopotentials (Troullier & Martins, 1991). A plane wave basis set with an energy cutoff of 70 Ry was employed. For the MM part, the force field parameters were used in addition to the MD analysis (see Protocol 6.17.3 in detail). The ionic temperature is controlled by velocity rescaling and retained at 300 ± 40 K. An integration time step of 0.097 fs and a fictitious electron mass of 400 au ensure good control of the conserved quantities and preservation of the Born-Oppenheimer adiabaticity. The CPMD 3.15 program package was used for QM/MM Car-Parrinello molecular dynamics (CPMD) (Car & Parrinello, 1985) (CPMD; http://www.cpmd.org/) and Metadynamics simulations (Laio & Parrinello, 2002). Since the C—N bond cleavage in Ald and nucleophilic attack of O_γ of Ser^{112} to the amide C atom in Ald were considered in the analysis, these two distances were adopted as collective variables (CVs) in MetaD simulations. For both collective variables, $M_\alpha = 20.0$ au and $k_\alpha = 0.24$. A new Gaussian penalty potential was added every 13.6 fs, with width and height sampled in the intervals of 0.06–0.36 Å and 0.4–3.8 kcal/mol, respectively. 2980 MetaD steps were needed to obtain a well-converged free-energy surface.

18.3 Unbiased CPMD simulations were performed for 8 ps to investigate the stability of the model structures and to compare them with the X-ray structures. MetaD CPMD simulations were also performed for approximately 40 ps to estimate the free energy landscape for amide-bond cleavage in Ald.

7. Summary

Conversion of synthetic polymers is recognized to be important from industrial and environmental perspectives. Especially, synthetic polyamides are generally highly resistant to biological degradation. A technical barrier to achieve the effective bioconversion of the polymer to monomer is to isolate the responsible enzymes which possess high activity on the targeted polymer. This chapter describes our approaches, i.e., strategy and protocols for enzyme assays and preparation of model substrates, to explore novel enzymes from the natural environment, and structural analysis of the enzymes to reveal the catalytic mechanisms and to improve the functions by protein engineering.

Acknowledgments

This work was supported in part by a grant-in-aid for scientific research (Japan Society for Promotion of Science) and grants from A-STEP (Adaptable & Seamless Technology Transfer Program through Target-driven R&D) (Japan Science and Technology Agency).

References

Baba, T., Boero, M., Kamiya, K., Ando, H., Negoro, S., Nakano, M., et al. (2015). Unraveling the degradation of artificial amide bonds in nylon oligomer hydrolase: From induced-fit to acylation processes. *Physical Chemistry Chemical Physics*, *17*, 4492–4504.

Baba, T., Kamiya, K., Matsui, T., Shibata, N., Higuchi, Y., Kobayashi, T., et al. (2011). Molecular dynamics studies on the mutational structures of a nylon-6 byproduct-degrading enzyme. *Chemical Physics Letters*, *507*, 157–161.

Berendsen, C. H. J., Postma, M. P. J., Gunsteren, F. W., Dinola, A., & Haak, R. J. (1984). Molecular dynamics with coupling to an external bath. *The Journal of Chemical Physics*, *81*, 3684–3690.

Brünger, A. T., Adams, P. D., Clore, G. M., DeLano, W. L., Gros, P., Grosse-Kunstleve, R. W., et al. (1998). Crystallography & NMR system (CNS), a new software suite for macromolecular structure determination. *Acta Crystallographica Section D*, *54*, 905–921.

Bujnicki, J. M. (2008). *Prediction of protein structures, functions and interactions*. Wiley.

Car, R., & Parrinello, M. (1985). Unified approach for molecular dynamics and density-functional theory. *Physical Review Letters*, *55*, 2471–2474.

Cornell, W. D., Cieplak, P., Bayly, C. I., Gould, I. R., Merz, K. M., Ferguson, D. M., et al. (1995). A second generation force field for the simulation of proteins, nucleic acids, and organic molecules. *Journal of the American Chemical Society*, *117*, 5179–5197.

de La Fortelle, E., & Bricogne, G. (1997). Maximum-likelihood heavy-atom parameter refinement for multiple Isomorphous replacement and multiwavelength anomalous diffraction methods. *Methods in Enzymology*, *276*, 472–494.

Deguchi, T., Kitaoka, Y., Kakezawa, M., & Nishida, T. (1998). Purification and characterization of a nylon-degrading enzyme. *Applied and Environmental Microbiology*, *64*, 1366–1371.

Drenth, J. (2007). *Principles of protein X-ray crystallography*. New York: Springer.

Emsley, P., & Cowtan, K. (2004). Coot: Model-building tools for molecular graphics. *Acta Crystallographica Section D*, *60*, 2126–2132.

Frisch, M. J., Trucks, G. W., Schlegel, H. B., Scuseria, G. E., Robb, M. A., Cheeseman, J. R., et al. (2004). *Gaussian 03, Revision E* (p. 01). Wallingford CT: Gaussian Inc.

Fuxreiter, M. (2014). *Computational approaches to protein dynamics: From quantum to coarse-grained methods*. CRC Press.

Hamprecht, F. A., Cohen, A. J., Tozer, D. J., & Handy, N. C. (1998). Development and assessment of new exchange-correlation functionals. *The Journal of Chemical Physics*, *109*, 6264–6271.

Igarashi, K., Uchihashi, T., Koivula, A., Wada, M., Kimura, S., Penttilä, M., et al. (2012). Visualization of cellobiohydrolase I from *Trichoderma reesei* moving on crystalline cellulose using high-speed atomic force microscopy. *Methods in Enzymology*, *510*, 169–182.

Kakudo, S., Negoro, S., Urabe, I., & Okada, H. (1995). Characterization of endotype 6-aminohexanoate-oligomer hydrolase from *Flavobacterium* sp. *Journal of Fermentation and Bioengineering*, *80*, 12–17.

Kamiya, K., Baba, T., Boero, M., Matsui, T., Negoro, S., & Shigeta, Y. (2014). Nylon-oligomer hydrolase promoting cleavage reactions in unnatural amide compounds. *Journal of Physical Chemistry Letters*, *5*, 1210–1216.

Kohn, W., & Sham, L. J. (1965). Self-consistent equations including exchange and correlation effects. *Physics Review*, *140*, A1133–A1138.

Laio, A., & Parrinello, M. (2002). Escaping free-energy minima. *Proceedings of the National Academy of Sciences of the United States of America*, *99*, 12562–12566.

Langer, G., Cohen, S. X., Lamzin, V. S., & Perrakis, A. (2008). Automated macromolecular model building for X-ray crystallography using ARP/wARP version 7. *Nature Protocols*, *3*, 1171–1179.

McRee, D. E. (1993). *Practical protein crystallography*. San Diego: Academic Press.

Nagai, K., Iida, K., Shimizu, K., Kinugasa, R., Izumi, M., Kato, D., et al. (2014). Enzymatic hydrolysis of nylons: Quantification of the reaction rate of nylon hydrolase for thin-layered nylons. *Applied Microbiology and Biotechnology*, *98*, 8751–8761.

Negoro, S. (2000). Biodegradation of nylon oligomers. *Applied Microbiology and Biotechnology*, *54*, 461–466.

Negoro, S. (2002). Biodegradation of nylon and other synthetic polyamides. In Steinbuechel (Ed.), *9. Biopolymers* (pp. 395–415). Berlin: Springer-Verlag.

Negoro, S., Kakudo, S., Urabe, I., & Okada, H. (1992). A new nylon oligomer degradation gene (*nylC*) on plasmid pOAD2 from a *Flavobacterium* sp. *Journal of Bacteriology*, *174*, 7948–7953.

Negoro, S., Kawashima, Y., Shibata, N., Kobayashi, T., Baba, T., Lee, Y. H., et al. (2016). Mutations affecting the internal equilibrium of the reaction catalyzed by 6-aminohexanoate-dimer hydrolase. *FEBS Letters*, *590*, 3133–3143.

Negoro, S., Ohki, T., Shibata, N., Mizuno, N., Wakitani, Y., Tsurukame, J., et al. (2005). X-ray crystallographic analysis of 6-aminohexanoate-dimer hydrolase: Molecular basis for the birth of a nylon oligomer degrading enzyme. *Journal of Biological Chemistry*, *280*, 39644–39652.

Negoro, S., Ohki, T., Shibata, N., Sasa, K., Hayashi, H., Nakano, H., et al. (2007). Nylon-oligomer degrading enzyme/substrate complex: Catalytic mechanism of 6-aminohexanoate-dimer hydrolase. *Journal of Molecular Biology*, *370*, 142–156.

Negoro, S., Shibata, N., Lee, Y. H., Takehara, I., Kinugasa, R., Nagai, K., et al. (2018). Structural basis of the correct subunit assembly, aggregation, and intracellular degradation of nylon hydrolase. *Scientific Reports*, *8*, 9725.

Negoro, S., Shibata, N., Tanaka, Y., Yasuhira, K., Shibata, H., Hashimoto, H., et al. (2012). Three-dimensional structure of nylon hydrolase and mechanism of nylon-6 hydrolysis. *Journal of Biological Chemistry*, *287*, 5079–5090.

Ohki, T., Mizuno, N., Shibata, N., Takeo, M., Negoro, S., & Higuchi, Y. (2005). Crystallization and X-ray diffraction analysis of 6-aminohexanoate-dimer hydrolase from *Arthrobacter* sp. KI72. *Acta Crystallographica. Section F*, *61*, 928–930.

Otwinowski, Z., & Minor, W. (1997). Processing of X-ray diffraction data collected in oscillation mode. *Methods in Enzymology*, *276*, 307–326.

Pearlman, D. A., Case, D. A., Caldwell, J. W., Ross, W. S., Cheatham, W. S. T. E., III, DeBolt, S., et al. (1995). AMBER, a package of computer programs for applying molecular mechanics, normal mode analysis, molecular dynamics and free energy calculations to simulate the structural and energetic properties of molecules. *Computer Physics Communications*, *91*, 1–41.

Ryckaert, J. P., Ciccotti, G., & Berendsen, H. J. C. (1977). Numerical integration of the Cartesian equations of motion of a system with constraints: Molecular dynamics of *n*-alkanes. *Journal of Computational Physics*, *23*, 321–341.

Takehara, I., Fujii, T., Tanimoto, Y., Kato, D., Takeo, M., & Negoro, S. (2018). Metabolic pathway of 6-aminohexanoate in the nylon oligomer-degrading bacterium *Arthrobacter* sp. KI72: Identification of the enzymes responsible for the conversion of 6-aminohexanoate to adipate. *Applied Microbiology and Biotechnology, 102*, 801–814.

Troullier, N., & Martins, J. L. (1991). Efficient pseudopotentials for plane-wave calculations. *Physical Review B, 43*, 1993–2006.

Turon, X., Rojas, O. J., & Deinhammer, R. S. (2008). Enzymatic kinetics of cellulose hydrolysis: A QCM-D study. *Langmuir, 24*, 3880–3887.

Wang, J., Cieplak, J., & Kollman, P. A. (2000). How well does a restrained electrostatic potential (RESP) model perform in calculating conformational energies of organic and biological molecules? *Journal of Computational Chemistry, 21*, 1049–1074.

Wang, J., Wang, W., Kollman, P. A., & Case, D. A. (2006). Automatic atom type and bond type perception in molecular mechanical calculations. *Journal of Molecular Graphics & Modelling, 25*, 247260.

Wang, J., Wolf, R. M., Caldwell, J. W., Kollman, P. A., & Case, D. A. (2004). Development and testing of a general AMBER force field. *Journal of Computational Chemistry, 25*, 1157–1174.

Warshel, A., & Levitt, M. (1976). Theoretical studies of enzymic reactions: Dielectric, electrostatic and steric stabilization of the carbonium ion in the reaction of lysozyme. *Journal of Molecular Biology, 103*, 227–249.

Yasuhira, K., Shibata, N., Mongami, G., Uedo, Y., Atsumi, Y., Kawashima, Y., et al. (2010). X-ray crystallographic analysis of the 6-aminohexanoate cyclic dimer hydrolase: Catalytic mechanism and evolution of an enzyme responsible for nylon-6 byproduct degradation. *Journal of Biological Chemistry, 285*, 1239–1248.

Yasuhira, K., Tanaka, Y., Shibata, H., Kawashima, Y., Ohara, A., Kato, D., et al. (2007). 6-Aminohexanoate oligomer hydrolases from the alkalophilic bacteria *Agromyces* sp. strain KY5R and *Kocuria* sp. strain KY2. *Applied and Environmental Microbiology, 73*, 7099–7102.

CHAPTER EIGHTEEN

Upcycling of hydrolyzed PET by microbial conversion to a fatty acid derivative

Gina Welsing, Birger Wolter, Henric M.T. Hintzen, Till Tiso, and Lars M. Blank*
iAMB—Institute of Applied Microbiology, ABBt—Aachen Biology and Biotechnology, RWTH Aachen University, Aachen, Germany
*Corresponding author: e-mail address: lars.blank@rwth-aachen.de

Contents

Abstract

The enzymatic degradation of polyethylene terephthalate (PET) results in a hydrolysate consisting almost exclusively of its two monomers, ethylene glycol and terephthalate. To biologically valorize the PET hydrolysate, microbial upcycling into high-value products is proposed. Fatty acid derivatives hydroxyalkanoyloxy alkanoates (HAAs) represent such valuable target molecules. HAAs exhibit surface-active properties and can be exploited in the catalytical conversion to drop-in biofuels as well as in the polymerization to bio-based poly(amide urethane). This chapter presents the genetic engineering methods of pseudomonads for the metabolization of PET monomers and the biosynthesis of HAAs with detailed protocols concerning product purification.

Methods in Enzymology, Volume 648
ISSN 0076-6879
https://doi.org/10.1016/bs.mie.2020.12.025

1. Introduction

One of the main challenges for modern society is to shift the current plastic industry towards a sustainable one. The development of modern plastic began in the 19th century with the invention of vulcanized rubber as well as polystyrene (PS) and polyvinyl chloride (Andrady & Neal, 2009). The production of plastic on industrial scale expanded in the early 1950s (Geyer, Jambeck, & Law, 2017). Since then, further synthetic polymers have been developed and made their way into nearly every aspect of our life. Their omnipresence in our daily life has led to the suggestion that we are living in the "plastic age" (Thompson, Moore, vom Saal, & Swan, 2009). Humankind undoubtedly benefits from the versatile properties of plastics like being water, gas, and shock-resistant, thermally, and electrically insulating, lightweight yet strong, and resilient (Li, 2020; PlasticsEurope, 2019; Thompson et al., 2009; Venkatachalam et al., 2012). These properties result in, for instance, the reduction of packaging weight, increase in food shelf life, or improved insulation of homes and thus contributing to a more resource-efficient economy. While all these plastic properties favor environmentally beneficial solutions, also the tremendous environmental damages caused by plastic waste must be addressed.

In 2018, 360 mt of new plastic were produced worldwide, and in total, 390 mt of plastic was converted into products (Conversio, 2018; PlasticsEurope, 2019). The production is continuously growing at a rate of approximately 3% per annum (Geyer et al., 2017). Also in 2018 around 250 mt of plastic waste were generated worldwide, of which 20% was recycled, 20% incinerated for energy recovery, 30% deposited in landfills, and the whereabouts of another 30% is unknown, so probably released into the environment due to improper disposal and leakage (Conversio, 2018). In the EU, better rates are reported with 32% of the plastic waste recycled, 43% incinerated for energy recovery, and 25% of the post-consumer plastic waste accumulates in landfills (PlasticsEurope, 2019). These numbers can differ depending on the data source and how the term "recycling" is defined. For example, 19% of the 32% recycled plastic waste was indeed exported to low-income countries (Lau et al., 2020) and nevertheless counted as recycled (Plastics Europe, 2019). Especially the direct release into the environment but also mismanaged landfills and old incineration plants pose a burden to the environment. Mismanaged landfills can release plastic into the environment, especially during extreme weather conditions, such as flooding. While in

Europe, tire abrasion is the main plastic entry into the environment, mismanaged post-consumer plastic packaging is a major source elsewhere, e.g., in Asia and Africa (Bertling, Bertling, & Hamann, 2018; Conversio, 2018). The resulting microplastics can accumulate in soil, and marine environments and might thus influence the entire ecosystem (Barnes, Galgani, Thompson, & Barlaz, 2009; Espinosa et al., 2020; Rochman et al., 2015). The impact of plastic waste, including microplastics, on the environment is still not fully understood with all its consequences and a subject of ongoing research.

These two sides of plastic, the beneficial product characteristics, on the one hand, and the adverse environmental effects, on the other hand, show the urgent need for sustainable plastic waste management and upcycling concepts.

Plastic recycling can be performed via mechanical or chemical methods or a combination thereof. On industrial scale, the vast majority of recycling is performed via mechanical methods, with chemical recycling only emerging recently on pilot scale (PlasticsEurope, 2019). Mechanical technologies depend on sorting steps, followed by remodeling of individual polymer types. The direct use of recyclate from mechanical recycling is often hindered, as recycling usually causes quality reduction (e.g., loss of molecular weight). Thus, blending with virgin plastic might be required for its reutilization in the same application. Alternatively, applications accepting plastics with lower quality need to be chosen, which results in continuous downcycling (Hestin et al., 2017; Van Eygen, Laner, & Fellner, 2018).

An alternative way to increase plastic recycling is to add additional value to plastic waste, not aiming for the same material or consumer good, but rather upcycling to higher-value chemicals and materials. As an economically valuable raw material, the interest in collecting and upcycling plastic waste would automatically increase to such an extent that the release of plastic waste into the environment would decrease significantly. The concept of upcycling has already been demonstrated using chemical methods such as different solvolysis (Carta, Cao, & D'Angeli, 2003; Vollmer et al., 2020), hydrogenolysis (Vollmer et al., 2020; Westhues, Idel, & Klankermayer, 2018), and pyrolysis processes (Vollmer et al., 2020).

However, these chemical upcycling methods will cause new environmental burdens, for instance, high energy costs (Garcia & Robertson, 2017; Geyer et al., 2017; PlasticsEurope, 2019). In Meys et al. (2020), a life cycle assessment (LCA) has shown that chemical recycling has the potential to reduce the environmental impact of plastic recycling compared to the current benchmark recycling methods, including mechanical recycling and incineration. However, it is clearly stated that every process pathway needs to be analyzed, and only some can compete with current benchmark recycling methods. In general, chemical recycling to monomers and value-added products tends to reduce the environmental impact, whereas chemical recycling to feedstocks and fuel tends to increase it.

Bio-upcycling of plastics represents a promising alternative. It follows the same principle giving plastic waste an additional value, whereby the chemical conversion methods are replaced by biological methods (Wei et al., 2020). The carbon-rich plastic waste streams are adopted as a substrate for biotechnological processes. This process resembles the natural degradation of lignocellulosic material, starting with the enzymatic hydrolysis of polymer substrates, followed by the metabolization of the resulting hydrolysates by microorganisms, which then secrete valuable molecules (Tiso et al., 2020). Potentially, the conversion from plastic to value-added chemicals should be even easier than from lignocellulosic biomass since the substrate is less complex, consisting of simple repeating monomers (Tiso, Narancic, et al., 2020). For instance, PET is a polymer consisting of ethylene glycol (EG) and terephthalic acid (TPA), with only minor fractions of copolymers like plasticizers, fillers, and colorants, depending on the application (e.g., isophthalic acid, diethylene glycol, or 1,4-cyclohexanedimethanol) (de Ilarduya & Muñoz-Guerra, 2014).

Yoshida et al. (2016) isolated a wild-type bacterium (*Ideonella sakaiensis*), able to grow on and assimilate amorphous PET. It was suggested that ester-bond hydrolyzing PETase and mono-(2-hydroxyethyl)TPA (MHET)ase were produced and hydrolyzed the PET. The molecular basis of the enzymes of this strain was also reported in other publications (Austin et al., 2018; Palm et al., 2019). This has been an exciting discovery, accelerating the hope of natural bioremediation of sites contaminated with plastic waste. However, it is not suggested to wait for natural evolution to occur (Wei et al., 2020). Plastic degradation seems to be slow at best, and the introduction of new plastic into the environment by far exceeds its decay (Jacquin et al., 2019).

The degradation of lignocellulosic material has had an evolutionary head start since hundreds of millions of years. Therefore, it is necessary to accelerate the natural evolution for plastic biodegradation with molecular biological methods. The biocatalytic upcycling of plastics must be optimized to such an extent that enzymes and microbes can rapidly and efficiently degrade and transform synthetic polymers previously considered non-biodegradable.

Proofs of principle for bio-upcycling of selected plastics are already available. For example, plastic monomers from pyrolyzed PET have been successfully converted to the intracellular biopolymer polyhydroxyalkanoate (PHA) by *Pseudomonas umsongensis* GO16 (previously classified as *Pseudomonas putida* GO16) (Kenny et al., 2008, 2012), and to five different aromatics and aromatic-derived compounds by engineered *Escherichia coli* (Kim et al., 2019). Similarly, pyrolyzed PE and PS have been converted by *Pseudomonas* spp. to PHA (Guzik et al., 2014; Ward, Goff, Donner, Kaminsky, & O'Connor, 2006). Furthermore, PET previously considered as recalcitrant to biological degradation has been depolymerized to its monomers by hydrolytic enzymes produced by bacteria and fungi (Tournier et al., 2020; Wei, Breite, et al., 2019; Wei, Song, et al., 2019; Wei & Zimmermann, 2017). In a recent publication, PET films were hydrolyzed with a thermostable polyester hydrolase resulting in nearly 100% EG and TPA. The hydrolysate was then utilized as carbon and energy feedstock for engineered *P. umsongensis* GO16 strains to synthesize the extracellular fatty acid derivative hydroxyalkanoyloxy-alkanoate (HAA) and the intracellular biopolymer PHA (Tiso, Narancic, et al., 2020). This chapter presents the workflow for converting EG and terephthalate (TA) to HAA by the evolved *P. umsongensis* GO16 KS3, including genetic engineering, cultivation, and analytical methods (Fig. 1).

Pseudomonas spp. are ubiquitous Gram-negative proteobacteria, which have been isolated from many habitats, including various soils, rhizospheres, fresh and marine water, animal tissue, and buildings (Remold, Purdy-Gibson, France, & Hundley, 2015). They are promising hosts for biotechnological applications since they entail a highly versatile metabolism, which enables the metabolization of a variety of substrates, including glycerol, xylose, and aromatic compounds (Blank, Ionidis, Ebert, Buhler, & Schmid, 2008; Köhler, Blank, Frick, & Schmid, 2015; Tiso, Wierckx, & Blank, 2014; Wierckx, Ballerstedt, de Bont, & Wery, 2005). Pseudomonads can synthesize industrially relevant chemicals like furandicarboxylic acid and the biodegradable plastic PHA from these substrates (Koopman, Wierckx, de Winde, & Ruijssenaars, 2010; Sun, Ramsay, Guay, & Ramsay, 2007). Thanks to their flexible and robust metabolism, *Pseudomonas* spp. can withstand stressful conditions like organic solvents or oxidative stress (Li, 2020; Nikel & de Lorenzo, 2018; Tiso et al., 2014; Zobel et al., 2015).

P. umsongensis strain GO16 is natively able to metabolize TA (Kenny et al., 2008). Tiso, Narancic, et al. (2020) and Tiso, Sauer, et al. (2020) applied adaptive laboratory evolution (ALE) to enable the additional

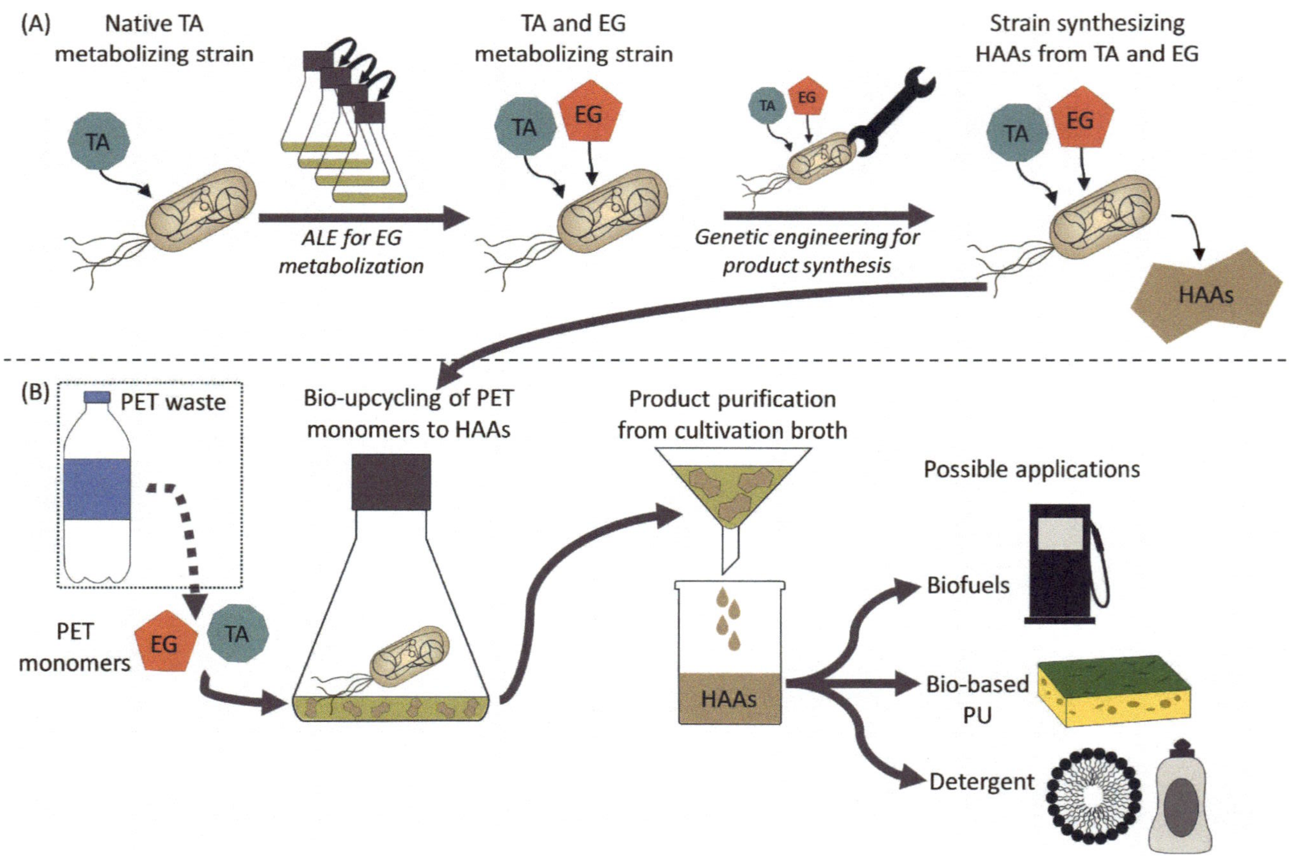

Fig. 1 Overview of the valorization of hydrolyzed polyethylene terephthalate (PET) by microbial transformation. (A) Bacterial strain development for metabolizing terephthalate (TA) and ethylene glycol (EG) for the production of hydroxyalkanoyloxy alkanoates (HAAs). (B) Biosynthesis, purification, and possible applications of HAAs obtained by bio-upcycling of PET monomers, potentially obtained from enzymatic hydrolysis of waste PET. PU: poly(amide urethane).

metabolization of EG as a second monomer from PET. The ALE procedure is described in the method section (see Section 2.1) of this chapter.

HAAs were originally described as the hydrophobic precursor of surface-active rhamnolipids. Recently, HAAs were also classified as biosurfactants themselves since they exhibit surface-active properties (Deziel, Lepine, Milot, & Villemur, 2003; Tiso et al., 2017). Moreover, they have been suggested as a renewable carbon-based molecule for diverse biotechnological applications. For instance, the use of HAAs as starting molecules for drop-in biofuels as well as their application as monomers for the chemo-catalytic synthesis of a novel bio-based poly(amide urethane) (Fig. 1B) have recently been reported (Mensah et al., 2020; Meyers et al., 2019; Tiso, Narancic, et al., 2020). Furthermore, some congeners are suited as precursors for ethenolysis to synthesize 1-octene (Tiso, Sauer, et al., 2020). To date, HAAs can be bio-derived from glucose, xylose, and ethanol (Bator, Karmainski, Tiso, & Blank, 2020; Bator, Wittgens, Rosenau, Tiso, & Blank, 2020; Meyers et al., 2019). Tiso, Narancic, et al. (2020) described the biosynthesis of HAAs from EG and TA with an evolved *P. umsongensis* GO16. Here, we present the methods employed for the bio-upcycling of PET to HAAs, including the determination of substrate consumption, genetic engineering for HAA synthesis, biosynthesis of HAA, as well as quantification and purification of HAAs.

2. Protocols

General laboratory safety procedures and personal protective materials (incl. lab coat, gloves, and safety glasses) should be used during the experiments. Solvents utilized for analytical purposes should be analytical grade. Water for analytics should be ultra-purified (e.g., using a MilliQ Direct, Merck, Darmstadt, Germany).

2.1 Adaptive laboratory evolution (ALE)

ALE is a useful strategy to adapt strains to clearly defined conditions. The procedure is methodically straightforward and enables the selection of mutated strains with desired properties as improved growth rates, reduced lag-phases, or increased chemical tolerances. ALE procedures can be performed with different microorganisms and strains to establish, for instance, the metabolism of a specific substrate.

The strain *P. umsongensis* GO16 can natively metabolize TA. The ALE procedure aims to extend this strain's metabolic spectrum by enabling the conversion of the non-native substrate EG as the second PET hydrolysis product. With the constant selective pressure of EG as the sole carbon source, repetitive batch cultivation should be carried out for many (>100) generations. The period of time chosen for ALE experiments depends on the growth rate, the observed mutations, and adaptions of the evolved strain. When ALE is applied to enhance growth, the basic prerequisite is minimal growth of the microorganism on the target substrate.

2.1.1 Materials, equipment, and reagents

2.1.1.1 Materials

- *Pseudomonas umsongensis* GO16 (GenBank: CP044409.1)

2.1.1.2 Equipment

- Incubation shaker (e.g., Infors HT Multitron shaker (50mm shaking diameter) by Infors AG, Bottmingen, Switzerland)
- Shake flasks with metal caps (250mL for pre-cultures, 500mL for main cultures)
- Centrifuge (e.g., Centrifuge 5920 R by Eppendorf, Hamburg, Germany)
- Swinging bucket rotor (e.g., S-4 × 750 by Eppendorf, Hamburg, Germany)
- 15 and 50mL centrifuge tubes (e.g., Falcon Conical Centrifuge Tubes by Thermo Fisher Scientific, Waltham, MA, United States)
- Photometer (e.g., Ultrospec 10 Cell Density Meter by Biochrom Ltd., Holliston, MA, USA)
- Microscale cultivation system (e.g., Growth Profiler 960 by EnzyScreen B.V., Heemstede, The Netherlands)

2.1.1.3 Reagents

- Mineral salts medium (MSM) adapted from Hartmans, Smits, van der Werf, Volkering, and de Bont (1989)
  - 3.88g/L K_2HPO_4, 2.12g/L $NaH_2PO_4 \bullet 2H_2O$, 2.00g/L $(NH_4)_2SO_4$, 10mg/L EDTA,
  - 0.1g/L $MgCl_2 \bullet 6H_2O$, 2mg/L $ZnSO_4 \bullet 7H_2O$, 1mg/L $CaCl_2 \bullet 2H_2O$, 5mg/L $FeSO_4 \bullet 7H_2O$,
  - 0.2mg $Na_2MoO_4 \bullet 2H_2O$, 0.2mg/L $CuSO_4 \bullet 5H_2O$, 0.4mg/L $CoCl_2 \bullet 6H_2O$,
  - 1.22mg $MnCl_2 \bullet 4H_2O$, bring pH to 7.0
- 1M glucose stock solution

- 1 M ethylene glycol stock solution
- MSM agar plates (1.5% agar)
- Genomic DNA isolation kit (e.g., High Pure PCR Template Preparation Kit by Roche Holding, Basel, Switzerland)
- PCR master mix (e.g., OneTaq 2X Master Mix with Standard Buffer by New England Biolabs, Ipswich, MA, United States)

2.1.2 Method

2.1.2.1 Cultivation

1. For the preculture, inoculate 25 mL of MSM supplemented with 20 mM glucose in a 250 mL shake flask with *P. umsongensis* GO16 from a cryopreserved glycerol stock. Cultivate at 30 °C with 200 rpm shaking speed before it reaches the stationary phase (between 12 and 16 h).
2. Harvest the pre-culture by centrifugation at 4800 × *g* for 2 min.
3. Resuspend the cell pellet in 5 mL MSM containing 30 mM EG as sole carbon source.
4. Use the resuspended cells to inoculate 50 mL MSM (in a 500 mL shake flask for sufficient aeration) containing 30 mM EG with a starting OD_{600} of 0.1.
5. Before the culture reaches the end of the exponential growth phase, re-inoculate in fresh medium using the same conditions as in step 4.
6. Repeat step 5 for repetitive batch cultivation until the termination criterion of enhanced growth is reached.
7. Once a distinctly improved growth rate is achieved, streak culture on MSM agar plates containing 30 mM EG to isolate single colonies.
8. To ensure the appearance of stable mutations, repetitively re-streak the isolated colonies on MSM agar plates containing EG. Then again, test clones for growth on selective medium containing EG. Choose clones for genome sequencing if growth on the selective medium can be reproduced.

To simultaneously assess the growth behavior of multiple single colonies in MSM with EG as substrate, cultivation can be performed in a microscale cultivation system (30 °C, 225 rpm). To monitor EG and TA consumption, sample the cultivation broth after cultivation is terminated. Store samples (1 mL) at −80 °C for further analysis by high-performance liquid chromatography coupled to a refractive index detector (EG) or an ultraviolet light detector (TA) (HPLC-RI/UV), as described in Sections 2.2 and 2.3.

According to the observed growth on the substrate, often ranked according to their superior growth rate, single evolved colonies can be selected to identify mutations responsible for improved fitness.

2.1.2.2 Genome (re-)sequencing

Genome analysis of the evolved strain is required to identify the mutations, which occurred during ALE cultivation.

1. Isolate genomic DNA of the starting and the evolved strains for re-sequencing using, e.g., the commercial kit.
2. Sequence the purified DNA using the Illumina technology as pair-end reads of 150 base pairs (e.g., by Eurofins Genomics, Ebersberg, Germany) and map the evolved strain genome sequence against the reference. To map the reference against the database use the Burrows-Wheeler Alignment tool (BWA) with default parameters (Li & Durbin, 2009).
3. Analyze single nucleotide polymorphisms and insertion or deletion polymorphisms with the GATK's Haplotype Caller (DePristo et al., 2011; McKenna et al., 2010) and visualize the results (e.g., Integrative Genomics Viewer) (Robinson et al., 2011).
4. Identify mutations, which are putatively responsible for the enhanced growth of the organism based on the genome annotation or any information from the literature or databases (e.g., gene annotation, synteny). Mutations in open reading frames, and in their upstream regions, especially in promoters, are of key interest. Gene function deletions and changes in the transcription rate are two common outcomes of ALE experiments.

2.1.2.3 Reverse engineering

Reverse engineering can reproduce the improved phenotype in the starting strain and ensure that a certain mutation is indeed responsible for the observed phenotype. The longer an ALE experiment is conducted, the more mutations can emerge, with the risk of mutation accumulation unrelated to the improved growth phenotype. While reverse engineering generates a deeper understanding of the investigated metabolism, highly mutated strains are difficult to be reversely engineered. The effect of specific mutations on the phenotype in *Pseudomonas* can be determined by recreating the mutations in the starting strain, applying the pEMG system by Martínez-García and de Lorenzo (2011). The method is a tool to acquire targeted gene modifications within gram-negative genomes. For pEMG vector design, proceed as described by Martínez-García and de Lorenzo (2011) and continue as follows:

1. Transfer the designed vector via conjugation into the starting strain and perform the mating procedure according to the modified protocol from Wynands et al. (2018).
2. Verify the mutants by colony PCR, using the alkaline polyethylene glycol-based method (Chomczynski & Rymaszewski, 2006) followed by a standardized PCR procedure (e.g., Woodman, 2008).
3. Test the growth behavior through growth comparison experiments on selective media (e.g., EG as carbon source) between the reverse engineered, the reference/starting, and the evolved ALE strains.

2.1.3 Alternative procedures

Depending on the desired properties for the evolved strain, the ALE method can be adapted. Serial re-inoculation can be performed with increasing selective pressure instead of constant conditions. A raised concentration, e.g., aiming for increased substrate tolerant mutants, leads to more targeted mutations, presuming the cells can bear higher selective pressure.

To evaluate the success of ALE experiments and to determine when to end the re-inoculations, a method established by LaCroix et al. (2015) can be used. For this, a determined parameter, such as the growth rate, is computed for each of the serial cultures. Observing this parameter over the experimental progress provides an overview of when beneficial mutations occur and are distributed in the population. These events are visible as jumps in the corresponding growth curves (Fig. 2) and can be applied as end criteria or to start further actions (e.g., genome re-sequencing) (LaCroix et al., 2015).

Furthermore, the automation of ALE experiments can enormously promote the experimental capabilities, throughput, and clarity in the outcome, compared to manually executed ALE experiments. However, implementing an automated system requires specialized equipment (Sandberg, Salazar, Weng, Palsson, & Feist, 2019). The automated system, established by Sandberg et al., tracks the culture's growth rate steadily and propagates cultures in constant exponential growth. By passing cultures to fresh medium when nutrients are still available in excess, growth limitation through carbon depletion is prevented (Sandberg et al., 2014). Automation accelerates the fixation of desired mutations since a precise selection pressure can be maintained even during long-term evolution experiments, consequently reducing the required time for fitness gains (Sandberg et al., 2019).

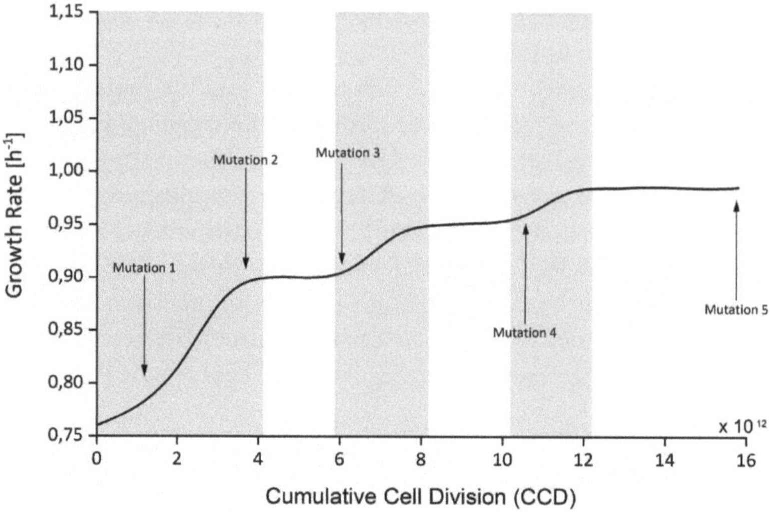

Fig. 2 Exemplary progression (smoothed data) of the fitness increase as a function of cumulative cell divisions (CCD) and highlighted jump regions (gray boxes).

2.2 Determination of substrate consumption: EG quantification

To evaluate the consumption of the substrate EG, its concentration in the fermentation broth can be analyzed by HPLC using a column for separation based mainly on ion-exchange together with size-exclusion and nonpolar interactions. A typical chromatogram of an EG standard obtained using the described method is depicted in Fig. 3.

2.2.1 Materials, equipment, and reagents

2.2.1.1 Material

- Centrifuge (e.g., 5418 R centrifuge by Eppendorf, Hamburg, Germany)
- 1 mL syringe (e.g., Inject-F fine dosage syringe by Braun, Melsungen, Germany)
- Syringe filters suited for HPLC sample preparation in aqueous and mixed organic solutions (e.g., Phenex-regenerated cellulose syringe filter (⌀ 4 mm, pore size 0.2 μm) by Phenomenex, Torrance, USA)
- HPLC vials compatible with the HPLC autosampler (e.g., Screw neck vial, N 9, 11.6 × 32.0 mm, 1.5 mL, flat bottom, clear by Macherey-Nagel GmbH & Co. KG, Düren, Germany)
- Screw caps for HPLC vials (e.g., Screw closure, N 9, PP, blue, center hole, Red Rubber/FEP colorless, 1.0 mm by Macherey-Nagel GmbH & Co. KG, Düren, Germany)

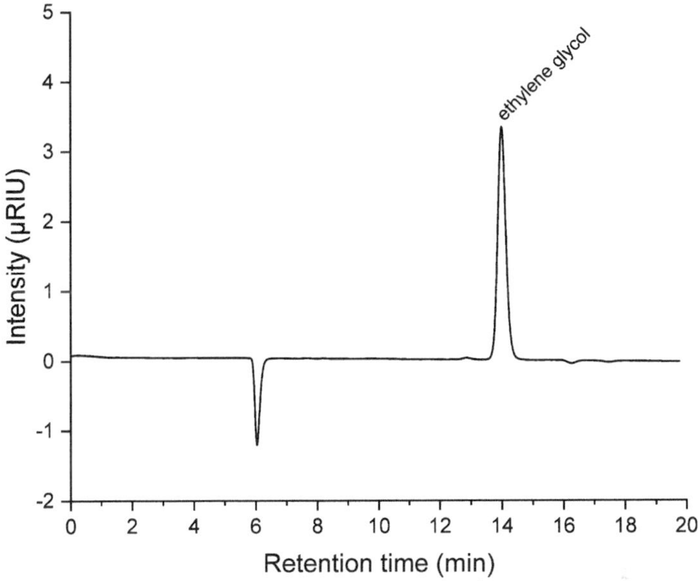

Fig. 3 Chromatogram of an ethylene glycol standard (30 mM) analyzed by HPLC-RI. Ethylene glycol elutes after about 14 min. µRIU: micro-refractive index units.

2.2.1.2 Equipment

- HPLC system composed of:
  - HPLC pump (e.g., pump LC-126 by Beckmann Coulter, Krefeld, Germany)
  - HPLC autosampler (e.g., autosampler LC-508 by Beckmann Coulter, Krefeld, Germany)
  - HPLC column oven (e.g., Jetstream 2 Plus Column Thermostat by Merck, Darmstadt, Germany)
- Refractive index detector (e.g., Smartline RI detector 2300 by Knauer, Berlin, Germany)
- HPLC column for analysis of alcohols in fermentation broth (e.g., Metab-AAC BF Series HPLC Column (300 × 7.8 mm)) together with a pre-column (e.g., Metab AAC BF Series Guard Column (10 × 7.8 mm) both by ISERA GmbH, Düren, Germany)

2.2.1.3 Reagents

- Ultra-purified water containing 5 mM sulfuric acid
- Ethylene glycol, analytical standard

2.2.2 *Method*

1. For quantification, prepare standard curves of EG in a suitable concentration range (e.g., 1–30 mM).
2. Separate cells and cell residues by centrifugation at 13,000 × *g* for 2 min and filter the supernatant with a syringe filter.
3. Apply ultra-purified water with 5 mM sulfuric acid as running buffer, pumping it isocratically at a flow rate of 0.7 mL/min and a column oven temperature of 40 °C.
4. Inject 10 μL of the sample and measure for 30 min to detect EG.

2.3 Determination of substrate consumption: TA quantification

To assess the consumption of TA, UV-absorbing TA, and associated low-molecular-weight esters (e.g., bis(2-hydroxyethyl) terephthalate (BHET) and MHET)—typical intermediates in enzymatic PET hydrolysis—can be measured by reversed-phase (RP-)HPLC using UV detection (Barth et al., 2015). For example, Barth et al. (2015) detected TA and BHET at a wavelength of 241 nm. Depending on the method sensitivity and whether additional molecules should be detected, it might be useful to choose another wavelength (see Fig. 4).

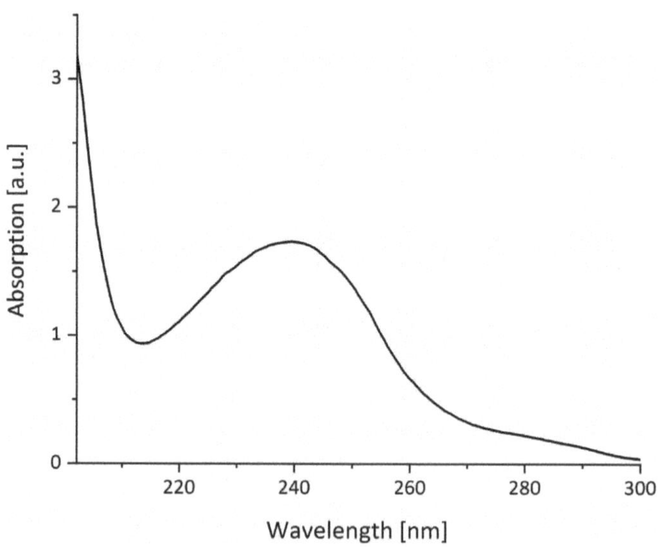

Fig. 4 UV spectrum of terephthalate in aqueous NaOH solution (100 μM terephthalic acid). AU: absorbance units.

Here, we describe a RP-HPLC method coupled with a diode array detector (DAD) set to a wavelength of 254 nm for terephthalate detection. A typical chromatogram obtained when measuring a terephthalate standard is shown in Fig. 5.

2.3.1 Materials, equipment, and reagents

2.3.1.1 Material

- See Section 2.2.1

2.3.1.2 Equipment

For HPLC set-up, see Section 2.2.1 with the following adaptations:

- UV detector (e.g., System Gold 168 Detector by Beckman Coulter, Brea, California, USA)
- HPLC column (e.g., ISAspher 100-5 C18 (250 × 4 mm) by ISERA GmbH, Düren, Germany)

2.3.1.3 Reagents

- Acetonitrile
- Ultra-purified water with 1% (v/v) formic acid
- Terephthalate, analytical standard

2.3.2 Method

1. For quantification, prepare standard curves of TA in a suitable concentration range (e.g., 1–30 mM).
2. Centrifuge 1 mL fermentation broth samples at 13,000 × *g* for 2 min and filter the supernatant with a syringe filter to remove any precipitates.
3. Perform the separation of TA with a column oven temperature set to 40 °C and an injection volume of 10 μL. Set the flow rate to 0.8 mL/min and detect the separated products at a wavelength of e.g., 254 nm.
4. Use acetonitrile and ultra-purified water containing 1% formic acid as mobile phase and apply the solvent gradient as follows: 2 min of 10% acetonitrile and 90% ultra-purified water containing 1% formic acid, increase acetonitrile to 90% within 12 min using a linear gradient, maintain the gradient for 2 min; linearly decrease acetonitrile to 10% between 16 and 18 min, stop the measurement after 20 min.

2.4 Genetic engineering for HAA synthesis

Here, we describe an approach to further engineer pseudomonads for the biosynthesis of a particular product (here HAAs as an example) from a given

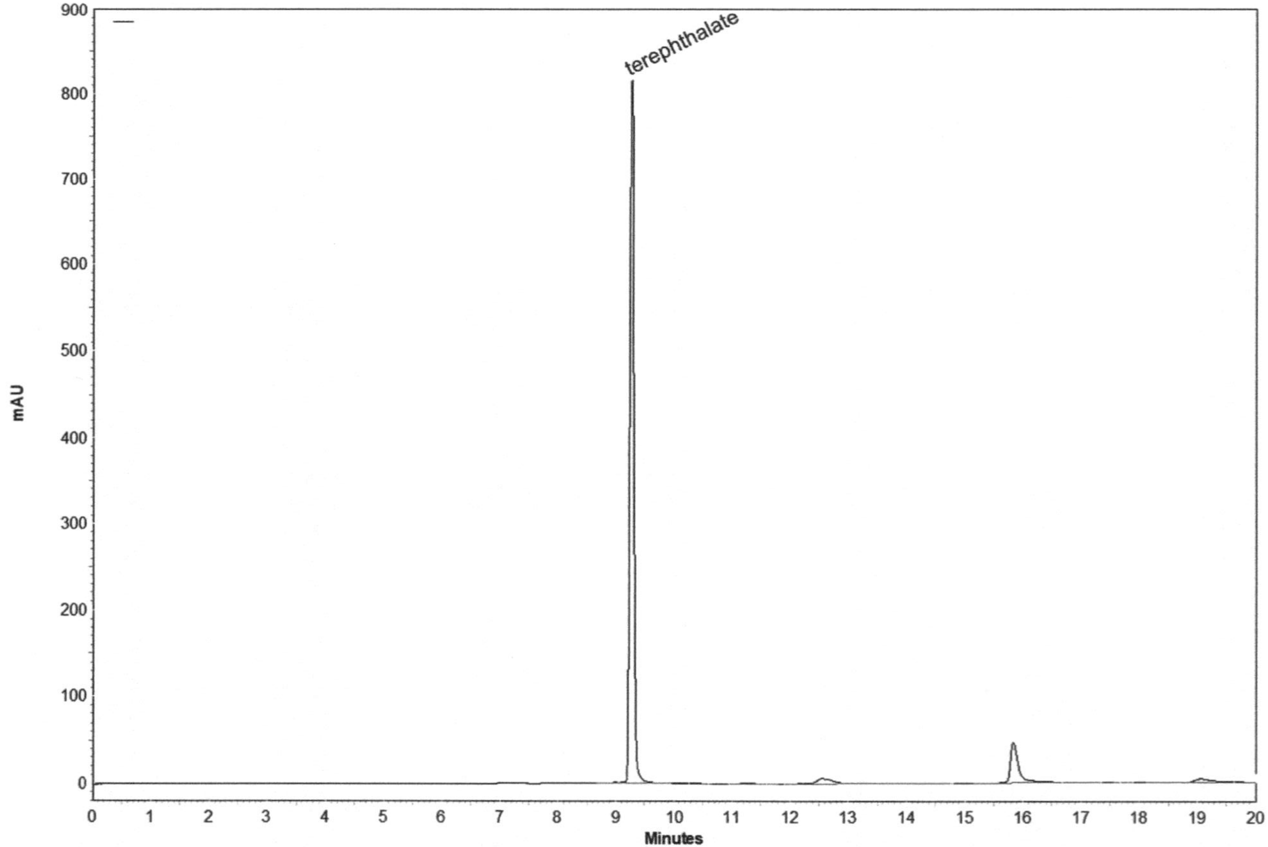

Fig. 5 Chromatogram obtained by HPLC-UV analysis of a terephthalate standard (5 mM terephthalic acid). The chromatogram shows one peak at 9.3 min corresponding to terephthalate. mAU: milli-absorbance units.

substrate such as plastic monomers. Expression of heterologous gene(s) required for the biosynthesis of the desired product can be either performed via a plasmid-based system or by genomic integration. For pseudomonads, the Tn7-based system has shown to provide the stable expression of heterologous genes upon chromosomal integration into the *att*Tn7 site downstream of the *glmS* gene (Köbbing, Blank, & Wierckx, 2020; Zobel et al., 2015). However, a plasmid-based expression can be suitable for fine-tuning the expression level of the gene of interest. The Standard European Vector Architecture (SEVA) is based on a simple, straightforward design to facilitate the construction of optimized vectors and other tools for the genetic engineering of microorganisms, including non-typical laboratory strains such as *P. umsongensis* GO16 (Martínez-García et al., 2019; Silva-Rocha et al., 2013).

2.4.1 Materials, equipment, and reagents

2.4.1.1 Strain and plasmid

- Cryopreserved glycerol stock of the desired production host (e.g., *P. umsongensis* GO16 KS03)
- Vector suitable for plasmid-based or chromosomal expression of *rhlA* or (an)other gene(s) required for target molecule synthesis (e.g., pSB01)
- 1.5 mL microcentrifuge tubes
- Electroporation cuvettes with a 0.2 cm gap for the electroporation of bacteria

2.4.1.2 Equipment

- Incubation shaker (Infors HT Multitron shaker (50 mm shaking diameter) by Infors AG, Bottmingen, Switzerland)
- Centrifuge (e.g., Centrifuge 5920 R by Eppendorf, Hamburg, Germany)
- Electroporator (e.g., Gene Pulser Xcell Microbial Electroporation System by Biorad Laboratories, Inc., Hercules, CA, USA)

2.4.1.3 Reagents

- Lysogeny broth (LB) medium: 10 g/L peptone, 5 g/L yeast extract, and 10 g/L NaCl
- 300 mM sucrose solution
- LB agar plates: LB medium is solidified by addition of 1.5 g/L agar
- Antibiotic stock solution (e.g., 50 mg/mL kanamycin)

2.4.2 Method

P. umsongensis GO16 KS3 is an evolved strain that was naturally able to metabolize TA and was subjected to ALE for growth on EG (see Section 2.1). Since *P. umsongensis* GO16 KS3 cannot synthesize HAAs, the introduction of a heterologous gene into the strain is required to produce the target molecule. The heterologous expression can, for instance, be achieved using a plasmid-based expression system. The plasmid pSB01, a derivative of pSEVA241, has successfully been utilized for the biosynthesis of HAAs in the laboratory strain *P. putida* KT2440 (Tiso et al., 2017). The vector, designed for plasmid-based expression, carries *rhlA* from *Pseudomonas aeruginosa* PAO1 (NCBI GeneID: 878955) and is equipped with the synthetic promotor SynPro8 for constitutive expression (Tiso et al., 2016). The *rhlA* gene codes for an acyltransferase, which catalyzes the esterification of two activated hydroxy fatty acids to HAA (Zhu & Rock, 2008). Here, we provide a protocol for the preparation of electrocompetent pseudomonads (e.g., *P. umsongensis* GO16 KS3) and subsequent electroporation with a suitable plasmid (e.g., pSB01) based on a method by Choi, Kumar, and Schweizer (2006). This procedure yields an engineered strain, *P. umsongensis* GO16 KS03 pSB01, that synthesizes HAAs from EG and TA.

2.4.2.1 Preparation of electrocompetent cells

1. Inoculate 5 mL of LB medium with the production strain from a cryopreserved glycerol stock. Incubate at 30 °C and shaking at 200 rpm overnight.
2. Transfer overnight culture into a 15 mL tube and centrifuge at 16,000 × *g* for 2 min. Discard the supernatant.
3. Resuspend the cell pellet in 2 mL of 300 mM sucrose solution and centrifuge again at 16,000 × g for 2 min.
4. Discard supernatant and repeat the centrifugation step.
5. Finally, resuspend cells in 100 μL of 300 mM sucrose.

2.4.2.2 Electroporation of electrocompetent *Pseudomonas* cells

1. Add 1 μL of plasmid DNA (1–10 ng) to the electrocompetent cells and mix gently by tapping the tube.
2. Transfer DNA/cell suspension into an electroporation cuvette (0.2 cm gap). Make sure that the fluid has contact to both electrodes. Wipe dry cuvette and position in an electroporator. Apply a pulse with the following settings: 2.5 kV, 400 Ω, and 25 μF.

3. Immediately add 1 mL LB medium directly to the electroporation cuvette. Transfer suspension to a 1.5 mL microcentrifuge tube and incubate for 2 h at 30 °C with shaking at 200 rpm.
4. Spread 50 µL of the LB culture on selective LB agar plates.
5. Centrifuge the microcentrifuge tube for 30 s at 16,000 × *g* and discard the supernatant.
6. Resuspend cells in the remaining supernatant and spread 50 µL on selective LB agar plates.
7. Incubate agar plates at 30 °C overnight and screen transformants by colony PCR to confirm the transformation of the plasmid before further analyses are performed.

2.5 Biosynthesis of HAA

For biosynthesis of HAAs, pseudomonads can be grown in shake flasks to study the general performance regarding growth rate, substrate consumption, production rate, and yield. Cultivation in Fernbach flasks (Tiso, Sauer, et al., 2020) or a bioreactor is recommended to obtain more product and to evaluate the strain performance at a larger scale. For *P. putida* KT2440, the synthesis of HAAs in a bioreactor has been successfully demonstrated (Blesken et al., 2020). Here, we describe the cultivation in shake flasks for the general assessment of a pseudomonad engineered and/or evolved to produce the platform molecule HAA from PET monomers, as shown earlier (Tiso, Narancic, et al., 2020). To avoid foam formation and to ensure reproducibility, the utilization of shake flasks without baffles is recommended.

2.5.1 Materials, equipment, and reagents

2.5.1.1 Strain

- Production host capable of metabolizing PET monomers engineered to synthesize HAAs (e.g., *P. umsongensis* GO16 KS03 pSB01)

2.5.1.2 Equipment

- 500 mL shake flasks
- 1.5 mL microcentrifuge tubes for sampling
- Incubation shaker (Infors HT Multitron shaker (50 mm shaking diameter) by Infors AG, Bottmingen, Switzerland)

2.5.1.3 Reagents

- MSM medium (see Section 2.1.1)
- 1 M ethylene glycol stock solution

- Terephthalate stock solution (400 mM TPA)

  TPA is poorly soluble in water (17–19 mg/L), but the addition of NaOH increases the solubility of TPA in water due its deprotonation resulting in the formation of terephthalate (e.g., 33 g/L at 20 °C with 2% NaOH) (Kim, Lee, & Chung, 2002; Park & Sheehan, 1996). The best dissolution can be achieved at pH 7 (Kim et al., 2002). Prepare 100 mL of a TA stock solution (400 mM TPA) as follows:

  (a) Dissolve 3.2 g NaOH in 80 mL ultra-purified water and add 6.64 g TPA.

  (b) Stir for 10 min and fill to 95 ml with ultra-purified water.

  (c) Adjust pH to 7 by *dropwise* adding 6 M NaOH until TPA is completely dissolved. Upon addition of NaOH the pH will increase temporarily. As more TPA goes into solution, the pH will decrease again and more NaOH can be added. It might take several hours until the pH stabilizes and all TPA is dissolved.

  (d) Fill up to 100 ml with ultra-purified water and autoclave.
- Antibiotic stock solution (e.g., kanamycin (50 mg/mL))

2.5.2 Method

1. Prepare MSM medium supplemented with 20 mM of EG and 20 mM of TPA.
2. Inoculate 50 mL of the prepared MSM supplemented with kanamycin (50 µg/mL) in 500 mL shake flasks with *P. umsongensis* GO16 KS03 pSB01.
3. Perform the cultivation for 24 h in an incubation shaker at 30 °C and shaking at 250 rpm. Maintain humidity at 80%. Sample fermentation broth (1 mL) every 2 h for subsequent EG, TA, and HAA quantification. Store samples at −80 °C for further analysis.

2.6 Quantification of HAA

The HAA concentration in the fermentation broth can be determined via an RP-HPLC coupled to a charged aerosol detector (CAD) (Bator, Wittgens, et al., 2020; Tiso et al., 2017). The expression of the *rhlA* from *P. aeruginosa* PAO1 in other pseudomonads results in the production of four HAA congeners, with 3-hydroxydecanoate being the main hydroxy fatty acid (Tiso et al., 2017; Tiso, Narancic, et al., 2020). The 3-hydroxydecanoate dimer is accordingly the main congener produced, followed by $C_{12:1}$-C_{10}, C_{12}-C_{10}, and C_8-C_{10} (Fig. 6). Congener quantification can be performed using individual standards (Blesken et al., 2020). The method presented here reliably separates all four HAA congeners produced (Fig. 6).

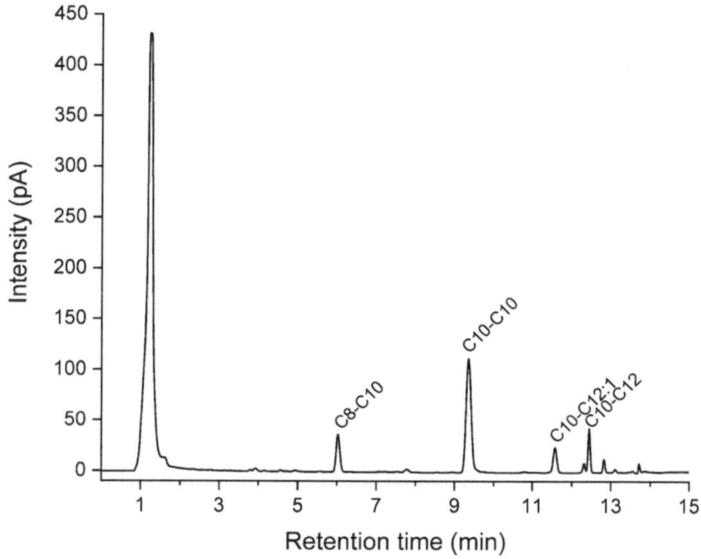

Fig. 6 Chromatogram of a culture supernatant containing HAA. Peaks corresponding to the HAA congeners C_8-C_{10}, C_{10}-C_{10}, C_{10}-$C_{12:1}$, and C_{10}-C_{12} are indicated. pA: picoampere.

2.6.1 Materials, equipment, and reagents

2.6.1.1 Material

- See Section 2.2.1

2.6.1.2 Equipment

For HPLC set-up see Section 2.2.1 with the following adaptations:

- Charged Aerosol Detector (e.g., Corona Veo Charged Aerosol Detector)
- Nitrogen generator (e.g., Parker Balston NitroVap-1LV by Parker Hannifin GmbH, Kaarst, Germany)
- HPLC column (e.g., NUCLEODUR C18 Gravity column (150 × 4.6 mm; particle size of 3 μm)) protected by a pre-column ((4 × 4.6 mm, particle size 3 μm) both by Macherey-Nagel GmbH & Co. KG, Düren, Germany)

2.6.1.3 Reagents

- Ultra-purified water with 0.2% (v/v) formic acid
- Acetonitrile
- HAA standards of all four HAA congeners: C_8-C_{10}, C_{10}-C_{10}, C_{10}-$C_{12:1}$, and C_{10}-C_{12}

2.6.2 Method

1. Prepare standard solutions of all four HAA congeners in a suitable concentration range (e.g., 0.1–5 g/L) in acetonitrile/water (1:1; v/v).
2. Separate the cells and cell residues by centrifugation at 13,000 × *g* for 2 min.
3. Mix 500 μL of the supernatant with 500 μL of 100% acetonitrile and vortex.
4. Incubate samples for at least 4 h at 4 °C to precipitate proteins that might clog the column.
5. Centrifuge the samples again and filter the supernatant with a syringe filter.
6. Analyze samples on an RP-HPLC-CAD system supplied with a continuous nitrogen stream.
7. Perform the separation of HAA species on an HPLC column protected by a pre-column with a column oven temperature set to 40 °C, an injection volume of 5 μL, and a flow rate of 1 mL/min. Use acetonitrile and ultra-purified water containing 0.2% formic acid as the mobile phase and apply the solvent gradient as follows:
   (a) Apply 70% of acetonitrile and 30% ultra-purified water containing 0.2% formic acid for 1 min.
   (b) Increase acetonitrile to 80% within 8 min using a linear gradient.
   (c) Linearly increase to 100% acetonitrile between 9 and 10 min.
   (d) Decrease linearly to 70% of acetonitrile between 11 and 12.5 min.
   (e) Stop the measurement after 15 min.

2.6.3 Alternative procedure

Instead of initially separating the cells from the supernatant, acetonitrile and the fermentation broth sample can immediately be mixed in a 1:1 ratio. Afterward, continue with step 4. This procedure results in the measurement of about 10% higher HAA concentrations.

2.7 HAA purification

HAAs are a highly diverse group of β-hydroxy-fatty acids differing in their chain length. As many as 53 congeners were recently obtained by expressing seven bacterial RhlA acyltransferases in *Escherichia coli* (Germer et al., 2020). *P. putida* KT2440 synthesizes a mixture of four main congeners upon heterologous expression of *rhlA* from *P. aeruginosa* (Tiso et al., 2017). *P. umsongensis* GO16 KS03 pSB01 produces the same congeners when grown on PET monomers (Tiso, Narancic, et al., 2020). To isolate the product HAA from the cultivation broth, purification can be performed by adsorption/desorption (Tiso et al., 2017). We use an LC glass column packed with hydrophobic octadecylsilyl-A (see Section 2.7.1) to adsorb HAA from the fermentation broth. The adsorption capacity of the adsorption column is dependent on the flow behavior and the proportion of other hydrophobic components such as pyoverdines. Therefore, it is recommendable to determine the adsorption capacity for every technical set-up and every adsorption material. For octadecylsilyl-A, we determined an adsorption capacity of 1 g HAA per 10 g packaging material. Here, we provide a protocol for adsorption/desorption to separate HAA from the supernatant. The described procedure results in up to 90% purity of HAA produced in *P. putida* KT2440 (Tiso et al., 2017). For further purification and congener separation, we describe a method for preparative HPLC (Tiso, Sauer, et al., 2020).

2.7.1 Materials, equipment, and reagents

2.7.1.1 Material

- Tubes to collect fractions (e.g., 15 and 50 mL Falcon Conical Centrifuge Tubes by Thermo Fisher Scientific, Waltham, MA, United States)
- 250 mL centrifuge bottles (e.g., Nalgene Wide-Mouth Lab Quality HDPE Bottles by Thermo Fisher Scientific Inc., Waltham, MA, USA)
- Hydrophobic packing material for open chromatography (e.g., Octadecylsilyl-A AA12SA5 (150 μm particle size; 120 nm pore size) by YMC Co., Ldt., Kyoto, Japan)

2.7.1.2 Equipment

- Centrifuge (e.g., 5920 R centrifuge by Eppendorf, Hamburg, Germany)
- Fixed-angle rotor (e.g., FA-6x250 by Eppendorf, Hamburg, Germany)
- Constant airflow or rotary evaporator
- LC glass column (e.g., Bioline glass column, by Knauer GmbH, Berlin, Germany)
- High-pressure dosing pump (e.g., BlueShadow Pump P40 by Knauer GmbH, Berlin, Germany)
- Preparative HPLC system consisting of:
  - Binary high-pressure gradient pump (e.g., Azura P6.1L pump by Knauer GmbH, Berlin, Germany)
  - HPLC autosampler (e.g., AZURA autosampler 3950 by Knauer GmbH, Berlin, Germany)

- HPLC column (e.g., VP250/21 NUCLEODUR C18 HTec column by Macherey-Nagel GmbH & Co. KG, Düren, Germany)
- Suitable detector; we chose a low-temperature evaporative light-scattering detector for HPLC (e.g., SEDEX 58 LT-ELSD detector by SEDERE, Olivet, France)

- Fraction collector (e.g., Foxy R1 fraction collector by Teledyne ISCO, Lincoln, United States)

2.7.1.3 Reagents

- Acetonitrile
- Ethanol
- Ultra-purified water
- Ultra-purified water containing 0.2% (v/v) formic acid

2.7.2 Method

2.7.2.1 Separation of HAA from the supernatant

1. Mix 500 mL culture broth with 250 mL acetonitrile to sterilize and precipitate protein.
2. Evaporate acetonitrile under constant airstream overnight.
3. Centrifuge the suspension for 30 min at 10,000 × *g*.
4. Pack LC glass column with 30 g of adsorbent packaging material.
5. Prime LC glass column with ethanol for 15 min with 2–6 bed volumes to remove possible hydrophobic impurities of the adsorbent packaging material.
6. For adsorption, pump the prepared supernatant onto the column using a flow rate of 2–5 mL/min.
7. Once the entire suspension has passed the column, wash with 2 L ultra-purified water to remove hydrophilic contaminants.
8. For desorption of HAA, apply the following ethanol gradient with a flow rate of, e.g., 1 mL/min and continuously collect 10 mL fractions. Use pure ethanol and ultra-purified water for desorption of HAA.
   - **(a)** Apply 100% water for 5 min.
   - **(b)** Linearly increase ethanol to 30% between min 5 and 35.
   - **(c)** Increase ethanol to 50% until min 65.
   - **(d)** Further increase ethanol to 75% until min 95.
   - **(e)** Until min 105, increase ethanol to 95%.
   - **(f)** Finally, increase ethanol to 100%.
   - **(g)** Stop the elution after 180 min or when no more HAA is eluted from the column.
9. Flush the column with 2 L ultra-purified water to clean the column.

10. Analyze the collected fractions regarding their HAA content by RP-HPLC-CAD, as described in Section 2.6.
11. Pool the fractions with a high HAA concentration and evaporate ethanol and water under a constant air stream or in a rotary evaporator (60 °C, −0.8 bar) to obtain purified HAAs. (If preparative HPLC is subsequently used, only reduce the volume to a suitable concentration.)
12. To establish the purity, weigh in some dried HAA mixture. Dissolve in acetonitrile: water (1:1, v/v) to a concentration of 1 g/L. Determine the true HAA concentration by RP-HPLC-CAD (see Section 2.6) and compare to the 1 g/L of the HAA mixture to assess purity.

2.7.2.2 Purification of HAA by preparative HPLC

To obtain HAA with a purity higher than 90%, further purify the HAAs by preparative HPLC. This method is also suitable to separate different HAA congeners (C_8-C_{10}, C_{10}-C_{10}, C_{10}-$C_{12:1}$, and C_{10}-C_{12}).

1. Set the flow rate to 10 mL/min and choose the injection volume depending on the HAA concentration obtained by adsorption/desorption (e.g., 3 mL sample per run).
2. Apply acetonitrile and ultra-purified water containing 0.2% (v/v) formic acid and collect all peaks detected. After the first run, perform RP-HPLC-CAD (see Section 2.6) to assign peaks to the four HAA congeners. Based on the peak assignment, the fraction collector can be programmed to individually collect the congeners in the following runs. Apply the gradient as follows:
   (a) Apply 70% acetonitrile for 5 min.
   (b) Linearly increase acetonitrile to 80% between 5 and 35 min.
   (c) Further increase acetonitrile to 100% until 37 min.
   (d) Decreased acetonitrile to 70% from 42 to 52 min.
   (e) Terminate the measurement after 60 min.
3. Evaporate acetonitrile and water from the solutions of the separated congeners.
4. The purity of the individual congeners can be established as described in Section 2.7.2 step 12.

3. Conclusion

The outlined method for the metabolic engineering of pseudomonads represents a powerful tool for the bioupcycling of TPA and EG to value-added chemicals. Through a tandem upcycling approach, combining this methodology with enzymatic plastic hydrolysis, PET waste can be fully upcycled by entirely biotechnological means.

Acknowledgments

The authors have received funding from the European Union's Horizon 2020 research and innovation program under grant agreement no. 633962 for the project P4SB and no. 870294 for the project MIX-UP. HMTH has been funded by the German Federal Ministry of Education and Research under the grant agreement No. 031B0899 in the project MIPLACE within the framework of the ERA-Cobiotech program of the EU. TT and LMB have been partially funded by the Deutsche Forschungsgemeinschaft (DFG, German Research Foundation) under Germany's Excellence Strategy—Exzellenzcluster 2186 "The Fuel Science Center" ID: 390919832.

References

Andrady, A. L., & Neal, M. A. (2009). Applications and societal benefits of plastics. *Philosophical Transactions of the Royal Society of London. Series B, Biological Sciences, 364*(1526), 1977–1984. https://doi.org/10.1098/rstb.2008.0304.

Austin, H. P., Allen, M. D., Donohoe, B. S., Rorrer, N. A., Kearns, F. L., Silveira, R. L., et al. (2018). Characterization and engineering of a plastic-degrading aromatic polyesterase. *Proceedings of the National Academy of Sciences, 115*(19), E4350–E4357.

Barnes, D. K., Galgani, F., Thompson, R. C., & Barlaz, M. (2009). Accumulation and fragmentation of plastic debris in global environments. *Philosophical Transactions of the Royal Society of London. Series B, Biological Sciences, 364*(1526), 1985–1998. https://doi.org/10.1098/rstb.2008.0205.

Barth, M., Oeser, T., Wei, R., Then, J., Schmidt, J., & Zimmermann, W. (2015). Effect of hydrolysis products on the enzymatic degradation of polyethylene terephthalate nanoparticles by a polyester hydrolase from *Thermobifida fusca. Biochemical Engineering Journal, 93*, 222–228. https://doi.org/10.1016/j.bej.2014.10.012.

Bator, I., Karmainski, T., Tiso, T., & Blank, L. M. (2020). Killing two birds with one stone—Strain engineering facilitates the development of a unique rhamnolipid production process. *Frontiers in Bioengineering and Biotechnology, 8*, 899. https://doi.org/10.3389/fbioe.2020.00899.

Bator, I., Wittgens, A., Rosenau, F., Tiso, T., & Blank, L. M. (2020). Comparison of three xylose pathways in *Pseudomonas putida* KT2440 for the synthesis of valuable products. *Frontiers in Bioengineering and Biotechnology*, 7, 480. https://doi.org/10.3389/fbioe.2019.00480.

Bertling, J., Bertling, R., & Hamann, L. (2018). *Kunststoffe in der Umwelt: Mikro-und Makroplastik. Ursachen, Mengen, Umweltschicksale, Wirkungen, Lösungsansätze, Empfehlungen* (p. 2018). Frauenhofer-Institut für Umwelt-, Sicherheits-und Energietechnik UMSICHT.

Blank, L. M., Ionidis, G., Ebert, B. E., Buhler, B., & Schmid, A. (2008). Metabolic response of *Pseudomonas putida* during redox biocatalysis in the presence of a second octanol phase. *The FEBS Journal, 275*(20), 5173–5190. https://doi.org/10.1111/j.1742-4658.2008.06648.x.

Blesken, C. C., Bator, I., Eberlein, C., Heipieper, H. J., Tiso, T., & Blank, L. M. (2020). Genetic cell-surface modification for optimized foam fractionation. *Frontiers in Bioengineering and Biotechnology, 8*. https://doi.org/10.3389/fbioe.2020.572892, 572892.

Carta, D., Cao, G., & D'Angeli, C. (2003). Chemical recycling of poly (ethylene terephthalate)(PET) by hydrolysis and glycolysis. *Environmental Science and Pollution Research, 10*(6), 390–394.

Choi, K.-H., Kumar, A., & Schweizer, H. P. (2006). A 10-min method for preparation of highly electrocompetent *Pseudomonas aeruginosa* cells: Application for DNA fragment transfer between chromosomes and plasmid transformation. *Journal of Microbiological Methods*, *64*(3), 391–397. https://doi.org/10.1016/j.mimet.2005.06.001.

Chomczynski, P., & Rymaszewski, M. (2006). Alkaline polyethylene glycol-based method for direct PCR from bacteria, eukaryotic tissue samples, and whole blood. *BioTechniques*, *40*(4), 454, 456, 458. https://doi.org/10.2144/000112149.

Conversio. (2018). Global plastic flow 2018. Mainaschaff, Germany: Conversio Market & Strategy GmbH.

de Ilarduya, A. M., & Muñoz-Guerra, S. (2014). Chemical structure and microstructure of poly(alkylene terephthalate)s, their copolyesters, and their blends as studied by NMR. *Macromolecular Chemistry and Physics*, *215*(22), 2138–2160. https://doi.org/10.1002/macp.201400239.

DePristo, M. A., Banks, E., Poplin, R., Garimella, K. V., Maguire, J. R., Hartl, C., et al. (2011). A framework for variation discovery and genotyping using next-generation DNA sequencing data. *Nature Genetics*, *43*(5), 491–498. https://doi.org/10.1038/ng.806.

Deziel, E., Lepine, F., Milot, S., & Villemur, R. (2003). *rhlA* is required for the production of a novel biosurfactant promoting swarming motility in *Pseudomonas aeruginosa*: 3-(3-hydroxyalkanoyloxy)alkanoic acids (HAAs), the precursors of rhamnolipids. *Microbiology*, *149*(Pt. 8), 2005–2013. https://doi.org/10.1099/mic.0.26154-0.

Espinosa, M. J. C., Blanco, A. C., Schmidgall, T., Atanasoff-Kardjalieff, A. K., Kappelmeyer, U., Tischler, D., et al. (2020). Toward biorecycling: Isolation of a soil bacterium that grows on a polyurethane oligomer and monomer. *Frontiers in Microbiology*, *11*, 404. https://doi.org/10.3389/fmicb.2020.00404.

Garcia, J. M., & Robertson, M. L. (2017). The future of plastics recycling. *Science*, *358*(6365), 870. https://doi.org/10.1126/science.aaq0324.

Germer, A., Tiso, T., Müller, C., Behrens, B., Vosse, C., Scholz, K., et al. (2020). Exploiting the natural diversity of RhlA acyltransferases for the synthesis of the rhamnolipid precursor 3-(3-hydroxyalkanoyloxy)alkanoic acid. *Applied and Environmental Microbiology*, *86*(6), e02317–e02319. https://doi.org/10.1128/AEM.02317-19.

Geyer, R., Jambeck, J. R., & Law, K. L. (2017). Production, use, and fate of all plastics ever made. *Science Advances*, *3*(7), e1700782. https://doi.org/10.1126/sciadv.1700782.

Guzik, M. W., Kenny, S. T., Duane, G. F., Casey, E., Woods, T., Babu, R. P., et al. (2014). Conversion of post-consumer polyethylene to the biodegradable polymer polyhydroxyalkanoate. *Applied Microbiology and Biotechnology*, *98*(9), 4223–4232. https://doi.org/10.1007/s00253-013-5489-2.

Hartmans, S., Smits, J. P., van der Werf, M. J., Volkering, F., & de Bont, J. A. M. (1989). Metabolism of styrene oxide and 2-phenylethanol in the styrene-degrading *Xanthobacter* strain 124X. *Applied and Environmental Microbiology*, *55*(11), 2850. https://doi.org/10.1128/AEM.55.11.2850-2855.

Hestin, M., Mitsios, A., Said, S. A., Fouret, F., Berwald, A., & Senlis, V. (2017). *Blueprint for plastics packaging waste: Quality sorting & recylcing*. Retrieved from https://www2.deloitte.com/my/en/pages/risk/articles/blueprint-for-plastics-packaging-waste.html.

Jacquin, J., Cheng, J., Odobel, C., Pandin, C., Conan, P., Pujo-Pay, M., et al. (2019). Microbial ecotoxicology of marine plastic debris: A review on colonization and biodegradation by the "plastisphere". *Frontiers in Microbiology*, *10*, 865.

Kenny, S. T., Runic, J. N., Kaminsky, W., Woods, T., Babu, R. P., Keely, C. M., et al. (2008). Up-cycling of PET (polyethylene terephthalate) to the biodegradable plastic PHA (polyhydroxyalkanoate). *Environmental Science & Technology*, *42*(20), 7696–7701. https://doi.org/10.1021/es801010e.

Kenny, S. T., Runic, J. N., Kaminsky, W., Woods, T., Babu, R. P., & O'Connor, K. E. (2012). Development of a bioprocess to convert PET derived terephthalic acid and biodiesel derived glycerol to medium chain length polyhydroxyalkanoate. *Applied Microbiology and Biotechnology*, *95*(3), 623–633. https://doi.org/10.1007/s00253-012-4058-4.

Kim, H. T., Kim, J. K., Cha, H. G., Kang, M. J., Lee, H. S., Khang, T. U., et al. (2019). Biological valorization of poly(ethylene terephthalate) monomers for upcycling waste PET. *ACS Sustainable Chemistry & Engineering*, 7(24), 19396–19406. https://doi.org/10.1021/acssuschemeng.9b03908.

Kim, Y.-B., Lee, K., & Chung, J.-H. (2002). Optimum cleaning-in-place conditions for stainless steel microfiltration membrane fouled by terephthalic acid solids. *Journal of Membrane Science*, *209*(1), 233–240. https://doi.org/10.1016/S0376-7388(02)00347-2.

Köbbing, S., Blank, L. M., & Wierckx, N. (2020). Characterization of context-dependent effects on synthetic promoters. *Frontiers in Bioengineering and Biotechnology*, *8*, 551. https://doi.org/10.3389/fbioe.2020.00551.

Köhler, K. A., Blank, L. M., Frick, O., & Schmid, A. (2015). D-Xylose assimilation via the Weimberg pathway by solvent-tolerant *Pseudomonas taiwanensis* VLB120. *Environmental Microbiology*, *17*(1), 156–170. https://doi.org/10.1111/1462-2920.12537.

Koopman, F., Wierckx, N., de Winde, J. H., & Ruijssenaars, H. J. (2010). Efficient whole-cell biotransformation of 5-(hydroxymethyl)furfural into FDCA, 2,5-furandicarboxylic acid. *Bioresource Technology*, *101*(16), 6291–6296. https://doi.org/10.1016/j.biortech.2010.03.050.

LaCroix, R. A., Sandberg, T. E., O'Brien, E. J., Utrilla, J., Ebrahim, A., Guzman, G. I., et al. (2015). Use of adaptive laboratory evolution to discover key mutations enabling rapid growth of *Escherichia coli* K-12 MG1655 on glucose minimal medium. *Applied and Environmental Microbiology*, *81*(1), 17–30. https://doi.org/10.1128/AEM.02246-14.

Lau, W. W. Y., Shiran, Y., Bailey, R. M., Cook, E., Stuchtey, M. R., Koskella, J., et al. (2020). Evaluating scenarios toward zero plastic pollution. *Science*, *369*, 1455–1461. https://doi.org/10.1126/science.aba9475.

Li, W.-J. (2020). *Plastic monomer degradation: Engineering Pseudomonas putida KT2440 for plastic monomer utilization* (PhD thesis). Aachen: RWTH Aachen University.

Li, H., & Durbin, R. (2009). Fast and accurate short read alignment with Burrows-Wheeler transform. *Bioinformatics*, *25*(14), 1754–1760. https://doi.org/10.1093/bioinformatics/btp324.

Martínez-García, E., & de Lorenzo, V. (2011). Engineering multiple genomic deletions in Gram-negative bacteria: Analysis of the multi-resistant antibiotic profile of *Pseudomonas putida* KT2440. *Environmental Microbiology*, *13*(10), 2702–2716. https://doi.org/10.1111/j.1462-2920.2011.02538.x.

Martínez-García, E., Goñi-Moreno, A., Bartley, B., McLaughlin, J., Sánchez-Sampedro, L., Pascual del Pozo, H., et al. (2019). SEVA 3.0: An update of the Standard European Vector Architecture for enabling portability of genetic constructs among diverse bacterial hosts. *Nucleic Acids Research*, *48*(D1), D1164–D1170. https://doi.org/10.1093/nar/gkz1024.

McKenna, A., Hanna, M., Banks, E., Sivachenko, A., Cibulskis, K., Kernytsky, A., et al. (2010). The genome analysis toolkit: A MapReduce framework for analyzing next-generation DNA sequencing data. *Genome Research*, *20*(9), 1297–1303. https://doi.org/10.1101/gr.107524.110.

Mensah, J. B., Hergesell, A. H., Brosch, S., Golchert, C., Artz, J., & Palkovits, R. (2020). Catalytic deoxygenation of bio-based 3-hydroxydecanoic acid to secondary alcohols and alkanes. *Green Chemistry*, *22*(11), 3522–3531. https://doi.org/10.1039/d0gc00691b.

Meyers, J., Mensah, J. B., Holzhäuser, F. J., Omari, A., Blesken, C. C., Tiso, T., et al. (2019). Electrochemical conversion of a bio-derivable hydroxy acid to a drop-in oxygenate diesel fuel. *Energy & Environmental Science*, *12*(8), 2406–2411. https://doi.org/10.1039/C9EE01485C.

Meys, R., Frick, F., Westhues, S., Sternberg, A., Klankermayer, J., & Bardow, A. (2020). Towards a circular economy for plastic packaging wastes—The environmental potential of chemical recycling. *Resources, Conservation and Recycling*, *162*, 105010. https://doi.org/10.1016/j.resconrec.2020.105010.

Nikel, P. I., & de Lorenzo, V. (2018). *Pseudomonas putida* as a functional chassis for industrial biocatalysis: From native biochemistry to trans-metabolism. *Metabolic Engineering*, *50*, 142–155. https://doi.org/10.1016/j.ymben.2018.05.005.

Palm, G. J., Reisky, L., Böttcher, D., Müller, H., Michels, E. A. P., Walczak, M. C., et al. (2019). Structure of the plastic-degrading *Ideonella sakaiensis* MHETase bound to a substrate. *Nature Communications*, *10*(1), 1717. https://doi.org/10.1038/s41467-019-09326-3.

Park, C., & Sheehan, J. R. (1996). In J. I. Korschwite, & M. Howe-Grant (Eds.), *Vol. 18. Encylopedia of chemical technology* (4th ed.). New York: John Wiley and Sons.

PlasticsEurope. (2019). *Plastics—The Facts 2019. An analysis of European plastics production, demand and waste data. The Facts.* Retrieved from www.plasticseurope.org/en/resources/publications/1804-plastics-facts-2019.

Remold, S. K., Purdy-Gibson, M. E., France, M. T., & Hundley, T. C. (2015). *Pseudomonas putida* and *Pseudomonas fluorescens* species group recovery from human homes varies seasonally and by environment. *PLoS One*, *10(5)*. https://doi.org/10.1371/journal.pone.0127704.

Robinson, J. T., Thorvaldsdóttir, H., Winckler, W., Guttman, M., Lander, E. S., Getz, G., et al. (2011). Integrative genomics viewer. *Nature Biotechnology*, *29*(1), 24–26. https://doi.org/10.1038/nbt.1754.

Rochman, C. M., Tahir, A., Williams, S. L., Baxa, D. V., Lam, R., Miller, J. T., et al. (2015). Anthropogenic debris in seafood: Plastic debris and fibers from textiles in fish and bivalves sold for human consumption. *Scientific Reports*, *5*, 14340. https://doi.org/10.1038/srep14340.

Sandberg, T. E., Pedersen, M., LaCroix, R. A., Ebrahim, A., Bonde, M., Herrgard, M. J., et al. (2014). Evolution of *Escherichia coli* to 42°C and subsequent genetic engineering reveals adaptive mechanisms and novel mutations. *Molecular Biology and Evolution*, *31*(10), 2647–2662. https://doi.org/10.1093/molbev/msu209.

Sandberg, T. E., Salazar, M. J., Weng, L. L., Palsson, B. O., & Feist, A. M. (2019). The emergence of adaptive laboratory evolution as an efficient tool for biological discovery and industrial biotechnology. *Metabolic Engineering*, *56*, 1–16. https://doi.org/10.1016/j.ymben.2019.08.004.

Silva-Rocha, R., Martínez-García, E., Calles, B., Chavarria, M., Arce-Rodriguez, A., de Las Heras, A., et al. (2013). The Standard European Vector Architecture (SEVA): A coherent platform for the analysis and deployment of complex prokaryotic phenotypes. *Nucleic Acids Research*, *41*(Database issue), D666–D675. https://doi.org/10.1093/nar/gks1119.

Sun, Z., Ramsay, J. A., Guay, M., & Ramsay, B. A. (2007). Carbon-limited fed-batch production of medium-chain-length polyhydroxyalkanoates from nonanoic acid by *Pseudomonas putida* KT2440. *Applied Microbiology and Biotechnology*, *74*(1), 69–77. https://doi.org/10.1007/s00253-006-0655-4.

Thompson, R. C., Moore, C. J., vom Saal, F. S., & Swan, S. H. (2009). Plastics, the environment and human health: Current consensus and future trends. *Philosophical Transactions of the Royal Society of London. Series B, Biological Sciences*, *364*(1526), 2153–2166. https://doi.org/10.1098/rstb.2009.0053.

Tiso, T., Narancic, T., Wei, R., Pollet, E., Beagan, N., Schröder, K., et al. (2020). *Bio-upcycling of polyethylene terephthalate.* https://doi.org/10.1101/2020.03.16.993592.

Tiso, T., Sabelhaus, P., Behrens, B., Wittgens, A., Rosenau, F., Hayen, H., et al. (2016). Creating metabolic demand as an engineering strategy in *Pseudomonas putida* – Rhamnolipid synthesis as an example. *Metabolic Engineering Communications*, *3*, 234–244. https://doi.org/10.1016/j.meteno.2016.08.002.

Tiso, T., Sauer, D. F., Beckerle, K., Blesken, C. C., Okuda, J., & Blank, L. M. (2020). A combined bio-chemical synthesis route for 1-octene sheds light on rhamnolipid structure. *Catalysts*, *10*(8). https://doi.org/10.3390/catal10080874.

Tiso, T., Wierckx, N., & Blank, L. M. (2014). In P. Grunwald (Ed.), *Non-pathogenic Pseudomonas as platform hosts for industrial biocatalysis* Industrial Biocatalysis Pan Stanford Publishing Pte Ltd.

Tiso, T., Zauter, R., Tulke, H., Leuchtle, B., Li, W.-J., Behrens, B., et al. (2017). Designer rhamnolipids by reduction of congener diversity: Production and characterization. *Microbial Cell Factories*, *16*(1), 225. https://doi.org/10.1186/s12934-017-0838-y.

Tournier, V., Topham, C. M., Gilles, A., David, B., Folgoas, C., Moya-Leclair, E., et al. (2020). An engineered PET depolymerase to break down and recycle plastic bottles. *Nature*, *580*(7802), 216–219. https://doi.org/10.1038/s41586-020-2149-4.

Van Eygen, E., Laner, D., & Fellner, J. (2018). Circular economy of plastic packaging: Current practice and perspectives in Austria. *Waste Management*, *72*, 55–64. https://doi.org/10.1016/j.wasman.2017.11.040.

Venkatachalam, S., Nayak, S. G., Labde, J. V., Gharal, P. R., Rao, K., & Kelkar, A. K. (2012). Degradation and recyclability of poly(ethylene terephthalate). In H. E.-D. M. Saleh (Ed.), *Polyester* IntechOpen.

Vollmer, I., Jenks, M. J., Roelands, M. C., White, R. J., van Harmelen, T., de Wild, P., et al. (2020). Beyond mechanical recycling: Giving new life to plastic waste. *Angewandte Chemie International Edition*, *59*, 15402–15423. https://doi.org/10.1002/anie.201915651.

Ward, P. G., Goff, M., Donner, M., Kaminsky, W., & O'Connor, K. E. (2006). A two step chemo-biotechnological conversion of polystyrene to a biodegradable thermoplastic. *Environmental Science & Technology*, *40*(7), 2433–2437. https://doi.org/10.1021/es0517668.

Wei, R., Breite, D., Song, C., Grasing, D., Ploss, T., Hille, P., et al. (2019). Biocatalytic degradation efficiency of postconsumer polyethylene terephthalate packaging determined by their polymer microstructures. *Advanced Science*, *6*(14), 1900491. https://doi.org/10.1002/advs.201900491.

Wei, R., Song, C., Grasing, D., Schneider, T., Bielytskyi, P., Böttcher, D., et al. (2019). Conformational fitting of a flexible oligomeric substrate does not explain the enzymatic PET degradation. *Nature Communications*, *10*(1), 5581. https://doi.org/10.1038/s41467-019-13492-9.

Wei, R., Tiso, T., Bertling, J., O'Connor, K., Blank, L. M., & Bornscheuer, U. T. (2020). Possibilities and limitations of biotechnological plastic degradation and recycling. *Nature Catalysis*, *3*(11), 867–871. https://doi.org/10.1038/s41929-020-00521-w.

Wei, R., & Zimmermann, W. (2017). Biocatalysis as a green route for recycling the recalcitrant plastic polyethylene terephthalate. *Microbial Biotechnology*, *10*(6), 1302–1307. https://doi.org/10.1111/1751-7915.12714.

Westhues, S., Idel, J., & Klankermayer, J. (2018). Molecular catalyst systems as key enablers for tailored polyesters and polycarbonate recycling concepts. *Science Advances*, *4*(8), eaat9669. https://doi.org/10.1126/sciadv.aat9669.

Wierckx, N. J., Ballerstedt, H., de Bont, J. A., & Wery, J. (2005). Engineering of solvent-tolerant *Pseudomonas putida* S12 for bioproduction of phenol from glucose. *Applied and Environmental Microbiology*, *71*(12), 8221–8227. https://doi.org/10.1128/AEM.71.12.8221-8227.2005.

Woodman, M. E. (2008). Direct PCR of intact bacteria (colony PCR). *Current Protocols in Microbiology*, Appendix 3. https://doi.org/10.1002/9780471729259.mca03ds9.

Wynands, B., Lenzen, C., Otto, M., Koch, F., Blank, L. M., & Wierckx, N. (2018). Metabolic engineering of *Pseudomonas taiwanensis* VLB120 with minimal genomic modifications for high-yield phenol production. *Metabolic Engineering*, *47*, 121–133. https://doi.org/10.1016/j.ymben.2018.03.011.

Yoshida, S., Hiraga, K., Takehana, T., Taniguchi, I., Yamaji, H., Maeda, Y., et al. (2016). A bacterium that degrades and assimilates poly (ethylene terephthalate). *Science*, *351*(6278), 1196–1199. https://doi.org/10.1126/science.aad6359.

Zhu, K., & Rock, C. O. (2008). RhlA converts β-hydroxyacyl-acyl carrier protein intermediates in fatty acid synthesis to the β-hydroxydecanoyl-β-hydroxydecanoate component of rhamnolipids in *Pseudomonas aeruginosa*. *Journal of Bacteriology*, *190*(9), 3147. https://doi.org/10.1128/JB.00080-08.

Zobel, S., Benedetti, I., Eisenbach, L., de Lorenzo, V., Wierckx, N., & Blank, L. M. (2015). Tn7-based device for calibrated heterologous gene expression in *Pseudomonas putida*. *ACS Synthetic Biology*, *4*(12), 1341–1351. https://doi.org/10.1021/acssynbio.5b00058.

CHAPTER NINETEEN

Screening and cultivating microbial strains able to grow on building blocks of polyurethane

María José Cárdenas Espinosa, Andrea Colina Blanco, Hermann J. Heipieper, and Christian Eberlein*
Helmholtz Centre for Environmental Research—UFZ, Department of Environmental Biotechnology, Leipzig, Germany
*Corresponding author: e-mail address: christian.eberlein@ufz.de

Contents

Abstract

The diverse benefits of synthetic polymers is overshadowed by the amount of plastic waste and its whereabouts. The problem can only be tackled by reducing and recycling of plastics. In this respect, investigating the (microbial) degradation of each type of polymer currently used may provide further understanding that fosters the development of new feasible recycling technologies. Here, we present a strategy to isolate bacteria from environmental samples that are able to degrade hydrolysis products and building blocks of polyurethane (PUR). Protocols are presented to enrich bacteria on the primary diamines 2,4-diaminotoluene (TDA) and 4,4′-diaminodiphenylmethane (MDA) as well as an oligomeric PUR (Sigma Aldrich, proprietary composition). For TDA and the oligomeric PUR, methods are suggested to monitor their concentration in bacterial enrichment cultures.

Methods in Enzymology, Volume 648
ISSN 0076-6879
https://doi.org/10.1016/bs.mie.2020.12.008

1. Introduction

Synthetic polymers (plastics) are typically derived from petroleum oil and are an important pillar of our modern society. Among plastics, polyurethanes (PUR) are used as basic material (often as foams) in medical, textile, packaging and construction industries. In 2018 only, the demand for PUR in Europe increased to about 4 million tons (PlasticsEurope, 2019). As the building blocks for polyurethane synthesis are numerous and diverse, there is a huge range of different polyurethane polymers. The properties of PUR highly depend on the structure of the polymer backbone and they can be tailored to have a certain rigidity, strength, flexibility and toughness. However, the common characteristic that all PUR share is the presence of urethane bonds (also known as carbamate bonds). Precursors used to synthesize PUR are diisocyanates and polyols together with additives such as catalysts, cross linkers, and chain extenders, among others. Besides the mentioned urethane bonds, the polyols used for synthesis can additionally contain polyester (polyester PUR) or polyether bonds (polyether PUR). Regarding the diisocyanates employed for the PUR polymerization process, aliphatic and aromatic compounds are common. Fig. 1 shows the diisocyanates mainly used for PUR production: 4,4′-methylenediphenyl diisocyanate (MDI) and toluene-2,4-diisocyanate (TDI).

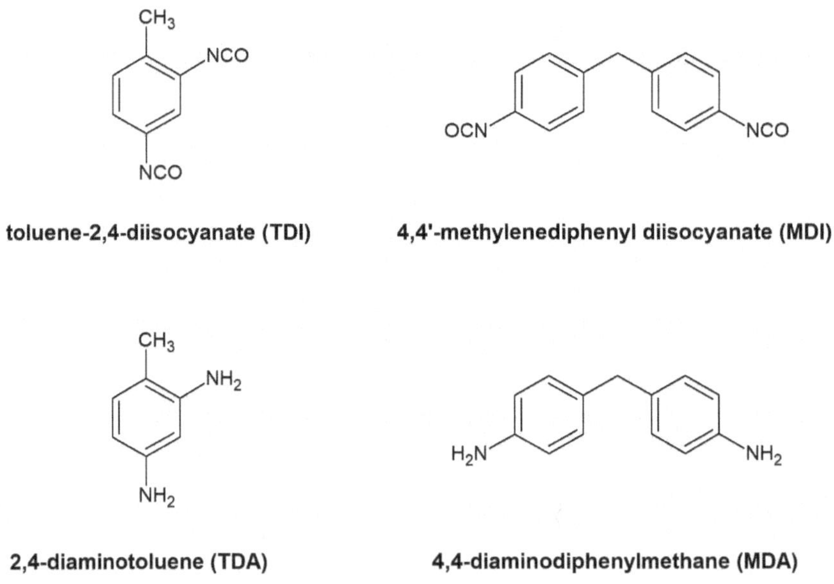

Fig. 1 Common aromatic diisocyanates used for PU synthesis and their corresponding diamines.

In order to investigate PUR biodegradation, two approaches are conceivable: to investigate the change in properties of bulk polymers or the release of degradation products including oligomers and monomeric building blocks. The first studies of PUR polymer degradation by microbial communities have been reported approximately 50 years ago with the exposure of polyester and polyether PUR to fungal activity (Kanavel, Koons, & Lauer, 1966; Ossefort & Testroet, 1966). The purpose of these initial studies was to improve the durability and quality of the materials rather than solving an environmental concern of plastic waste management. Since the 1990s, many studies focused their efforts on analyzing PUR degradation using Gram-positive and Gram-negative bacteria and are summarized in comprehensive review papers (Cregut, Bedas, Durand, & Thouand, 2013; Mahajan & Gupta, 2015; Wierckx et al., 2018). Some of them aimed to improve PUR properties. Other studies focused on the increasing accumulation of plastic waste in landfills or on the decreasing raw material supply (e.g., crude oil). Regarding the latter, a recycling strategy for polyethylene terephthalate (PET) to produce novel PUR polymers was recently accomplished. Enzymatic hydrolysis of PET, subsequent degradation of the building blocks and hydroxyalkanoyloxy-alkanoates (HAAs) production in a *Pseudomonas* strain led to the chemo-catalytic synthesis of a partially bio-based PUR (Tiso et al., 2020).

Generally, polymer degradation is caused by oxidation and hydrolysis reactions. However, many factors such as the chemical structure of the polymer (crystallinity, cross-linking, chemical groups in the molecular chains, etc.) and the abiotic conditions (pH, redox conditions, humidity, temperature, etc.) have an impact on the biodegradation reactions. The protocols presented in this chapter refer to standardized lab conditions in batch cultures of microbial (pure) cultures isolated from the environment. The following chapter describes the investigation of the biodegradation of building blocks of polyurethane polymers by isolating strains from the environment that are capable to use them as sole source of carbon and energy. However, diisocyanates are not stable in the environment due to their immediate reaction with water and the subsequent formation of the corresponding primary amine after decarboxylation (Kreye, Mutlu, & Meier, 2013; Ozaki, 1972). Therefore, diisocyanates such as TDI and MDI are not suitable as model substances to investigate PUR building block degradation in aqueous environments. Instead, their corresponding primary diamines 2,4-diaminotoluene (TDA) and 4,4′-diaminodiphenylmethane (MDA) were chosen. In addition, they are the environmentally relevant PUR hydrolysis products because these diamines were identified as intermediates in the biodegradation of PUR material (Akutsu-Shigeno et al., 2006; Matsumiya, Murata, Tanabe, Kubota, & Kubo, 2010) and are common precursors in polyurethane production processes. In summary, due to the immediate reaction of PU monomers MDI and TDI with water, these monomers were replaced by MDA and TDA to investigate the PUR monomer metabolism (Utomo et al., 2020). Also, an oligomeric PUR was chosen as a model compound for PUR biodegradation experiments. The isolation of a bacterial strain on the oligomeric PUR was reported earlier (Mukherjee et al., 2011).

2. Methods

2.1 Isolation and characterization of bacteria from soil samples capable of degrading PU building blocks

2.1.1 Rationale

So far no pure culture was isolated that uses TDA or MDA as sole source of carbon and energy. Thus, nature's diversity was mined for a bacterial TDA/MDA degrader on a site rich in brittle plastic waste. This site was chosen to increase the chances to find a PUR building block consumer. To avoid the availability of an additional carbon source a mineral medium was chosen for isolation (Hartmans, Smits, van der Werf, Volkering, & de Bont, 1989).

2.1.2 Materials, equipment and reagents

UV/visible spectrophotometer
TDA (Sigma Aldrich, molar mass = 122.17 g/mol; 98% purity)
MDA (Sigma Aldrich, molar mass = 198.26 g/mol; purity ≥97.0%)
PU diol solution (Sigma Aldrich; 88% m/m; proprietary composition), aliphatic, dihydroxy-functional oligomeric PUR
Salts (Table 1) and trace elements (Table 2) to prepare mineral medium (Table 3)

2.1.3 Protocol

1. Prepare agar plates (Table 3) containing 10 mM of TDA or 10 mM MDA as well as agar plates not containing any additional carbon source
2. Take about 10 g of soil, preferably from a soil rich in brittle plastic waste and fill it into a 50 mL test tube
3. Proceed directly or store the sample at 4 °C

Table 1 Mineral salts solution, 10×.

Substance	Concentration (g/L distilled H_2O)
$Na_2HPO_4 \times 2H_2O$	70
KH_2PO_4	28
NaCl	5
NH_4Cl	10

Table 2 Trace elements solution, 500×.

Substance	Concentration (g/L distilled H_2O)	Volume (mL/L)
$MgSO_4 \times 7H_2O$	50.0	
$FeSO_4 \times 7H_2O$	5.0	
$MnSO_4 \times H_2O$	2.5	
$ZnCl_2$	3.2	
$CaCl_2 \times 6H_2O$	0.5	
$BaCl_2$	0.3	
$CoSO_4 \times 7H_2O$	0.18	
$CuSO_4 \times 5H_2O$	0.18	
H_3BO_3	3.25	
EDTA	5.0	
HCl (37%)	—	73

Table 3 Mineral medium used to isolate bacteria.

Substance	Quantity
Mineral salts, 10x	50 mL
Trace elements, 500x	1 mL
Distilled water	449 mL
Agar-agar (3.5% final concentration)	17.5 g

Carbon source was either TDA or MDA (10 mM in agar plates, 2 mM in liquid cultures) or PU diol solution (3 g/L).

4. Resuspend 1 g of each soil sample in 9 mL of 0.9% NaCl (m/v) and then dilute it 1:10 with 0.9% NaCl (m/v), this solution can be stored at 4 °C
5. Prepare a dilution series (10^{-1}, 10^{-2} and 10^{-3}) of the stock solution (prepared in number 4) and use 100 µL for inoculation of agar plates prepared in number 1
6. Store the plates at 30 °C
7. Check the plates after 5 days of incubation, discard the plates completely covered with fungi. From the plates with fungi and bacteria, transfer the latter to fresh plates containing the same carbon source
8. Check after a few days if the transferred bacteria grew and repeat the transfer two or three times more
9. Choose bacteria that grew after each transfer and check their growth on agar plates not containing a carbon source to avoid isolation of autotrophic strains
10. For growth in liquid culture prepare the same media as mentioned above without the addition of agar-agar (Table 3)
11. Prepare stock solutions of 200 mM for TDA and MDA in 10 mL methanol. These solutions must be prepared and used the same week that you inoculate the cultures. In addition, it should be made sure beforehand that methanol cannot be used as a carbon source for the strains isolated
12. Use 250 mL sterile flasks with 50 mL mineral medium containing 2 mM of TDA or MDA (5 mL from stock solution prepared in number 11 in 50 mL total culture volume) or 3 g/L PU diol solution
13. Pick a colony of a strain of interest and inoculate a liquid culture. Perform the experiment in triplicates
14. Incubate the cultures at 30 °C and 106 rpm
15. Measure optical density (OD) daily in a UV/VIS spectrophotometer at a wavelength of 560 nm during 14 days
16. To test if TDA or MDA are not only carbon and energy but also nitrogen source for the isolated bacteria, prepare mineral media as described above but without the addition of NH_4Cl
17. Compare the obtained optical densities between N-deficient media and N-containing media

2.1.4 Safety considerations and standards

TDA and MDA are toxic substances. Avoid contact with skin and eyes as well as formation of dust and aerosols.

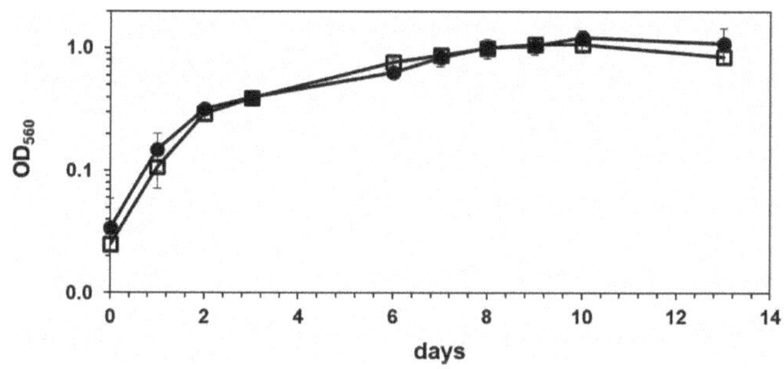

Fig. 2 Example for growth of isolated strain on 2mM TDA in mineral medium containing an additional nitrogen source (filled circles) and in nitrogen-deficient mineral medium (empty squares). n=3.

2.1.5 Analysis and statistics

In Fig. 2, optical density data from an isolated strain growing on 2mM TDA in nitrogen-deficient and nitrogen-containing mineral media is shown (Cárdenas Espinosa et al., 2020). The growth yields are the same and from that, it can be derived that TDA not only serves as a carbon and energy but also as a nitrogen source for the isolated bacterial strain.

2.1.6 Alternative methods/procedures

In order to support the hypothesis that a bacterial strain was successfully enriched on a specific compound, a substrate concentration dependent growth yield test can be performed additionally to the protocol above. Here, the correlation between the concentration of TDA or MDA and the growth yields (OD values) of the isolated bacteria in the same phase of growth were compared. For example, when comparing 1 and 2mM of TDA as substrate concentrations, the growth yield on 2mM should roughly be twice the yield on 1mM. Take into account that concentrations much higher than those used for isolating the strain could be toxic for the cells. For example lower OD values on high concentrations ($\geq$5mM) of TDA or MDA in liquid cultures indicate that these concentrations are already toxic for the isolated bacteria.

2.2 Quantification of TDA degradation

2.2.1 Rationale

A further way to confirm the degradation of the PUR model compounds in the isolated strains is to measure their consumption via chromatographic methods. Here, a protocol is presented where TDA degradation was monitored by high performance liquid chromatography (HPLC) once the media were inoculated with the bacterial strains.

2.2.2 Materials, equipment and reagents

TDA (Sigma Aldrich, molar mass = 122.17 g/mol; 98% purity)
Methanol, Trimethylamine
Equipment for cultivation of bacterial isolates on TDA (see Section 2.1)
0.2 μm filters, for example, polyethersulfone membrane from Whatman™-GE Healthcare
Shimadzu HPLC with CBM-20A (communication bus module), SPD-M20A (photodiode array detector), CTO-20AC (column oven), DGU-20A3 (degasser), LC- 20AB (liquid chromatograph) or equivalent
C18 column (LiChroCART® 125–4, RP-18e, 5 μm, Merck KGaA) or equivalent

2.2.3 Protocol

1. Inoculate 50 mL of 2 mM TDA media with a selected strain (see Section 2.1) and also prepare a sterile control
2. Collect 1 mL of the cultures daily for 14 days and perform $OD_{560\ nm}$ measurements as well to monitor the biomass growth
3. Prepare standards with TDA concentrations between 0.1 and 3 mM in mineral media to obtain a calibration curve
4. Centrifuge all samples (7 min at 15700 × *g*) and filter through a 0.2 μm polyethersulfone membrane
5. Quantify TDA concentrations by HPLC using a C18 column
6. HPLC program: Isocratic elution within 5 min with a mixture of 39.5% methanol, 59.5% distilled water, and 1% trimethylamine (flow rate: 0.65 mL/min; injection volume: 75 μL); temperature of the column constant at 25 °C. Detection was done with a photodiode array detector at 278 nm
7. Calculate TDA concentrations according to the calibration curve

2.2.4 Safety considerations and standards

TDA is a toxic and carcinogenic substance. Avoid contact with skin and eyes as well as formation of dust and aerosols.

2.2.5 Analysis and statistics

An example for TDA consumption in course of the cultivation is depicted in Fig. 3 and shows a depletion of 85% of the initial TDA concentration within 4 days. In contrast to that, the sterile control shows only a minor decrease in TDA concentration.

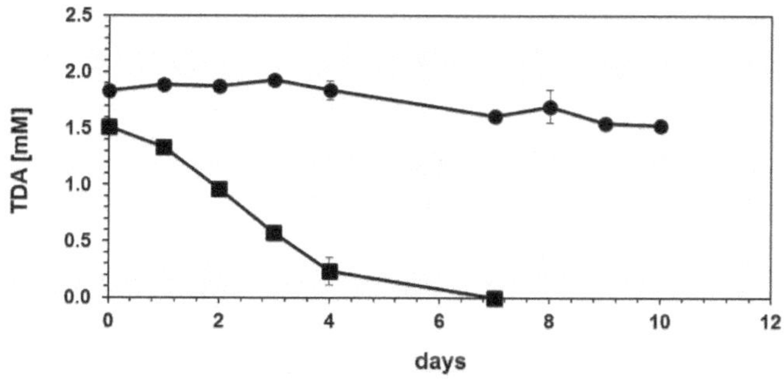

Fig. 3 Consumption of TDA measured via HPLC during the cultivation of an isolated strain on TDA as sole source of carbon and energy in mineral medium (filled squares) and in a sterile control containing TDA (filled circles). n = 3.

2.3 Quantification of PU diol degradation

2.3.1 Rationale

The proprietary composition of the commercially available PU diol solution makes it difficult to assess changes in its concentration during the course of cultivation with isolated bacteria. However, two methods were reported to solve this problem to a certain extent (Howard, Vicknair, & MacKie, 2001; Mukherjee et al., 2011), the determination of the concentration of the oligomeric PUR with high performance thin layer chromatography (HPTLC) and an agar plate assay.

2.3.2 Materials, equipment and reagents

PU diol solution (Sigma Aldrich; 88% m/m; proprietary composition), aliphatic, dihydroxy-functional oligomeric PUR
Ethyl acetate
Hexane
5% Ethanolic sulfuric acid
Equipment for cultivation of bacterial isolates on PU diol solution (see Section 2.1)
Aluminum-backed HPTLC sheets of 0.2 mm layers of silica gel 60 F_{254} (Merck) or equivalent
CAMAG® Linomat 5 (semi-automatic sample application system for bandwise spray-on application of the sample onto the chromatographic layer, including one dosing syringe 100 μL, 90–230 V) or equivalent
CAMAG® Twin Trough Chamber for 20 × 20 cm plates or equivalent
CAMAG® TLC Scanner 4 (densitometric evaluation of HPTLC plates, spectral range 190–900 nm, plate sizes up to 20 × 20 cm, absorbance and fluorescence mode, Software: CAMAG visionCATS) or equivalent
Separatory funnel

2.3.3 Protocol

1. Inoculate 50 mL of mineral media containing 3 g/L PU diol solution with a selected strain (see Section 2.1).
2. Collect 1 mL of the culture at different time points
3. Centrifuge the samples (7 min at 15700 × *g*).
4. Mix the supernatant equivoluminar with ethyl acetate in a separatory funnel, shake well, separate and continue with the ethyl acetate phase
5. Apply the extract of the supernatant (10 μL) on aluminum-backed HPTLC sheets of 0.2 mm layers of silica gel 60 F_{254} using the CAMAG® Linomat 5 sample applicator
6. Perform the chromatographic run in a CAMAG® Twin Trough Chamber, saturated with ethyl acetate and hexane (3:1).
7. Evaporate the solvent from the plates and incubated into freshly prepared solution of 5% ethanolic sulfuric acid and heat for 3 min at 150 °C
8. Wait for color development on the plates and then scan them in absorbance/transmittance mode at 450 nm with a scan speed of 40 mm/s using CAMAG® TLC Scanner 4

2.3.4 Analysis and statistics

In case of a successful PU diol degradation silica plates with samples of a later time point in the course of cultivation show lower values at 450 nm than those at the beginning of the cultivation. Examples can be seen in Mukherjee et al. (2011).

2.3.5 Alternative method

1. Prepare agar plates containing 50 mM phosphate buffer (pH 7), PU diol solution (3 g/L) and 0.001% (w/v) rhodamine B (results in an opaque, pink colored, solid medium).
2. Cut wells of 1 cm diameter in the agar plates and load either with cell suspensions or cell extracts
3. Incubate the agar plates at 37 °C for up to 72 h
4. Irradiate plates with UV light at 350 nm. PU cleavage activity can be detected as orange fluorescent halos surrounding the wells

3. Outlook

Further studies on the isolated strains like differential gene expression studies and transcriptomics as well as proteomics should be conducted to elucidate the degradation pathway of PUR monomers as well as the cleavage of PUR oligomers and polymers. Information on the (extracellular) enzymes and metabolites involved could provide valuable information to foster biotechnical approaches to treat plastic waste.

Acknowledgment

The Financial support by the European Union's Horizon 2020 Research and Innovation Program under grant agreement no. 633962 for the project P4SB is greatly appreciated.

References

Akutsu-Shigeno, Y., Adachi, Y., Yamada, C., Toyoshima, K., Nomura, N., Uchiyama, H., et al. (2006). Isolation of a bacterium that degrades urethane compounds and characterization of its urethane hydrolase. *Applied Microbiology and Biotechnology*, *70*(4), 422–429. https://doi.org/10.1007/s00253-005-0071-1.

Cárdenas Espinosa, M. J., Blanco, A. C., Schmidgall, T., Atanasoff-Kardjalieff, A. K., Kappelmeyer, U., Tischler, D., et al. (2020). Toward biorecycling: Isolation of a soil bacterium that grows on a polyurethane oligomer and monomer. *Frontiers in Microbiology*, *11*, 404. https://doi.org/10.3389/fmicb.2020.00404.

Cregut, M., Bedas, M., Durand, M. J., & Thouand, G. (2013). New insights into polyurethane biodegradation and realistic prospects for the development of a sustainable waste recycling process. *Biotechnology Advances*, *31*(8), 1634–1647. https://doi.org/10.1016/j.biotechadv.2013.08.011.

Hartmans, S., Smits, J. P., van der Werf, M. J., Volkering, F., & de Bont, J. A. M. (1989). Metabolism of styrene oxide and 2-phenylethanol in the styrene-degrading Xanthobacter strain 124X. *Applied and Environmental Microbiology*, *55*(11), 2850–2855.

Howard, G. T., Vicknair, J., & MacKie, R. I. (2001). Sensitive plate assay for screening and detection of bacterial polyurethanase activity. *Letters in Applied Microbiology*, *32*(3), 211–214. https://doi.org/10.1046/j.1472-765x.2001.00887.x.

Kanavel, G. A., Koons, P. A., & Lauer, R. E. (1966). Fungus resistance of millable urethanes. *Rubber Chemistry and Technology*, *39*(4), 1338–1346.

Kreye, O., Mutlu, H., & Meier, M. A. R. (2013). Sustainable routes to polyurethane precursors. *Green Chemistry*, *15*(6), 1431–1455. https://doi.org/10.1039/c3gc40440d.

Mahajan, N., & Gupta, P. (2015). New insights into the microbial degradation of polyurethanes. *RSC Advances*, *5*(52), 41839–41854. https://doi.org/10.1039/c5ra04589d.

Matsumiya, Y., Murata, N., Tanabe, E., Kubota, K., & Kubo, M. (2010). Isolation and characterization of an ether-type polyurethane-degrading micro-organism and analysis of degradation mechanism by Alternaria sp. *Journal of Applied Microbiology*, *108*(6), 1946–1953. https://doi.org/10.1111/j.1365-2672.2009.04600.x.

Mukherjee, K., Tribedi, P., Chowdhury, A., Ray, T., Joardar, A., Giri, S., et al. (2011). Isolation of a Pseudomonas aeruginosa strain from soil that can degrade polyurethane diol. *Biodegradation*, *22*(2), 377–388. https://doi.org/10.1007/s10532-010-9409-1.

Ossefort, Z. T., & Testroet, F. B. (1966). Hydrolytic stability of urethane elastomers. *Rubber Chemistry and Technology*, *39*, 1308–1327.

Ozaki, S. (1972). Recent advances in isocyanate chemistry. *Chemical Reviews*, *72*(5), 457–496. https://doi.org/10.1021/cr60279a002.

PlasticsEurope. (2019). *Plastics – The facts 2019: An analysis of European plastics production, demand and waste data*. Retrieved from Belgium: Brussels.

Tiso, T., Narancic, T., Wei, R., Pollet, E., Beagan, N., Schröder, K., et al. (2020). Bio-upcycling of polyethylene terephthalate. *bioRxiv*. https://doi.org/10.1101/2020.03.16.993592.

Utomo, R. N. C., Li, W. J., Tiso, T., Eberlein, C., Doeker, M., Heipieper, H. J., et al. (2020). Defined microbial mixed culture for utilization of polyurethane monomers. *ACS Sustainable Chemistry & Engineering*. https://doi.org/10.1021/acssuschemeng.0c06019.

Wierckx, N., Narancic, T., Eberlein, C., Wei, R., Drzyzga, O., Magnin, A., et al. (2018). Plastic biodegradation: Challenges and opportunities. In R. Steffan (Ed.), *consequences of microbial interaction with hydrocarbons, oils and lipids: Biodegradation and bioremediation. Handbook of hydrocarbon and lipid microbiology* (pp. 333–361). Cham: Springer. https://doi.org/10.1007/978-3-319-50433-9_23.

CHAPTER TWENTY

Engineering microalgae as a whole cell catalyst for PET degradation

Daniel Moog[a,b,*,†], Jan Zarzycki[c], Karl-Heinz Rexer[d], Tobias J. Erb[b,c], and Uwe G. Maier[a,b]

[a]Laboratory for Cell Biology, University of Marburg, Marburg, Germany
[b]SYNMIKRO Research Center, Marburg, Germany
[c]Max Planck Institute for Terrestrial Microbiology, Marburg, Germany
[d]Department for Evolutionary Ecology of Plants, University of Marburg, Marburg, Germany
*Corresponding author: e-mail address: daniel.moog@biologie.uni-marburg.de

Contents

† Current address: Max Planck Institute for Terrestrial Microbiology, Karl-von-Frisch-Str. 10, 35043 Marburg, Germany

Methods in Enzymology, Volume 648
ISSN 0076-6879
https://doi.org/10.1016/bs.mie.2020.12.023

Abstract

Plastic pollution has become a serious issue on Earth. Although efficient industrial recycling processes exist, a significant fraction of plastic waste still ends up in nature, where it can endure for centuries. Slow mechanical and chemical decay lead to the formation of micro- and nanoplastics, which are washed from land into rivers and finally end up in the oceans. As such particles cannot be efficiently removed from the environment, biological degradation mechanisms are highly desirable. Several enzymes have been described that are capable of degrading certain plastic materials such as polyethylene terephthalate (PET). Such enzymes have a huge potential for future biotechnology applications. However, they require model systems that can be efficiently adapted to very specific conditions. Here, we present detailed instructions, how to convert the model diatom *Phaeodactylum* into a solar-fueled microbial cell factory for PETase expression, resulting in a whole cell catalyst for PET degradation at moderate temperatures under saltwater conditions.

1. Introduction

Efficient management of plastic waste is one of the major economic and ecological challenges of the 21st century. Globally, plastic production rates are steadily increasing, and a functional circular economy is yet to be established. Plastic waste is dumped into landfills or washed into the sea via rivers and from wastewater, often in the form of micro- and nanoplastics. Released into the environment, plastics are taken up by animals, and accumulate in the trophic cycle. Most synthetic petroleum-based plastic materials are artificial polymers foreign to nature, and thus being inaccessible for biological degradation. Nevertheless, certain polyester-based plastics, such as polyethylene terephthalate (PET) can be attacked enzymatically and thereby disassembled into their constituents (terephthalic acid, TPA, and ethylene glycol, EG, as well as mono(2-hydroxyethyl) terephthalic acid, MHET, in the case of PET), which can serve some microorganisms as energy source (Yoshida et al., 2016). The most prominent PET degrading enzymes are bacterial and fungal cutinases, such as leaf-branch compost cutinase (LCC; Sulaiman et al., 2012) and the recently discovered bacterial PET hydrolase (PETase) from *Ideonella sakaiensis* (Yoshida et al., 2016). The primary substrates of cutinases are components of the plant cuticle; however, due to similarities in the chemical structure cutinases accept PET for catalysis. Therefore, cutinases and PETase have a high potential for use in biological PET degradation processes, not only for industrial recycling, but also for wastewater treatment and bioremediation approaches.

Whereas LCC and cutinases in general are adapted to higher temperatures between 50 °C and 80 °C (Sulaiman et al., 2012; Tournier et al., 2020), PETase is a mesophilic enzyme being most active at temperatures between 30 °C and 40 °C (Yoshida et al., 2016). This means that LCC is only poorly active at mesophilic conditions while PETase activity is significantly reduced at temperatures exceeding 45 °C (Yoshida et al., 2016). However, so far, in vitro enzymatic PET degradation is most efficient near the glass transition temperature (T_g) of the polymer, which for PET is around 75 °C (60–65 °C in aqueous solution (Kawai et al., 2014)), rendering the polymer chains most accessible for enzymatic attack. Temperature, together with PET crystallinity, molecular size, surface hydrophobicity and topology are crucial factors for biodegradability of the polymer (Kawai, Kawabata, & Oda, 2019). As a consequence, the varying PET-hydrolyzing enzymes are differently well suited for certain approaches: LCC is much more suitable for biotechnological industrial (high-throughput) applications (e.g., PET recycling), whereas PETase might be the better choice for use in wastewater treatment and (microbial) bioremediation approaches, where temperatures in the mesophilic range or below are prevalent.

Genes encoding LCCs and PETases, have been expressed in various bacterial as well as a few eukaryotic hosts (see e.g., Huang et al., 2018; Moog et al., 2019; Shirke et al., 2018; Sulaiman et al., 2012; Tournier et al., 2020; Yoshida et al., 2016). Established bacterial expression platforms, such as *Escherichia coli* or *Bacillus subtilis*, are highly reliable and productive. However, they lack the ability to generate complex post-translational modifications (PTMs) on proteins, which, even in case the original versions do not carry PTMs, might provide useful features to recombinant plastic degrading enzymes (see below). PTMs include the addition of sugar residues (*N*- and *O*-glycosylation) to a protein's primary structure as well as phosphorylation, acetylation and ubiquitinylation, to name a few, and have an influence on enzyme folding, stability and activity (Hempel & Maier, 2016). For recombinant protein production, glycosylation in particular can be highly important. It is known to stabilize proteins, making them more resistant to higher temperatures and thereby significantly prolonging their half-life (Shental-Bechor & Levy, 2008; Sola & Griebenow, 2009). In case of LCC, this phenomenon has been observed when cutinase was synthesized by a eukaryotic host (*Pichia pastoris*) (Shirke et al., 2018). Moreover, the majority of established bacterial expression systems requires an organic carbon source for growth in a bioreactor, which can make the process of enzyme production costly. A simple solution to that problem is to use a photoautotrophic host that can efficiently use the greenhouse gas CO_2 as a carbon source. Microalgae combine the benefits of photoautotrophy with a fast growth rate (e.g., in a contained bioreactor or even open pond) and the ability to generate eukaryotic PTMs on (recombinant) proteins. In addition, methods for efficient genetic modification and genome editing are well established for many microalgae, making them excellent model systems for synthetic biological and biotechnological approaches.

Recently, a microalga was developed via synthetic biology as a whole cell catalyst for PET degradation (Moog et al., 2019). In this work, the diatom *Phaeodactylum tricornutum* was genetically modified to synthesize and secrete PETase into the surrounding culture medium. Diatom-produced extracellular PETase was shown to be a functional PET degradation enzyme in saltwater at mesophilic conditions (Moog et al., 2019). This chapter provides detailed instructions for the successful production of transgenic diatoms capable of producing and secreting PET degrading enzymes such as PETase.

2. Microalgae as phototrophic cell factories for recombinant protein production

Microalgae combine several advantages as recombinant gene expression systems. These include photoautotrophy, high growth rates, easy scalability, robustness, inexpensive cultivation, rapid genetic manipulation and the potential to achieve high protein yields, among others; thus, microalgae can be considered as high-potential low-cost protein/enzyme production platforms (Baier, Kros, et al., 2018; Hempel & Maier, 2016). Several microalgal species have proven suitable for production of recombinant proteins: among them the green algae *Chlamydomonas reinhardtii*, *Dunaliella salina*, *Haematococcus pluvialis* and several *Chlorella* spp. strains, but also stramenopiles such as the diatom *P. tricornutum* and *Nannochloropsis* spp., the latter belonging to the eustigmatophytes, have shown great potential (Brasil, Gonçalves de Siqueira, Chan Salum, Zanette, & Spier, 2017). Recently, also red microalgae came into focus as hosts for recombinant protein production (Li & Bock, 2018) and the list of potential microalgae expression platforms is constantly expanding, as are the available molecular genetic tools. A prerequisite for recombinant protein production, is a sequenced genome, genetic transformation methods (ideally stable, with a wide range of available selection markers) and knowledge on functional genetic elements that can be used to drive expression of genes (e.g., inducible promoters).

Moreover, the use of known targeting signals would enable the trafficking of proteins to subcellular locations/compartments or even outside of the cells (secretion). The latter can be beneficial for downstream processing, as the recombinant product can easily be separated from the microalgal cells (Baier, Kros, et al., 2018).

So far, the green alga *C. reinhardtii* and the diatom *P. tricornutum* are among the most developed eukaryotic microalgal model systems for recombinant protein production (Vavitsas et al., 2019). Both microalgae can easily be manipulated by different genetic tools (see Section 2.2). For example, advanced genome editing techniques are available and even rapid high-throughput approaches via modular cloning (MoClo) exist (Crozet et al., 2018; D'Adamo et al., 2019). A wide range of recombinant proteins including therapeutics and feed supplements has already been produced with *C. reinhardtii* and *P. tricornutum*. Some of these proteins could even be directed out of the cell via the secretory system (for details see, e.g., Hempel & Maier, 2016). For the diatom *P. tricornutum* this, e.g., was the case for a fully assembled functional human IgG antibody against the hepatitis B virus surface antigen (Hempel & Maier, 2012). Recently, also a functional bacterial enzyme (PETase), which was post-translationally modified, could be produced and secreted into the culture medium by a transgenic *P. tricornutum* cell line (Moog et al., 2019). Depending on the prerequisites of the secreted recombinant enzyme, the composition of the culture medium is important and might influence enzymatic activity, especially if the microalgal expression host is to be used as whole cell catalyst. In case of *P. tricornutum*, the medium is based on brackish water (usually 1.66% salinity) and allows PET degradation by PETase (Moog et al., 2019).

2.1 Design of the genetic constructs

For successful genetic transformation of a model organism and production of a recombinant protein, certain requirements have to be met. First, for efficient expression of non-endogenous genes by an individual model system, it is often necessary to adapt/optimize the codon usage to the requirements of the specific host. Especially when a bacterial gene is to be expressed in a eukaryote model organism (or vice versa), which in some cases might have a considerably different GC-content than the donor organism, codon usage adaption/optimization of the gene might be crucial for success. Thus, knowing the genome sequence, GC- and codon-biases of the host organism is a huge advantage. However, in case of the diatom *P. tricornutum*

(GC-content of 48%) it has been shown that bacterial genes can be readily integrated into the cell nucleus without prior codon optimization, resulting in efficient expression (Hempel, Bozarth, et al., 2011).

Second, depending on where in the cell the recombinant protein should be finally located, it is possible to equip the amino acid sequence with native or host-compatible targeting signals. These signal sequences (peptides) can direct the recombinant protein into the desired subcellular localization or even mediate protein transport out of the cell into the culture medium (secretion) in case an enzyme should function outside the cell, or for an easy isolation of the protein itself from the medium. Protein secretion can be achieved by adding the (codon optimized) sequence of a secretory signal peptide to the 5′-end of the target gene (Moog et al., 2019). Depending on the nature of the signal sequence it is either cleaved off or remains part of the mature protein after transport.

Third, there might be the requirement for recombinant proteins to be purified or analyzed via immunodetection to assess production efficiency (see also Section 2.3). In case no specific antibody for the expressed recombinant protein exists, it might be expedient to add a genetically encoded tag (common tags are His, FLAG or HA) to the protein sequence. This modification will allow immunodetection via Western blot, as well as other applications such as protein localization via immunofluorescence or affinity purification. When choosing the tag position (N-/C-terminal, or internal) and size (short peptide vs. long sequence or entire protein) it is important that the tag does not impair the function or activity of the produced protein or enzyme.

Moreover, in some model systems, including the green alga *C. reinhardtii*, it might be beneficial for expression of recombinant genes to introduce "artificial" intron sequences from native genes into the sequences or to generate transcript fusions with selection marker genes that can be cleaved after protein synthesis (Baier, Wichmann, Kruse, & Lauersen, 2018). This procedure might prevent gene silencing or other mechanisms causing inefficient expression.

2.2 Genomic transformation

As mentioned earlier, *C. reinhardtii* and *P. tricornutum* are two of the most prominent microalgal model systems used in biotechnology. A whole palette of molecular tools is already established and more advanced techniques are currently being developed. Genomic transformation of microalgae is commonly achieved by two different techniques: biolistic bombardment using DNA-coated particles and electroporation (see e.g., Mayfield & Kindle, 1990; Miyahara, Aoi, Inoue-Kashino, Kashino, & Ifuku, 2013; Niu et al., 2012; Shimogawara, Fujiwara, Grossman, & Usuda, 1998; Zhang & Hu, 2014). These conventional transformation methods usually lead to random integration of gene constructs in the nuclear genome (via non-homologous end joining, NHEJ). In *C. reinhardtii* and *P. tricornutum* even functional protein complexes, such as human antibodies (IgG) consisting of two light and two heavy chains (encoded by two separate genes) that are connected via disulfide bridges, can be synthesized (Hempel, Lau, Klingl, & Maier, 2011; Hempel & Maier, 2012; Hempel et al., 2017). In case of the diatom *P. tricornutum*, for example, this is possible as multiple genes can be introduced into the diatom at the same time or via sequential transformations (Hempel & Maier, 2016). To this end, plasmids with two multiple cloning sites (MCSs) such as the pPha-DUAL-[2xNR] vector (GenBank: JN180664.1), optionally equipped with different selection marker genes, can be used enabling the (simultaneous) genomic insertion and expression of six different genes or more. Notably, in case of *P. tricornutum* (and the related diatom *Thalassiosira pseudonana*) via bacterial conjugation, even entire artificial chromosomes (episomes) can be introduced into and maintained within the diatom cell nucleus. This allows the expression of whole gene clusters encoding complex (synthetic) biochemical pathways (Diner, Bielinski, Dupont, Allen, & Weyman, 2016; Karas et al., 2015). Also, targeted genomic integration, e.g., via homologous recombination, or even DNA-free transformation is possible in certain microalgae.

For an efficient production of recombinant proteins, it is essential that the plasmids carrying the designed genetic constructs (Section 2.1) are equipped with the necessary genetic elements to specifically drive transcription. In microalgae, several suitable promoter and terminator elements are established for gene expression—some of them even enable a controlled expression via light or supplementation of certain molecules such as nitrate or ammonium (nitrate reductase promoter) into the culture medium. It is also possible to induce transcription in dependence of a lack of specific nutrients in the medium such as phosphate (for chemically inducible *P. tricornutum* promoters see, e.g., Chu, Ewe, Rio Bartulos, Kroth, & Gruber, 2016; Dell'Aquila et al., 2020). The selective regulation of gene expression enables the coordinated synthesis of the protein of interest, which is especially important for biotechnological applications using model organisms as whole cell biocatalysts.

2.3 Screening and clone selection

After successful transformation of the microalgal host via the biolistic particle bombardment, electroporation or one of the more recently developed methods (e.g., bacterial conjugation), positive clones have to be selected from those that did not integrate the genetic material. This is usually done on selective medium employing an antibiotic to which positive clones are resistant after uptake and genomic integration of the constructed shuttle plasmid. Thus, only clones carrying the resistance cassette can grow. These clones need to be further characterized for production of the recombinant protein, as especially those clones transformed by conventional methods (biolistic bombardment and electroporation) can exhibit strong differences with respect to the level of produced recombinant protein. These differences can result from the location of the random insertion of genes into certain genetic contexts surrounding the integration sites. This is in contrast to expression of episomal genes, for which different clones usually show quite uniform levels of protein yield (Karas et al., 2015). Protein production levels of different microalgal clones can be analyzed, e.g., via immunodetection using a specific antibody against the produced recombinant protein via SDS-PAGE and Western blot. Depending on the desired localization of the recombinant protein (intracellular vs. extracellular/secreted) specific fractions of the culture (medium or cell pellet fraction) or a cellular subfractionation (e.g., in case of membrane proteins) have to be analyzed and compared for protein abundance. This may require a normalization of the protein abundance between different fractions. If the target protein is meant to be exported to the extracellular milieu, secretion efficiency has to be determined by comparing recombinant protein abundance in the medium and the cell fractions. Only those clones showing the highest yield of protein in the desired fraction will be selected for further use in downstream experiments.

2.4 Determining optimal expression conditions

As a next step, optimal expression conditions have to be investigated using the best performing clones (see Section 2.3). Especially, parameters such as duration of expression (induction time) and stability of the protein/enzyme have to be determined. In case of *P. tricornutum*, 2–3 days induction of the nitrate reductase promoter via supplementation of nitrate into the culture medium already seem enough to produce and secrete significant amounts of recombinant protein and even fully assembled protein complexes (Hempel & Maier, 2012; Moog et al., 2019). In the case of bacterial

PETase produced by the diatom, the secreted enzyme successively accumulated in the culture medium after expression was induced at day 0 and was stable for at least 7 days (Moog et al., 2019). Besides testing induction time and duration, additional factors such as light intensity, temperature and the composition of the culture medium (e.g., salinity, CO_2 bubbling, pH, etc.) might influence microalgal growth and thus expression as well as enzyme activity and stability.

2.5 Plastic degradation assays

When a suitable microalgal cell line effectively producing the protein of interest is selected, plastic degradation assays can be performed in order to test enzyme functionality. This can be done in two different ways: either the whole culture consisting of microalgal cells secreting a plastic degrading enzyme can be used as a whole cell catalyst or the produced recombinant enzymes are separated from the living cells prior plastic degradation experiments. The most prominent advantage of using a cell-free plastic degradation assay is that the reaction conditions, especially with respect to temperature, can be easily adapted to the optimum. For the enzymatic breakdown of PET film with PETase, this would be a temperature between 30 °C and 40 °C (Yoshida et al., 2016), conditions in which several microalgae including the diatom *P. tricornutum* are unable to thrive. For a more thermostable enzyme, such as LCC, the temperature in a cell-free assay can be increased even further and adapted to both the enzyme optimum and the glass transition temperature of PET where the plastic is most susceptible to enzymatic degradation (Wei et al., 2019). A main disadvantage of using an enzyme solution without cells is that enzyme stability has to be very high as after loss of function, no further protein is supplied to the assay (Tournier et al., 2020).

In contrast, when using a whole cell catalyst, reaction conditions are largely depending on the conditions required for optimal culture growth. In case of *P. tricornutum*, this would be a moderate temperature around 21 °C (26 °C max.) (Bojko et al., 2013), which is quite distant from optimal reaction conditions of PETase (temperature optimum between 30 °C and 40 °C) (Yoshida et al., 2016). Although, the enzymatic reaction is much slower at 21 °C, if expression is induced permanently, there is a constant supply of new PETase enzyme to the reaction. Especially in non-industrial approaches where incubation time does not play a decisive role (e.g., in long-term bioremediation approaches) the PET hydrolase-producing microalgal whole cell catalyst might be a suitable system.

To initially test the functionality of a secreted PET degrading enzyme, assays using the medium fraction (with or without cells) of a “positive” clone (see Sections 2.3 and 2.4) can be performed. Ideally, this can be done in a small reaction tube (1.5 mL), which is not made of enzymatically degradable material (e.g., polypropylene, PP), using 1 mL of enzyme solution (algal culture/medium fraction) and a piece of PET film (see below) of a defined size. After incubation for 10–14 days at 30–40 °C, the plastic film can be analyzed for traces of surface degradation via scanning electron microscopy (SEM), whereas the liquid fraction can be investigated for the presence of PET degradation products (MHET, TPA, EG) via high-performance liquid chromatography (HPLC). Both methods will provide information on whether the enzyme present in the solution was capable of degrading PET. For the plastic degradation assay, it is highly important to use a negative control (for example, the medium fraction of a wild type culture). As an alternative to using the culture/medium fraction itself, the assay can be performed with purified enzyme, ensuring that the expressed protein is solely responsible for PET degradation and enabling fine tuning of the enzyme concentration in the reaction.

3. Introducing the *PETase* gene into the genome of a diatom

To produce a plastic degrading enzyme in a microalgal host, the gene of interest has to be integrated into the genome of the model organism. Alternatively, some model systems, including *P. tricornutum*, allow maintenance of an expression vector (plasmid) carrying the gene of interest episomally. The following section will provide a detailed list of equipment and chemicals necessary for converting the diatom *P. tricornutum* into a PETase producing platform that can be used as a whole cell catalyst for PET degradation. Please note that genetic transformation is performed by the conventional method of DNA delivery via particle bombardment resulting in random genomic integration(s) of the DNA fragment. For further details see below and Moog et al. (2019).

3.1 Equipment

1. Computer and software for gene design/engineering
2. Codon usage adapted (expression host-specific)/engineered *petase* gene
3. Cloning material, including expression vector (pPha_NR, NCBI accession number: JN180663), suitable cloning kit and competent bacterial cell (*Escherichia coli* TOP10) for plasmid transformation

4. Plasmid isolation kit to gain high purity plasmid for algal transformation (Mini and Midi kit)
5. *Phaeodactylum tricornutum* UTEX 646 strain (Bohlin) for genetic transformation and expression of PETase
6. Vortex
7. Gene gun for DNA delivery via biolistic particle bombardment (e.g., BIO-RAD PDS-1000/He™ System) including all accessory equipment (tungsten M10 micro-particles, helium gas, microcarrier disks, 1350 psi rupture disks, stopping screens, forceps, vacuum pump)
8. Parafilm
9. Petri dishes (92 × 16 mm)
10. Centrifuges and sterile tubes (1.5–50 mL)
11. Micropipettes and sterile tips (0.1–1000 μL)
12. Light microscope
13. Cell counting chamber
14. UV/Vis-spectrophotometer and plastic cuvettes
15. Culturing material, including Erlenmeyer flasks (50–500 mL), orbital laboratory shaker and light source (10,000 lx)
16. Centrifugal filter units with holding proteins larger than 10 kDa (e.g., Amicon® Ultra-15)
17. Western blot material, including SDS-PAGE equipment, blotting instrument, nitrocellulose membrane, Whatman paper, and analogous/digital signal detection system
18. Specific antibodies for protein/tag detection
19. Autoclave for preparation of culture medium
20. Plastic substrate (Polyethylene terephthalate film, e.g., GoodFellow ES301445 amorphous PET film, thickness: 0.25 mm)
21. Non-PET reaction tubes (e.g., 1.5 mL)
22. Incubation chamber or heat block with adjustable temperature (37 °C)
23. Cutting tool or whole-puncher for plastic film
24. High performance liquid chromatography system (and accessory equipment)
25. Scanning electron microscope, sample stubs and sputter coater

3.2 Chemicals

1. Algal growth medium (f/2) components: Tropic Marin salt (Dr. Biener GmbH), Tris/HCl (pH 8.0), NaH_2PO_4, NH_4Cl, $NaNO_3$, vitamins and trace elements, agar agar (solid medium)
2. Bacterial growth medium (LB)

3. Antibiotics (Zeocin, Ampicillin)
4. Spermidine (0.1 M)
5. Calciumchloride (2.5 M)
6. Ethanol (abs., HPLC grade)
7. Phosphate buffered saline (PBS; 137 mM NaCl, 2.7 mM KCl, 10 mM Na_2HPO_4, 1.8 mM KH_2PO_4; pH 7.4)
8. Trichloracetic acid (TCA)
9. Acetone
10. SDS-urea loading dye (8 M urea, 1% (*v*/v) β-mercaptoethanol)
11. β-Mercaptoethanol
12. Sodium hydroxide
13. Sodium dodecyl sulfate (SDS)
14. Authentic HPLC standards: terephthalic acid (TPA), mono(2-hydroxyethyl) terephthalate (MHET) and bis(2-hydroxyethyl) terephthalate (BHET)

4. Protocol

4.1 Generating the genetic construct for transformation

1.1. Design the *PETase* gene sequence for optimal expression in the host system. In case the diatom *P. tricornutum* is used for expression, the codon usage of the bacterial gene should be adapted to the one of the host organism (see e.g., (Montsant, Jabbari, Maheswari, & Bowler, 2005) for codon usage table of *P. tricornutum*).

Tip: This service is usually also offered by gene synthesis companies. For efficient secretion of the protein by the diatom, substitute the coding sequence of the bacterial signal peptide with one encoding a signal peptide of a secreted *P. tricornutum* protein, for example from an alkaline phosphatase (first 30 aa of Phatr2: 49678). Moreover, for protein immunodetection, add the coding sequence of a short amino acid tag to the 3′-end of the gene (e.g., FLAG-tag: DYKDDDDK) followed by a stop-codon. Depending on the cloning strategy, equip the gene with suitable 5′ and 3′ restriction sites. Get the gene synthesized.

1.2. Clone the modified synthetic gene sequence into a vector with an appropriate expression cassette such as the pPha_NR plasmid (which contains an inducible nitrate reductase promoter to drive gene expression) using suitable restriction sites or via advanced cloning methods.

1.3. Transform the construct into *E. coli* (TOP10) for plasmid amplification. Isolate the plasmid (Mini kit) and verify the construct by sequencing.

1.4. Prepare a 50 mL culture of a positive *E. coli* (TOP10) clone carrying the correct plasmid and isolate high purity plasmids via Midi preparation for algal transformation.

4.2 Transformation of the diatom

2.1. Grow *P. tricornutum* at 21 °C in continuous light (10,000 lx) for 7–10 days in 150 mL liquid f/2 medium (pH 8.0) containing 1.5 mM NH_4Cl as a nitrogen source (non-inducing conditions).

2.2. Count and harvest cells by centrifugation (1500 × *g*, 5 min, 21 °C). Adjust cell number to 10^9 cells/mL and carefully resuspend the cells in the required medium volume.

2.3. Plate 10^8 cells on three sterile f/2 agar plates (circular form, diameter of 3–4 cm in the middle of the plate) without antibiotics and let them dry for approximately 30 min.

2.4. Coat tungsten particles (M10) with plasmid DNA by combining 5 μg DNA (in 5–10 μL) with 50 μL of M10 particles, 50 μL $CaCl_2$ (2.5 M) and 20 μL spermidine (0.1 M).

2.5. Vortex for 1 min and allow the particles to sediment for 10 min. Remove the supernatant and resuspend the pellet in 200 μL ethanol (abs., HPLC grade).

2.6. Again, vortex for 1 min and allow sedimentation of the DNA-coupled particles for 5 min. Remove the supernatant and resuspend the particles in 50 μL of ethanol (abs., HPLC grade).

2.7. Prepare 15 μL of the resuspended DNA-coated particles for diatom transformation with the BIO-RAD PDS-1000/He™ system by pipetting the mixture on a microcarrier membrane that is subsequently air-dried (evaporation of ethanol).

2.8. All parts of the gene gun have to be sterilized with ethanol (70% *v*/v) prior use. After assembly of the different gene gun components and insertion of the f/2 agar plate containing wild type cells, close the instrument chamber and apply a vacuum of −23 to −25 psi. Bombard the cells with the DNA-coupled particles using a pressure of 1350 psi (specific rupture disk). Re-pressurize the chamber with air, remove the plate and seal it with parafilm. Do three parallel transformation approaches using three different plates with wild type cells in the center.

2.9. Incubate the transformed diatom cells for 24 h in continuous light (10,000 lx) at 21 °C.

2.10. Gently wash the cells off with 1 mL f/2-medium and spread the volume to equal parts on three new f/2 plates containing a selection marker (e.g., 75 μg/mL Zeocin).

2.11. Let the cells grow for 2–3 weeks in an incubation chamber (21 °C, 10,000 lx) until positive clones are visual (brownish colonies). The clones can be cultured further on agar plates or in liquid f/2.

4.3 Testing the expression and secretion of the recombinant PET hydrolase

3.1 Grow positive clones under non-inducing conditions (NH_4Cl as nitrogen source) in 50 mL of f/2 medium containing Zeocin for 7 days.

3.2 Measure the optical density of the culture at 600 nm (OD_{600}) using a photometer.

3.3 Harvest cells (corresponding to an OD_{600} of 1.0 for culturing in 50 mL fresh medium) by centrifugation (1500 × *g*, 5 min., 21 °C), remove the supernatant, wash cells once with 10 mL f/2 medium containing 0.9 mM $NaNO_3$ (inducing conditions) and culture the cells in 50 mL f/2 with nitrate for 3 days in which the recombinant gene expression is activated on an orbital shaker at 21 °C.

3.4 Separate the medium fraction, which should contain the overproduced secreted protein, from the cells by centrifugation (1500 × *g*, 5 min., 21 °C). Tip: To remove residual intact cells, filtrate the medium fraction using a 0.22 μm pore size filter.

3.5 Concentrate the protein solution to a volume of less than 1 mL using an Amicon® Ultra-15 centrifugal filter unit withholding proteins larger than 10 kDa.

3.6 Precipitate the proteins via addition of TCA (10% *v*/v final conc.) via centrifugation (20,000 × *g*, 15 min, 4 °C). Remove the supernatant.

3.7 Wash the pellet (usually not visible) with 1 mL of 80% (*v*/v) ice-cold acetone by centrifugation (20,000 × *g*, 15 min, 4 °C). Remove the supernatant, air-dry the pellet and dissolve proteins in 20 μL of SDS-urea buffer at 60 °C for 20 min.

3.8 Wash the cell pellet from step 3.4 once with 5–10 mL PBS by centrifugation (1500 × *g*, 5 min., 21 °C) and remove the supernatant. Produce a total protein extract by resuspending the cells in

200 µL alkaline lysis buffer containing 1.7 M NaOH, 7.5% (v/v) β-mercaptoethanol and protease inhibitor cocktail (PIC, 1:200). Incubate on ice for 30 min and precipitate proteins as described in steps 3.6 and 3.7 (with TCA and acetone). Extend acetone wash to two to three repeats. Dissolve air-dried protein pellet in 100 µL of SDS-urea buffer at 60 °C for 20 min. Centrifuge (≥16,000 × *g*, 1 min., RT) and transfer the protein-containing supernatant into a new reaction tube for further analysis.

3.9 Prepare an SDS-gel and load the total medium fraction as well as 10 µg of total protein extract from the cell pellet and separate proteins via electrophoresis. Transfer separated proteins to a nitrocellulose membrane using Western blot. For immunodetection of the PETase equipped with a FLAG-tag use the α-FLAG-tag antibody from Rockland at a dilution of 1:2000. As a control for non-secreted proteins, an antibody against α-tubulin (Sigma-Aldrich, 1:2000) can be used.

3.10 After signal detection, identify those clones on the Western blot showing highest PETase abundance in the medium fraction. These cell lines will be subjected to further plastic degradation experiments.

4.4 Analyzing the functionality of the PETase against its substrate

4.1 Using transgenic diatom clones with the strongest production and secretion of PETase, perform steps 3.1–3.4 (Section 4.3).

4.2 Take 1 mL of the medium fraction containing the PETase enzyme and transfer the solution into a non-PET 1.5 mL reaction tube (e.g., consisting of polypropylene). For a control experiment, set up the same approach with 1 mL of a wild type (WT) *P. tricornutum* culture medium fraction (devoid of PETase).

4.3 Add a defined piece—for example, generated using a whole puncher—of PET film (e.g., from GoodFellow) into the reaction tube and incubate the sample for 14 days at 30–40 °C (conditions optimal for PETase enzyme function). Remove the PET film and wash it with 1% (*w/v*) SDS to stop PETase reaction. Dry the PET film (by air or under vacuum) and store the liquid fraction at −80 °C for further (HPLC) analysis.

4.4 Analyze the PET film incubated with PETase-containing and PETase-lacking (wild type) medium fractions via scanning electron microscopy (SEM) for surface alterations of the plastic structure. To this end samples must be coated with gold (or an alternative conductive material). If degradation was successful the surface of the PET film should be covered with holes clearly visible at a magnification of 4000 ×.

4.5 Use the liquid fraction of the sample for HPLC analysis to investigate the presence of PET degradation products (at least: TPA and MHET).

4.6 As HPLC standards, use pure authentic MHET and TPA solutions. For a detailed HPLC protocol see Moog et al. (2019).

4.7 The MHET and TPA formation in the sample can be calculated by using the detected peak areas of the HPLC chromatograms.

4.5 PET degradation experiments

5.1 For using a transgenic PETase producing and secreting *P. tricornutum* cell line as a whole cell catalyst for PET degradation, perform steps 3.1–3.3 (Section 4.3). You can grow the cells in any suitable volume of f/2-medium in a sterile glass container or a photobioreactor.

5.2 Add PET substrate to the culture at the time PETase gene expression is induced. Since *P. tricornutum* grows optimally at 21 °C, PETase performance will decrease compared to previously used reaction temperatures between 30 °C and 40 °C (see Section 4.4, step 4.3). Thus, the experimental setup might require adaption of the PET substrate from film to shredded or powder PET (nano-/microplastics), which is much more susceptible to enzymatic degradation under these conditions (Moog et al., 2019).

5.3 Run the experiment for several days/weeks/months. Although diatom produced PETase is quite stable and active for several weeks, it is necessary to add more/fresh nitrate to the culture medium or to provide fresh medium to maintain constant algal growth and to keep the PETase gene expression induced.

5.4 Perform necessary control experiments to follow PET biodegradation in the diatom culture (see steps 4.4–4.7, Section 4.4), for example by removing PET particles for SEM or fractions of the liquid culture for HPLC analysis. For the latter, filter the culture using a 0.22 μm pore size filter prior further processing (to remove cells and larger particles).

5. Analysis of enzymatic plastic degradation

5.1 Scanning electron microscopy

A common method to prove enzymatic plastic degradation is SEM. Via this technique, surface alterations of plastic (PET) film can be visualized at high resolution after the sample has been coated with a conductive material such as gold. An essential component of this experiment is the analysis of a negative control, that is a plastic film that has not been in contact with a plastic degradative enzyme, in parallel. Such plastic film can also possess scratches or marks from handling and it represents a crucial element for comparing the effects of enzymatic polymer hydrolysis vs. potential mechanical damages of the plastic film. A direct comparison of a degradation positive and negative PET film sample is shown in Fig. 1. Here, clear traces of enzymatic degradation (wholes and clefts) are visible on the surface of the amorphous PET film (PETA, Goodfellow) on the left side, which was treated with 1 mL of the sterile filtered (0.22 μm) supernatant of PETase producing diatom culture for 42 days at 37 °C (Fig. 1). In contrast, only dust particles and slight scratches from handling are visible on the surface of the negative control—an amorphous PET film sample incubated with the filtered supernatant of a wild type diatom culture for 42 days at 37 °C.

As an exact quantitative analysis of the degree of enzymatic plastic degradation is not possible via SEM, additional HPLC analyses of the liquid fractions of the samples can be very helpful to estimate the amounts of monomers (TPA, MHET, and EG) formed by PETase (see Section 5.2).

5.2 High performance liquid chromatography

Full degradation of PET film should result in the monomers TPA and EG, which should be detectable in the soluble fraction of the samples. However, as reported from studies using bacterial expressed PETase (see e.g., Yoshida

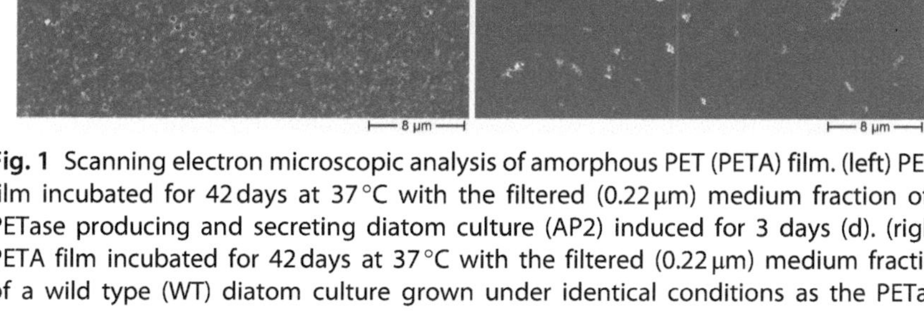

Fig. 1 Scanning electron microscopic analysis of amorphous PET (PETA) film. (left) PETA film incubated for 42 days at 37 °C with the filtered (0.22 μm) medium fraction of a PETase producing and secreting diatom culture (AP2) induced for 3 days (d). (right) PETA film incubated for 42 days at 37 °C with the filtered (0.22 μm) medium fraction of a wild type (WT) diatom culture grown under identical conditions as the PETase producing strain. Magnification: 4000 ×.

et al., 2016), the enzymatic PET breakdown by PETase is mainly resulting in the formation of MHET and rather low amounts of TPA (and EG). This seems to be different in the diatom system, in which PETase-containing culture supernatant, depending on the reaction conditions, mainly produces TPA and (if at all) only a minor fraction of MHET (Moog et al., 2019). The reaction products (monomers) present in the liquid fraction of the sample can be detected via HPLC (see e.g., Moog et al., 2019). In addition to a qualitative identification, another benefit of HPLC is the possibility of a quantitative sample analysis, in which the amount of formed PET degradation products can be calculated. This can be done by using the detected peak areas of HPLC chromatograms and comparing those values to chromatograms of standards of known concentrations. Fig. 2 shows a the HPLC analysis of the aqueous fractions of amorphous PET degradation samples (PETA) shown in Fig. 1 (for HPLC protocol details see Moog et al. (2019)). TPA (calculated concentration: 291 μM) and MHET (calculated concentration: 38 μM) could be identified in the samples containing PETase, whereas no degradation products were detected in the negative control (wild type culture supernatant devoid of PETase).

A significant benefit of HPLC is the possibility of a continuous monitoring of the enzymatic catalytic performance, thus giving information on the degree of biological plastic degradation (i.e., the formation of degradation products) taking place in the sample at a given time. At different time points, parts of the aqueous fraction can be easily removed from the reaction and analyzed via HPLC, while the sample is further incubated in the remaining

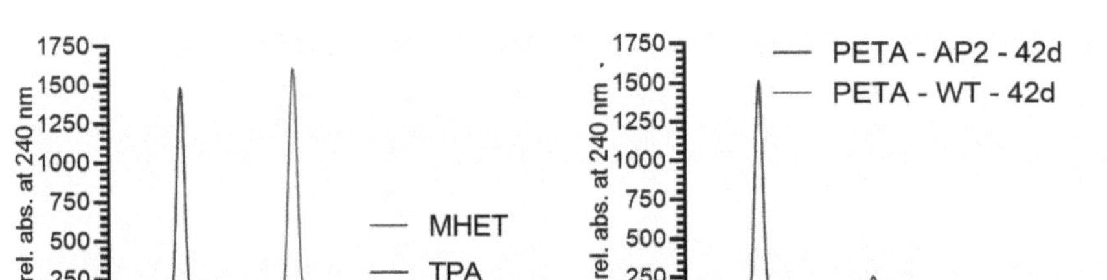

Fig. 2 HPLC chromatograms with specific peaks for TPA and MHET. (left) Chromatograms of the standards for TPA and MHET (15 μL of 60 μM solutions). (right) HPLC chromatograms for samples PETA - AP2 - 42d (degradation positive) and PETA - WT - 42d (negative control) incubated for 42 days at 37C°. 3 μL of each sample was analyzed. Abbreviation: MHET: mono(2-hydroxyethyl) terephthalic acid; TPA: terephthalic acid; rel.: relative, abs.: absorption; min: minutes; nm: nanometers.

enzyme-containing solution or active culture. This allows a constant insight into the PETase activity in a cell free approach as well as in an experimental setup, in which the diatom is used as a whole cell catalyst for PET degradation.

6. Summary

Microalgae are an aspiring chassis for the production of extracellular recombinant proteins. Especially for the efficient secretion of recombinant enzymes (even such of bacterial origin) into the surrounding (saltwater) medium, the diatom *P. tricornutum* has proven to be a very useful model system. This chapter provides detailed instruction how to convert *P. tricornutum* into a solar-fueled microbial cell factory for PETase production, resulting in a whole cell catalyst for PET degradation at moderate temperatures under saltwater conditions. The presented diatom system can be easily adapted for the synthesis and secretion of other soluble plastic degrading enzymes. Although the culture conditions of *P. tricornutum* may not be completely optimal for efficient and rapid PET degradation, the culture supernatant might have the potential to be used for bioremediation approaches and wastewater treatment with the aim of breaking down nano- and microplastic PET particles at moderate temperatures.

Acknowledgments

This work was supported by the University of Marburg, the SYNMIKRO Research Center and the Max Planck Institute for Terrestrial Microbiology.

References

Baier, T., Kros, D., Feiner, R. C., Lauersen, K. J., Muller, K. M., & Kruse, O. (2018). Engineered fusion proteins for efficient protein secretion and purification of a human growth factor from the green microalga Chlamydomonas reinhardtii. *ACS Synthetic Biology*, *7*, 2547–2557.

Baier, T., Wichmann, J., Kruse, O., & Lauersen, K. J. (2018). Intron-containing algal transgenes mediate efficient recombinant gene expression in the green microalga Chlamydomonas reinhardtii. *Nucleic Acids Research*, *46*, 6909–6919.

Bojko, M., Brzostowska, K., Kuczynska, P., Latowski, D., Olchawa-Pajor, M., Krzeszowiec, W., et al. (2013). Temperature effect on growth, and selected parameters of Phaeodactylum tricornutum in batch cultures. *Acta Biochimica Polonica*, *60*, 861–864.

Brasil, B. S. A. F., Gonçalves de Siqueira, F., Chan Salum, T. F., Zanette, C. M., & Spier, M. R. (2017). Microalgae and cyanobacteria as enzyme biofactories. *Algal Research*, *25*, 76–89.

Chu, L., Ewe, D., Rio Bartulos, C., Kroth, P. G., & Gruber, A. (2016). Rapid induction of gfp expression by the nitrate reductase promoter in the diatom Phaeodactylum tricornutum. *PeerJ*, *4*, e2344.

Crozet, P., Navarro, F. J., Willmund, F., Mehrshahi, P., Bakowski, K., Lauersen, K. J., et al. (2018). Birth of a photosynthetic chassis: A moclo toolkit enabling synthetic biology in the microalga Chlamydomonas reinhardtii. *ACS Synthetic Biology*, 7, 2074–2086.
D'Adamo, S., Schiano di Visconte, G., Lowe, G., Szaub-Newton, J., Beacham, T., Landels, A., et al. (2019). Engineering the unicellular alga Phaeodactylum tricornutum for high-value plant triterpenoid production. *Plant Biotechnology Journal*, *17*, 75–87.
Dell'Aquila, G., Zauner, S., Heimerl, T., Kahnt, J., Samel-Gondesen, V., Runge, S., et al. (2020). Mobilization and cellular distribution of phosphate in the diatom Phaeodactylum tricornutum. *Frontiers in Plant Science*, *11*, 579.
Diner, R. E., Bielinski, V. A., Dupont, C. L., Allen, A. E., & Weyman, P. D. (2016). Refinement of the diatom episome maintenance sequence and improvement of conjugation-based DNA delivery methods. *Frontiers in Bioengineering and Biotechnology*, *4*, 65.
Hempel, F., Bozarth, A. S., Lindenkamp, N., Klingl, A., Zauner, S., Linne, U., et al. (2011). Microalgae as bioreactors for bioplastic production. *Microbial Cell Factories*, *10*, 81.
Hempel, F., Lau, J., Klingl, A., & Maier, U. G. (2011). Algae as protein factories: Expression of a human antibody and the respective antigen in the diatom Phaeodactylum tricornutum. *PLoS One*, *6*, e28424.
Hempel, F., & Maier, U. G. (2012). An engineered diatom acting like a plasma cell secreting human IgG antibodies with high efficiency. *Microbial Cell Factories*, *11*, 126.
Hempel, F., & Maier, U. G. (2016). Microalgae as solar-powered protein factories. *Advanced Technologies for Protein Complex Production and Characterization*, *896*, 241–262.
Hempel, F., Maurer, M., Brockmann, B., Mayer, C., Biedenkopf, N., Kelterbaum, A., et al. (2017). From hybridomas to a robust microalgal-based production platform: Molecular design of a diatom secreting monoclonal antibodies directed against the Marburg virus nucleoprotein. *Microbial Cell Factories*, *16*, 131.
Huang, X., Cao, L., Qin, Z., Li, S., Kong, W., & Liu, Y. (2018). Tat-independent secretion of polyethylene terephthalate hydrolase PETase in Bacillus subtilis 168 mediated by its native signal peptide. *Journal of Agricultural and Food Chemistry*, *66*, 13217–13227.
Karas, B. J., Diner, R. E., Lefebvre, S. C., McQuaid, J., Phillips, A. P., Noddings, C. M., et al. (2015). Designer diatom episomes delivered by bacterial conjugation. *Nature Communications*, *6*, 6925.
Kawai, F., Kawabata, T., & Oda, M. (2019). Current knowledge on enzymatic pet degradation and its possible application to waste stream management and other fields. *Applied Microbiology and Biotechnology*, *103*, 4253–4268.
Kawai, F., Oda, M., Tamashiro, T., Waku, T., Tanaka, N., Yamamoto, M., et al. (2014). A novel ca(2)(+)-activated, thermostabilized polyesterase capable of hydrolyzing polyethylene terephthalate from Saccharomonospora viridis ahk190. *Applied Microbiology and Biotechnology*, *98*, 10053–10064.
Li, Z., & Bock, R. (2018). Replication of bacterial plasmids in the nucleus of the red alga Porphyridium purpureum. *Nature Communications*, *9*, 3451.
Mayfield, S. P., & Kindle, K. L. (1990). Stable nuclear transformation of Chlamydomonas reinhardtii by using a C. reinhardtii gene as the selectable marker. *Proceedings of the National Academy of Sciences*, *87*, 2087–2091.
Miyahara, M., Aoi, M., Inoue-Kashino, N., Kashino, Y., & Ifuku, K. (2013). Highly efficient transformation of the diatom Phaeodactylum tricornutum by multi-pulse electroporation. *Bioscience, Biotechnology, and Biochemistry*, 77, 874–876.
Montsant, A., Jabbari, K., Maheswari, U., & Bowler, C. (2005). Comparative genomics of the pennate diatom Phaeodactylum tricornutum. *Plant Physiology*, *137*, 500–513.
Moog, D., Schmitt, J., Senger, J., Zarzycki, J., Rexer, K. H., Linne, U., et al. (2019). Using a marine microalga as a chassis for polyethylene terephthalate (PET) degradation. *Microbial Cell Factories*, *18*, 171.

Niu, Y. F., Yang, Z. K., Zhang, M. H., Zhu, C. C., Yang, W. D., Liu, J. S., et al. (2012). Transformation of diatom Phaeodactylum tricornutum by electroporation and establishment of inducible selection marker. *BioTechniques*, *52*. https://doi.org/10.2144/000113881.

Shental-Bechor, D., & Levy, Y. (2008). Effect of glycosylation on protein folding: A close look at thermodynamic stabilization. *Proceedings of the National Academy of Sciences*, *105*, 8256–8261.

Shimogawara, K., Fujiwara, S., Grossman, A., & Usuda, H. (1998). High-efficiency transformation of Chlamydomonas reinhardtii by electroporation. *Genetics*, *148*, 1821–1828.

Shirke, A. N., White, C., Englaender, J. A., Zwarycz, A., Butterfoss, G. L., Linhardt, R. J., et al. (2018). Stabilizing leaf and branch compost cutinase (LCC) with glycosylation: Mechanism and effect on PET hydrolysis. *Biochemistry*, *57*, 1190–1200.

Sola, R. J., & Griebenow, K. (2009). Effects of glycosylation on the stability of protein pharmaceuticals. *Journal of Pharmaceutical Sciences*, *98*, 1223–1245.

Sulaiman, S., Yamato, S., Kanaya, E., Kim, J. J., Koga, Y., Takano, K., et al. (2012). Isolation of a novel cutinase homolog with polyethylene terephthalate-degrading activity from leaf-branch compost by using a metagenomic approach. *Applied Environmental Microbiology*, *78*, 1556–1562.

Tournier, V., Topham, C. M., Gilles, A., David, B., Folgoas, C., Moya-Leclair, E., et al. (2020). An engineered PET depolymerase to break down and recycle plastic bottles. *Nature*, *580*, 216–219.

Vavitsas, K., Crozet, P., Vinde, M. H., Davies, F., Lemaire, S. D., & Vickers, C. E. (2019). The synthetic biology toolkit for photosynthetic microorganisms. *Plant Physiology*, *181*, 14–27.

Wei, R., Breite, D., Song, C., Grasing, D., Ploss, T., Hille, P., et al. (2019). Biocatalytic degradation efficiency of postconsumer polyethylene terephthalate packaging determined by their polymer microstructures. *Advanced Science*, *6*, 1900491.

Yoshida, S., Hiraga, K., Takehana, T., Taniguchi, I., Yamaji, H., Maeda, Y., et al. (2016). A bacterium that degrades and assimilates poly(ethylene terephthalate). *Science*, *351*, 1196–1199.

Zhang, C., & Hu, H. (2014). High-efficiency nuclear transformation of the diatom Phaeodactylum tricornutum by electroporation. *Marine Genomics*, *16*, 63–66.

CHAPTER TWENTY-ONE

Yeast cell surface display of bacterial PET hydrolase as a sustainable biocatalyst for the degradation of polyethylene terephthalate

Zhuozhi Chen[a,†], Yunjie Xiao[a,†], Gert Weber[b], Ren Wei[c], and Zefang Wang[a,*]

[a]School of Life Sciences, Tianjin University, Tianjin, China
[b]Macromolecular Crystallography, Helmholtz-Zentrum Berlin, Berlin, Germany
[c]Junior Research Group Plastic Biodegradation, Institute of Biochemistry, University of Greifswald, Greifswald, Germany
*Corresponding author: e-mail address: zefangwang@tju.edu.cn

Contents

† These authors contributed equally.

Methods in Enzymology, Volume 648
ISSN 0076-6879
https://doi.org/10.1016/bs.mie.2020.12.030

Abstract

Enzymatic hydrolysis of polyethylene terephthalate (PET) is considered to be an environmentally friendly method for the recycling of plastic waste. Recently, a bacterial enzyme named *Is*PETase was found in *Ideonella sakaiensis* with the ability to degrade amorphous PET at ambient temperature suggesting its possible use in recycling of PET. However, applying the purified *Is*PETase in large-scale PET recycling has limitations, i.e., a complicated production process, high cost of single-use, and instability of the enzyme. Yeast cell surface display has proven to be an effectual alternative for improving enzyme degradation efficiency and realizing industrial applications. This chapter deals with the construction and application of a whole-cell biocatalyst by displaying *Is*PETase on the surface of yeast (*Pichia pastoris*) cells.

1. Introduction

PET is currently the most widely used synthetic polyester in the world. PET is formed by polycondensation of terephthalic acid (terephthalate, TPA) and ethylene glycol (EG) and possesses good mechanical, chemical and thermal properties (Rieckmann & Völker, 2004; Webb, Arnott, Crawford, & Ivanova, 2013). In 2018, global PET production reached approximately 360 million tonnes. PET is mainly used in the production of packaging materials (e.g., beverage bottles) and synthetic fiber (Global PET Resin Production Capacity, 2016). Currently, large amounts of PET debris are entering landfills or natural environment by uncontrolled littering (Kawai, Kawabata, & Oda, 2019). Given that PET exhibits a pronounced durability, it can accumulate in the environment and persist for hundreds of years, the uncontrolled disposal of waste PET becomes a global problem. At present, many studies have detected plastic fragments and particles (e.g., micro- and nanoplastics) in the bodies of more than 690 marine organisms (Carbery, O'Connor, & Palanisami, 2018; Toussaint et al., 2019). Thus, the continuous enrichment of plastic particles via the food chain may pose a potential threat to human health (Bouwmeester, Hollman, & Peters, 2015). Recalcitrant PET waste accumulates rapidly in ecosystems worldwide, and has caused serious environmental pollution (Casabianca et al., 2020; Cressey, 2016; Huang et al., 2020). In 2016, only about 7% of post-consumer PET bottles were recycled (The New Plastics Economy: Rethinking the Future of Plastics, 2016), which led to a significant waste of resources. By chemical or biocatalytic approaches, PET can be dissembled to its monomers TPA and EG which can be recovered, purified and reused for the synthesis of virgin polymers (Rahimi & García, 2017; Wei & Zimmermann, 2017).

PET exhibits good structural regularity and strong crystallization ability, but the rigidity of its molecular chain and high glass transition temperature hinder the flexibility of its molecular chain. Therefore, PET exists mostly in a semicrystalline form with both amorphous and crystalline fractions (Marten, Müller, & Deckwer, 2003). Regularly arranged regions are called crystalline regions, and randomly arranged regions are called amorphous regions. PET can crystallize as the temperature increases to above its glass transition temperature (T_g) at >70 °C. The melting point of PET is in the temperature range around 260 °C. If the melt cools quickly to far below T_g, it will form a transparent amorphous structure.

The percentage degree of crystallinity is usually used to describe the fraction of crystalline regions in a semi-crystalline polymers like PET. PET with different crystallinities can be used for different application scenarios (Kawai et al., 2019). Most PET used in the production of beverage bottles and textiles has a high crystallinity of 30%–40%. However, the crystallinity of PET used for food packaging is low, for example, at approximately 8% (Kawai et al., 2014; Oda et al., 2018). Commercially available low-crystalline PET films provided by Goodfellow Ltd. (PET-GF; thickness 0.25 mm) with a crystallinity of lower than 7% are widely used in the investigations on biological PET degradation (Ronkvist, Xie, Lu, & Gross, 2009; Tournier et al., 2020; Wei, Song, et al., 2019). Many studies have indicated that PET hydrolase preferentially hydrolyzes the amorphous region of PET, and the crystalline region is minimally decomposed by enzymes (Brueckner, Eberl, Heumann, Rabe, & Guebitz, 2008; Gamerith et al., 2017; Wei, Braite, et al., 2019). Additionally, the increase in crystallinity will limit the mobility of polymer chains and reduce their accessibility to the hydrolases (Kawai et al., 2019; Wei, Braite, et al., 2019). Therefore, the higher the crystallinity of PET, the more resistant it is to microbial and enzymatic degradation (Table 1).

1.1 The current status of enzymatic PET depolymerization

Up to now, several strategies have been developed to degrade and recycle waste PET, including chemical, mechanical and biological methods. The mechanical method involves a relatively simple process including sorting, cutting, heating, and extrusion. Thereby, PET pellets from the mechanical recycling are used to directly produce new end product. Although mechanical recycling has been widely used in industry, the performance of recycled PET will be greatly reduced after this process (Anadón et al., 2011).

Table 1 Application of cutinase in the degradation of low-crystalline PET.

Enzyme	Strain	Crystallinity of PET/%	Temperature/°C	Days	Weight lose/%	References
FsC	*Fusarium solani f. pisi*	7	40	4	5	Ronkvist et al. (2009)
HiC	*Humicola Insolens*	7	70	4	97 ± 3	Ronkvist et al. (2009)
TfH	*Thermobifida fusa* DSM43793	10	55	21	≈ 50	Müller, Schrader, Profe, Dresler, and Deckwer (2005)
		–	65	2	12.9 ± 1.2	Then et al. (2015)
TfCut2	*Thermobifida fusa* KW3	–	65	2	12.6 ± 0.2	Then et al. (2015)
TfCut2 Variant	*Thermobifida fusa* KW3	–	70	2	25.0 ± 0.8	Then et al. (2016)
LCC Variant	Leaf branch compost	10	72	0.42	90	Tournier et al. (2020)
Cut190 Variant	*Saccharomonospora viridis* AHK190	6.3	63	3	13.5 ± 0.5	Kawai et al. (2014)
		8.4			27.0 ± 1.0	Kawai et al. (2014)

In contrast, chemical recycling is aiming at the conversion of waste PET polymers into the monomers TPA and EG or partially depolymerized oligomers which can be used to synthesize virgin PET without significant performance loss compared to those of the waste PET before the recycling procedure (Lin & Yang, 2007). However, chemical depolymerization through hydrolysis, alcoholysis, ammonolysis, and other methods is based on the use of harmful chemical reagents or at extreme reaction conditions. Although this method is a typical closed-loop recycling process, the application of this method is also limited due to secondary environmental pollution, high energy consumption, and high monetary cost (Rahimi & García, 2017).

In recent years, the biocatalytic method has attracted widespread attention due to its advantages of simple operation, environmental friendliness, and energy-saving features and has opened up a "green route" for PET recycling (Shoda, Uyama, Kadokawa, Kimura, & Kobayashi, 2016; Taniguchi et al., 2019; Wei et al., 2020; Wei & Zimmermann, 2017; Zhang, Fevre, Jones, & Waymouth, 2018). A variety of PET hydrolases have been identified, including esterases, lipases, and cutinases (Doris et al., 2013; Groß, Hamacher, Schmitz, & Jager, 2017; Joo et al., 2018). These enzymes can depolymerize amorphous or low-crystalline PET (Wei and Zimmermann, 2017). Unfortunately, none of these enzymes can effectively degrade high-crystalline PET (Kawai et al., 2019).

1.2 *Is*PETase as a promising biocatalyst for PET depolymerization

In 2016, Japanese scientists discovered a new PET hydrolase (*Is*PETase) from a bacterium called *Ideonella sakaiensis* 201-F6 that can degrade amorphous PET at ambient temperature and assimilate its degradation monomers, albeit very slowly (Yoshida et al., 2016). The main degradation product was mono(2-hydroxyethyl) terephthalic acid (MHET). In addition, a small amount of TPA and bis(2-hydroxyethyl)-TPA (BHET) was produced. Further studies on enzyme activity revealed that recombinant *Is*PETase exhibited increased enzyme activity at pH 9.0. MHET can be potentially internalized by *Ideonella sakaiensis* 201-F6, hydrolyzed by the MHETase (Palm et al., 2019) and further metabolized by this bacterium. At a reaction temperature below 40 °C, *Is*PETase showed higher degradation performance on selected PET samples than with other fungal (e.g., FsC from *Fusarium solani*) and bacterial cutinases (e.g., TfH from *Thermobifida fusca* and LCC derived from a plant metagenome) (Yoshida et al., 2016). This feature makes *Is*PETase a promising candidate for the degradation and recycling of PET at ambient temperature. Afterward, many studies have reported the crystal structure of *Is*PETase and corresponding protein engineering investigations towards the reaction mechanisms of enzymatic PET hydrolysis as well as *Is*PET mutants with improved catalytic properties (Austin et al., 2018; Fecker et al., 2018; Han et al., 2017; Joo et al., 2018).

By comparing the structure of *Is*PETase and other PET hydrolases, the substrate binding pocket of *Is*PETase appears to be much wider than those of other PET hydrolases. For example, the binding pocket widths of TfCut2 and LCC are 3.0 and 3.6 Å, respectively, whereas that of *Is*PETase reached 8.5 Å. Similar findings were also reported by an earlier study (Austin et al., 2018). These special structural features may explain why *Is*PETase can potentially accommodate and degrade high-crystalline PET at a higher rate at ambient temperature, at which condition the high-crystalline polymer chains are indeed in a less flexible status (Falkenstein et al., 2020; Wei, Song, et al., 2019). Nevertheless, there is still a need to further improve the efficiency of *Is*PETase in the degradation of high-crystalline PET which is currently still far from the requirements of industrial applications (Fecker et al., 2018; Liu et al., 2018).

1.3 Brief introduction of whole-cell biocatalysts

To realize industrial applications, enzymes are typically immobilized because the stability and reusability of free enzymes are not ideal. However, immobilization will greatly increase the application costs (De Carvalho, 2011). The traditional methods of enzyme immobilization include covalent binding of the enzyme to a solid support, cross-linking the enzyme to polymers, and encapsulation of the enzyme in a polymer matrix. However, the chemical reagents used in these methods may denature the enzyme, resulting in a loss of enzyme activity. Additionally, the enzyme purification and preparation processes are time consuming and costly (Sheldon & Van Pelt, 2013).

To avoid these shortcomings, researchers have developed cell surface display systems, which utilize synthetic biology approaches to design renewable biocatalysts by expressing target enzymes on the surface of microbial cells (Dwevedi & Kayastha, 2012). The basic principle of the cell surface display techniques involves an anchor protein (typically a cellular surface protein) tethered to the target protein, which is transported to the cell surface as fused cargo of the anchor protein. After surface display, the function and activity of the enzyme are maintained in the cell surface display system since the enzyme is fixed on the supporting cells, which enhances its stability and reusability. In addition, the target enzyme is synthesized in microbial cells

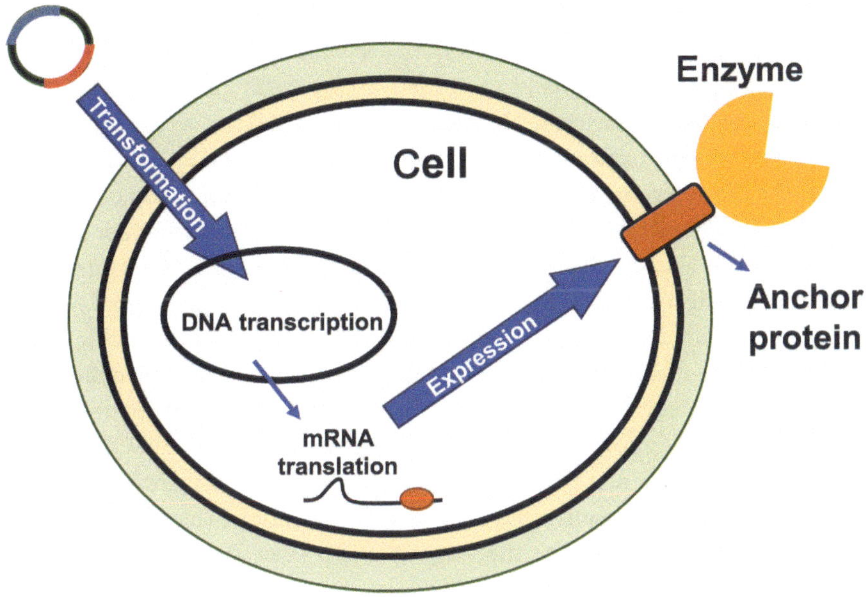

Fig. 1 Schematic diagram of the whole-cell biocatalyst with surface-displayed recombinant enzyme.

and then automatically transferred to the cell surface through the cellular secretion system (Fig. 1). This technology avoids the cumbersome and expensive process of enzyme separation and purification, thereby enabling a broad application prospect (Nakatani & Hori, 2017). Cell surface display technology has been applied to the development of biosensors for the detection of organic pollutants and the development of whole-cell biocatalysts for the production of renewable biofuels (Fan, Hou, Liang, Feng, & Liu, 2015; Liang et al., 2019; Liu, Feng, & Liang, 2016; Liu, Lang, Liang, & Shi, 2017).

1.4 Brief introduction of the yeast surface display system

Cell surface display is categorized into prokaryotic cell surface display and eukaryotic cell surface display. The prokaryotic cell surface display system cannot express complex eukaryote-derived proteins as they are often glycosylated or require disulfide bond isomerization, for activation after translation. Eukaryotic surface display systems have been successfully used to express and display several complex proteins in the recent decades (Park, 2020).

The commonly used eukaryotic hosts for cell surface display are yeasts such as *Pichia pastoris* and *Saccharomyces cerevisiae*. The target protein and the anchor protein are usually expressed in the form of a fusion protein. During the secretion process, the anchor protein interacts with the cell wall through a covalent or non-covalent bond. Yeast surface display systems are categorized into two types according to the fusion method of the target protein and the anchor protein. Depending on the location of target protein relative to the anchor protein, these can be divided into N-terminal and C-terminal fusion variants. For N-terminal fusion, a secretion signal peptide needs to be added to the N-terminus of the target protein to achieve the transport of the fusion protein to the cell surface. The choice of anchor protein is crucial for surface display systems (Annie Gai & Dane Wittrup, 2007). For GPI (glycosylphosphatidyl alcohol) anchor proteins, the GPI site of the cell wall is combined with the GPI signal peptide of the anchor protein. The fixed protein is connected to the cell wall by covalent bonding to achieve the surface display of the target protein (Der Vaart et al., 1997).

1.5 Application of whole-cell biocatalyst in lipase and cutinase

At present, numerous studies on the surface display of lipase and cutinase have been published (Lee, Choi, Han, Choi, & Sang, 2005; Lee, Lee, & Park, 2005; Liu, Zhao, Jia, Xu, & Yan, 2010; Ma et al., 2008). Researchers used Sed1p anchored protein to display lipase derived from *Rhizopus oryzae* on the surface of yeast cells, improving the optimal temperature and pH of lipase and other properties (Li et al., 2015). Using YlPir1p anchored protein to display lipase Lip2p derived from *Yarrowia lipolytica* on the surface of yeast cells, the whole-cell biocatalyst exhibits improved thermal and organic solvent stability compared to free enzymes (Yuzbasheva et al., 2015). Lpp'OmpA anchor protein was used to display the lipase derived from *Bacillus licheniformis* on the surface of *Escherichia coli* cells. Through a series of optimizations, the optimal pH, temperature, and substrate conditions of the whole-cell biocatalyst reaction were determined (Jo, Han, Kim, Kwon, & Lee, 2014). Cutinase FsC showed superior degradation performance against phthalate-based plasticizers after being displayed on the surface of yeast cells (Zhu & Wei, 2019).

The ability of cutinases to degrade PET depends largely on the crystallinity of the polymer and the reaction temperature applied. Effective PET

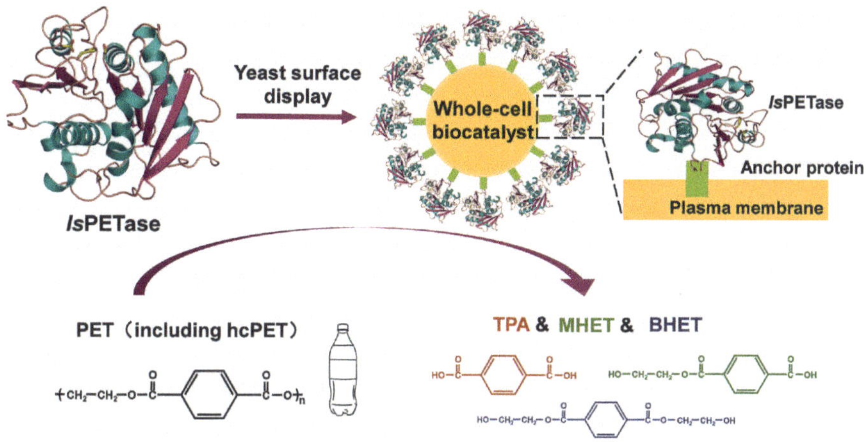

Fig. 2 Scheme of the PET degradation using surface displayed *Is*PETase. hcPET: high-crystalline PET (Chen et al., 2020).

hydrolysis only occurs below a certain critical crystallinity (Wei, Breite, et al., 2019). A whole-cell biocatalyst expressing recombinant LCC has been shown with industrial promise in low-crystalline PET degradation (Wei et al., 2020; Yan, Wei, Cui, Bornscheuer, & Liu, 2020). However, the vast majority of PET used in beverage bottle manufacturing and other industries is highly crystalline, and how to degrade high-crystalline PET (hcPET) using biocatalysis remains still a difficult task. Therefore, we constructed several engineered yeast cells displaying *Is*PETase to improve its degradation efficiency against hcPET as detailed in this chapter (Fig. 2).

2. Protocols

2.1 Construction of a whole-cell biocatalyst by displaying *Is*PETase on the surface of yeast (*Pichia pastoris*) cells

Materials

- *Pichia pastoris* (*P. pastoris*) strain GS115
- *Escherichia coli* (*E. coli*) DH5α
- pPIC9 plasmid
- *petase* gene
- *gcw21* gene (NCBI Reference Sequence: XM_002491407)
- *gcw51* gene (NCBI Reference Sequence: XP_002493782)
- *gcw61* gene (NCBI Reference Sequence: XP_002494322)
- Primers (Table 2)

Table 2 List of the primers used for plasmid construction.

21-F	GGTGGTGGAGGTAGTGGAGGAGGTGGTAGTGA CCAGCGAATCTCTGTCAC
21-R	CCGGAATTCTTATAACAAAGCAGCGGCGGCAATT
51-F	GGTGGTGGAGGTAGTGGAGGAGGTGGTAGT GATGACGATGACTCATTAC
51-R	CCGGAATTCCTAGATCAATAGGGCAATGG
61-F	GGTGGTGGAGGTAGTGGAGGAGGTGGTAGT AACAACCTATCAAACGAGAGT
61-R	ATAGTTTAGCGGCCGCTTAAATCAATAGAG CAACACCGGC
P-F	CCGCTCGAGAAAAGAGATTACAAGGATGACGAC GATAAGCAAACCAATCCTTATGCCCGTG
P-R	ACTACCACCTCCTCCACTACCTCCACCACC GCTACAGTTGGCGGTACGAA

In this chapter, the *petase* gene (GenBank number NZ_BBYR01000074.1) was first sequence-optimized and synthesized by BGI Group (China).

Equipment

- Polymerase chain reaction (PCR) cycler (Bio-Rad, USA)
- MicroPulser Electroporator (Bio-Rad, USA)

Reagents

- LB medium
- Ampicillin
- YPD medium (1%, w/v yeast extract, 2%, w/v peptone, and 2%, w/v glucose)
- BMGY medium (1%, w/v yeast extract, 2%, w/v peptone, 100 mM potassium phosphate pH 6.0, 1.34%, w/v yeast nitrogen base (YNB), and 1%, v/v glycerol)
- BMMY medium (1%, w/v yeast extract, 2%, w/v peptone, 100 mM potassium phosphate pH 6.0, 1.34%, w/v yeast nitrogen base (YNB), and 1%, v/v methanol)
- *EcoRI* (Thermo Fisher Scientific, USA)
- *XhoI* (Thermo Fisher Scientific, USA)
- *StuI* (Thermo Fisher Scientific, USA)
- T4 ligase (Thermo Fisher Scientific, USA)

Protocol

1. Amplify the *gcw21*, *gcw51*, and *gcw61* genes by PCR from the genomic DNA of *P. pastoris* GS115 using primers 21/51/61-F and 21/51/61-R, and amplify the *petase* gene by PCR with primers P-F and P-R (Table 2).
2. Use gel to purify PCR products. Then, use the overlap PCR method to amplify the fusion fragments PETase-GCW21/51/61, using primers 21/51/61-F and P-R (Table 2).
3. Use *XhoI* and *EcoRI* to digest PETase-GCW21/51/61 and pPIC9 at 37 °C.
4. Ligate the digested fragments PETase-GCW21/51/61 along with plasmid pPIC9 digested with *XhoI* and *EcoRI* at 22 °C for 30 min and transform into *E. coli* DH5α. Select several single clonies and culture it in LB medium at 37 °C for 12 h, and extract the plasmids.
5. Purify the resulting recombinant plasmids from positive clones named pPIC9-PETase-GCW21/51/61 and verify by restriction enzyme digestion and sequencing.
6. Linearize the recombinant plasmids were linearized with *StuI* and then transformed into *P. pastoris* GS115 by electroporation. In detail, mix *P. pastoris* GS115 competent cells with 5–20 μg of linearized plasmid, transfer them to a pre-cooled electroporation cup, and place them on ice for 5 min. After a 1.5 kV electric shock, immediately add 1 mL of pre-cooled 1 M sorbitol to the conversion cup. Transfer the contents to a sterile centrifuge tube, place it in a 30-degree incubator for 1 h, divide it into 200–600 μL, and plate on an MD plate.
7. Finally, use the genome of the positive transformants selected from YPD plates containing 100 μg/mL ampicillin as the selective marker was used as a template, and perform PCR using the target gene primers and AOX primers, to identify true positive transformants, which are named GS115/PETase-GCW21/51/61.
8. Culture *P. pastoris* GS115 was cultured at 30 °C in the following media: YPD for sub-cultivation, BMGY for cell growth, and BMMY for recombinant protein production.

2.2 Verification of *Is*PETase expression

2.2.1 Western Blot

Materials

- GS115/PETase-GCW21/51/61
- Nitrocellulose membrane

Equipment

- Chemiluminescence multicolor fluorescent gel imaging system, Alliance Q9 (UVITEC)

Reagents

- PBS (phosphate-buffered saline, dissolve 8 g NaCl, 0.2 g KCl, 1.44 g Na_2HPO_4, 0.24 g KH_2PO_4 in 800 mL of double-distilled H_2O. Adjust the pH to 7.4 with HCl, and then add H_2O to 1 L)
- Primary anti-FLAG antibody
- Goat anti-rabbit IgG antibody

Protocol

1. Before SDS-PAGE sample preparation, the OD_{600} of the yeast culture solution is measured.
2. After 48 h, 2×10^7 cells from the three recombinant strains are obtained. Centrifuge the cells at 12,000 rpm for 1 min, wash twice with PBS, and add 20 μL of ddH_2O (double-distilled) to resuspend. Then, add 4 μL of loading buffer, and subject to metal bath treatment at 100 °C for 15 min. Then, centrifuge samples at 12,000 rpm for 2 min. The supernatant is obtained and subjected to SDS-PAGE.
3. Subsequently, blot the gel at 100 V for 1 h to the nitrocellulose membrane under ice bath conditions.
4. After transfer, incubate the membrane in a blocking solution (PBS with 5%, w/v nonfat dry milk) at room temperature for 1 h with shaking.
5. Incubate on a 4 °C shaker overnight in PBS supplemented with primary anti-FLAG antibody at a concentration of 1:1000.
6. Incubate with PBS supplemented with a goat anti-rabbit secondary antibody at a concentration of 1:5000 on a shaker at room temperature for 1 h.
7. Lay a layer of cling film on the imaging sample plate of the photoresist, put the nitrocellulose (NC) film on the cling film, mix the two coloring solutions (SuperSignal West Pico Chemiluminescent Substrate, product number 34080, Thermo Fisher Scientific, USA) at a 1:1 ratio, add the mixture to the NC film, and incubate in the dark for 5 min.
8. Take pictures with the chemiluminescence multicolor fluorescent gel imaging system. Briefly, select the appropriate filter, start the exposure, wait for the imaging to be completed, and save the picture.

2.2.2 Immunofluorescence microscopy analysis

Materials

- *P. pastoris* GS115/PETase-GCW21/51/61

Equipment

- Fluorescence microscope (PerkinElmer, USA)

Reagents

- 70% (v/v) ethanol & 1% (v/v) HCl
- HEPES (20 mM, pH 7.0)
- PBS
- Primary anti-FLAG antibody
- Goat anti-rabbit IgG antibody

Protocol

1. After 48 h, 4×10^7 cells from the three recombinant strains are obtained. Wash the cells twice with HEPES, pH 7.0. The supernatant is discarded after centrifugation and the cells are fixed with 37% formaldehyde for 1 h.
2. Incubate on a 4 °C shaker overnight in PBS supplemented with primary anti-FLAG antibody at a concentration of 1:1000.
3. Incubate with PBS supplemented with a goat anti-rabbit secondary antibody at a concentration of 1:5000 on a shaker at room temperature for 1 h.
4. To confirm the expression and localization of PETase-GCW21/51/61 fusion protein on the yeast cell surface, yeast cells are visualized using a fluorescence microscope.

2.3 Detection of release products by high-performance liquid chromatography (HPLC)

Materials

- PET film (Goodfellow Ltd., biaxially oriented, product number ES301440, thickness 0.175 mm, diameter 6 mm)
- *P. pastoris* GS115/PETase-GCW51

Equipment

- Waters e2695 equipped with a HyPURITY C18 (Thermo Fisher Scientific, No 22105-254630) column (4.6 × 250 mm)

Reagents

- 0.5% Triton X-100, 10 mM Na_2CO_3
- 50 mM glycine-NaOH (pH 9.0)
- 18 mM phosphate buffer (pH 2.5)
- 18 mM phosphate buffer (pH 2.5) containing 10% (v/v) DMSO
- 16 mM phosphate buffer (pH 2.5) containing 20% (v/v) DMSO
- Methanol

Protocol

1. Before the reaction, soak the PET film separately in 0.5% Triton X-100, 10 mM Na_2CO_3, and distilled water for 30 min each and then air-dry for the reaction.
2. Incubate GS115/PETase-GCW21/51/61 cells (4×10^7/mL, induced for 48 h) with PET film in buffer containing 50 mM glycine-NaOH (pH 9.0) for 18 h at 30 °C. The relative enzyme activity of different cells is calculated by measuring the amount of MHET degradation product using HPLC.
3. HPLC is performed on a Waters e2695 equipped with a HyPURITY C18 (Thermo Fisher Scientific, No 22105-254630) column (4.6 × 250 mm). As mobile phase, use methanol/18 mM phosphate buffer (pH 2.5) at a flow rate of 0.5 mL/min, and monitor the effluent at a wavelength of 240 nm. The typical elution condition is as follows: 0 to 30 min, 25% (v/v) methanol; 30–50 min, 25%–100% methanol linear gradient.
4. Termination Reaction: All PET hydrolysis reactions are terminated by dilution of the aqueous solution with 18 mM phosphate buffer (pH 2.5) containing 10% (v/v) DMSO followed by heat treatment (85 °C, 10 min). All BHET hydrolysis reactions are terminated by dilution of the aqueous solution with 16 mM phosphate buffer (pH 2.5) containing 20% (v/v) DMSO followed by heat treatment (80 °C, 10 min). The supernatant obtained by centrifugation (13,500 × *g*, 10 min) is analyzed by HPLC.
5. We normalize the amount (moles) of MHET produced in each experiment to the amount (moles) of *Is*PETase present in each experiment. Then, the resulting ratio is further divided by enzymatic degradation time (18 × 3600 s) to obtain an "ad hoc" value of the turnover rate." By doing this, the turnover rate can be determined by HPLC and is used to evaluate PETase enzyme activity.

2.4 Optimization of the surface display system

2.4.1 Materials

- PET film (Goodfellow, biaxially oriented, product number ES301440, thickness 0.175 mm, diameter 6 mm)
- BHET (Sigma-Aldrich)
- *para*-nitrophenol (pNP)-acetate (Sigma-Aldrich)
- *p*NP-butyrate (Sigma-Aldrich)
- *p*NP-octanoate (Sigma-Aldrich)
- *P. pastoris* GS115/PETase-GCW51

Equipment

- Waters e2695 equipped with a HyPURITY C18 (Thermo Fisher Scientific, No 22105-254630) column (4.6 × 250 mm)

Reagents

- 50 mM NaH_2PO_4-NaOH (pH 5.0)
- 50 mM Na_2HPO_4-HCl (pH 6.0–8.0)
- 50 mM glycine-NaOH (pH 9.0–10)
- Methanol
- 40 mM Na_2HPO_4-HCl (pH 7.0), 80 mM NaCl, and 20% (v/v) DMSO
- 45 mM Na_2HPO_4-HCl (pH 7.0), 90 mM NaCl, and 10% (v/v) DMSO

Protocols

1. To optimize the induction time, *P. pastoris* GS115/PETase-GCW51 is induced for 0, 24, 48, 72, and 96 h for enzyme activity assays.
2. To optimize pH, the reactions are in performed 50 mM NaH_2PO_4-NaOH (pH 5.0), 50 mM Na_2HPO_4-HCl (pH 6.0–8.0), or 50 mM glycine-NaOH (pH 9.0–10).
3. To optimize temperature, the reactions are performed at 20, 30, 40, and 50 °C, separately.
4. Standardization of cell number to protein concentration: To compare the enzyme activities of the surface-displayed recombinant *P. pastoris* and PETase protein, the amount of the two should be standardized based on the protein content. Use Image J to analyze the western blot band of PETase according to the gray value and generate a standard curve (Fig. 3).
5. Optimization of protein concentration for the PET hydrolysis reaction: The PETase protein concentration is 0.5, 1, 5, 10, 20, 40, 70, and 100 μg/mL. The MHET quality of PETase reaction with PET at 10 μg/mL is defined as 100%, and the relative enzyme activity is calculated. The *P. pastoris* GS115/PETase-GCW51 cell concentration is standardized to 2×10^6/mL, 4×10^6/mL, 2×10^7/mL, 4×10^7/mL, 1.2×10^8/mL, 2×10^8/mL, 2.8×10^8/mL, and 4×10^8/mL. The MHET quality of the GS115/PETase-GCW51 cellular reaction with PET at 4×10^7/mL is defined as 100%, and the relative enzyme activity is calculated. All reactions are performed in 50 mM glycine-NaOH (pH 9.0) at 30 °C for 18 h.
6. Degradation of BHET and *p*NP-aliphatic Esters with the Whole-Cell Biocatalyst: Cells (4×10^7/mL) and purified PETase (0.303 μg/mL) are separately incubated with 0.9 mM BHET (Sigma-Aldrich) in 40 mM Na_2HPO_4-HCl (pH 7.0), 80 mM NaCl, and 20% (v/v) DMSO for 18 h at 30 °C and then separately incubated with 1 mM *p*NP-acetate, *p*NP-butyrate, and *p*NP-octanoate (Sigma-Aldrich) in

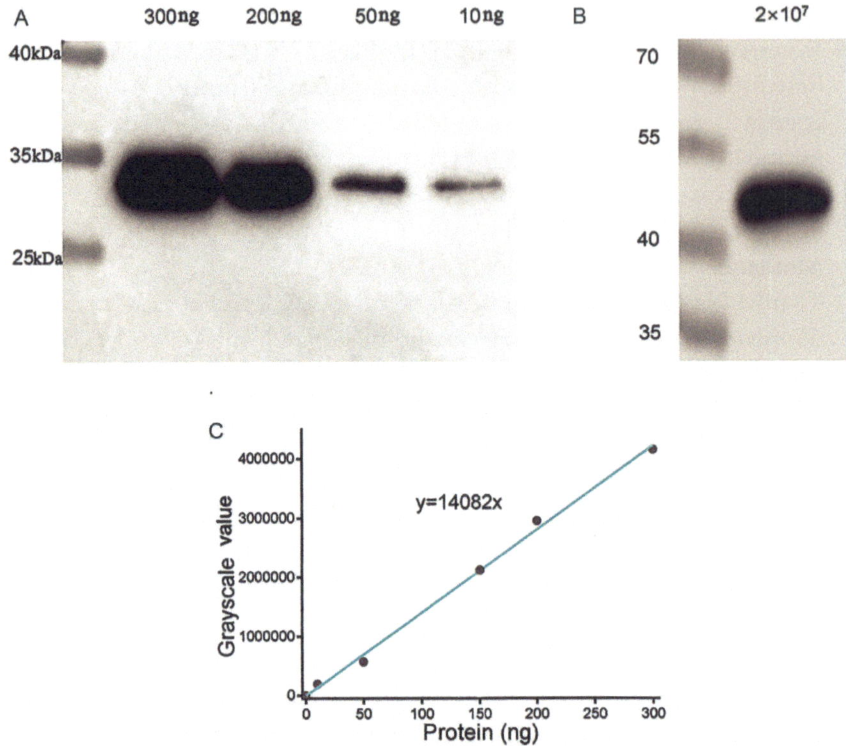

Fig. 3 Standardization of cell number to protein concentration. (A) Western blot analysis of the quality of *Is*PETase in 300, 200, 50, and 10 ng. (B) Western blot analysis of the expression of PETase-GCW51 under the induction time of 48 h in the amount of 2×10^7 GS115/PETase-GCW51 cells. (C) The protein-grayscale analysis standard curve (Chen et al., 2020).

45 mM Na_2HPO_4-HCl (pH 7.0), 90 mM NaCl, and 10% (v/v) DMSO for 5 min at 30 °C. The production of *p*NP is monitored at a wavelength of 415 nm.

2.5 The application of the whole-cell biocatalyst

2.5.1 Materials

- PET film (Goodfellow Ltd., biaxially oriented, product number ES301440, thickness 0.175 mm, diameter 6 mm)
- Commercial PET bottles (brands: Coca-cola, Nestlé, and Pepsi-cola)

Equipment

- Waters e2695 equipped with a HyPURITY C18 (Thermo Fisher Scientific, No 22105-254630) column (4.6 × 250 mm)

Reagents

- 50 mM glycine-NaOH buffer (pH 9.0)
- 10% (v/v) methanol
- 10% (v/v) ethanol
- 0.1% (v/v) Triton X-100

Protocols

1. To assess the thermostability of PETase displayed on yeast, incubate *P. pastoris* GS115/PETase-GCW51 cells at 30 °C for 1–7 days in 50 mM glycine-NaOH buffer (pH 9.0). Initiate the PET degradation reaction after 3 washes with 50 mM glycine-NaOH (pH 9.0) at 3500 × *g* for 5 min. Quantify the production of MHET by HPLC.
2. To assess the chemical/solvent conditions for PETase stability displayed on yeast, incubate *P. pastoris* GS115/PETase-GCW51 cells in 50 mM glycine-NaOH (pH 9.0) supplemented with methanol (final concentration, 10%, v/v), ethanol (final concentration, 10%, v/v), and Triton X-100 (final concentration, 0.1%, v/v) for 18 h at 30 °C. Initiate the PET degradation reaction after 3 washes with 50 mM glycine-NaOH (pH 9.0) at 3500 × *g* for 5 min. Quantify the production of MHET by HPLC.
3. To assess the activity of PETase after freeze-dried, it is measured after incubation at 4 °C overnight in 50 mM glycine-NaOH buffer (pH 9.0). The OD_{600} value was measured to ensure that the cells count is the same as that prior to freeze-drying. Initiate the PET degradation reaction and quantify the production of MHET by HPLC.
4. Degradation of commercial PET bottles with the whole-cell biocatalyst: *P. pastoris* GS115/PETase-GCW51 cells are respectively incubated with 6 mm diameter hcPET cut out from different commercial PET bottles (brands: Coca-cola, Nestlé, and Pepsi-cola) in 50 mM glycine-NaOH buffer (pH 9.0) for 18 h at 30 °C. The production of MHET is quantified by HPLC.

Acknowledgment

This work was supported by the National Natural Science Foundation of China (grant numbers 81601593 & 31970048).

References

Anadón, A., Binderup, M. L., Bursch, W., Castle, L., Crebelli, R., & Engel, K. H. (2011). Scientific opinion on the criteria to be used for safety evaluation of a mechanical recycling process to produce recycled PET intended to be used for manufacture of materials and articles in contact with food. *EFSA Journal*, *9*(7), 2184.

Annie Gai, S., & Dane Wittrup, K. (2007). Yeast surface display for protein engineering and characterization. *Current Opinion in Structural Biology*, *17*(4), 467–473.

Austin, H. P., Allen, M. D., Donohoe, B. S., Rorrer, N. A., Kearns, F. L., Silveira, R. L., et al. (2018). Characterization and engineering of a plastic-degrading aromatic polyesterase. *Proceedings of the National Academy of Sciences of the United States of America*, *115*(19), 4350–4357.

Bouwmeester, H., Hollman, P. C. H., & Peters, R. J. B. (2015). Potential health impact of environmentally released micro- and nanoplastics in the human food production chain: Experiences from nanotoxicology. *Environmental Science & Technology*, *49*(15), 8932–8947.

Brueckner, T., Eberl, A., Heumann, S., Rabe, M., & Guebitz, G. M. (2008). Enzymatic and chemical hydrolysis of poly(ethylene terephthalate) fabrics. *Journal of Polymer Science Part A*, *46*(19), 6435–6443.

Carbery, M., O'Connor, W. O., & Palanisami, T. (2018). Trophic transfer of microplastics and mixed contaminants in the marine food web and implications for human health. *Environment International*, *115*, 400–409.

Casabianca, S., Capellacci, S., Penna, A., Cangiotti, M., Fattori, A., Corsi, I., et al. (2020). Physical interactions between marine phytoplankton and PET plastics in seawater. *Chemosphere*, *238*, 124560.

Chen, Z., Wang, Y., Cheng, Y., Wang, X., Tong, S., Yang, H., et al. (2020). Efficient biodegradation of highly crystallized polyethylene terephthalate through cell surface display of bacterial PETase. *Science of the Total Environment*, *709*, 136138.

Cressey, D. (2016). Bottles, bags, ropes and toothbrushes: The struggle to track ocean plastics. *Nature*, *536*(7616), 263–265.

De Carvalho, C. C. C. R. (2011). Enzymatic and whole cell catalysis: Finding new strategies for old processes. *Biotechnology Advances*, *29*(1), 75–83.

Der Vaart, J. M. V., Biesebeke, R. T., Chapman, J. W., Toschka, H. Y., Klis, F. M., & Verrips, C. T. (1997). Comparison of cell wall proteins of *Saccharomyces cerevisiae* as anchors for cell surface expression of heterologous proteins. *Applied and Environmental Microbiology*, *63*(2), 615–620.

Doris, R., Antonio Orcal, Y., Sabine, Z., Jing, W., Susanne, N., Georg, S., et al. (2013). Fusion of binding domains to *Thermobifida cellulosilytica* cutinase to tune sorption characteristics and enhancing PET hydrolysis. *Biomacromolecules*, *14*(6), 1769–1776.

Dwevedi, A., & Kayastha, A. M. (2012). *Enzyme immobilization: A breakthrough in enzyme technology and boon to enzyme based industries. Vol. 3*. Nova Science Publishers.

Falkenstein, P., Gräsing, D., Bielytskyi, P., Zimmermann, W., Matysik, J., Wei, R., et al. (2020). UV pretreatment impairs the enzymatic degradation of polyethylene terephthalate. *Frontiers in Microbiology*, *11*, 689.

Fan, S., Hou, C., Liang, B., Feng, R., & Liu, A. (2015). Microbial surface displayed enzymes based biofuel cell utilizing degradation products of lignocellulosic biomass for direct electrical energy. *Bioresource Technology*, *192*(1), 821–825.

Fecker, T., Galaz-Davison, P., Engelberger, F., Narui, Y., Sotomayor, M., Parra, L. P., et al. (2018). Active site flexibility as a hallmark for efficient PET degradation by *I. sakaiensis* PETase. *Biophysical Journal*, *114*(6), 1302–1312.

Gamerith, C., Zartl, B., Pellis, A., Guillamot, F., Marty, A., Acero, E. H., et al. (2017). Enzymatic recovery of polyester building blocks from polymer blends. *Process Biochemistry*, *59*, 58–64.

Global PET. (2016). *Global PET resin production capacity*. Retrieved from https://www.plasticsinsight.com/global-pet-resin-production-capacity/.

Groß, C., Hamacher, K., Schmitz, K., & Jager, S. (2017). Cleavage product accumulation decreases the activity of cutinase during PET hydrolysis. *Journal of Chemical Information and Modeling*, *57*(2), 243–255.

Han, X., Liu, W., Huang, J. W., Ma, J., Zheng, Y., Ko, T. P., et al. (2017). Structural insight into catalytic mechanism of PET hydrolase. *Nature Communications*, *8*(1), 2106.

Huang, J. S., Koongolla, J. B., Li, H. X., Lin, L., Pan, Y. F., Liu, S., et al. (2020). Microplastic accumulation in fish from Zhanjiang mangrove wetland, South China. *Science of the Total Environment*, *708*, 134839.

Jo, J. H., Han, C.-W., Kim, S.-H., Kwon, H.-J., & Lee, H.-H. (2014). Surface display expression of *Bacillus licheniformis* lipase in *Escherichia coli* using Lpp'OmpA chimera. *Journal of Microbiology*, *52*(10), 856–862.

Joo, S., Cho, I. J., Seo, H., Son, H. F., Sagong, H.-Y., Shin, T. J., et al. (2018). Structural insight into molecular mechanism of poly(ethylene terephthalate) degradation. *Nature Communications*, *9*(1), 382.

Kawai, F., Kawabata, T., & Oda, M. (2019). Current knowledge on enzymatic PET degradation and its possible application to waste stream management and other fields. *Applied Microbiology and Biotechnology*, *103*(11), 4253–4268.

Kawai, F., Oda, M., Tamashiro, T., Waku, T., Tanaka, N., Yamamoto, M., et al. (2014). A novel Ca^{2+}-activated, thermostabilized polyesterase capable of hydrolyzing polyethylene terephthalate from *Saccharomonospora viridis* AHK190. *Applied Microbiology and Biotechnology*, *98*(24), 10053–10064.

Lee, S. H., Choi, J. I., Han, M. J., Choi, J. H., & Sang, Y. L. (2005). Display of lipase on the cell surface of *Escherichia coli* using OprF as an anchor and its application to enantioselective resolution in organic solvent. *Biotechnology and Bioengineering*, *90*(2), 223–230.

Lee, S. H., Lee, S. Y., & Park, B. C. (2005). Cell surface display of lipase in *Pseudomonas putida* KT2442 using OprF as an anchoring motif and its biocatalytic applications. *Applied and Environmental Microbiology*, *71*(12), 8581–8586.

Li, W., Shi, H., Ding, H., Wang, L., Zhang, Y., Li, X., et al. (2015). Cell surface display and characterization of *Rhizopus oryzae* lipase in *Pichia pastoris* using Sed1p as an anchor protein. *Current Microbiology*, *71*(1), 150–155.

Liang, B., Wang, G., Yan, L., Ren, H., Feng, R., Xiong, Z., et al. (2019). Functional cell surface displaying of acetylcholinesterase for spectrophotometric sensing organophosphate pesticide. *Sensors and Actuators B: Chemical*, *279*, 483–489.

Lin, Y. H., & Yang, M. H. (2007). Catalytic conversion of commingled polymer waste into chemicals and fuels over spent FCC commercial catalyst in a fluidized-bed reactor. *Applied Catalysis B: Environmental*, *69*(3–4), 145–153.

Liu, A., Feng, R., & Liang, B. (2016). Microbial surface displaying formate dehydrogenase and its application in optical detection of formate. *Enzyme and Microbial Technology*, *91*, 59–65.

Liu, B., He, L., Wang, L., Li, T., Li, C., Liu, H., et al. (2018). Protein crystallography and site-direct mutagenesis analysis of the poly(ethylene terephthalate) hydrolase PETase from *Ideonella sakaiensis*. *Chembiochem*, *19*(14), 1471–1475.

Liu, A., Lang, Q., Liang, B., & Shi, J. (2017). Sensitive detection of maltose and glucose based on dual enzyme-displayed bacteria electrochemical biosensor. *Biosensors and Bioelectronics*, *87*, 25–30.

Liu, W., Zhao, H., Jia, B., Xu, L., & Yan, Y. (2010). Surface display of active lipase in *Saccharomyces cerevisiae* using Cwp2 as an anchor protein. *Biotechnology Letters*, *32*(2), 255–260.

Ma, Y., Tao, X., Ren, R., Bei, G., Jiang, Z., & Wei, D. (2008). Efficient display of active lipase LipB52 with a *Pichia pastoris* cell surface display system and comparison with the LipB52 displayed on *Saccharomyces cerevisiae* cell surface. *BMC Biotechnology*, *8*(1), 4.

Marten, E., Müller, R.-J., & Deckwer, W.-D. (2003). Studies on the enzymatic hydrolysis of polyesters. II. Aliphatic–aromatic copolyesters. *Polymer Degradation and Stability*, *88*, 371–381.

Müller, R., Schrader, H., Profe, J., Dresler, K., & Deckwer, W. (2005). Enzymatic degradation of poly(ethylene terephthalate): Rapid hydrolyse using a hydrolase from *T. fusca*. *Macromolecular Rapid Communications*, *26*(17), 1400–1405.

Nakatani, H., & Hori, K. (2017). Cell surface protein engineering for high-performance whole-cell catalysts. *Frontiers of Chemical Engineering in China, 11*(1), 46–57.
Oda, M., Yamagami, Y., Inaba, S., Oida, T., Yamamoto, M., Kitajima, S., et al. (2018). Enzymatic hydrolysis of PET: Functional roles of three Ca^{2+} ions bound to a cutinase-like enzyme, Cut190*, and its engineering for improved activity. *Applied Microbiology and Biotechnology, 102*(23), 10067–10077.
Palm, G. J., Reisky, L., Böttcher, D., Müller, H., Michels, E. A. P, Walczak, M. C., et al. (2019). Structure of the plastic-degrading *Ideonella sakaiensis* MHETase bound to a substrate. *Nature Communications, 10*, 1717.
Park, M. (2020). Surface display technology for biosensor applications: A review. *Sensors (Basel), 20*, 2775.
Rahimi, A., & García, J. M. (2017). Chemical recycling of waste plastics for new materials production. *Nature Reviews Chemistry, 1*(6), 0046.
Rieckmann, T., & Völker, S. (2004). Poly(ethylene terephthalate) polymerization–mechanism, catalysis, kinetics, mass transfer and reactor design. In J. Scheirs, & T. E. Long (Eds.), *Modern polyesters: Chemistry and technology of polyesters and copolyesters* John Wiley & Sons, Ltd.
Ronkvist, A. M., Xie, W., Lu, W., & Gross, R. A. (2009). Cutinase-catalyzed hydrolysis of poly(ethylene terephthalate). *Macromolecules, 42*(14), 5128–5138.
Sheldon, R. A., & Van Pelt, S. (2013). Enzyme immobilisation in biocatalysis: Why, what and how. *Chemical Society Reviews, 42*(15), 6223–6235.
Shoda, S., Uyama, H., Kadokawa, J., Kimura, S., & Kobayashi, S. (2016). Enzymes as green catalysts for precision macromolecular synthesis. *Chemical Reviews, 116*(4), 2307–2413.
Taniguchi, I., Yoshida, S., Hiraga, K., Miyamoto, K., Kimura, Y., & Oda, K. (2019). Biodegradation of PET: Current status and application aspects. *ACS Catalysis, 9*(5), 4089–4105.
Then, J., Wei, R., Oeser, T., Barth, M., Belisarioferrari, M. R., Schmidt, J., et al. (2015). Ca^{2+} and Mg^{2+} binding site engineering increases the degradation of polyethylene terephthalate films by polyester hydrolases from *Thermobifida fusca*. *Biotechnology Journal, 10*(4), 592–598.
Then, J., Wei, R., Oeser, T., Gerdts, A., Schmidt, J., Barth, M., et al. (2016). A disulfide bridge in the calcium binding site of a polyester hydrolase increases its thermal stability and activity against polyethylene terephthalate. *FEBS Open Bio, 6*(5), 425–432.
Tournier, V., Topham, C., Gilles, A., David, B., Folgoas, C., Moya-Leclair, E., et al. (2020). An engineered PET depolymerase to break down and recycle plastic bottles. *Nature, 580*, 216–219.
Toussaint, B., Raffael, B., Angers-Loustau, A., Gilliland, D., Kestens, V., Petrillo, M., et al. (2019). Review of micro- and nanoplastic contamination in the food chain. *Food Additives and Contaminants Part A-Chemistry Analysis Control Exposure & Risk Assessment, 36*, 639–673.
Webb, H. K., Arnott, J., Crawford, R. J., & Ivanova, E. P. (2013). Plastic degradation and its environmental implications with special reference to poly(ethylene terephthalate). *Polymers, 5*(1), 1–18.
Wei, R., Breite, D., Song, C., Gräsing, D., Ploss, T., Hille, P., et al. (2019). Biocatalytic degradation efficiency of postconsumer polyethylene terephthalate packaging determined by their polymer microstructures. *Advanced Science, 6*(14), 1900491.
Wei, R., Song, C., Gräsing, D., Schneider, T., Bielytskyi, P., Böttcher, D., et al. (2019). Conformational fitting of a flexible oligomeric substrate does not explain the enzymatic PET degradation. *Nature Communications, 10*, 5581.
Wei, R., Tiso, T., Bertling, J., Connor, K. E. O., Blank, L. M., & Bornscheuer, U. T. (2020). Possibilities and limitations of biotechnological plastic degradation and recycling. *Nature Catalysis, 3*, 867–871.

Wei, R., & Zimmermann, W. (2017). Biocatalysis as a green route for recycling the recalcitrant plastic polyethylene terephthalate. *Microbial Biotechnology*, *10*(6), 1302–1307.

World Economic Forum. (2016). *The new plastics economy: Rethinking the future of plastics*. World Economic Forum. Retrieved from www3.weforum.org/docs/WEF_The_New_Plastics_Economy.pdf.

Yan, F., Wei, R., Cui, Q., Bornscheuer, U. T., & Liu, Y.-J. (2020). Thermophilic whole-cell degradation of polyethylene terephthalate using engineered *Clostridium thermocellum*. *Microbial Biotechnology*. https://doi.org/10.1111/1751-7915.13580.

Yoshida, S., Hiraga, K., Takehana, T., Taniguchi, I., Yamaji, H., Maeda, Y., et al. (2016). A bacterium that degrades and assimilates poly(ethylene terephthalate). *Science*, *351*(6278), 1196–1199.

Yuzbasheva, E. Y., Yuzbashev, T. V., Perkovskaya, N. I., Mostova, E. B., Vybornaya, T. V., Sukhozhenko, A. V., et al. (2015). Cell surface display of *Yarrowia lipolytica* lipase Lip2p using the cell wall protein YlPir1p, its characterization, and application as a whole-cell biocatalyst. *Applied Biochemistry and Biotechnology*, *175*(8), 3888–3900.

Zhu, B., & Wei, N. (2019). Biocatalytic degradation of parabens mediated by cell surface displayed cutinase. *Environmental Science and Technology*, *53*(1), 354–364.

Zhang, X., Fevre, M., Jones, G. O., & Waymouth, R. M. (2018). Catalysis as an enabling science for sustainable polymers. *Chemical Reviews*, *118*, 839–885.